STUDENT SOLUTIONS MANUAL
Laurel Technical Services

EXPERIENCING INTRODUCTORY ALGEBRA

JoAnne Thomasson
Bob Pesut

PRENTICE HALL, Upper Saddle River, NJ 07458

Senior Editor: Karin Wagner
Supplement Editor: Kate Marks
Special Projects Manager: Barbara A. Murray
Production Editor: Barbara A. Till
Supplement Cover Manager: Paul Gourhan
Supplement Cover Designer: Liz Nemeth
Manufacturing Buyer: Alan Fischer

Printed in the United States of America

10 9 8 7 6 5 4 3 2 1

ISBN 0-13-799958-5

Prentice-Hall International (UK) Limited, *London*
Prentice-Hall of Australia Pty. Limited, *Sydney*
Prentice-Hall Canada, Inc., *London*
Prentice-Hall Hispanoamericana, S.A., *Mexico*
Prentice-Hall of India Private Limited, *New Delhi*
Prentice-Hall of Japan, Inc., *Tokyo*
Simon & Schuster Asia Pte. Ltd., *Singapore*
Editora Prentice-Hall do Brazil, Ltda., *Rio de Janeiro*

Table of Contents

Discovery Answers

Chapter 1

Discovery 1

1. a. 8
   b. 8
2. a. 12
   b. 12
3. a. 3.7
   b. 3.7
4. a. $\dfrac{3}{4}$
   b. $\dfrac{3}{4}$

The sum of two positive rational numbers is a positive number that is the sum of the absolute values of the addends.

Discovery 2

1. a. −8
   b. 8
2. a. −12
   b. 12
3. a. −3.7
   b. 3.7
4. a. $-\dfrac{3}{4}$
   b. $\dfrac{3}{4}$

The sum of two negative rational numbers is a negative number that is the opposite of the sum of the absolute values of the addends.

Discovery 3

1. a. 4
   b. 4
2. a. −6
   b. 6
3. a. −1.3
   b. 1.3
4. a. $\dfrac{1}{4}$
   b. $\dfrac{1}{4}$

The sum of a positive and a negative rational number has the same sign as the addend with the larger absolute value and is the difference of the absolute values of the addends.

Discovery 4

1. a. −4
   b. 4
2. a. 6
   b. 6
3. a. 1.3
   b. 1.3
4. a. $-\dfrac{1}{4}$

b. $\dfrac{1}{4}$

b. $-\dfrac{1}{4}$

The sum of a negative and a positive rational number has the same sign as the addend with the larger absolute value and is the difference of the absolute values of the addends.

The difference of two negative rational numbers is the sum of the minuend and the opposite of the subtrahend.

Discovery 5

1. **a.** 4

   b. 4

2. **a.** −6

   b. −6

3. **a.** −1.3

   b. −1.3

4. **a.** $\dfrac{1}{4}$

   b. $\dfrac{1}{4}$

The difference of two positive rational numbers is the sum of the minuend and the opposite of the subtrahend.

Discovery 6

1. **a.** −4

   b. −4

2. **a.** 6

   b. 6

3. **a.** 1.3

   b. 1.3

4. **a.** $-\dfrac{1}{4}$

Discovery 7

1. **a.** 8

   b. 8

2. **a.** −12

   b. −12

3. **a.** 3.7

   b. 3.7

4. **a.** $-\dfrac{3}{4}$

   b. $-\dfrac{3}{4}$

The difference of a positive and negative rational number is the sum of the minuend and the opposite of the subtrahend.

Discovery 8

1. **a.** 12

   b. 12

2. **a.** 12

   b. 12

3. **a.** 3

   b. 3

4. **a.** $\dfrac{1}{8}$

   b. $\dfrac{1}{8}$

The product of two rational numbers with like signs is a positive number that is the product of the absolute values of the factors.

Discovery 9

1. a.   −12

   b.   12

2. a.   −12

   b.   12

3. a.   −3

   b.   3

4. a. $-\dfrac{1}{8}$

   b. $\dfrac{1}{8}$

The product of two rational numbers with unlike signs is a negative number that is the opposite of the product of the absolute values of the factors.

Discovery 10

1. a.   −72

   b.   72

2. a.   72

   b.   72

3. a.   −72

   b.   72

4. a.   72

   b.   72

The product of three or more rational numbers is a positive number when the number of negative factors is even. The product of three or more rational numbers is a negative number when the number of negative factors is odd. The absolute value of the product is the product of the absolute values of the factors.

Discovery 11

1. 0

2. 0

3. 0

The product of zero and any rational number (or numbers) is zero.

Discovery 12

1. a.   4

   b.   4

2. a.   4

   b.   4

3. a.   0.48

   b.   0.48

4. a.   2

   b.   2

The quotient of two rational numbers with like signs is a positive number that is the quotient of the absolute values of the dividend and divisor.

Discovery 13

1. a.   −4

   b.   4

2. a.   −4

   b.   4

3. a.   −0.48

   b.   0.48

4. a.   −2

   b.   2

The quotient of two rational numbers with unlike signs is a negative number that is the opposite of the quotient of the absolute values of the dividend and divisor.

Discovery 14

1. This results in a calculator error.

2. This results in a calculator error.

3. This results in a calculator error.

4. This results in a calculator error.

The quotient of a rational number divided by zero is undefined.

Discovery 15

1. 0

2. 0

3. 0

4. 0

5. This results in a calculator error.

The quotient of zero divided by any rational number other than zero is zero. The quotient of zero divided by zero is indeterminate.

Discovery 16

1. 4

2. –8

3. 16

4. –32

The sign of an exponential expression with a negative base is negative if the exponent is an odd number (such as 3 and 5) and is positive if the exponent is an even number (such as 2 and 4).

Discovery 17

1. 3

2. –3

3. 2.4

4. $\dfrac{1}{2}$

5. 0

An exponential expression with an exponent of one results in the base number.

Discovery 18

1. 1

2. 1

3. 1

4. 1

5. indeterminate

An exponential expression with an exponent of zero and a nonzero base will be 1 and with a zero base will be indeterminate.

Discovery 19

1. a. 0.5

   b. 0.5

2. a. 0.25

   b. 0.25

3. a. 0.125

   b. 0.125

4. a. –0.5

   b. –0.5

5. a. −0.25

  b. −0.25

6. a. −0.125

  b. −0.125

An exponential expressions with a nonzero base and a negative integer exponent is equal to an exponential expression having a base that is the reciprocal of the original base and an exponent that is the opposite of the original exponent.

Discovery 20

  1. a. 2

    b. 2

  2. a. 4

    b. 4

  3. a. 8

    b. 8

  4. a. −2

    b. −2

  5. a. 4

    b. 4

  6. a. −8

    b. −8

An exponential expression with a fraction base and a negative integer exponent is equal to an exponential expression having a base that is the reciprocal of the original base and an exponent that is the opposite of the original exponent.

Discovery 21

  1. $(1.1)^2 = 1.21$
     $(1.2)^2 = 1.44$
     $(1.3)^2 = 1.69$
     $(1.4)^2 = 1.96$
     $(1.5)^2 = 2.25$

Therefore, $\sqrt{2}$ must be between 1.4 and 1.5 .

  2. $(1.41)^2 = 1.9881$
     $(1.42)^2 = 2.0164$

Therefore, $\sqrt{2}$ must be between 1.41 and 1.42 .

Discovery 22

  1. a. 1.732050808

    b. 1.732050808

  2. a. 1.44224957

    b. 1.44224957

  3. a. 1.316074013

    b. 1.316074013

  4. a. error

    b. error

  5. a. −1.44224957

    b. −1.44224957

  6. a. error

    b. error

A radical expression is equal to an exponential expression having a base that is the radicand of the radical expression and an exponent that is the reciprocal of the index of the radical expression.

Discovery 23

  1. a. −8

    b. −8

  2. a. −4

    b. −4

  3. a. −4

    b. 4

4. **a.** −8

  b. 8

5. **a.** 12

  b. 12

6. **a.** −12

  b. −12

7. **a.** 3

  b. $\dfrac{1}{3}$

8. **a.** −3

  b. $-\dfrac{1}{3}$

Changing the order of real numbers when addition and multiplication operations are performed results in equivalent values. However, changing the order of real numbers when subtraction and division operations are performed results in different values.

Discovery 24

1. **a.** −5

  b. −5

2. **a.** −1

  b. −1

3. **a.** −7

  b. −1

4. **a.** −5

  b. −11

5. **a.** 36

  b. 36

6. **a.** −36

  b. −36

7. **a.** 1

  b. 9

8. **a.** −9

  b. −1

Changing the grouping of real numbers when addition and multiplication operations are performed results in equivalent values. However, changing the grouping of real numbers when subtraction and division operations are performed results in different values.

Discovery 25

1. **a.** 20

  b. 20

2. **a.** 4

  b. 4

The product of a real number factor and a sum of real number addends is the same as the sum of the products of the factor and each addend.

Discovery 26

1. 7

2. −4

3. 1

4. 1

Perform all multiplication and division in order from left to right before performing addition and subtraction in order left to right.

Chapter 3

Discovery 1

Ordered pairs may vary.
In quadrant I, both coordinates are positive; in quadrant II, the *x*-coordinate is negative and the *y*-coordinate is positive; in quadrant III, both coordinates are negative; and in quadrant IV, the

x-coordinate is positive and the *y*-coordinate is negative.

Discovery 2

Ordered pairs may vary.
On the *x*-axis, the *y*-coordinate is always 0, and on the *y*-axis, the *x*-coordinate is always 0.

Discovery 3

1.

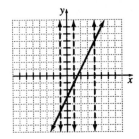

2.

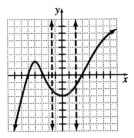

3.

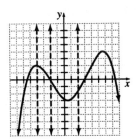

A vertical line does not cross the graph of a function more than once.

4.

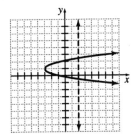

5.

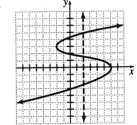

Discovery 4

1.

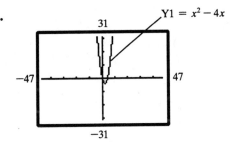

2. falls; rises

3. 12; 5; 0; –3; –4; –3; 0; 5; 12

4. decrease; increase

A function is increasing if the graph is rising. A function is decreasing if the graph is falling.

Chapter 4

Discovery 1

1.

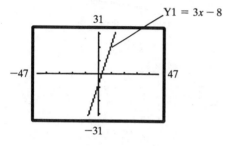

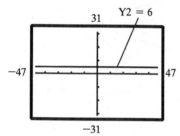

2.

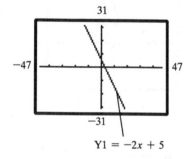

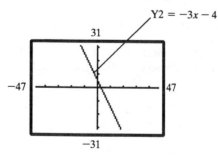

The graphs of the functions Y1 and Y2 are straight lines.

3.

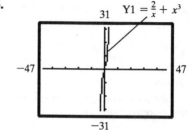

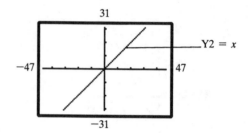

The graph of the function Y1 is not a straight line but a curve.
The graph of the function Y2 is a straight line.

Discovery 2

x	$2x + 3 = x + 5$		
0	3	5	$3 < 5$
1	5	6	$5 < 6$
2	7	7	$7 = 7$
3	9	8	$9 > 8$
4	11	9	$11 > 9$

The solution of an equation is the value for the variable that determines equal values for the expression on the left and the expression on the right of the equation.

Discovery 3

x	$(5x + 4) - 2(3x + 1) = 2(x - 7)$			
3	−1		−8	−1 > −8
4	−2		−6	−2 > −6
5	−3		−4	−3 > −4
6	−4		−2	−4 < −2
7	−5		0	−5 < 0

The solution of an equation is between two integer values when the order of the values for the expression on the left and the expression on the right changes from less than to greater than or greater than to less than.

Discovery 4

x	$2x + 5 = 2x + 10$			
0	5	10	5 ≠ 10	10 − 5 = 5
1	7	12	7 ≠ 12	12 − 7 = 5
2	9	14	9 ≠ 14	14 − 9 = 5
3	11	16	11 ≠ 16	16 − 11 = 5

If the values of the expression on the left and the expression on the right are never equal and their differences remain the same, then the equation has no solution.

Discovery 5

x	$2x + 5 = (x + 3) + (x + 2)$		
0	5	5	5 = 5
1	7	7	7 = 7
2	9	9	9 = 9
3	11	11	11 = 11

If the values of the expression on the left and the expression on the right are always equal, then the equation has many solutions.

Discovery 6

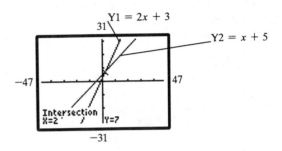

The value of the x-coordinate of the point of intersection of the two graphs is the solution of the equation.
The value of the y-coordinate of the point of intersection of the two graphs is the value obtained when both expressions are evaluated with the solution.

Discovery 7

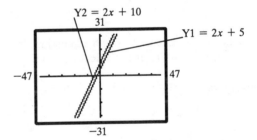

If the graphs of the functions defined by the expression on the left and the expression on the right do not intersect (are parallel), then the equation has no solution.

Discovery 8

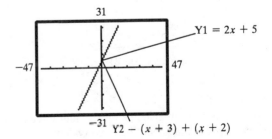

If the graphs of the functions defined by the expression on the left and the expression on the right intersect at all times (are coinciding), then the equation has many solutions.

Discovery 9

1.

$$7 = 7$$

$7 + (-2)$	$7 + (-2)$
5	5

2.

$$6 + 1 = 4 + 3$$

$6 + 1 + 2$	$4 + 3 + 2$
9	9

3.

$$6 + 1 = 4 + 3$$

$6 + 1 + (-2)$	$4 + 3 + (-2)$
5	5

If the same number is added to both expressions in an equation, the result is an equivalent equation.

Discovery 10

1.

$$7 = 7$$

$7(-2)$	$7(-2)$
-14	-14

2.

$$6 + 1 = 4 + 3$$

$(6 + 1)2$	$(4 + 3)2$
14	14

3.

$$6 + 1 = 4 + 3$$

$(6 + 1)(-2)$	$(4 + 3)(-2)$
-14	-14

If both expressions in an equation are multiplied by the same number, the result is an equivalent equation.

Discovery 11

$$2x + 5 = 2x + 10$$
$2x + 5 - 2x = 2x + 10 - 2x$ Subtract $2x$ from both sides.
$$5 = 10 \qquad \text{Simplify.}$$

When isolating the variable results in an equation without a variable and the equation is false (a contradiction), then the equation has no solution.

Discovery 12

$$2x + 5 = (x + 3) + (x + 2)$$
$2x + 5 = 2x + 5$ Simplify the right side.
$2x + 5 - 2x = 2x + 5 - 2x$ Subtract $2x$ from both sides.
$$5 = 5 \qquad \text{Simplify.}$$

When isolating the variable results in an equation without a variable and the equation is true (an identity), then the equation has many solutions.

Discovery 13

1.

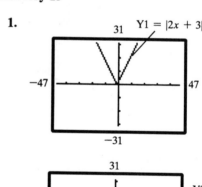

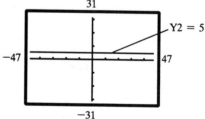

2.

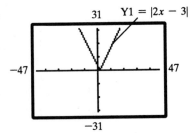

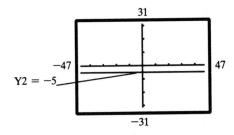

The graph of the function defined by the absolute value expression is a V-shaped graph made up of two lines and the graph of the constant function is a line.

3.

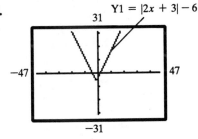

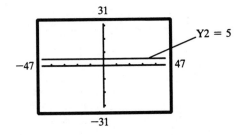

4. $Y1 = 4|2x + 3|$

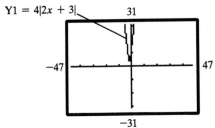

The graph of the function defined by the expression containing the absolute value expression is a V-shaped graph made up of lines, and the graph of the constant function is a line.

Discovery 14

1.

x	$|x|$	3
−5	5	3
−4	4	3
−3	3	3
−2	2	3
−1	1	3
0	0	3
1	1	3
2	2	3
3	3	3
4	4	3
5	5	3

solution: −3 and 3

2.

x	$|x|$	−3
−5	5	−3
−4	4	−3
−3	3	−3
−2	2	−3
−1	1	−3
0	0	−3
1	1	−3
2	2	−3
3	3	−3
4	4	−3
5	5	−3

solution: no solution

3.

x	$|x|$	0
−5	5	0
−4	4	0
−3	3	0
−2	2	0
−1	1	0
0	0	0
1	1	0
2	2	0
3	3	0
4	4	0
5	5	0

solution: 0

The number of solutions that can be found for a linear absolute value equation is determined by the number the absolute value expression is equated to. That is, if the number is positive, two solutions can be found. If the number is negative, no solution can be found. If the number is zero, then one solution can be found.

Discovery 15

1.

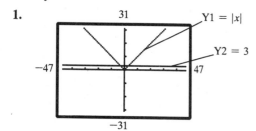

solution: −3 and 3

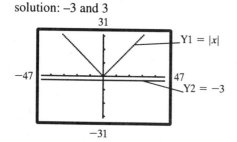

solution: no solution

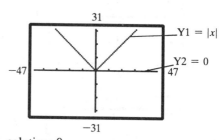

solution: 0

The number of solutions that can be found for a linear absolute value equation is determined by the number the absolute value expression is equated to. That is, if the number is positive, two solutions can be found. If the number is negative, no solution can be found. If the number is zero, then one solution can be found.

Chapter 5

Discovery 1

1.

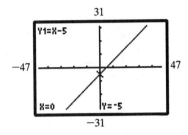

2.

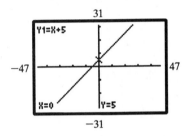

3.

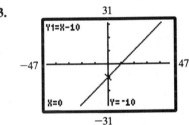

4.

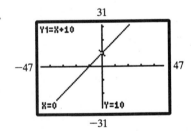

Given an equation solved for y, the y-coordinate of the y-intercept is the constant term of the expression in the equation.

Discovery 2

1.

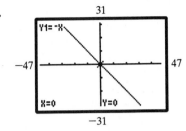

2.

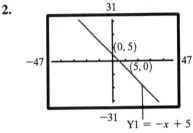

3.

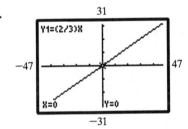

4.

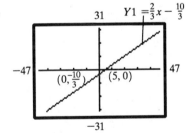

The graph of a linear equation in two variables written in standard form has one point that is both the x-intercept and y-intercept if the constant term is zero. The intercept is the origin, $(0, 0)$.

Discovery 3

1. $\dfrac{1}{2}$

2. 2

3. $-\dfrac{1}{3}$

4. -3

5. 0

6. undefined

7. positive; rise

8. negative; fall

9. zero; horizontal

10. undefined; vertical

11. more

Discovery 4

1.

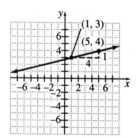

2. 1

3. 4

4. 1

5. 4

6. $\dfrac{1}{4}$

The slope of the graph is the ratio of the difference of the ordered pairs' *y*-coordinates to the difference in the ordered pairs' *x*-coordinates.

Discovery 5

Possible integer coordinate points are labeled.

1. a.

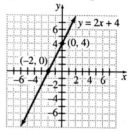

b.

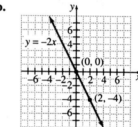

c.

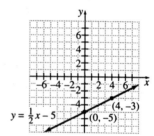

2. a. 2

b. -2

c. $\dfrac{1}{2}$

3 a. 2

b. -2

c. $\dfrac{1}{2}$

Given a linear equation solved for *y*, the slope of its graph is the coefficient of the *x*-term.

Discovery 6

1. a.

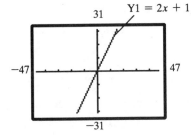

$Y1 = 2x + 1$

b.

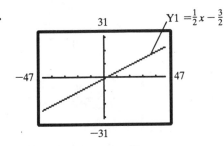

$Y1 = \frac{1}{2}x - \frac{3}{2}$

c.

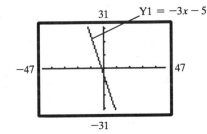

$Y1 = -3x - 5$

2. a. $m = 2$ $b = 1$ $m = 2$ $b = 1$

b. $m = \frac{1}{2}$ $b = -\frac{3}{2}$ $m = \frac{1}{2}$

       $b = -\frac{3}{2}$

c. $m = -3$ $b = -5$ $m = -3$ $b = -5$

3. coinciding

4. equal

5. equal

The graphs of two linear equations in two variables are coinciding if their corresponding equations written in slope-intercept form have equal slopes (*m*) and equal *y*-intercepts (*b*).

Discovery 7

1. a.

$Y2 = 2x + 10$

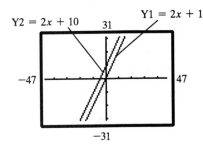

$Y1 = 2x + 1$

b.

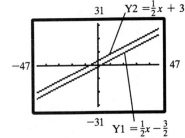

$Y2 = \frac{1}{2}x + 3$

$Y1 = \frac{1}{2}x - \frac{3}{2}$

c. $Y1 = -3x - 5$

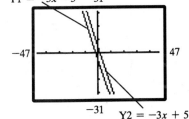

$Y2 = -3x + 5$

2 a. $m = 2$ $b = 1$ $m = 2$ $b = 10$

b. $m = \frac{1}{2}$ $b = -\frac{3}{2}$ $m = \frac{1}{2}$ $b = 3$

c. $m = -3$ $b = -5$ $m = -3$ $b = 5$

3. parallel

4. equal

5. not equal

The graphs of two linear equations in two variables are parallel if their corresponding equations written in slope-intercept form have equal slopes (*m*) and non-equal *y*-intercepts (*b*).

Discovery 8

1. a.

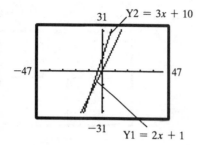

b.

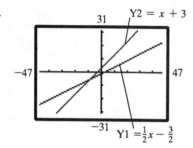

c.

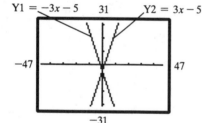

2. a. $m = 2$ $b = 1$ $m = 3$ $b = 10$

b. $m = \dfrac{1}{2}$ $b = -\dfrac{3}{2}$ $m = 1$ $b = 3$

c. $m = -3$ $b = -5$ $m = 3$ $b = -5$

3. intersecting

4. not equal

The graphs of two linear equations in two variables are intersecting if their corresponding equations written in slope-intercept form have non-equal slopes (m).

Discovery 9

1. a.

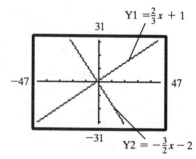

b.

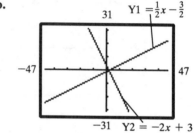

c.

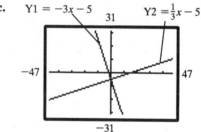

2. a. $m = \dfrac{2}{3}$ $b = 1$ $m = -\dfrac{3}{2}$ $b = -2$

b. $m = \dfrac{1}{2}$ $b = -\dfrac{3}{2}$ $m = -2$ $b = 3$

c. $m = -3$ $b = -5$ $m = \dfrac{1}{3}$ $b = -5$

3. intersecting and perpendicular

4. not equal

5. opposite

The graphs of two linear equations in two variables are intersecting and perpendicular if their corresponding equations written in slope-intercept form have non-equal slopes (m) which are opposite reciprocals.

Chapter 6

Discovery 1

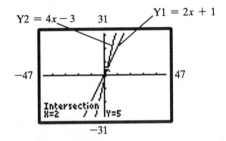

1. $(2, 5)$

2. $(2, 5)$

If the graphs of the two equations of a system intersect, then the coordinates of the point of intersection is the ordered pair solution of the system.

Discovery 2

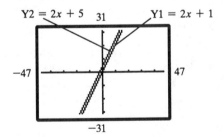

If the graphs of the equations of a system do not intersect, then the system of linear equations is inconsistent or has no solution.

Discovery 3

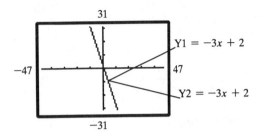

If the graphs of the equations of a system are coinciding, then the system of linear equations consists of dependent equations or has an infinite number of solutions.

Discovery 4

Solve the first equation for y.
$y = 2x + 1$
Substitute $(2x + 1)$ for y in the second equation and solve for x.
$-4x + 2(2x + 1) = 3$
$\quad -4x + 4x + 2 = 3$
$\qquad\qquad\qquad 2 = 3$ (contradiction)

If the substitution method yields a contradiction, then the system of linear equations is inconsistent or has no solution.

Discovery 5

Substitute $(-3x + 2)$ for y in the first equation.
$\quad -3x - (-3x + 2) = -2$
$\quad\;\; -3x + 3x - 2 = -2$
$\qquad\qquad\qquad -2 = -2$ (identity)
If the substitution method yields an identity, then the system of linear equations has dependent equations or an infinite number of solutions.

Discovery 6

Multiply both sides of the first equation by -2.
$\quad -2(-2x + y) = -2(1)$
The new system is:
$\qquad 4x - 2y = -2$
$\qquad -4x + 2y = 3$
Add corresponding members
$\qquad\qquad 0 = 1$ (contradiction)

If the elimination method yields a contradiction, then the system of linear equations is inconsistent or has no solution.

Discovery 7

Write both equations in standard form.
$$-3x - y = -2$$
$$3x + y = 2$$
Add corresponding members.
$$0 = 0 \text{ (identity)}$$
If the elimination method yields an identity, then the system of linear equations has dependent equations or an infinite number of solutions.

Chapter 7

Discovery 1

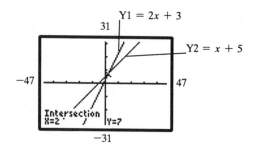

1. $(2, 7)$

2. 2

3. below; left; less than

4. $x \le 2$

Discovery 2

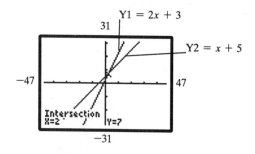

1. $(2, 7)$

2. 2

3. above; right; greater than

4. $x \ge 2$

Discovery 3

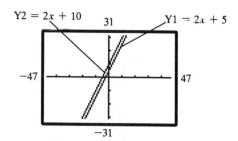

A "less than" inequality Y1 < Y2 has no solution if the graphs of the two functions defined by Y1 and Y2 coincide or if the graphs do not intersect and the graph of Y1 lies above the graph of Y2.

A "greater than" inequality Y1 > Y2 has no solution if the graphs of the two functions defined by Y1 and Y2 coincide or if the graphs do not intersect and the graph of Y1 lies below the graph of Y2.

A "less than or equal to" inequality Y1 ≤ Y2 has no solution if the graphs of the two functions defined by Y1 and Y2 do not intersect and the graph of Y1 lies above the graph of Y2.

A "greater than or equal to" inequality Y1 ≥ Y2 has no solution if the graphs of the two functions defined by Y1 and Y2 do not intersect and the graph of Y1 lies below the graph of Y2.

Discovery 4

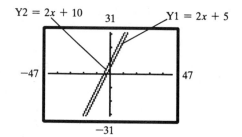

A "less than" inequality Y1 < Y2 has a solution set of all real numbers if the graphs of the two functions defined by Y1 and Y2 do not intersect and the graph of Y1 lies below the graph of Y2.

A "greater than" inequality Y1 > Y2 has a solution set of all real numbers if the graphs of the two functions defined by Y1 and Y2 do not intersect and the graph of Y1 lies above the graph of Y2.

Also, the inequality Y1 ≤ Y2 or Y1 ≥ Y2 has a solution set of all real numbers if the graphs of the two functions defined by Y1 and Y2 are the same.

Discovery 5

1.
$$10 < 12$$

$10 + 2$	$12 + 2$
12	14

2.
$$10 < 12$$

$10 + (-2)$	$12 + (-2)$
8	10

3.
$$10 < 12$$

$10 - 2$	$12 - 2$
8	10

4.
$$10 < 12$$

$10 - (-2)$	$12 - (-2)$
12	14

If a number is added to (or subtracted from) both expressions in a true inequality the resulting inequality is also true.

Discovery 6

1.
$$10 < 12$$

$10 \cdot 2$	$12 \cdot 2$
20	24

2.
$$10 < 12$$

$10 \cdot (-2)$	$12 \cdot (-2)$
−20	−24

3.
$$10 < 12$$

$10 \div 2$	$12 \div 2$
5	6

4.
$$10 < 12$$

$10 \div (-2)$	$12 \div (-2)$
−5	−6

If the expressions in a true inequality are both multiplied (or divided) by a positive number the resulting inequality is also true.
If the expressions in a true inequality are both multiplied (or divided) by a negative number the resulting inequality is not true.

Discovery 7

$$2x + 5 > 2x + 10$$
$2x + 5 - 2x > 2x + 10 - 2x$ Subtract $2x$ from both
sides.
$$5 > 10 \qquad \text{Simplify.}$$

A linear inequality has no solution if in the process of solving the variable term is deleted and the result is a false inequality.

Discovery 8

$$2x + 5 < 2x + 10$$
$2x + 5 - 2x < 2x + 10 - 2x$ Subtract $2x$ from both
sides.
$$5 < 10 \qquad \text{Simplify.}$$

A linear inequality has a solution set of all real numbers if in the process of solving the variable term is deleted and the result is a true inequality.

Discovery 9

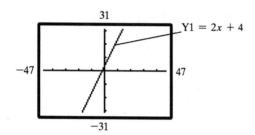

1. a. An infinite number of ordered pairs can be found.

  b. yes

2. a. An infinite number of ordered pairs can be found.

  b. no

3. a. An infinite number of ordered pairs can be found.

  b. yes

The solution set of a "less than or equal to" inequality that is solved for y may be illustrated by graphing the line for the related equation and shading all points below the line.

Discovery 10

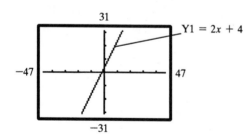

1. a. An infinite number of ordered pairs can be found.

  b. no

2 a. An infinite number of ordered pairs can be found.

  b. yes

3. a. An infinite number of ordered pairs can be found.

  b. no

The solution set of a "greater than" inequality that is solved for y may be illustrated by graphing the line for the related equation with a dashed line and shading all points above the line.

Discovery 11

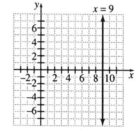

1. a. An infinite number of ordered pairs can be found.

  b. yes

  c. no

  d. yes

  e. no

2. a. An infinite number of ordered pairs can be found.

  b. yes

  c. yes

  d. no

  e. no

3. a. An infinite number of ordered pairs can be found.

  b. no

  c. no

d. yes

e. yes

The solution set of a linear inequality containing no y-term that is solved for x may be illustrated by graphing the line for the related equation. If the inequality symbol is <, then the line is dashed and all points to the left of the line are shaded. If the inequality symbol is >, then the line is dashed and all points to the right of the line are shaded. If the inequality symbol is ≤, then the line is solid and all points to the left of the line is shaded. If the inequality symbol is ≥, then the line is solid and all points to the right of the line is shaded.

Chapter 8

Discovery 1

1.

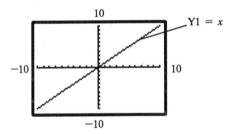

2.

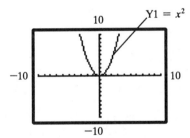

3.

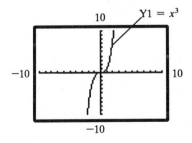

4.

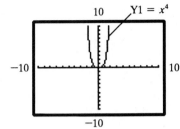

Each polynomial relation passes the vertical line test.

Discovery 2

1.–6.

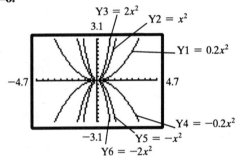

7. positive; upward

8. negative; downward

9. narrower

10. wider

Discovery 3

1.

x	y	
-3	9	
-2	4	
-1	1	
0	0	← vertex
1	1	
2	4	
3	9	

2.

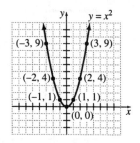

3. 1

4. 4

5. 9

Discovery 4

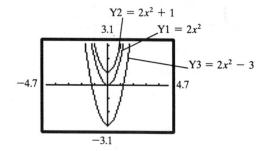

The y-coordinate of the y-intercept is the constant term.

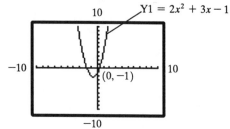

Chapter 9

Discovery 1

  1. a. 1024

    b. 1024

2. a. 64

  b. 64

3. a. $\dfrac{27}{64}$

  b. $\dfrac{27}{64}$

To multiply exponential expressions with like bases, add the exponents.

  4. x^7

Discovery 2

  1. a. 64

    b. 64

  2. a. 4

    b. 4

  3. a. $\dfrac{9}{16}$

    b. $\dfrac{9}{16}$

  4. a. indeterminate

    b. indeterminate

To divide exponential expressions with like bases, subtract the exponent in the denominator from the exponent in the numerator.

  5. x^4

Discovery 3

  1. a. 4096

    b. 4096

  2. a. 256

    b.   256

3. a. $\dfrac{729}{4096}$

    b. $\dfrac{729}{4096}$

To raise an exponential expression to a power, multiply the exponents.

    4. x^{12}

Discovery 4

    1. a.   512

       b.   512

    2. a.   36

       b.   36

    3. a. $\dfrac{27}{1000}$

       b. $\dfrac{27}{1000}$

Raising a product to a power is equivalent to the product of the factors raised to a power.

    4. $x^4 y^4$

Discovery 5

    1. $x^2 - 25$

    2. $9x^2 - 1$

    3. $x^2 - y^2$

The product of the sum and difference of the same two terms is the difference of the square of the first term and the square of the second term.

Discovery 6

    1. $x^2 + 10x + 25$

    2. $9x^2 - 6x + 1$

    3. $x^2 + 2xy + y^2$

    4. $x^2 - 2xy + y^2$

The square of a binomial that is a sum of two terms is the square of the first term, plus two times the product of the first and last terms, plus the square of the last term.

The square of a binomial that is a difference of two terms is the square of the first term, minus two times the product of the first and last terms, plus the square of the last term.

Discovery 7

    1. $x \cdot x \cdot x; \quad x \cdot x \cdot y; \quad x \cdot x \cdot x \cdot x$

    2. x

    3.   2

    4. x^2

The GCF for a set of monomials with variable factors is the common variable with the smallest exponent common in the set of factors.

Discovery 8

    1. a. $x^2 + 5x + 6$

       b.   6; 5

    2. a. $x^2 - 5x + 6$

       b.   6; −5

The first term in each binomial factor is x, or the square root of the quadratic term, x^2. The second term in each of the binomial factors must be the factors of the constant term, c. The sum of the factors is the coefficient b.

Chapter 10

Discovery 9

1. $(x + 3)(x + 2)$; positive

2. $(x - 3)(x - 2)$; negative

3. $(x + 3)(x - 2)$; positive

4. $(x - 3)(x + 2)$; negative

The signs of the factors are determined in the following way:

> If c is positive, then the factors both have the same sign as b.
>
> If c is negative, then the factors have different signs (the factor with the larger absolute value has the same sign as b).

Discovery 10

1. $x^3 + 125$

2. $x^3 - 125$

The first product is a sum of two cubes. The second product is a difference of two cubes. In order to factor these polynomials, we turn these results around and write a binomial and trinomial factor.

Discovery 1

x	$4x^2 - x - 3$	$3x$	
−3	36	−9	$36 > -9$
−2	15	−6	$15 > -6$
−1	2	−3	$2 > -3$
0	−3	0	$-3 < 0$
1	0	3	$0 < 3$
2	11	6	$11 > 6$
3	30	9	$30 > 9$

The expression on the left is greater than the expression on the right for x-values of −3, −2, and −1. The expression on the left is less than the expression on the right for x-values of 0 and 1. Since each function is defined for all real numbers between −1 and 0, the expression on the left is equal to the expression on the right at some x-value between −1 and 0.

Also, the expression on the left is less than the expression on the right for x-values of 0 and 1 and the expression on the left is greater than the expression on the right for x-values of 2 and 3. Since each function is defined for all real numbers between 1 and 2, the expression on the left is equal to the expression on the right at some x-value between 1 and 2.

Discovery 2

x	$x^2 + x + 10$	$x^2 + x - 10$		
0	10	−10	$10 \neq -10$	$10 - (-10) = 20$
1	12	−8	$12 \neq -8$	$12 - (-8) = 20$
2	16	−4	$16 \neq -4$	$16 - (-4) = 20$
3	22	2	$22 \neq 2$	$22 - 2 = 20$

The expression on the right is always 20 less than the expression on the left. Therefore, it appears that the two expressions are never equal. The equation does not have a solution.

Discovery 3

x	$2x^3 + 4x^2 + 5$	$(x^3 + 3x^2 + 4) + (x^3 + x^2 + 1)$	
0	5	5	$5 = 5$
1	11	11	$11 = 11$
2	37	37	$37 = 37$
3	95	95	$95 = 95$

The two expressions are equal for all values of the independent variable in the table. If we could view this table with all real-number values for x, the expressions would always be equal. Therefore, the solution of the equation is the set of all real numbers (the permissible replacements for the independent variable).

Discovery 4

x	$x^2 + x + 1$	$4x^2 + 4x + 4$
0	1	4
1	3	12
2	7	28
3	13	52

The expression on the left is always less than the expression on the right. Therefore, the two expressions are never equal. It appears the equation does not have a solution.

Discovery 5

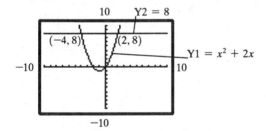

The values of the x-coordinates of the points of intersection of the two graphs are the solutions of the equation.

The values of the y-coordinates of the points of intersection of the two graphs are the values obtained when both expressions are evaluated with the solution.

Discovery 6

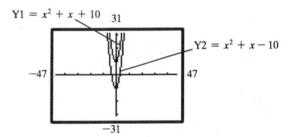

The two graphs do not appear to intersect. Therefore, there is no ordered pair common to both functions.

Discovery 7

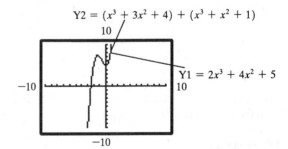

There is only one graph on the screen. (Actually, both graphs are the same.) Therefore, all ordered pairs on the graph are common to both functions, or all
x-coordinates in the domain of the functions are solutions of the equation.

Discovery 8

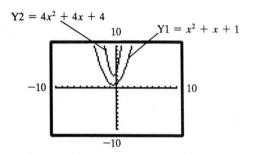

Y2 = $4x^2 + 4x + 4$

10

Y1 = $x^2 + x + 1$

−10   10

−10

The two graphs do not appear to intersect. Therefore, there is no ordered pair common to both graphed functions. We need more information to determine the solution.

Discovery 9

  1. a. 4.583

    b. 4.583

  2. a. 2.45

    b. 2.45

To multiply square roots, we multiply the radicands and then take the square root of the product.

Discovery 10

  1. a. 1.732

    b. 1.732

  2. a. 1.414

    b. 1.414

To divide square roots, we divide the radicands and then take the square root of the quotient.

Discovery 11

  1. 49

  2. 41

  3. 0

  4. −7

The radicand, $b^2 - 4ac$, determines the characteristics of the quadratic equation solutions. If the radicand is a perfect square ($\neq 0$), there will be two rational-number solutions. If the radicand is 0, there will be one rational-number solution. If the radicand is a positive number that is not a perfect square, then there will be two irrational-number solutions. If the radicand is a negative number, there will be no real-number solutions.

Discovery 12

  1.

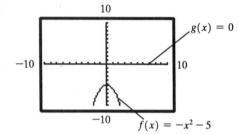

10

$g(x) = 0$

−10   10

−10

$f(x) = -x^2 - 5$

    a. no solution

    b. no solution

    c. all real numbers

    d. all real numbers

  2. $h(x) = x^2 + 5x + 8$

10

−10   10

$j(x) = 0$

−10

    a. all real numbers

    b. all real numbers

    c. no solution

    d. no solution

3.

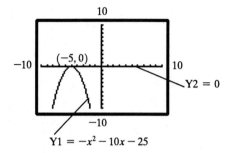

$Y1 = -x^2 - 10x - 25$

  a. no solution

  b. $x = -5$

  c. $x \neq -5$

  d. all real numbers

4.

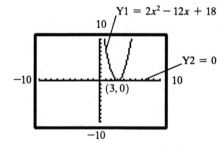

  a. $x \neq 3$

  b. all real numbers

  c. no solution

  d. $x = 3$

Chapter 1

1.1 Experiencing Algebra the Exercise Way

1. a. -15: integer, rational

   b. 29: natural, whole, integer, rational

   c. 1 million: natural, whole, integer, rational

   d. $\dfrac{3}{7}$: rational

3. a. 0: whole, integer, rational

   b. 12.75: rational

   c. $2\dfrac{4}{9}$: rational

   d. -8.35: rational

5. wins \$3 = 3
wins \$1.75 = 1.75
loses \$2.00 = -2
loses \$1.25 = -1.25

7. in the red \$345.67 = -345.67

9. correct = 5
incorrect = -5
unanswered = -2

11. $\dfrac{17}{3} = 5\dfrac{2}{3}$
Graph between 5 and 6.

13.

$$-3.5 \quad -2 \; -1\tfrac{1}{4} \; -\tfrac{4}{5} \quad \tfrac{1}{2} \quad 1.5 \; \tfrac{9}{4} \qquad 4$$

15. $-9 < -5$

17. $0 > -6$

19. $5 > 3$

21. $-\dfrac{1}{2} < -\dfrac{2}{5}$

23. $\dfrac{3}{7} < \dfrac{7}{10}$

25. $2\dfrac{3}{5} > -2\dfrac{3}{5}$

27. $1\dfrac{4}{5} = 1.8$

29. $-3.7 > -5.8$

31. $1.7 < 3.2$

33. T

35. F

37. F

39. F

41. F

43. F

45. F

47. T

49. F

51. $0.295 < 0.7$

53. $-1 > -5$

55. $0.4 = \dfrac{2}{5}$

57. $|15.34| = 15.34$

59. $|-15.34| = 15.34$

61. $\left|-\left(-3\dfrac{1}{3}\right)\right| = \left|3\dfrac{1}{3}\right| = 3\dfrac{1}{3}$

63. $\left|-\left(-\left(-3\frac{1}{3}\right)\right)\right| = \left|-3\frac{1}{3}\right| = 3\frac{1}{3}$

65. $-|23| = -23$

67. $-|-23| = -|23| = -23$

69. $-(15) = -15$

71. $-(-25) = 25$

73. $-\left(-\left(-\frac{3}{2}\right)\right) = -\left(\frac{3}{2}\right) = -\frac{3}{2}$

75. loss of $712 million = -712

1.1 Experiencing Algebra the Calculator Way

1. F

2. F

3. F

```
-193987<-194879
                 0
123/579>296/978
                 0
15+11/19<15+9/16
                 0
```

4. F

5. F

```
119/250>0.476
                 0
-149/275<-356/56
7
                 0
```

6. F

7. F

```
-13.0987>-13.087
9
                 0
-129/500=-0.2585
                 0
```

8. T

9. T

10. F

```
12.368=1546/125
                 1
abs(-1.5)=1.5
                 1
abs(-1.5)=-1.5
                 0
```

11. F

12. T

```
abs(-3/2)=-1.5
                 0
abs(-3/2)=1.5
                 1
```

13. F

14. T

```
abs(-4)=-4
                 0
-abs(4)=-4
                 1
```

1.2 Experiencing Algebra the Exercise Way

1. $-7 + 9 = 2$

3. $0 + (-13) = -13$

5. $-19 + (-12) = -31$

7. $15 + (-15) = 0$

9. $52 + (-13) = 39$

11. $32 + (-579) = -547$

13. $2.7 + 3.96 = 6.66$

15. $1.2 + (-2.5) = -1.3$

17. $-2.73 + 4.1 = 1.37$

19. $-1.1 + (-2.27) = -3.37$

21. $-1.25 + 0 = -1.25$

23. $1.23 + (-1.23) = 0$

25. $-\dfrac{3}{5} + \left(-\dfrac{1}{2}\right)$

$= -\dfrac{6}{10} + \left(-\dfrac{5}{10}\right)$

$= -\dfrac{11}{10}$

27. $-\dfrac{7}{9} + \dfrac{1}{6}$

$= -\dfrac{14}{18} + \dfrac{3}{18}$

$= -\dfrac{11}{18}$

29. $\dfrac{2}{3} + \left(-\dfrac{2}{9}\right)$

$= \dfrac{6}{9} + \left(-\dfrac{2}{9}\right)$

$= \dfrac{4}{9}$

31. $\dfrac{2}{3} + 0 = \dfrac{2}{3}$

33. $\dfrac{1}{3} + \left(-\dfrac{1}{3}\right) = 0$

35. $-2\dfrac{3}{4} + 3\dfrac{2}{3}$

$= -2\dfrac{9}{12} + 3\dfrac{8}{12}$

$= -\dfrac{33}{12} + \dfrac{44}{12}$

$= \dfrac{11}{12}$

37. $1 + \left(-4\dfrac{3}{4}\right) = -3\dfrac{3}{4}$

39. $-13 + (-25) + (-26)$
$= -38 + (-26)$
$= -64$

41. $17 + (-23) + (-16) + 13$
$= -6 + (-16) + 13$
$= -22 + 13$
$= -9$

43. $1 + (-2.3) + (-5.71)$
$= -1.3 + (-5.71)$
$= -7.01$

45. $-\dfrac{1}{2} + \dfrac{1}{3} + \left(-\dfrac{1}{4}\right)$

$= -\dfrac{6}{12} + \dfrac{4}{12} + \left(-\dfrac{3}{12}\right)$

$= -\dfrac{2}{12} + \left(-\dfrac{3}{12}\right)$

$= -\dfrac{5}{12}$

47. $1124 + (-923) + 2305 + (-1156) + (-109)$
$= 201 + 2305 + (-1156) + (-109)$
$= 2506 + (-1156) + (-109)$
$= 1350 + (-109)$
$= 1241$

49. $1500 + 150 + 150 + 150 + (-75) + (-500) + (-200) + 12$
$= 1650 + 150 + 150 + (-75) + (-500) + (-200) + 12$
$= 1800 + 150 + (-75) + (-500) + (-200) + 12$
$= 1950 + (-75) + (-500) + (-200) + 12$
$= 1875 + (-500) + (-200) + 12$
$= 1375 + (-200) + 12$
$= 1175 + 12$
$= 1187$
The net balance of Lindsay's account is $1187.

51. $-148{,}561 + 5{,}253{,}628 = 5{,}105{,}067$
The census for 1990 of Cook County, Illinois was 5,105,067.

53. $8 + \left(-4\dfrac{3}{4}\right) + 22\dfrac{1}{2}$

$= 8 + \left(-4\dfrac{3}{4}\right) + 22\dfrac{2}{4}$

$= \dfrac{32}{4} + \left(-\dfrac{19}{4}\right) + \dfrac{90}{4}$

$= \dfrac{13}{4} + \dfrac{90}{4}$

$= \dfrac{103}{4}$

$= 25\dfrac{3}{4}$

Heath had a net gain of $25\dfrac{3}{4}$ yards.

55. $-255 + 375 + (-575) + 1525$
$= 120 + (-575) + 1525$
$= -455 + 1525$
$= 1070$
Karin is \$1070 above her quota, because 1070 is a positive number.

57. $-2.5 + 1.25 + (-1.8) + (-2.5)$
$= -1.25 + (-1.8) + (-2.5)$
$= -3.05 + (-2.5)$
$= -5.55$
Beth lost 5.5 pounds in four weeks.

59. $378 + 322 + 218$
$= 700 + 218$
$= 918$
Liu drove 918 miles

61. $25 + 8 + (-12) + 17$
$= 33 + (-12) + 17$
$= 21 + 17$
$= 38$
Julia had 38 chips at the end of the third hand.

63. $\dfrac{1}{2} + \dfrac{1}{4} + 1 + 1 + \dfrac{1}{2}$

$= \dfrac{2}{4} + \dfrac{1}{4} + \dfrac{4}{4} + \dfrac{4}{4} + \dfrac{2}{4}$

$= \dfrac{2 + 1 + 4 + 4 + 2}{4}$

$= \dfrac{13}{4}$

$= 3\dfrac{1}{4}$

The recipe requires $3\dfrac{1}{4}$ cups of dry ingredients.

1.2 Experiencing Algebra the Calculator Way

A.

1. $\dfrac{55}{7} = 7\dfrac{6}{7}$

```
55/7
          7.857142857
Ans-7
          .8571428571
Ans▶Frac
                   6/7
```

2. $-\dfrac{295}{113} = -2\dfrac{69}{113}$

```
-295/113
          -2.610619469
Ans+2
          -.610619469
Ans▶Frac
               -69/113
```

3. $\dfrac{1227}{487} = 2\dfrac{253}{487}$

```
1227/487
          2.519507187
Ans-2
          .5195071869
Ans▶Frac
               253/487
```

4. $-\dfrac{108}{19} = -5\dfrac{13}{19}$

```
-108/19
         -5.684210526
Ans+5
         -.6842105263
Ans▶Frac
              -13/19
```

B.

1. $-\dfrac{171}{92} + \left(-\dfrac{170}{33}\right) = -7\dfrac{31}{3036}$

```
-(171/92)+(-170/
33)
       -7.010210804
Ans+7
       -.0102108037
Ans▶Frac
            -31/3036
```

2. $\dfrac{171}{92} + \left(-\dfrac{170}{33}\right) = -3\dfrac{889}{3036}$

```
(171/92)+(-170/3
3)
       -3.292819499
Ans+3
       -.2928194993
Ans▶Frac
           -889/3036
```

3. $-\dfrac{171}{92} + \left(\dfrac{170}{33}\right) = 3\dfrac{889}{3036}$

```
-(171/92)+(170/3
3)
        3.292819499
Ans-3
        .2928194993
Ans▶Frac
            889/3036
```

4. $\dfrac{171}{92} + \left(\dfrac{170}{33}\right) = 7\dfrac{31}{3036}$

```
(171/92)+(170/33
)
        7.010210804
Ans-7
        .0102108037
Ans▶Frac
             31/3036
```

5. $15\dfrac{17}{19} + 21\dfrac{13}{17} = 37\dfrac{213}{323}$

```
(15+17/19)+(21+1
3/17)
        37.65944272
Ans-37
        .6594427245
Ans▶Frac
             213/323
```

6. $15\dfrac{17}{19} + \left(-21\dfrac{13}{17}\right) = -5\dfrac{281}{323}$

```
(15+17/19)-(21+1
3/17)
        -5.86996904
Ans+5
        -.8699690402
Ans▶Frac
            -281/323
```

7. $-15\dfrac{17}{19} + 21\dfrac{13}{17} = 5\dfrac{281}{323}$

```
-(15+17/19)+(21+
13/17)
        5.86996904
Ans-5
        .8699690402
Ans▶Frac
            281/323
```

8. $-15\dfrac{17}{19} + \left(-21\dfrac{13}{17}\right) = -37\dfrac{213}{323}$

```
-(15+17/19)-(21+
13/17)
        -37.65944272
Ans+37
        -.6594427245
Ans▶Frac
            -213/323
```

9. $437.925 + (-108.0065) = 329.9185$

10. $-437.925 + (-108.0065) = -545.9315$

```
437.925+-108.006
5
          329.9185
-437.925+-108.00
65
         -545.9315
```

11. $-437.925 + 108.0065 = -329.9185$

12. $437.925 + 108.0065 = 545.9315$

```
-437.925+108.006
5
         -329.9185
437.925+108.0065
          545.9315
```

13. $893,475 + 1,093,007 = 1,986,482$

14. −893,475 + 1,093,007= 199,532

```
893475+1093007
        1986482
-893475+1093007
         199532
```

15. 893,475 + (−1,093,007) = −199,532

16. −893,475 + (−1,093,007) = −1,986,482

```
893475+-1093007
        -199532
-893475+-1093007

       -1986482
```

1.3 Experiencing Algebra the Exercise Way

1. −7 −9 = −7 + (−9) = −16

3. 0 − (−13) = 0 + 13 = 13

5. −19 − (−12) = −19 + 12 = −7

7. 15 − (−15) = 15 + 15 = 30

9. 52 − (−13) = 52 + 13 = 65

11. 32 − 579 = 32 + (−579) = −547

13. 1.2 − (−2.5) = 1.2 + 2.5 = 3.7

15. −2.73 − 4.1 = −2.73 + (−4.1) = −6.83

17. −1.1 − (−2.27) = −1.1 + 2.27 = 1.17

19. −1.25 − 0 = −1.25

21. 1.23 − (−1.23) = 1.23 + 1.23 = 2.46

23. 2.7 − 3.96 = 2.7 + (−3.96) = −1.26

25. $-\dfrac{3}{5}-\left(-\dfrac{1}{2}\right)=-\dfrac{3}{5}+\dfrac{1}{2}=-\dfrac{6}{10}+\dfrac{5}{10}=-\dfrac{1}{10}$

27. $-\dfrac{7}{9}-\dfrac{1}{6}$

$=-\dfrac{7}{9}+\left(-\dfrac{1}{6}\right)$

$=-\dfrac{14}{18}+\left(-\dfrac{3}{18}\right)$

$=-\dfrac{17}{18}$

29. $\dfrac{2}{3}-\left(-\dfrac{7}{9}\right)=\dfrac{2}{3}+\dfrac{7}{9}=\dfrac{6}{9}+\dfrac{7}{9}=\dfrac{13}{9}$

31. $\dfrac{2}{3}-0=\dfrac{2}{3}$

33. $\dfrac{1}{3}-\left(-\dfrac{1}{3}\right)=\dfrac{1}{3}+\dfrac{1}{3}=\dfrac{2}{3}$

35. $\dfrac{3}{7}-3=\dfrac{3}{7}+(-3)=\dfrac{3}{7}+\left(-\dfrac{21}{7}\right)=-\dfrac{18}{7}$

37. $1\dfrac{5}{6}-\dfrac{1}{3}=1\dfrac{5}{6}+\left(-\dfrac{1}{3}\right)=\dfrac{11}{6}+\left(-\dfrac{2}{6}\right)=\dfrac{9}{6}=\dfrac{3}{2}$

39. $-5-\left(-1\dfrac{4}{5}\right)=-5+1\dfrac{4}{5}=-\dfrac{25}{5}+\dfrac{9}{5}=-\dfrac{16}{5}$

41. $3\dfrac{2}{3}-5=3\dfrac{2}{3}+(-5)=\dfrac{11}{3}+\left(-\dfrac{15}{3}\right)=-\dfrac{4}{3}$

43. −13 − (−25) − (−26)
   = −13 + 25 + 26
   = 12 + 26
   = 38

45. 17 − (−23) − (−16) − 13
   = 17 + 23 + 16 + (−13)
   = 40 + 16 + (−13)
   = 56 + (−13)
   = 43

47. 1.2 − (−2.31) − (−5.7)
   = 1.2 + 2.31 + 5.7
   = 3.51 + 5.7
   = 9.21

49. $5 - (-7) - (-2.3)$
$= 5 + 7 + 2.3$
$= 12 + 2.3$
$= 14.3$

51. $-\dfrac{1}{2} - \dfrac{1}{3} - \left(-\dfrac{1}{4}\right)$
$= -\dfrac{1}{2} + \left(-\dfrac{1}{3}\right) + \dfrac{1}{4}$
$= -\dfrac{6}{12} + \left(-\dfrac{4}{12}\right) + \dfrac{3}{12}$
$= -\dfrac{10}{12} + \dfrac{3}{12}$
$= -\dfrac{7}{12}$

53. $1124 - (-924) - 2305 - (-1156) - (-109)$
$= 1124 + 924 + (-2305) + 1156 + 109$
$= 2048 + (-2305) + 1156 + 109$
$= -257 + 1156 + 109$
$= 899 + 109$
$= 1008$

55. $23 + 56 - 34 + (-12) - 68 - (-31)$
$= 23 + 56 + (-34) + (-12) + (-68) + 31$
$= 79 + (-34) + (-12) + (-68) + 31$
$= 45 + (-12) + (-68) + 31$
$= 33 + (-68) + 31$
$= -35 + 31$
$= -4$

57. $1\dfrac{1}{5} - 2\dfrac{3}{10} + \dfrac{4}{5} - \left(-\dfrac{7}{10}\right) + \left(-\dfrac{3}{5}\right)$
$= 1\dfrac{1}{5} + \left(-2\dfrac{3}{10}\right) + \dfrac{4}{5} + \dfrac{7}{10} + \left(-\dfrac{3}{5}\right)$
$= 1\dfrac{2}{10} + \left(-2\dfrac{3}{10}\right) + \dfrac{8}{10} + \dfrac{7}{10} + \left(-\dfrac{6}{10}\right)$
$= \dfrac{12}{10} + \left(-\dfrac{23}{10}\right) + \dfrac{8}{10} + \dfrac{7}{10} + \left(-\dfrac{6}{10}\right)$
$= -\dfrac{11}{10} + \dfrac{8}{10} + \dfrac{7}{10} + \left(-\dfrac{6}{10}\right)$
$= -\dfrac{3}{10} + \dfrac{7}{10} + \left(-\dfrac{6}{10}\right)$
$= \dfrac{4}{10} + \left(-\dfrac{6}{10}\right)$
$= -\dfrac{2}{10}$
$= -\dfrac{1}{5}$

59. $3.75 - 1.2 + (-1.09) - (-0.76) + 13.13$
$= 3.75 + (-1.2) + (-1.09) + 0.76 + 13.13$
$= 2.55 + (-1.09) + 0.76 + 13.13$
$= 1.46 + 0.76 + 13.13$
$= 2.22 + 13.13$
$= 15.35$

61. $98 - (-90) = 98 + 90 = 188$
The range is 188°F.

63. $130 - (-180) = 130 + 180 = 310$
The change of the mean surface temperature is 310°C.

65. $20,320 - (-282) = 20,320 + 282 = 20,602$
The difference is 20,602 feet.

67. $420.35 + 185.00 + 75.00 + (-50.00) + (-120.00) + (-12.55) + (-110.76) + (-5.50)$
$= 680.35 + (-298.81)$
$= 381.54$
Rolanda's current balance is \$381.54.

69. $20 + \left(-2\frac{1}{2}\right) + \left(-1\frac{2}{3}\right) + (-1) + \left(-2\frac{1}{4}\right)$

$= \frac{240}{12} + \left(-2\frac{6}{12}\right) + \left(-1\frac{8}{12}\right) + \left(-\frac{12}{12}\right) + \left(-2\frac{3}{12}\right)$

$= \frac{240}{12} + \left(-\frac{30}{12}\right) + \left(-\frac{20}{12}\right) + \left(-\frac{12}{12}\right) + \left(-\frac{27}{12}\right)$

$= \frac{210}{12} + \left(-\frac{20}{12}\right) + \left(-\frac{12}{12}\right) + \left(-\frac{27}{12}\right)$

$= \frac{190}{12} + \left(-\frac{12}{12}\right) + \left(-\frac{27}{12}\right)$

$= \frac{178}{12} + \left(-\frac{27}{12}\right)$

$= \frac{151}{12}$

$= 12\frac{7}{12}$

Betty has $12\frac{7}{12}$ cups of flour left.

71. $25 + (-5.75) + 15 + (-12.50) + (-4.50) + 2.35$
$= 19.25 + 15 + (-12.50) + (-4.50) + 2.35$
$= 34.25 + (-12.50) + (-4.50) + 2.35$
$= 21.75 + (-4.50) + 2.35$
$= 17.25 + 2.35$
$= 19.60$
There is $19.60 left in the petty cash fund.

73. $19.95 + 19.95 + 19.95 + (-25) + (-59.27) + (-19.95)$
$= 59.85 + (-104.22)$
$= -44.37$
Rosie lost $44.37 that morning.

1.3 Experiencing Algebra the Calculator Way

1. $-\frac{171}{92} - \left(-\frac{170}{33}\right) = 3\frac{889}{3036}$

```
-171/92--170/33
      3.292819499
Ans-3
      .2928194993
Ans▶Frac
      889/3036
```

2. $\frac{171}{92} - \left(-\frac{170}{33}\right) = 7\frac{31}{3036}$

```
171/92--170/33
      7.010210804
Ans-7
      .0102108037
Ans▶Frac
      31/3036
```

3. $-\dfrac{171}{92} - \left(\dfrac{170}{33}\right) = -7\dfrac{31}{3036}$

```
-171/92-170/33
      -7.010210804
Ans+7
      -.0102108037
Ans▶Frac
           -31/3036
```

4. $\dfrac{171}{92} - \dfrac{170}{33} = -3\dfrac{889}{3036}$

```
171/92-170/33
      -3.292819499
Ans+3
      -.2928194993
Ans▶Frac
          -889/3036
```

5. $15\dfrac{17}{19} - 21\dfrac{13}{17} = -5\dfrac{281}{323}$

```
(15+17/19)-(21+1
3/17)
      -5.86996904
Ans+5
      -.8699690402
Ans▶Frac
          -281/323
```

6. $15\dfrac{17}{19} - \left(-21\dfrac{13}{17}\right) = 37\dfrac{213}{323}$

```
(15+17/19)-(-21-
13/17)
      37.65944272
Ans-37
      .6594427245
Ans▶Frac
          213/323
```

7. $-15\dfrac{17}{19} - 21\dfrac{13}{17} = -37\dfrac{213}{323}$

```
-(15+17/19)-(21+
13/17)
      -37.65944272
Ans+37
      -.6594427245
Ans▶Frac
          -213/323
```

8. $-15\dfrac{17}{19} - \left(-21\dfrac{13}{17}\right) = 5\dfrac{281}{323}$

```
-(15+17/19)-(-21
-13/17)
      5.86996904
Ans-5
      .8699690402
Ans▶Frac
          281/323
```

9. $437.925 - (-108.0065) = 545.9315$

10. $-437.925 - (-108.0065) = -329.9185$

```
437.925--108.006
5
         545.9315
-437.925--108.00
65
         -329.9185
```

11. $-437.925 - 108.0065 = -545.9315$

12. $437.925 - 108.0065 = 329.9185$

```
-437.925-108.006
5
         -545.9315
437.925-108.0065
          329.9185
```

13. $893,475 - 1,093,007 = -199,532$

14. $-893,475 - 1,093,007 = -1,986,482$

```
893475-1093007
         -199532
-893475-1093007
         -1986482
```

15. $893,475 - (-1,093,007) = 1,986,482$

16. $-893,475 - (-1,093,007) = 199,532$

```
893475--1093007
         1986482
-893475--1093007
          199532
```

1.4 Experiencing Algebra the Exercise Way

1. $45 \cdot (-3) = -135$

3. $-32 \cdot (-4) = 128$

5. $-25 \cdot 5 = -125$

7. $51 \cdot 3 = 153$

9. $0 \cdot (-15) = 0$

11. $(0.88)(-1.1) = -0.968$

13. $(-1.7)(-0.2) = 0.34$

15. $(24.3)(0.3) = 7.29$

17. $(-0.25)(50) = -12.5$

19. $\left(\dfrac{2}{5}\right)\left(\dfrac{25}{48}\right) = \dfrac{50}{240} = \dfrac{5}{24}$

21. $\left(1\dfrac{2}{3}\right)\left(-\dfrac{3}{4}\right) = \left(\dfrac{5}{3}\right)\left(-\dfrac{3}{4}\right) = -\dfrac{15}{12} = -\dfrac{5}{4}$

23. $\left(-\dfrac{1}{3}\right)\left(-\dfrac{3}{7}\right) = \dfrac{3}{21} = \dfrac{1}{7}$

25. $\left(-\dfrac{4}{7}\right)\left(\dfrac{3}{16}\right) = -\dfrac{12}{112} = -\dfrac{3}{28}$

27. $(0)\left(-3\dfrac{5}{19}\right) = 0$

29. $(-1.11)(0) = 0$

31. four negative factors: positive product

33. five negative factors: negative product

35. two negative factors: positive product

37. three negative factors: negative product

39. $(-2)(-3)(-4)(-10)(20) = 4800$

41. $\left(-\dfrac{1}{5}\right)\left(-\dfrac{2}{3}\right)\left(-\dfrac{4}{5}\right)\left(\dfrac{1}{2}\right) = -\dfrac{8}{150} = -\dfrac{4}{75}$

43. $(5.2)(-0.1)(-2.2) = 1.144$

45. $\left(\dfrac{1}{3}\right)(-2)(4.2)(5) = -\dfrac{42}{3} = -14$

47. $(5)(-6.8)\left(\dfrac{1}{2}\right)(0) = 0$

49. $(14)(0)(-35)(0)(-312) = 0$

51. $12 \cdot 225 = 2700$
Steve pays $2700 in rent each year.

53. 22% of 645
$(0.22)(-645) = -141.9$
Sara has $141.90 deducted every 2 weeks.
$(6)(-141.9) = -851.4$
This amounts to $851.40 every 12 weeks.

55. $(4)\left(-4\dfrac{1}{2}\right) = -18$
Heath lost 18 yards on these four sacks.

57. $(4.5)(4)(7) = 126$
The track athlete will run 126 miles in four weeks.

59. $(20)(25) = 500$
There are 500 soldiers in each group.

61. $(10)(3)(4) = 120$
There are 120 people riding in the cars.

63. $(0.05)(40)\left(1\dfrac{1}{2}\right)(3) = 9$
Sammy earns $9 for his work.

65. $(4)(3)(20) = 240$
Nellie will be able to store 240 CD's.

67. $(50)\left(8\dfrac{1}{2}\right)(4) = 1700$
George drove approximately 1700 miles.

69. $(24)(0.59)(14) = 198.24$
The total retail value for the vegetables is $198.24.

71. $(100)\left(2\dfrac{3}{4}\right)(12) = 3300$
Drucilla will pay her parents $3300.

73. $(8)(6)(4)(24) = 4608$
4608 ounces will be needed to fill the order.

1.4 Experiencing Algebra the Calculator Way

1. $\left(-\dfrac{171}{92}\right)\left(-\dfrac{170}{33}\right) = 9\dfrac{291}{506}$

```
(-171/92)*(-170/
33)
        9.575098814
Ans-9
        .5750988142
Ans▶Frac
             291/506
```

2. $\left(\dfrac{171}{92}\right)\left(-\dfrac{170}{33}\right) = -9\dfrac{291}{506}$

```
(171/92)*(-170/3
3)
         -9.575098814
Ans+9
         -.5750988142
Ans▸Frac
            -291/506
```

3. $\left(-\dfrac{171}{92}\right)\left(\dfrac{170}{33}\right) = -9\dfrac{291}{506}$

```
(-171/92)*(170/3
3)
         -9.575098814
Ans+9
         -.5750988142
Ans▸Frac
            -291/506
```

4. $\left(\dfrac{171}{92}\right)\left(\dfrac{170}{33}\right) = 9\dfrac{291}{506}$

```
(171/92)*(170/33
)
          9.575098814
Ans-9
          .5750988142
Ans▸Frac
             291/506
```

5. $\left(15\dfrac{17}{19}\right)\left(21\dfrac{13}{17}\right) = 345\dfrac{305}{323}$

```
(15+17/19)*(21+1
3/17)
         345.9442724
Ans-345
         .9442724458
Ans▸Frac
            305/323
```

6. $\left(15\dfrac{17}{19}\right)\left(-21\dfrac{13}{17}\right) = -345\dfrac{305}{323}$

```
(15+17/19)*(-21-
13/17)
        -345.9442724
Ans+345
         -.9442724458
Ans▸Frac
           -305/323
```

7. $\left(-15\dfrac{17}{19}\right)\left(21\dfrac{13}{17}\right) = -345\dfrac{305}{323}$

```
(-15-17/19)*(21+
13/17)
        -345.9442724
Ans+345
         .9442724458
Ans▸Frac
           -305/323
```

8. $\left(-15\dfrac{17}{19}\right)\left(-21\dfrac{13}{17}\right) = 345\dfrac{305}{323}$

```
(-15-17/19)*(-21
-13/17)
         345.9442724
Ans-345
         .9442724458
Ans▸Frac
            305/323
```

9. $(437.925)(-108.0065) = -47{,}298.74651$

10. $(-437.925)(-108.0065) = 47{,}298.74651$

```
437.925*-108.006
5
        -47298.74651
-437.925*-108.00
65
         47298.74651
```

11. $(-437.925)(108.0065) = -47{,}298.74651$

12. $(437.925)(108.0065) = 47{,}298.74651$

```
-437.925*108.006
5
        -47298.74651
437.925*108.0065
         47298.74651
```

13. $(893)(1{,}093) = 976{,}049$

14. $(-893)(1{,}093) = -976{,}049$

```
893*1093
            976049
-893*1093
           -976049
```

15. $(893)(-1{,}093) = -976{,}049$

16. $(-893)(-1{,}093) = 976{,}049$

```
893*-1093
           -976049
-893*-1093
            976049
```

1.5 Experiencing Algebra the Exercise Way

1. $45 \div (-3) = -15$

3. $-32 \div (-4) = 8$

5. $51 \div 3 = 17$

7. $0 \div (-15) = 0$

9. $18 \div (-18) = -1$

11. $26 \div (-0.13) = -200$

13. $-1.7 \div (-0.2) = 8.5$

15. $24.3 \div 0.3 = 81$

17. $-2.7 \div (-2.7) = 1$

19. $\dfrac{-25}{5} = -5$

21. $\dfrac{-16}{0} =$ undefined

23. $\dfrac{0.88}{-1.1} = -0.8$

25. $\dfrac{-5.7}{19} = -0.3$

27. $\dfrac{2}{5} \div \dfrac{25}{48} = \dfrac{2}{5} \cdot \dfrac{48}{25} = \dfrac{96}{125}$

29. $\left(1\dfrac{2}{3}\right) \div \left(-\dfrac{3}{4}\right) = \left(\dfrac{5}{3}\right)\left(-\dfrac{4}{3}\right) = -\dfrac{20}{9}$

31. $\dfrac{6}{7} \div \left(-\dfrac{18}{19}\right) = \dfrac{6}{7} \cdot \left(-\dfrac{19}{18}\right) = -\dfrac{114}{126} = -\dfrac{19}{21}$

33. $-\dfrac{1}{3} \div \left(-\dfrac{3}{7}\right) = -\dfrac{1}{3} \cdot \left(-\dfrac{7}{3}\right) = \dfrac{7}{9}$

35. $\left(-1\dfrac{1}{4}\right) \div \left(-\dfrac{4}{5}\right) = \left(-\dfrac{5}{4}\right) \cdot \left(-\dfrac{5}{4}\right) = \dfrac{25}{16}$

37. $\dfrac{4}{7} \div \dfrac{3}{16} = -\dfrac{4}{7} \cdot \dfrac{16}{3} = -\dfrac{64}{21}$

39. $0 \div \left(-3\dfrac{5}{19}\right) = 0$

41. $\dfrac{3}{5} \div 0 =$ undefined

43. $-\dfrac{2}{3} \div \left(-\dfrac{2}{3}\right) = -\dfrac{2}{3} \cdot \left(-\dfrac{3}{2}\right) = \dfrac{6}{6} = 1$

45. $\dfrac{8}{9} \div 4 = \dfrac{8}{9} \cdot \dfrac{1}{4} = \dfrac{8}{36} = \dfrac{2}{9}$

47. $14 \div \left(-\dfrac{1}{3}\right) = 14 \cdot \left(-\dfrac{3}{1}\right) = -42$

49. $\dfrac{0}{1.3} = 0$

51. $(-15)(4) \div (-3)(12) \div 3(-10)$
$= -60 \div (-3)(12) \div 3(-10)$
$= 20(12) \div 3(-10)$
$= 240 \div 3(-10)$
$= 80(-10)$
$= -800$

53. $(-3.3)(2.7) \div (-11)(0.6)$
$= -8.91 \div (-11)(0.6)$
$= 0.81(0.6)$
$= 0.486$

55. $\left(\dfrac{2}{3}\right)\left(-\dfrac{5}{8}\right) \div \left(-\dfrac{5}{16}\right)$
$= -\dfrac{10}{24} \div \left(-\dfrac{5}{16}\right)$
$= -\dfrac{10}{24} \cdot \left(-\dfrac{16}{5}\right)$
$= \dfrac{160}{120}$
$= \dfrac{4}{3}$

57. $\dfrac{20 + \left(-3\frac{1}{2}\right) + 8\frac{3}{4} + 12}{4} = \dfrac{37\frac{1}{4}}{4} = 9\dfrac{5}{16}$

Heath averaged $9\dfrac{5}{16}$ yards per play.

59. $\dfrac{3.5 + 2.0 + 0.5 + (-1.5) + (-2.5) + 3.5 + 4.0 + (-4.0) + 3.5 + 2.5 + 3.5}{11}$

$= \dfrac{15}{11}$

$= 1.\overline{36}$

Professor Chips' average rating was approximately 1.36 points.

61. $364 \div 26 = 14$

There will be 14 soldiers in each row.

63. $479 \div 60 = 7.98\overline{3}$

It will take Michaela approximately 8 hours to drive to New Orleans.

65. $3200 \div 12 = 266.\overline{6}$

The distributor will have 266 full packs.

67. $659 \div 19.4 \approx 33.97$

Al will use approximately 34 gallons of gas on his trip.

69. $47,500 \div 40 = 1187.5$

1187 bottles can be filled.

71. $16,000,000 \text{ seconds} \times \dfrac{1 \text{ minute}}{60 \text{ seconds}} \times \dfrac{1 \text{ hour}}{60 \text{ minutes}} \times \dfrac{1 \text{ day}}{24 \text{ hours}}$

$\approx 185.2 \text{ days}$

It will take a little over 185 days.

73. $850 \div 35 \approx 24.3$

It will take Anita 25 weeks to pay her mother.

75. $335 \div (3 \times 20) = 335 \div 60 \approx 5.6$

Billie will need 6 CD cabinets to store her collection.

77. $1,035,900,000 \div 4 = 258,975,000$

The estimated expenditure for a quarter was $258,975,000.

1.5 Experiencing Algebra the Calculator Way

1. $-\dfrac{171}{92} \div \left(-\dfrac{17}{3}\right) = \dfrac{513}{1564}$

2. $\dfrac{171}{92} \div \left(-\dfrac{17}{3}\right) = -\dfrac{513}{1564}$

```
(-171/92)/(-17/3
)►Frac
            513/1564
(171/92)/(-17/3)
►Frac
           -513/1564
```

3. $-\dfrac{171}{92} \div \dfrac{17}{3} = -\dfrac{513}{1564}$

4. $\dfrac{171}{92} \div \dfrac{17}{3} = \dfrac{513}{1564}$

```
(-171/92)/(17/3)
►Frac
           -513/1564
(171/92)/(17/3)►
Frac
            513/1564
```

5. $\left(15\dfrac{17}{19}\right) \div \left(21\dfrac{13}{17}\right) = \dfrac{2567}{3515}$

6. $\left(15\dfrac{17}{19}\right) \div \left(-21\dfrac{13}{17}\right) = -\dfrac{2567}{3515}$

```
(15+17/19)/(21+1
3/17)►Frac
           2567/3515
(15+17/19)/-(21+
13/17)►Frac
          -2567/3515
```

7. $\left(-15\dfrac{17}{19}\right) \div \left(21\dfrac{13}{17}\right) = -\dfrac{2567}{3515}$

8. $\left(-15\dfrac{17}{19}\right) \div \left(-21\dfrac{13}{17}\right) = \dfrac{2567}{3515}$

```
-(15+17/19)/(21+
13/17)►Frac
          -2567/3515
-(15+17/19)/-(21
+13/17)►Frac
           2567/3515
```

9. $\dfrac{437.925}{-108.0065} \approx -4.055$

10. $\dfrac{-437.925}{-108.0065} \approx 4.055$

```
437.925/-108.006
5
        -4.054617083
-437.925/-108.00
65
         4.054617083
```

11. $\dfrac{-437.925}{108.0065} \approx -4.055$

12. $\dfrac{437.925}{108.0065} \approx 4.055$

```
-437.925/108.006
5
        -4.054617083
437.925/108.0065
         4.054617083
```

13. $893 \div 1093 = 0.817$

14. $-893 \div 1093 \approx -0.817$

```
893/1093
        .8170173833
-893/1093
        -.8170173833
```

15. $893 \div (-1093) \approx -0.817$

16. $-893 \div (-1093) \approx 0.817$

```
893/-1093
        -.8170173833
-893/-1093
         .8170173833
```

1.6 Experiencing Algebra the Exercise Way

1. $15 \cdot 15 \cdot 15 \cdot 15 \cdot 15 \cdot 15 = 15^6$

3. $\left(-\dfrac{1}{5}\right)\left(-\dfrac{1}{5}\right)\left(-\dfrac{1}{5}\right)\left(-\dfrac{1}{5}\right) = \left(-\dfrac{1}{5}\right)^4$

5. $(3.7)(3.7)(3.7) = (3.7)^3$

7. $0 \cdot 0 \cdot 0 \cdot 0 \cdot 0 \cdot 0 \cdot 0 = 0^7$

9. $(-55)^8$, positive

11. -55^8, negative

13. $(-55)^5$, negative

15. -55^5, negative

17. $3^4 = 3 \cdot 3 \cdot 3 \cdot 3 = 81$

19. $(-3)^4 = (-3)(-3)(-3)(-3) = 81$

21. $-3^4 = -(3 \cdot 3 \cdot 3 \cdot 3) = -81$

23. $-(-3)^4 = -[(-3)(-3)(-3)(-3)] = -81$

25. $(-4)^3 = (-4)(-4)(-4) = -64$

27. $-(-4)^3 = -[(-4)(-4)(-4)] = -(-64) = 64$

29. $(-2.5)^2 = (-2.5)(-2.5) = 6.25$

31. $-(0.5)^3 = -[(0.5)(0.5)(0.5)] = -0.125$

33. $-\left(-\dfrac{3}{7}\right)^2 = -\left[\left(-\dfrac{3}{7}\right)\left(-\dfrac{3}{7}\right)\right] = -\dfrac{9}{49}$

35. $\left(1\dfrac{1}{3}\right)^3 = \left(\dfrac{4}{3}\right)^3 = \left(\dfrac{4}{3}\right)\left(\dfrac{4}{3}\right)\left(\dfrac{4}{3}\right) = \dfrac{64}{27}$

37. $0^8 = 0$

39. $1^{32} = 1$

41. $(-1)^{29} = -1$

43. $(-1)^{122} = 1$

45. $(-79)^4 = 38,950,081$

47. $(-1.08)^5 \approx -1.469$

49. $\left(-\dfrac{2}{3}\right)^6 = \dfrac{64}{729}$

51. $\left(\dfrac{5}{6}\right)^5 = \dfrac{3125}{7776}$

53. $(-37)^5 = -69,343,957$

55. $(2.24)^6 \approx 126.325$

57. $-\left(3\dfrac{2}{11}\right)^4 = -\left(\dfrac{35}{11}\right)^4 \approx -102.495$

59. $-\left(7\dfrac{21}{29}\right)^7 = -\left(\dfrac{224}{29}\right)^7 \approx -1,640,401.445$

61. $1256^1 = 1256$

63. $-(13.06)^1 = -13.06$

65. $1256^0 = 1$

67. $-4721^0 = -1$

69. $7^{-2} = \left(\dfrac{1}{7}\right)^2 = \dfrac{1}{49}$

71. $\left(\dfrac{2}{3}\right)^{-4} = \left(\dfrac{3}{2}\right)^4 = \dfrac{81}{16}$

73. $(-0.2)^{-2}$
$= \left(-\dfrac{2}{10}\right)^{-2}$
$= \left(-\dfrac{10}{2}\right)^2$
$= (-5)^2$
$= 25$

75. $-11^{-2} = -\left(\dfrac{1}{11}\right)^2 = -\dfrac{1}{121}$

77. $(-12)^{-1} = \left(-\dfrac{1}{12}\right)^1 = -\dfrac{1}{12}$

79. $-(-97)^{-1} = -\left(-\dfrac{1}{97}\right)^1 = \dfrac{1}{97}$

81. $-\left(-\dfrac{5}{6}\right)^{-4} = -\left(-\dfrac{6}{5}\right)^{4} = -\dfrac{1296}{625}$

83. $-(2.06)^{-4} = -\left(\dfrac{1}{2.06}\right)^{4} \approx 0.056$

85. $-\left(2\dfrac{2}{9}\right)^{-2} = -\left(\dfrac{20}{9}\right)^{-2} = -\left(\dfrac{9}{20}\right)^{2} = -\dfrac{81}{400}$

87. $\left(-\dfrac{2}{7}\right)^{-3} = \left(-\dfrac{7}{2}\right)^{3} = -\dfrac{343}{8}$

89. $\left(1\dfrac{1}{5}\right)^{-3} = \left(\dfrac{6}{5}\right)^{-3} = \left(\dfrac{5}{6}\right)^{3} = \dfrac{125}{216}$

91. $-\left(-5\dfrac{4}{11}\right)^{-3}$

$= -\left(-\dfrac{59}{11}\right)^{-3}$

$= -\left(-\dfrac{11}{59}\right)^{3}$

≈ -0.006

93. $(3.5)^{3} = 42.875$
The volume of the pit is 42.875 ft^{3}.

95. $(6.5)^{2} = 42.25$
Aladdin's magic carpet will cover 42.25 ft^{2}.

97. $(755)^{2} = 570,025$
The Great Pyramid covers 570,025 ft^{2} of ground.

1.6 Experiencing Algebra the Calculator Way

1. $17^{6} = 24,137,569$
```
17^6
        24137569
```

2. $-17^{6} = -24,137,569$
```
-17^6
        -24137569
```

3. $(-17)^{6} = 24,137,569$
```
(-17)^6
        24137569
```

4. $17^{5} = 1,419,857$
```
17^5
        1419857
```

5. $-17^{5} = -1,419,857$
```
-17^5
        -1419857
```

6. $(-17)^{5} = -1,419,857$
```
(-17)^5
        -1419857
```

7. $-(-24)^{6} = -191,102,976$
```
-(-24)^6
        -191102976
```

8. $-\left(\dfrac{2}{9}\right)^4 = -\dfrac{16}{6561}$

```
-(2/9)^4▶Frac
          -16/6561
```

9. $\left(-\dfrac{2}{9}\right)^4 = \dfrac{16}{6561}$

```
(-2/9)^4▶Frac
          16/6561
```

10. $(12.89)^4 \approx 27,606.520$

```
(12.89)^4
        27606.52033
```

11. $-(12.89)^4 \approx -27,606.520$

```
-(12.89)^4
        -27606.52033
```

12. $(-0.56)^4 \approx 0.098$

```
(-0.56)^4
        .09834496
```

13. $-(-0.56)^4 \approx -0.098$

```
-(-0.56)^4
        -.09834496
```

14. $-\left(3\dfrac{5}{6}\right)^4 = -\left(\dfrac{23}{6}\right)^4 = -\dfrac{279,841}{1296}$

```
-(23/6)^4▶Frac
        -279841/1296
```

15. $\left(-4\dfrac{1}{4}\right)^4 = \left(-\dfrac{17}{4}\right)^4 = \dfrac{83,521}{256}$

```
(-17/4)^4▶Frac
        83521/256
```

16. $-\left(-4\dfrac{1}{4}\right)^4 = -\left(-\dfrac{17}{4}\right)^4 = -\dfrac{83,521}{256}$

```
-(-17/4)^4▶Frac
        -83521/256
```

17. $\left(-\dfrac{4}{7}\right)^7 \approx -0.020$

```
(-4/7)^7
        -.0198945289
```

18. $-\left(-\dfrac{4}{7}\right)^7 \approx 0.020$

```
-(-4/7)^7
        .0198945289
```

19. $(-4.076)^7 \approx -18,691.288$

```
(-4.076)^7
        -18691.28792
```

20. $-(-4.076)^7 \approx 18,691.288$

```
-(-4.076)^7
       18691.28792
```

21. $\left(2\dfrac{2}{7}\right)^5 = \left(\dfrac{16}{7}\right)^5 \approx 62.389$

```
(16/7)^5
       62.38924258
```

22. $-\left(-3\dfrac{2}{3}\right)^7 = -\left(-\dfrac{11}{3}\right)^7 = \dfrac{19,487,171}{2187}$

```
-(-11/3)^7▶Frac
   19487171/2187
```

23. $\left(-7\dfrac{19}{21}\right)^3 = \left(-\dfrac{166}{21}\right)^3 = -\dfrac{4,574,296}{9261}$

```
(-166/21)^3▶Frac
   -4574296/9261
```

24. $\left(-17\dfrac{21}{37}\right)^6 = \left(-\dfrac{650}{37}\right)^6 \approx 29,394,751.66$

```
(-650/37)^6
      29394751.66
```

1.7 Experiencing Algebra the Exercise Way

1. $23,450,000,000 = 2.345 \times 10^{10}$

3. $-203,415,000,000,000 = -2.03415 \times 10^{14}$

5. $0.0000000176 = 1.76 \times 10^{-8}$

7. $-0.000006591 = -6.591 \times 10^{-6}$

9. $3.6943 = 3.6943 \times 10^{0}$

11. $-4.7502 = -4.7502 \times 10^{0}$

13. 6.3 E17
$= 6.3 \times 10^{17}$
$= 630,000,000,000,000,000$

15. $-7.1103 \text{ E}5 = -7.1103 \times 10^{5} = -711,030$

17. $-3.7 \text{ E}{-5} = -3.7 \times 10^{-5} = -0.000037$

19. $1.966 \text{ E}{-2} = 1.966 \times 10^{-2} = 0.01966$

21. $4.356 \text{ E}0 = 4.356 \times 10^{0} = 4.356$

23. $-9.95 \text{ E}0 = -9.95 \times 10^{0} = -9.95$

25. $2.7 \times 10^{7} = 27,000,000$

27. $-4.005 \times 10^{6} = -4,005,000$

29. $4.056 \times 10^{-7} = 0.0000004056$

31. $-3.0303 \times 10^{-4} = -0.00030303$

33. $1.26 \times 10^{0} = 1.26$

35. $-4.5 \times 10^{0} = -4.5$

37. $\$1.024769 \times 10^{12} = \$1,024,769,000,000$

39. $1.024769 \times 10^{12} - 2.71638 \times 10^{11}$
$= 7.53131 \times 10^{11}$
The GNP of the U.K. is $\$7.53131 \times 10^{11}$ or $\$753,131,000,000$ more than the GNP of India.

41. $14,438,000 = 1.4438 \times 10^7$ crimes

$255,458,000 = 2.55458 \times 10^8$ population

$$\frac{1.4438 \times 10^7}{2.55458 \times 10^8} \approx 0.056518$$

5.6518×10^{-2} ratio

In 1992, there were about 0.056518 crimes per person in the U.S.

43. $1,878,285$

$= 1.878285 \times 10^6$ American Indians

$248,700,000 = 2.487 \times 10^8$ total population

$$\frac{1,878,285}{248,700,000} = 0.00755$$

0.755% of the population were American Indians.

45. $258,245,000 \times 365$

$= 9.4259425 \times 10^{10}$

9.4259425×10^{10} or 94,259,425,000 cups of fruit juice were consumed in a year.

47. $2.65 \times 10^9 = 2,650,000,000$

2,650,000,000 Christmas cards are expected to be sold in 1996.

49. $6,000,000 \times 2000 = 12,000,000,000$

$= 1.2 \times 10^{10}$

Each pyramid at Giza would weigh over 1.2×10^{10} pounds.

51. $3,700 \times 5280 = 1.9536 \times 10^7$

The Great Wall of China is 1.9536×10^7 feet long.

1.7 Experiencing Algebra the Calculator Way

1. a. 6.023 E23

```
6.023E23
```

b. 602,300,000,000,000,000,000,000

c. scientific notation; takes up less room

d. 1.2046 E26

```
1.2046E26
```

e. 1.2046×10^{26}

2. a. 5.642151×10^9

b. 5.642151 E9

```
5.642151E9
```

c.
```
1.6926453*10^10
```

1.6926453×10^{10} people will live in the world.

d. 16,926,453,000

1.8 Experiencing Algebra the Exercise Way

1. $\sqrt{36} = 6$ because $6^2 = 36$.

3. $\sqrt{256} = 16$ because $16^2 = 256$.

5. $-\sqrt{25} = -5$ because $5^2 = 25$.

7. $\sqrt{0.64} = 0.8$ because $0.8^2 = 0.64$.

9. $\sqrt{\dfrac{49}{81}} = \dfrac{7}{9}$ because $7^2 = 49$ and $9^2 = 81$.

11. $-\sqrt{\dfrac{16}{9}} = -\dfrac{4}{3}$ because $4^2 = 16$ and $3^2 = 9$.

13. $\sqrt{1} = 1$ because $1^2 = 1$.

15. $-\sqrt{0} = 0$ because $0^2 = 0$.

17. $\sqrt{-16}$ is not a real number.

19. $\sqrt{10}$ lies between $\sqrt{9} = 3$ and $\sqrt{16} = 4$ or approximately 3.162.

21. $-\sqrt{3}$ lies between $-\sqrt{4} = -2$ and $-\sqrt{1} = -1$ or approximately -1.732.

23. $-\sqrt{1200} \approx -34.641$

25. $\sqrt{10.5} \approx 3.240$

27. $-\sqrt{\dfrac{1}{10}} \approx -0.316$

29. $\sqrt{5\dfrac{8}{13}} \approx 2.370$

31. $-\sqrt{2.5} \approx -1.581$

33. $\sqrt[3]{64} = 4$

35. $\sqrt[3]{1728} = 12$

37. $\sqrt[3]{1234} \approx 10.726$

39. $\sqrt[3]{-125} = -5$

41. $-\sqrt[4]{1296} = -6$

43. $\sqrt[6]{17} \approx 1.604$

45. $\sqrt[4]{-1296}$ is not a real number

47. $\sqrt[5]{-12.7} \approx -1.662$

49. $\sqrt[4]{11.3} \approx 1.833$

51. $\sqrt[3]{\dfrac{1}{8}} = \dfrac{1}{2}$

53. $\sqrt[5]{-\dfrac{32}{3125}} = -\dfrac{2}{5}$

55. $-\sqrt[3]{2\dfrac{5}{7}} \approx -1.395$

57. $\sqrt[5]{-7\dfrac{19}{32}} = \sqrt[5]{-\dfrac{243}{32}} = -\dfrac{3}{2} = -1.5$

59.

61. $A = s^2$
$182.25 = s^2$
$s = \sqrt{182.25} = 13.5$
Each side is 13.5 feet.

63. $l = 8$ ft
$w = 5$ ft
diagonal $= \sqrt{8^2 + 5^2}$
$= \sqrt{64 + 25}$
$= \sqrt{89}$
≈ 9.434
Jennie will need approximately 9.434 feet of logs.

65. Current $= \dfrac{120}{400\sqrt{3}} \approx 0.173$
The current will be approximately 0.173 ampere.

67. Speed $= 2\sqrt{5 \cdot 50}$
$= 2\sqrt{250}$
$= 2(15.8113883)$
≈ 31.623
The speed was approximately 32 mph.

69. Speed $= (3.25)\sqrt{300}$
$= (3.25)(17.32050808)$
≈ 56.29
The speed was approximately 56 mph.

71. flow rate $= (2.97)(1.5^2)\sqrt{40}$
$\approx (2.97)(2.25)(6.32455532)$
≈ 42.264
The flow rate is approximately
42.26 gallons per minute.

1.8 Experiencing Algebra the Calculator Way

1. $\sqrt{351,649} = 593$

```
√(351649)
            593
```

2. $\sqrt[3]{-13,312,053} = -237$

```
³√(-13312053)
            -237
```

3. $\sqrt{\dfrac{1681}{2601}} = \dfrac{41}{51}$

```
√(1681/2601)▶Fra
c
            41/51
```

4. $\sqrt{695.798884} = 26.378$

```
√(695.798884)
            26.378
```

5. $\sqrt[3]{-\dfrac{512}{729}} = -\dfrac{8}{9}$

```
³√(-512/729)▶Fra
c
            -8/9
```

6. $\sqrt[5]{1,160,290,625} = 65$

```
5×√1160290625
            65
```

7. $\sqrt{-24,025}$ is not a real number.

8. $-\sqrt{24,025} = -155$

```
-√(24025)
            -155
```

9. $\sqrt[5]{14539.33568} = 6.8$

```
5×√14539.33568
            6.8
```

10. $\sqrt{\pi} \approx 1.772$

```
√(π)
            1.772453851
```

11. $\sqrt{1999} \approx 44.710$

```
√(1999)
            44.71017781
```

12. $\sqrt{\dfrac{2}{3}} \approx 0.816$

```
√(2/3)
            .8164965809
```

1.9 Experiencing Algebra the Exercise Way

1. b

3. c

5. f

7. b

9. a

11. e

13. d

15. g

17. h

19. j

21. i

23.

Number	12	−3	−6.2	3.5	$\dfrac{5}{9}$	$-\dfrac{3}{5}$
Opposite	−12	3	6.2	−3.5	$-\dfrac{5}{9}$	$\dfrac{3}{5}$
Reciprocal	$\dfrac{1}{12}$	$-\dfrac{1}{3}$	$-\dfrac{1}{6.2}$	$\dfrac{1}{3.5}$	$\dfrac{9}{5}$	$-\dfrac{5}{3}$

25. $-4.33 + 1.9 = 1.9 + (-4.33)$

27. $58(-65) = -65(58)$

29. $5.6[(-3.7)(1.1)] = [5.6(-3.7)](1.1)$

31. $\left(\dfrac{5}{13} + \dfrac{8}{13}\right) + \dfrac{1}{13} = \dfrac{5}{13} + \left(\dfrac{8}{13} + \dfrac{1}{13}\right)$

33. $\left(\dfrac{3}{8}\right)\left(\dfrac{5}{7} - \dfrac{1}{9}\right) = \left(\dfrac{3}{8}\right)\left(\dfrac{5}{7}\right) - \left(\dfrac{3}{8}\right)\left(\dfrac{1}{9}\right)$

35. $2.7(-1.5 + 3.2) = 2.7(-1.5) + 2.7(3.2)$

37. $\dfrac{217 - 175}{7} = \dfrac{217}{7} - \dfrac{175}{7}$

39. $-(15 + 19.3) = -15 + (-19.3) = -15 - 19.3$

41. $-\left(-\dfrac{6}{7} - \dfrac{5}{9}\right) = -\left(-\dfrac{6}{7}\right) + \left[-\left(-\dfrac{5}{9}\right)\right] = \dfrac{6}{7} + \dfrac{5}{9}$

43. $-\left(1\dfrac{1}{7} - 2\dfrac{1}{5}\right)$

$= -1\dfrac{1}{7} + \left[-\left(-2\dfrac{1}{5}\right)\right]$

$= -1\dfrac{1}{7} + 2\dfrac{1}{5}$

45. $-(19.37 + 15.043)$

$= -19.37 + (-15.043)$

$= -19.37 - 15.043$

47. $15(17) + 15(21) = 15(17 + 21)$

49. $-3(14) - 3(21) = -3(14 + 21)$

51. $-6(12) - 6(-13) = -6(12 - 13)$

53. $-3(15) + (-3)(17) = -3(15 + 17)$

55. $-(6^2 - 12) + 5(-3 - 8)$

$= -(36 - 12) + 5(-3 - 8)$

$= -(24) + 5(-11)$

$= -24 + (-55)$

$= -79$

57. $[4(7 - 5) + 2] - 9$

$= [4(2) + 2] - 9$

$= (8 + 2) - 9$

$= 10 - 9$

$= 1$

59. $-\sqrt{36 + 64}$

$= -\sqrt{100}$

$= -10$

61. $\dfrac{18 - 4^2 + 7}{2 + 1}$

$= \dfrac{18 - 16 + 7}{2 + 1}$

$= \dfrac{2 + 7}{3}$

$= \dfrac{9}{3}$

$= 3$

63. $2.7 + 5.6 - 16 \div 4 - 3 \cdot 2$

$= 2.7 + 5.6 - 4 - 3 \cdot 2$

$= 2.7 + 5.6 - 4 - 6$

$= 8.3 - 4 - 6$

$= 4.3 - 6$

$= -1.7$

65. $3(-4) - (5)(6) + 4$

$= -12 - 30 + 4$

$= -42 + 4$

$= -38$

67. $(4.3)3 - 5(1.6) + 42.9 \div 3$

$= 12.9 - 5(1.6) + 42.9 \div 3$

$= 12.9 - 8 + 42.9 \div 3$

$= 12.9 - 8 + 14.3$

$= 4.9 + 14.3$

$= 19.2$

69. $-5(39 - 4^2) - 2(23 - 11)$

$= -5(39 - 16) - 2(23 - 11)$

$= -5(23) - 2(12)$

$= -115 - 24$

$= -139$

71. $4(15 - 8) + 31(14 - 11)$

$= 4(7) + 31(3)$

$= 28 + 93$

$= 121$

73. $-|5.2 - 31.3 + 3.95| = -|-22.15| = -22.15$

75. $6^2 + 12 \div (-2) - 12 \cdot (-4)$

$= 36 + 12 \div (-2) - 12 \cdot (-4)$

$= 36 + (-6) - (-48)$

$= 30 + 48$

$= 78$

77. $\left(\dfrac{2}{3}\right)^2 \div \left(\dfrac{1}{3} + \dfrac{1}{2}\right) \cdot \left(\dfrac{8}{9}\right)$

$= \dfrac{4}{9} \div \left(\dfrac{1}{3} + \dfrac{1}{2}\right) \cdot \left(\dfrac{8}{9}\right)$

$= \dfrac{4}{9} \div \left(\dfrac{5}{6}\right) \cdot \left(\dfrac{8}{9}\right)$

$= \dfrac{4}{9} \cdot \dfrac{6}{5} \cdot \dfrac{8}{9}$

$= \dfrac{64}{135}$

79. $\left(\dfrac{1}{5}\right) \cdot \left(\dfrac{15}{22}\right) \div \left(\dfrac{1}{11} + \dfrac{1}{33}\right) - \left(\dfrac{1}{3}\right)^2$

$= \left(\dfrac{1}{5}\right) \cdot \left(\dfrac{15}{22}\right) \div \left(\dfrac{1}{11} + \dfrac{1}{33}\right) - \left(\dfrac{1}{9}\right)$

$= \dfrac{3}{22} \div \left(\dfrac{4}{33}\right) - \left(\dfrac{1}{9}\right)$

$= \dfrac{3}{22} \cdot \left(\dfrac{33}{4}\right) - \dfrac{1}{9}$

$= \dfrac{9}{8} - \dfrac{1}{9}$

$= \dfrac{73}{72}$

81. $100 - (24 + 7^2 - 5) \cdot 3 + 102 \div 2$

$= 100 - (24 + 49 - 5) \cdot 3 + 102 \div 2$

$= 100 - (68) \cdot 3 + 102 \div 2$

$= 100 - 204 + 51$

$= -53$

83. $2\{3[5 - 2(3 + 4)] + 9 \div 3\} - 9(-8)$

$= 2\{3[5 - 2(7)] + 3\} + 72$

$= 2\{3[5 - 14] + 3\} + 72$

$= 2\{3[-9] + 3\} + 72$

$= 2\{-27 + 3\} + 72$

$= 2\{-24\} + 72$

$= -48 + 72$

$= 24$

85. $\dfrac{29 + 3 - 2^3}{2^2 + 2^3}$

$= \dfrac{29 + 3 - 8}{4 + 8}$

$= \dfrac{24}{12}$

$= 2$

87. $15 - \sqrt{2^2 + 3 \cdot 7}$

$= 15 - \sqrt{4 + 21}$

$= 15 - \sqrt{25}$

$= 15 - 5$

$= 10$

89. $\dfrac{4 + 3 \cdot 9 - 5^2 - 2 \cdot 3}{6^2 - 5}$

$= \dfrac{4 + 3 \cdot 9 - 25 - 2 \cdot 3}{36 - 5}$

$= \dfrac{4 + 27 - 25 - 6}{36 - 5}$

$= \dfrac{0}{31}$

$= 0$

91. $\dfrac{8^2 + 3 \cdot 12}{5 - 4 \cdot 6 + 2 \cdot 3^2 + 1^2}$

$= \dfrac{64 + 3 \cdot 12}{5 - 4 \cdot 6 + 2 \cdot 9 + 1}$

$= \dfrac{64 + 36}{5 - 24 + 18 + 1}$

$= \dfrac{100}{0}$

$=$ undefined

93. $\dfrac{87 + 80 + 92 + 76 + 100}{5}$

$= \dfrac{435}{5}$

$= 87$

Holly's average grade on the tests was 87.

95. $\dfrac{675 + 375 + 545 + 390}{4}$

$= \dfrac{1985}{4}$

$= \$496.25$

Estrelita's average weekly sales were $496.25.

97. $\dfrac{168 - 136}{12} = \dfrac{32}{12} = 2\dfrac{2}{3}$ lb

Marilyn lost on average $2\dfrac{2}{3}$ lb per week.

99. "Can't Buy Me Love" sold 7 million records
" Do They Know It's Christmas" sold
7 million records
"We Are the World" sold 7 million records

$x = 2(7) - \dfrac{1}{2}$

$= 13.5$

"I Will Always Love You" sold 13.5 million records

$x = 4(7) + 2$

$= 30$

"White Christmas" sold 30 million records
Total = 7 + 7 + 7 + 13.5 + 30 = 64.5 million
Total records sold was 64.5 million.

1.9 Experiencing Algebra the Calculator Way

1.
```
1085+57(273(480-
575)-(-1233))-(-
857+654)+56²
            -1403590
```

2.
```
15/17-(-19/23+7/
2)-5/3(8/15(2/3-
7/9)-14/17)+(7/3
)²
       5.124198794
```

3.
```
(11.98644/(-2.36
))²(5.76-3.1³)+(
15.75)^-2
      -619.9054362
```

4.
```
√((2.35)³-(11/19
)²)
    3.555656753
```

5.
```
(57²-31(15+17)-4
7(-53))/(14(31)-
29²+15³)▸Frac
           1187/742
```

6.
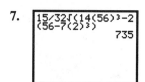
```
57^2-43(-59)
            5786
36^2-54(2^2)(6)
               0
■
```

When this expression is evaluated directly, a "Divide by 0" error message is returned. The numerator and denominator must then be evaluated separately to tell whether the fraction is indeterminate or undefined.

$\dfrac{5786}{0}$ is undefined.

7.
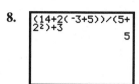
```
15/32√(14(56)³-2
(56-7(2)³)
             735
```

8.
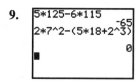
```
(14+2(-3+5))/(5+
2²)+3
               5
```

9.
```
5*125-6*115
            -65
2*7^2-(5*18+2^3)
               0
■
```

When this expression is evaluated directly, a "Divide by 0" error message is returned. The numerator and denominator must then be evaluated separately to tell whether the fraction is indeterminate or undefined.

$\dfrac{-65}{0}$ is undefined.

10.
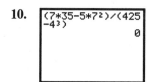
```
(7*35-5*7²)/(425
-4³)
               0
```

Chapter 1 Review

Reflections

1–11. Answers will vary.

Exercises

1. −15: integers, rational numbers
    0: whole numbers, integers, rational numbers
    13: natural numbers, whole numbers, integers, rational numbers
    −12.97: rational numbers

    $3\dfrac{5}{8}$: rational numbers

    $\dfrac{12}{4} = 3$: natural numbers, whole numbers, integers, rational numbers

2. 3 above par = 3
    even par = 0
    2 below par= −2
    1 above par = 1

3.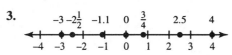

4. $-7 > -9$

5. $\dfrac{13}{45} < \dfrac{4}{9}$

6. $\dfrac{27}{75} = 0.36$

7. $2035 > 491$

8. $12.304 < 12.344$

9. $-28 < 13$

10. $-1.34 < -1.04$

11. $-\dfrac{3}{4} < -\dfrac{5}{8}$

12. $-4\dfrac{1}{3} > -5\dfrac{1}{2}$
    True

13. $\dfrac{13}{23} < \dfrac{51}{91}$
    False

14. $-408 > -513$
    True

15. $-\left|-\dfrac{17}{33}\right| = -\dfrac{17}{33}$

16. $\left|-(-67)\right| = |67| = 67$

17. $-(-(-(257))) = -257$

18. $35 + (-19) = 16$

19. $-1123 + (-3406) = -4529$

20. $82.56 + (-43.7) = 38.86$

21. $-5\dfrac{2}{9} + 4\dfrac{5}{9} = -\dfrac{47}{9} + \dfrac{41}{9} = -\dfrac{6}{9} = -\dfrac{2}{3}$

22. $18 + \left(-1\dfrac{1}{3}\right) = \dfrac{54}{3} + \left(-\dfrac{4}{3}\right) = \dfrac{50}{3}$

23. $\dfrac{102}{43} + \left(-\dfrac{29}{75}\right)$

    $= \dfrac{7650}{3225} + \left(-\dfrac{1247}{3225}\right)$

    $= \dfrac{6403}{3225}$

    $= 1\dfrac{3178}{3225}$

24. $235,407 + (-571,004) = -335,597$

25. $12.097 + 1.92 = 14.017$

26. $\dfrac{32}{77} + \dfrac{50}{99} = \dfrac{288}{693} + \dfrac{350}{693} = \dfrac{638}{693} = \dfrac{58}{63}$

27. $-49.071 + 105.399 = 56.328$

28. $-\dfrac{103}{345}+\left(-\dfrac{21}{115}\right)=-\dfrac{103}{345}+\left(-\dfrac{63}{345}\right)$

　　$=-\dfrac{166}{345}$

29. $0+(-123)=-123$

30. $53+(-97)+33+50+(-83)+(-101)$
　　$=-44+33+50+(-83)+(-101)$
　　$=-11+50+(-83)+(-101)$
　　$=39+(-83)+(-101)$
　　$=-44+(-101)$
　　$=-145$

31. $\dfrac{1}{8}+\dfrac{5}{6}+\left(-\dfrac{2}{3}\right)+\left(-\dfrac{7}{12}\right)+\left(-\dfrac{5}{2}\right)$

　　$=\dfrac{3}{24}+\dfrac{20}{24}+\left(-\dfrac{16}{24}\right)+\left(-\dfrac{14}{24}\right)+\left(-\dfrac{60}{24}\right)$

　　$=\dfrac{23}{24}+\left(-\dfrac{16}{24}\right)+\left(-\dfrac{14}{24}\right)+\left(-\dfrac{60}{24}\right)$

　　$=\dfrac{7}{24}+\left(-\dfrac{14}{24}\right)+\left(-\dfrac{60}{24}\right)$

　　$=-\dfrac{7}{24}+\left(-\dfrac{60}{24}\right)$

　　$=-\dfrac{67}{24}$

32. $-3.5+4.08+(-1.9)+7.36$
　　$=0.58+(-1.9)+7.36$
　　$=-1.32+7.36$
　　$=6.04$

33. $650+(-250)+1200+(-700)=900$
Willy was \$900 above quota at the end of four weeks.

34. $-35-(-61)=-35+61=26$

35. $-\dfrac{3}{8}-\dfrac{4}{7}=-\dfrac{3}{8}+\left(-\dfrac{4}{7}\right)$

　　$=-\dfrac{21}{56}+\left(-\dfrac{32}{56}\right)$

　　$=-\dfrac{53}{56}$

36. $15.6-18=15.6+(-18)=-2.4$

37. $0-3.97=0+(-3.97)=-3.97$

38. $-2.3-(-2.3)=-2.3+2.3=0$

39. $3\dfrac{5}{9}-5\dfrac{2}{9}=\dfrac{32}{9}+\left(-\dfrac{47}{9}\right)=-\dfrac{15}{9}=-\dfrac{5}{3}$

40. $\dfrac{23}{112}-\dfrac{19}{24}=\dfrac{69}{336}+\left(-\dfrac{266}{336}\right)=-\dfrac{197}{336}$

41. $553-(-392)=553+392=945$

42. $-\dfrac{34}{75}-\left(-\dfrac{203}{275}\right)=-\dfrac{374}{825}+\dfrac{609}{825}=\dfrac{235}{825}=\dfrac{47}{165}$

43. $3.5-4.7-(-8.2)-(-10.1)-4.9-(-16.7)$
　　$=3.5+(-4.7)+8.2+10.1+(-4.9)+16.7$
　　$=28.9$

44. $\dfrac{3}{2}-\dfrac{1}{4}-\left(-\dfrac{8}{3}\right)-\dfrac{5}{6}-\left(-\dfrac{7}{2}\right)-3$

　　$=\dfrac{18}{12}+\left(-\dfrac{3}{12}\right)+\dfrac{32}{12}+\left(-\dfrac{10}{12}\right)+\dfrac{42}{12}+\left(-\dfrac{36}{12}\right)$

　　$=\dfrac{43}{12}$

　　$=3\dfrac{7}{12}$

45. $31 - 16 + (-23) - 45 - (-37) + 52 + (-83)$
$= 31 + (-16) + (-23) + (-45) + 37 + 52 + (-83)$
$= -47$

46. $3.7 - 6.83 + 5.5 + (-9.02) - 0.8 - (-15.2)$
$= 3.7 + (-6.83) + 5.5 + (-9.02) + (-0.8) + 15.2$
$= 7.75$

47. $\dfrac{4}{5} - \dfrac{7}{3} + \dfrac{8}{9} - \left(-\dfrac{4}{15}\right) + \left(-\dfrac{2}{3}\right) + 3$

$= \dfrac{36}{45} + \left(-\dfrac{105}{45}\right) + \dfrac{40}{45} + \dfrac{12}{45} + \left(-\dfrac{30}{45}\right) + \dfrac{135}{45}$

$= \dfrac{88}{45}$

$= 1\dfrac{43}{45}$

48. $735.66 + (-276.12) + (-187.05) + (-68.57) + 75.00 + 185.00 + 50 + (-4.65) + (-12.00)$
$= 1045.66 + (-548.39)$
$= 497.27$
Cleta's closing balance was $497.27.

49. $14{,}494 - (-282) = 14{,}494 + 282 = 14{,}776$
The difference in elevation between these two points is 14,776 feet.

50. $28\dfrac{1}{4} - 22\dfrac{1}{2}$

$= \dfrac{113}{4} + \left(-\dfrac{45}{2}\right)$

$= \dfrac{113}{4} + \left(-\dfrac{90}{4}\right)$

$= \dfrac{23}{4}$

John's gain per share was $5\dfrac{3}{4}$.

51. $(-13)(-6) = 78$

52. $\left(-7\dfrac{2}{5}\right)\left(-\dfrac{5}{7}\right) = \left(-\dfrac{37}{5}\right)\left(-\dfrac{5}{7}\right) = \dfrac{185}{35} = \dfrac{37}{7}$

53. $\left(-3\dfrac{1}{5}\right)\left(\dfrac{5}{16}\right) = \left(-\dfrac{16}{5}\right)\left(\dfrac{5}{16}\right) = -\dfrac{80}{80} = -1$

54. $(23.05)(-0.04) = -0.922$

55. $0 \cdot (-11) = 0$

56. $(-34)(20) = -680$

57. $\left(\dfrac{23}{171}\right)\left(\dfrac{33}{230}\right) = \dfrac{759}{39{,}330} = \dfrac{11}{570}$

58. $(13.902)(4.387) \approx 60.988$

59. $(-32)(20)(-1)(5)(-2)(10) = -64{,}000$

60. $(-1.1)(0.2)(-4)(-10)(-0.8) = 7.04$

61. $\left(-\dfrac{1}{3}\right)\left(\dfrac{6}{7}\right)\left(-\dfrac{14}{15}\right)\left(-\dfrac{25}{8}\right) = -\dfrac{2100}{2520} = -\dfrac{5}{6}$

62. $(-54)(21)(0)(32)(0)(-25) = 0$

63. $(14)(36)(16) = 8064$
8064 ounces of cleanser will be required to process the order.

64. $(22.5)(45) = 1012.5$
The total bill for labor is \$1012.50.

65. $(-78) \div (-3) = 26$

66. $\dfrac{5}{9} \div \left(-\dfrac{2}{9}\right) = \dfrac{5}{9} \cdot \left(-\dfrac{9}{2}\right) = -\dfrac{45}{18} = -\dfrac{5}{2}$

67. $-10.557 \div 2.3 = -4.59$

68. $0 \div 25 = 0$

69. $-2 \div 0 = \text{undefined}$

70. $-\dfrac{23}{356} \div \dfrac{46}{89} = -\dfrac{23}{356} \cdot \dfrac{89}{46} = -\dfrac{2047}{16,376} = -\dfrac{1}{8}$

71. $\dfrac{143,883}{657} = 219$

72. $7.32864 \div 1.056 = 6.94$

73. $(-25) \div (-5)(12)(-3) \div (-9)(-5)$
$= 5(12)(-3) \div (-9)(-5)$
$= 60(-3) \div (-9)(-5)$
$= -180 \div (-9)(-5)$
$= 20(-5)$
$= -100$

74. $(4.2)(-3.2) \div (1.6)(0.2) \div (-0.4)(-2.2)$
$= -13.44 \div (1.6)(0.2) \div (-0.4)(-2.2)$
$= -8.4(0.2) \div (-0.4)(-2.2)$
$= -1.68 \div (-0.4)(-2.2)$
$= 4.2(-2.2)$
$= -9.24$

75. $\left(\dfrac{5}{12}\right)\left(-\dfrac{6}{25}\right) \div \left(-\dfrac{3}{10}\right)\left(\dfrac{15}{17}\right) \div \left(-\dfrac{5}{13}\right)$
$= \left(\dfrac{5}{12}\right)\left(-\dfrac{6}{25}\right)\left(-\dfrac{10}{3}\right)\left(\dfrac{15}{17}\right)\left(-\dfrac{13}{5}\right)$
$= -\dfrac{58,500}{76,500}$
$= -\dfrac{13}{17}$

76. $\dfrac{10 + (-5) + 8 + 1 + (-12)}{5} = \dfrac{2}{5} = 0.4$
David's average is 0.4 points above the class average.

77. $\dfrac{625 + 690 + 620 + 590 + 630 + 660}{6}$
$= \dfrac{3815}{6} \approx 635.83$
A fair monthly rent would be about \$636.

78. $345 \div 18.7 \approx 18.45$
Clarence's average mileage is approximately 18.45 mpg.

79. **a.** positive

   b. negative

   c. negative

   d. negative

   e. negative

   f. positive

80. $2^8 = 2 \cdot 2 \cdot 2 \cdot 2 \cdot 2 \cdot 2 \cdot 2 \cdot 2 = 256$

81. $(1.2)^2 = (1.2)(1.2) = 1.44$

82. $\left(1\dfrac{1}{3}\right)^3 = \left(1\dfrac{1}{3}\right)\left(1\dfrac{1}{3}\right)\left(1\dfrac{1}{3}\right) = 2\dfrac{10}{27}$

83. $0^{10} = 0$

84. $1^{15} = 1$

85. $(-3)^4 = (-3)(-3)(-3)(-3) = 81$

86. $(-0.2)^3 = (-0.2)(-0.2)(-0.2) = -0.008$

87. $\left(-1\dfrac{1}{3}\right)^2 = \left(-1\dfrac{1}{3}\right)\left(-1\dfrac{1}{3}\right) = \dfrac{16}{9}$

88. $(-1)^{21} = -1$

89. $(-33)^5 = -39,135,393$

90. $(0.47)^6 \approx 0.011$

91. $(-56)^4 = 9,834,496$

92. $\left(-\dfrac{3}{4}\right)^6 = \dfrac{729}{4096}$

93. $-\left(-2\dfrac{1}{3}\right)^4 = -\left(-\dfrac{7}{3}\right)^4 = -\dfrac{2401}{81}$

94. $-15^0 = -1$

95. $-15^1 = -15$

96. $1^0 = 1$

97. $0^0 = \text{indeterminate}$

98. $1^1 = 1$

99. $0^1 = 0$

100. $(229,384)^0 = 1$

101. $\left(\dfrac{13}{23}\right)^0 = 1$

102. $(229,384)^1 = 229,384$

103. $(-12)^{-2} = \dfrac{1}{(-12)^2} = \dfrac{1}{144}$

104. $-12^{-2} = -\dfrac{1}{12^2} = -\dfrac{1}{144}$

105. $-9^{-3} = \dfrac{1}{(-9)^3} = -\dfrac{1}{729}$

106. $\left(-2\dfrac{3}{8}\right)^{-1} = \dfrac{1}{\left(-2\dfrac{3}{8}\right)^1} = \dfrac{1}{\left(-\dfrac{19}{8}\right)^1} = -\dfrac{8}{19}$

107. $\left(1\dfrac{3}{4}\right)^{-2} = \dfrac{1}{\left(1\dfrac{3}{4}\right)^2} = \dfrac{1}{\left(\dfrac{7}{4}\right)^2} = \dfrac{16}{49}$

108. $1^{-9} = \dfrac{1}{1^9} = 1$

109. $(1.8)^{-3} = \dfrac{1}{1.8^3} \approx 0.171$

110. $\left(\dfrac{7}{8}\right)^{-3} = \left(\dfrac{8}{7}\right)^3 = \dfrac{512}{343}$

111. $A = s^2$
$A = (7.5)^2 = 56.25$
Yes, since 56.25 ft^2 is greater than 60 ft^2, there will be enough plants to fill the garden.

112. $A = s^3$
$A = (2.5)^3 = 15.625$
15.625 ft^3 of mulch will fill the bin.

113. $0.000000189 = 1.89 \times 10^{-7}$

114. $-27,085,000,000 = -2.7085 \times 10^{10}$

115. $4.02 \text{ E}{-7} = 4.02 \times 10^{-7} = 0.000000402$

116. $-1.3 \text{ E}7 = -1.3 \times 10^7 = -13,000,000$

117. $5,904,822,000,000 = \$5.904822 \times 10^{12}$

118. $2.37 \times 10^8 = \$237,000,000$
$5.904822 \times 10^{12} - 2.37 \times 10^8$
$= \$5.904585 \times 10^{12}$
The U.S. GNP exceeds the GNP of the Solomon Islands by $\$5.904585 \times 10^{12}$.

119. $8 \times 10^{-6} = 0.000008$ m

120. $\sqrt{81} = 9$

121. $\sqrt{0.64} = 0.8$

122. $\sqrt{\dfrac{9}{25}} = \dfrac{3}{5}$

123. $-\sqrt{49} = -7$

124. $\sqrt{-16}$ is not a real number

125. $-\sqrt{470.89} = -21.7$

126. $-\sqrt{\dfrac{576}{1369}} = -\dfrac{24}{37}$

127. $\sqrt{2\dfrac{6}{11}} \approx 1.595$

128. $\sqrt{15} \approx 3.873$

129. $\sqrt{-1.2}$ is not a real number

130. $\sqrt[4]{81} = 3$

131. $\sqrt[3]{0.125} = 0.5$

132. $\sqrt[4]{\dfrac{16}{81}} = \dfrac{2}{3}$

133. $\sqrt[3]{-27,000} = -30$

134. $\sqrt[5]{-1} = -1$

135. $-\sqrt[4]{13.0321} = -1.9$

136. $-\sqrt[5]{10} \approx -1.585$

137.

138. $A = s^2$
$729 = s^2$
$s = \sqrt{729} = 27$
The diamond is about 27 meters on a side.

139. $V = s^3$
$91\dfrac{1}{8} = s^3$
$s = \sqrt[3]{91\dfrac{1}{8}} = 4.5$
The box is 4.5 in. on a side.

140. e

141. c

142. d

143. a

144. b

145.

Number	8	–12	$\dfrac{3}{17}$	$-\dfrac{5}{3}$
Opposite	–8	12	$-\dfrac{3}{17}$	$\dfrac{5}{3}$
Reciprocal	$\dfrac{1}{8}$	$-\dfrac{1}{12}$	$\dfrac{17}{3}$	$-\dfrac{3}{5}$

146. $-33.05 + 12.4 = 12.4 + (-33.05)$

147. $\left(1\dfrac{9}{10}\right)\left(2\dfrac{4}{19}\right) = \left(2\dfrac{4}{19}\right)\left(1\dfrac{9}{10}\right)$

148. $132 + (-207 + 391) = [132 + (-207)] + 391$

149. $\left(\dfrac{3}{7}\cdot\dfrac{14}{15}\right)\cdot\dfrac{5}{9} = \dfrac{3}{7}\cdot\left(\dfrac{14}{15}\cdot\dfrac{5}{9}\right)$

150. $-2.6(-1.9 + 3.2) = -2.6(-1.9) - 2.6(3.2)$

151. $\dfrac{5}{6}\left(-\dfrac{3}{5} - \dfrac{4}{15}\right) = \dfrac{5}{6}\left(-\dfrac{3}{5}\right) + \dfrac{5}{6}\left(-\dfrac{4}{15}\right)$

152. $\dfrac{1687 - 1372}{7} = \dfrac{1687}{7} - \dfrac{1372}{7}$

153. $-(2.7 + 3.09) = -2.7 + (-3.09) = -2.7 - 3.09$

154. $-[32 + (-51)] = -32 - (-51) = -32 + 51$

155. $21(18) + 21(47) = 21(18 + 47)$

156. $5(19) - 5(17) = 5(19 - 17)$

157. $17 + 31 - 20 \div 2 - 15 \cdot (-3)$
$= 17 + 31 - 10 - (-45)$
$= 83$

158. $-(27 - 4^2) - 16(-5 - 3)$
$= -(27 - 16) - 16(-8)$
$= -11 + 128$
$= 117$

159. $\dfrac{191 + 104 - 11^2}{5^2 + 3^3 - 46}$
$= \dfrac{191 + 104 - 121}{25 + 27 - 46}$
$= \dfrac{174}{6}$
$= 29$

160. $[15 + 21(14 - 18)] - 7^2$
$= [15 + 21(-4)] - 49$
$= (15 - 84) - 49$
$= -69 - 49$
$= -118$

161. $22 - \sqrt{9^2 - 45}$
$= 22 - \sqrt{81 - 45}$
$= 22 - \sqrt{36}$
$= 22 - 6$
$= 16$

162. $\dfrac{8 \cdot 9 - 2 \cdot 6^2}{125 - 7^2}$
$= \dfrac{8 \cdot 9 - 2 \cdot 36}{125 - 49}$
$= \dfrac{72 - 72}{125 - 49}$
$= \dfrac{0}{76}$
$= 0$

163. $\dfrac{78 - 3 \cdot 17}{5^2 - 4 \cdot 6 - 1^3}$
$= \dfrac{78 - 51}{25 - 24 - 1}$
$= \dfrac{27}{0}$
$=$ undefined

164. Area $= 12(14) - 6^2$
$= 168 - 36$
$= 132$
132 ft^2 will not be covered by the rug.

165. $2(7) + 2(5) + 2\sqrt{7^2 + 5^2}$
$= 14 + 10 + 2\sqrt{49 + 25}$
$= 14 + 10 + 2\sqrt{74}$
≈ 41.205
Mary will need approximately 41.205 feet of lace.

166. $\dfrac{77 + (77 + 4) + 68 + (68 - 6) + 82}{5} = \dfrac{370}{5}$
$= 74$
The average daily high temperature was 74°F.

167. $\dfrac{185 - 227}{20} = \dfrac{-42}{20} = -2.1$
Richard lost 2.1 pounds per week on average.

Chapter 1 Mixed Reveiw

1. $13^2 = 169$

2. $0^1 = 0$

3. $(2,333,145)^0 = 1$

4. $(3.4)^7 \approx 5252.335$

5. $\left(2\dfrac{1}{5}\right)^2 = \left(\dfrac{11}{5}\right)^2 = \dfrac{121}{25}$

6. $0^{35} = 0$

7. $6^{-3} = \left(\dfrac{1}{6}\right)^3 = \dfrac{1}{216}$

8. $(-6)^{-3} = \left(-\dfrac{1}{6}\right)^3 = -\dfrac{1}{216}$

9. $-6^{-3} = -\left(\dfrac{1}{6}\right)^3 = -\dfrac{1}{216}$

10. $(-5)^3 = -125$

11. $(-5)^4 = 625$

12. $(-3.34)^5 \approx -415.654$

13. $\left(-1\dfrac{2}{5}\right)^3 = \left(-\dfrac{7}{5}\right)^3 = -\dfrac{343}{125}$

14. $(-1)^{24} = 1$

15. $(-1)^{35} = -1$

16. $3^4 = 81$

17. $(-3)^6 = 729$

18. $(-22)^0 = 1$

19. $-22^0 = -1$

20. $-22^1 = -22$

21. $(2,333,145)^1 = 2,333,145$

22. $\left(\dfrac{79}{95}\right)^0 = 1$

23. $\left(-\dfrac{11}{65}\right)^1 = -\dfrac{11}{65}$

24. $(-0.0004)^0 = 1$

25. $20^{-2} = \left(\dfrac{1}{20}\right)^2 = \dfrac{1}{400}$

26. $(-20)^{-2} = \left(-\dfrac{1}{20}\right)^2 = \dfrac{1}{400}$

27. $-20^{-2} = -\left(\dfrac{1}{20}\right)^2 = -\dfrac{1}{400}$

28. $-(-3)^6 = -729$

29. $-\left(\dfrac{1}{3}\right)^5 = -\dfrac{1}{243}$

30. $(-2.1)^4 = 19.4481$

31. $\left(-\dfrac{8}{21}\right)^{-1} = \left(-\dfrac{21}{8}\right)^1 = -\dfrac{21}{8}$

32. $\left(1\dfrac{3}{5}\right)^{-2} = \left(\dfrac{8}{5}\right)^{-2} = \left(\dfrac{5}{8}\right)^2 = \dfrac{25}{64}$

33. $1^{-7} = \left(\dfrac{1}{1}\right)^7 = 1$

34. $-1^0 = -1$

35. $0^0 = $ indeterminate

36. $-1^1 = -1$

37. $\sqrt{2500} = 50$

38. $\sqrt{-400}$ is not a real number

39. $-\sqrt{400} = -20$

40. $\sqrt{\dfrac{16}{289}} = \dfrac{4}{17}$

41. $-\sqrt{368.64} = -19.2$

42. $\sqrt{5\dfrac{4}{9}} = \sqrt{\dfrac{49}{9}} = \dfrac{7}{3} = 2\dfrac{1}{3}$

43. $\sqrt{1.44} = 1.2$

44. $-\sqrt{\dfrac{5}{400}} \approx -0.112$

45. $-\sqrt{2335.6} \approx -47.282$

46. $\sqrt[3]{-64} = -4$

47. $\sqrt[4]{\dfrac{81}{16}} = \dfrac{3}{2}$

48. $\sqrt[4]{3\dfrac{13}{81}} = \sqrt[4]{\dfrac{256}{81}} = \dfrac{4}{3} = 1\dfrac{1}{3}$

49. $-\sqrt[5]{25} \approx -1.904$

50. $\sqrt[4]{2.56} \approx 1.265$

51. $\sqrt[3]{-2\dfrac{5}{7}} \approx -1.395$

52. $\sqrt[5]{-32} = -2$

53. $\sqrt[5]{\dfrac{243}{1024}} = \dfrac{3}{4}$

54. $-\sqrt[4]{6.5536} = -1.6$

55. $-\left|\dfrac{9}{25}\right| = -\dfrac{9}{25}$

56. $\left|-(-102)\right| = |102| = 102$

57. $-(-(-(-7))) = -(-7) = 7$

58. $5.5 \div 2.2 = 2.5$

59. $-549 + (-908) = -1457$

60. $3.07 + (-2.9) = 0.17$

61. $(-767)(-11) = 8437$

62. $\left(\dfrac{17}{72}\right)\left(\dfrac{9}{34}\right) = \dfrac{153}{2448} = \dfrac{1}{16}$

63. $0 \div 0 = $ indeterminate

64. $-5.7 + 8.26 = 2.56$

65. $1260 + 1111 = 2371$

66. $-67,853 + 80,000 = 12,147$

67. $\dfrac{17}{65} + \dfrac{8}{91} = \dfrac{119}{455} + \dfrac{40}{455} = \dfrac{159}{455}$

68. $-9.678 + (-9.678) = -19.356$

69. $-576 - (-394) = -576 + 394 = -182$

70. $-\dfrac{4}{13} - \dfrac{7}{39} = -\dfrac{12}{39} + \left(-\dfrac{7}{39}\right) = -\dfrac{19}{39}$

71. $\dfrac{184,008}{902} = 204$

72. $\dfrac{2}{5} \div \dfrac{56}{75} = \dfrac{2}{5} \cdot \dfrac{75}{56} = \dfrac{150}{280} = \dfrac{15}{28}$

73. $0.52 - 3 = 0.52 + (-3) = -2.48$

74. $0 - \dfrac{3}{11} = 0 + \left(-\dfrac{3}{11}\right) = -\dfrac{3}{11}$

75. $\dfrac{13}{25} - \dfrac{2}{5} = \dfrac{13}{25} + \left(-\dfrac{2}{5}\right) = \dfrac{13}{25} + \left(-\dfrac{10}{25}\right) = \dfrac{3}{25}$

76. $1,000,000 - 1,000 = 1,000,000 + (-1,000)$
$= 999,000$

77. $15.07 - (-0.907) = 15.07 + 0.907 = 15.977$

78. $\left(-\dfrac{12}{55}\right)\left(-\dfrac{5}{6}\right) = \dfrac{60}{330} = \dfrac{2}{11}$

79. $\left(-4\dfrac{3}{7}\right)\left(\dfrac{14}{31}\right) = \left(-\dfrac{31}{7}\right)\left(\dfrac{14}{31}\right) = -\dfrac{434}{217} = -2$

80. $(121.3)(12.1) = 1467.73$

81. $(0.09)(-0.06) = -0.0054$

82. $0 \cdot (-8.05) = 0$

83. $(-45) \cdot 65 = -2925$

84. $(-1)(-88) = 88$

85. $\dfrac{-1175}{-5} = 235$

86. $\dfrac{4}{13} \div \left(-\dfrac{12}{13}\right) = \dfrac{4}{13} \cdot \left(-\dfrac{13}{12}\right) = -\dfrac{52}{156} = -\dfrac{1}{3}$

87. $-21 \div 0 = $ undefined

88. $4000 \cdot 1200 = 4,800,000$

89. $-1\dfrac{4}{13} + 2\dfrac{1}{2} = -1\dfrac{8}{26} + 2\dfrac{13}{26}$
$$= -\dfrac{34}{26} + \dfrac{65}{26}$$
$$z = \dfrac{31}{26} = 1\dfrac{5}{26}$$

90. $-\dfrac{1}{8}+\left(-1\dfrac{1}{4}\right)$

$=-\dfrac{1}{8}+\left(-1\dfrac{2}{8}\right)$

$=-\dfrac{1}{8}+\left(-\dfrac{10}{8}\right)$

$=-\dfrac{11}{8}=-1\dfrac{3}{8}$

91. $-9.02\div(-1.1)=8.2$

92. $-\dfrac{8}{55}\div\left(-\dfrac{4}{17}\right)=-\dfrac{8}{55}\cdot\left(-\dfrac{17}{4}\right)=\dfrac{136}{220}=\dfrac{34}{55}$

93. $-\dfrac{17}{19}+\dfrac{45}{57}=-\dfrac{51}{57}+\dfrac{45}{57}=-\dfrac{6}{57}=-\dfrac{2}{19}$

94. $-4706+(-11{,}237)=-15{,}943$

95. $-21-453=-21+(-453)=-474$

96. $-\dfrac{4}{9}-\left(-\dfrac{47}{81}\right)=-\dfrac{4}{9}+\dfrac{47}{81}=-\dfrac{36}{81}+\dfrac{47}{81}=\dfrac{11}{81}$

97. $2.56+(-3.78)+0.5+9.882+(-1.05)+(-0.009)$
$=8.103$

98. $\dfrac{5}{7}+\dfrac{3}{5}+\left(-\dfrac{1}{2}\right)+\left(-\dfrac{23}{35}\right)+\left(-\dfrac{7}{10}\right)$

$=\dfrac{50}{70}+\dfrac{42}{70}+\left(-\dfrac{35}{70}\right)+\left(-\dfrac{46}{70}\right)+\left(-\dfrac{49}{70}\right)$

$=-\dfrac{38}{70}$

$=-\dfrac{19}{35}$

99. $-45-88-(-74)-(-90)-65-71-(-27)-53-(-48)$
$=-45+(-88)+74+90+(-65)+(-71)+27+(-53)+48$
$=-83$

100. $\dfrac{5}{6}-\dfrac{11}{30}-\left(-\dfrac{3}{5}\right)-\dfrac{14}{15}-\left(-\dfrac{9}{2}\right)-1$

$=\dfrac{25}{30}+\left(-\dfrac{11}{30}\right)+\dfrac{18}{30}+\left(-\dfrac{28}{30}\right)+\dfrac{135}{30}+\left(-\dfrac{30}{30}\right)$

$=\dfrac{109}{30}$

$=3\dfrac{19}{30}$

101. $57-61+(-32)-54-(-73)+25+(-38)$
$=57+(-61)+(-32)+(-54)+73+25+(-38)$
$=-30$

102. $7.3 - 38.6 + 5.5 + (-2.09) - 8 - (-2.51)$
$= 7.3 + (-38.6) + 5.5 + (-2.09) + (-8) + 2.51$
$= -33.38$

103. $(-55)(12)(-2)(9)(-3)(1) = -35{,}640$

104. $(-3.2)(0.6)(-5)(-100)(-0.1) = 96$

105. $\left(-\dfrac{14}{22}\right)\left(\dfrac{8}{21}\right)\left(-\dfrac{11}{4}\right)\left(-\dfrac{3}{5}\right) = -\dfrac{3696}{9240} = -\dfrac{2}{5}$

106. $(-11.2)(3.1)(0)(-9.4)(-1)(-7.5) = 0$

107. $(-15) \div (-3)(22)(-4) \div (-8)(-7)$
$= 5(22)(-4) \div (-8)(-7)$
$= 110(-4) \div (-8)(-7)$
$= -440 \div (-8)(-7)$
$= 55(-7)$
$= -385$

108. $(13.1)(-4.2) \div (2.62)(0.5) \div (-0.7)(-1.1)$
$= -55.02 \div (2.62)(0.5) \div (-0.7)(-1.1)$
$= -21(0.5) \div (-0.7)(-1.1)$
$= -10.5 \div (-0.7)(-1.1)$
$= 15(-1.1)$
$= -16.5$

109. $\left(\dfrac{9}{14}\right)\left(-\dfrac{7}{18}\right) \div \left(-\dfrac{3}{4}\right)\left(\dfrac{6}{7}\right) \div \left(-\dfrac{1}{21}\right)$
$= \left(\dfrac{9}{14}\right)\left(-\dfrac{7}{18}\right)\left(-\dfrac{4}{3}\right)\left(\dfrac{6}{7}\right)\left(-\dfrac{21}{1}\right)$
$= -\dfrac{31{,}752}{5292}$
$= -6$

110. $25 + 63 - 48 \div 3 - 22 \cdot (-8)$
$= 25 + 63 - 16 - (-176)$
$= 248$

111. $-3.4(-1.1) - 7(2.6) + 14.3$
$= 3.74 - 18.2 + 14.3$
$= -14.46 + 14.3$
$= -0.16$

112. $-(34 - 7^2) - 21(-8 - 4)$
$= -(34 - 49) - 21(-12)$
$= -(-15) + 252$
$= 267$

113. $\dfrac{412 + 204 - 4^2}{5^2 + 35}$
$= \dfrac{412 + 204 - 16}{25 + 35}$
$= \dfrac{600}{60}$
$= 10$

114. $[651 + 18(21 - 19)] - 5^2$
$= [651 + 18(2)] - 25$
$= (651 + 36) - 25$
$= 687 - 25$
$= 662$

115. $96 - \sqrt{11^2 - 21}$
$= 96 - \sqrt{121 - 21}$
$= 96 - \sqrt{100}$
$= 96 - 10$
$= 86$

116. $\dfrac{10 \cdot 25 - 2 \cdot 5^3}{275 - 3^5}$
$= \dfrac{10 \cdot 25 - 2 \cdot 125}{275 - 243}$
$= \dfrac{250 - 250}{32}$
$= \dfrac{0}{32}$
$= 0$

117. $\dfrac{35 - 6 \cdot 22}{6^3 - 4 \cdot 50 - 4^2}$
$= \dfrac{35 - 132}{216 - 200 - 16}$
$= -\dfrac{97}{0}$
$= \text{undefined}$

118. $42(15) - 42(23) = 42(15 - 23)$

119. $-31(12) - 31(42) = -31(12 + 42)$

120. $75 > 59$

121. $3.54 < 3.65$

122. $\dfrac{31}{51} > \dfrac{10}{17}$

123. $-142 < -105$

124. $-\dfrac{21}{59} > -\dfrac{29}{59}$

125. $\dfrac{3}{8} = 0.375$

126. $-9.05 > -9.07$

127. $-\dfrac{6}{7} < \dfrac{3}{7}$

128.

129.

130. $0.00000475 = 4.75 \times 10^{-6} = 4.75\text{ E}{-6}$

131. $-100,505,000 = -1.00505 \times 10^{8}$
$= -1.00505 \text{ E}8$

132. $1.12 \text{ E}9 = 1.12 \times 10^{9} = 1,120,000,000$

133. $-2.35 \text{ E}{-7} = -2.35 \times 10^{-7} = -0.000000235$

134.

Number	24	−15	$\dfrac{4}{9}$	$-\dfrac{5}{8}$
Opposite	−24	15	$-\dfrac{4}{9}$	$\dfrac{5}{8}$
Reciprocal	$\dfrac{1}{24}$	$-\dfrac{1}{15}$	$\dfrac{9}{4}$	$-\dfrac{8}{5}$

135. a. $275,000 - 183,000$
$= 275,000 + (-183,000)$
$= 92,000$
The business made a profit of $92,000.

b. $695,000 - 710,000$
$= 695,000 + (-710,000)$
$= -15,000$
The business lost $15,000.

136. $\dfrac{5\frac{3}{8} + \left(-1\frac{5}{8}\right) + \left(-3\frac{3}{4}\right) + 0}{4}$

$= \dfrac{\frac{43}{8} + \left(-\frac{13}{8}\right) + \left(-\frac{30}{8}\right) + 0}{4}$

$= \dfrac{0}{4}$

$= 0$

The stock had neither an increase nor a decrease over the 4-week period.

137. $763 \div \left(6\dfrac{1}{2}\right)$

$= 763 \div \left(\dfrac{13}{2}\right)$

$= 763 \cdot \left(\dfrac{2}{13}\right)$

$= \dfrac{1526}{13} \approx 117.4$

Approximately 117 credits run each minute.

138. $10 \times \dfrac{3}{4} = \dfrac{30}{4} = 7\dfrac{1}{2}$

It takes $7\dfrac{1}{2}$ gallons of water to fill ten 10-gallon hats.

139. $1012 - 430 = 1012 + (-430) = 582$
Captain Picard had 582 more people under his command.

140. $24 \times 28 \times 8 = 5376$
5376 footballs are needed.

141. $A = s^{2}$
$A = (14)^{2} = 196$
The porch's area is 196 ft^{2}.

142. $V = s^{3}$
$V = (6)^{3} = 216$
The volume of the fruit cellar is 216 ft^{3}.

143. $1,176,946,000 = 1.176946 \times 10^9$

144. $9.87 \times 10^8 = 987,000,000$

145. $A = s^2$
$300 = s^2$
$s = \sqrt{300} \approx 17.321$
The dimensions of the tarpaulin are approximately 17.321 feet on each side.

146. $V = s^3$
$343 = s^3$
$s = \sqrt[3]{343} = 7$
The dimensions of the tree house are 7 feet on each side.

147. $\dfrac{35 + (35 + 8) + 52 + (52 - 16)}{4}$
$= \dfrac{166}{4}$
$= 41.5$
Katie's average cost per framed print was $41.50.

148. $\dfrac{32,800 - 22,675}{6}$
$= \dfrac{10,125}{6}$
$= 1687.5$
Cecilia's average annual salary increase was $1687.50.

Chapter 1 Test

1. $\dfrac{4}{9} < \dfrac{5}{6}$

2. $-18 > -23$

3. $\dfrac{17}{25} = 0.68$

4. $-(-20) = 20$

5. $-\left| -15 + 10 \right| = -\left| -5 \right| = -|5| = -5$

6. $-\dfrac{17}{95} + \dfrac{4}{19} = -\dfrac{17}{95} + \dfrac{20}{95} = \dfrac{3}{95}$

7. $2\dfrac{3}{7} \div 1\dfrac{3}{14} = \dfrac{17}{7} \div \dfrac{17}{14} = \dfrac{17}{7} \cdot \dfrac{14}{17} = \dfrac{238}{119} = 2$

8. $4.378 - 7.98 = 4.378 + (-7.98) = -3.602$

9. $\left(-\dfrac{3}{4} \right)\left(-\dfrac{8}{15} \right) = \dfrac{24}{60} = \dfrac{2}{5}$

10. $-6985 - (-2576) = -6985 + 2576 = -4409$

11. $-413.9 + (-597.65) = -1011.55$

12. $-819 \div (-9) = 91$

13. $-\dfrac{44}{57} \div \dfrac{11}{19} = -\dfrac{44}{57} \cdot \dfrac{19}{11} = -\dfrac{836}{627} = -\dfrac{4}{3}$

14. $15.9 \div 0 =$ undefined

15. $0 \div (-53) = 0$

16. $\dfrac{-609 + 928}{29}$
$= \dfrac{319}{29}$
$= 11$

17.

18. $500 + (-123.75) + (-56.80) + (-95.87) + 250$
$= 473.58$
Sally's current account balance is $473.58.

19. 24% of 675
$0.24 \times 675 = 162$
Regina will have $162 per week deducted.

20. Answers will vary. Possible answer: The sign will be positive for an even number of factors and negative for an odd number of factors.

21. $(1.5)^2 = 2.25$

22. $3^4 = 81$

23. $\left(\dfrac{4}{3}\right)^4 = \dfrac{256}{81}$

24. $1^9 = 1$

25. $0^{12} = 0$

26. $(-4)^3 = -64$

27. $0^0 =$ indeterminate

28. $\left(-\dfrac{3}{8}\right)^{-2} = \left(-\dfrac{8}{3}\right)^2 = \dfrac{64}{9}$

29. $(-3)^0 = 1$

30. $-\sqrt{196} = -14$

31. $\sqrt{\dfrac{36}{121}} = \dfrac{6}{11}$

32. $-\sqrt{3.6} \approx -1.897$

33. $\sqrt[4]{7\dfrac{58}{81}} = \sqrt[4]{\dfrac{625}{81}} = \dfrac{5}{3} = 1\dfrac{2}{3}$

34. $(-4.008)^1 = -4.008$

35. $-[51.3 - (-20.9)]$
$= -(51.3 + 20.9)$
$= -72.2$

36. $\dfrac{2(5^2 + 3^2) - 8^2 - 2^2}{3.65}$

$= \dfrac{2(25 + 9) - 64 - 4}{3.65}$

$= \dfrac{2(34) - 64 - 4}{3.65}$

$= \dfrac{68 - 64 - 4}{3.65}$

$= \dfrac{0}{3.65}$

$= 0$

37. $\left(-\dfrac{5}{6}\right)\left(-\dfrac{3}{7}\right)\left(\dfrac{1}{2}\right)\left(-\dfrac{2}{15}\right) = -\dfrac{30}{1260} = -\dfrac{1}{42}$

38. $\sqrt[3]{17^2 - 6 \cdot 12 - 1} + 126 \div 9$
$= \sqrt[3]{289 - 6 \cdot 12 - 1} + 126 \div 9$
$= \sqrt[3]{289 - 72 - 1} + 126 \div 9$
$= \sqrt[3]{216} + 14$
$= 6 + 14$
$= 20$

39. $(-23)(-4)(0)(-17)(0)(-45) = 0$

40. $4.3 + (-0.1) - (-2) + (-1.1)$
$= 4.3 + (-0.1) + 2 + (-1.1)$
$= 5.1$

41. $(-42)(22) \div 77(-4) \div (-16)$
$= -924 \div 77(-4) \div (-16)$
$= -12(-4) \div (-16)$
$= 48 \div (-16)$
$= -3$

42.

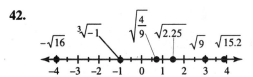

43. $5{,}239{,}000{,}000{,}000{,}000 = 5.239 \times 10^{15}$

44. $-1.04 \text{ E}{-}8$
$= -1.04 \times 10^{-8} = -0.0000000104$

45. $2.49 \times 10^{11} - 2.49 \times 10^{10} = 2.241 \times 10^{11}$
The GNP of Switzerland exceeds that of
Romania by \$2.241 $\times 10^{11}$ or
\$224,100,000,000.

46. Answers will vary. Possible answer: The set
of real numbers includes the set of rational
numbers, so all rational numbers are real
numbers. However, some real numbers, such
as π, are not rational numbers.

Chapter 2

2.1 Experiencing Algebra the Exercise Way

For exercises 1–19 let each x, y, z = some numbers.

1. $25 - x$

3. $\dfrac{3}{4}x$

5. $x \div 15$

7. $2.5x - 19.59 \div x$

9. $12x - 25$

11. $x + y + 2xy$

13. $(x + y + z) \div 3$

15. $2x - 5$

17. $\dfrac{1}{3}x + 6$

19. $2x \div (x + 5)$

21. Let a = number of adult tickets
c = number of children tickets
Total money = $6a + 2c$

23. Let
x = number of hours worked at \$9.50 per hr
y = number of hours worked at \$12.25 per hr
Total money = $9.5x + 12.25y$

For exercises 25–40 answers will vary. One possibility is given.

25. $x + 27$
The sum of a number and 27

27. $\dfrac{35}{a}$
The quotient of 35 divided by some number

29. $\dfrac{3}{5} \cdot z$
Three-fifths of a number

31. $t - 12.50$
12.50 subtracted from some number

33. $14 - 5z$
The difference of 14 and the product of a number times 5

35. $3.14d$
The product of 3.14 and some number

37. $\dfrac{1}{2}(L + H)$
One-half of the sum of two different numbers

39. $(a + b + c) \div 3$
The sum of three different grades divided by 3

41. $3x + 5$ for $x = -5$
$3x + 5 = 3(-5) + 5 = -15 + 5 = -10$

43. $3x + 5$ for $x = \dfrac{2}{3}$
$3x + 5 = 3\left(\dfrac{2}{3}\right) + 5 = 2 + 5 = 7$

45. $3x + 5$ for $x = -2.7$
$3x + 5 = 3(-2.7) + 5 = -8.1 + 5 = -3.1$

47. $18 - 3z$ for $z = -12.07$
$18 - 3z = 18 - 3(-12.07)$
$= 18 + 36.21$
$= 54.21$

49. $18 - 3z$ for $z = 23$
$18 - 3z = 18 - 3(23) = 18 - 69 = -51$

51. $18 - 3z$ for $z = -\dfrac{5}{6}$

$$18 - 3z = 18 - 3\left(-\dfrac{5}{6}\right)$$
$$= 18 + \dfrac{5}{2}$$
$$= \dfrac{36}{2} + \dfrac{5}{2}$$
$$= \dfrac{41}{2}$$

53. $\dfrac{1}{2}bh$ for $b = 56, h = 14$

$$\dfrac{1}{2}bh = \dfrac{1}{2}(56)(14) = 392$$

55. $\dfrac{1}{2}bh$ for $b = \dfrac{8}{5}, h = \dfrac{16}{3}$

$$\dfrac{1}{2}bh = \dfrac{1}{2}\left(\dfrac{8}{5}\right)\left(\dfrac{16}{3}\right) = \dfrac{64}{15}$$

57. $\dfrac{1}{2}bh$ for $b = 6.8, h = 4.2$

$$\dfrac{1}{2}bh = \dfrac{1}{2}(6.8)(4.2) = 14.28$$

59. $x^2 + 2x + 9$ for $x = -7$

$$x^2 + 2x + 9 = (-7)^2 + 2(-7) + 9$$
$$= 49 + (-14) + 9$$
$$= 44$$

61. $x^2 + 2x + 9$ for $x = 2.5$

$$x^2 + 2x + 9 = (2.5)^2 + 2(2.5) + 9$$
$$= 6.25 + 5 + 9 = 20.25$$

63. $\sqrt{4x^2 - 20x + 25}$ for $x = 3$

$$\sqrt{4x^2 - 20x + 25} = \sqrt{4(3)^2 - 20(3) + 25}$$
$$= \sqrt{36 - 60 + 25}$$
$$= \sqrt{1}$$
$$= 1$$

65. $\sqrt{4x^2 - 20x + 25}$ for $x = -4$

$$\sqrt{4x^2 - 20x + 25} = \sqrt{4(-4)^2 - 20(-4) + 25}$$
$$= \sqrt{64 + 80 + 25}$$
$$= \sqrt{169}$$
$$= 13$$

67. $|4.5 - 3.1b|$ for $b = 2$

$$|4.5 - 3.1b| = |4.5 - 3.1(2)|$$
$$= |4.5 - 6.2|$$
$$= |-1.7|$$
$$= 1.7$$

69. $|4.5 - 3.1b|$ for $b = -2$

$$|4.5 - 3.1b| = |4.5 - 3.1(-2)|$$
$$= |4.5 + 6.2|$$
$$= |10.7|$$
$$= 10.7$$

71. $\dfrac{x^2 + 2x + 1}{x + 1}$ for $x = -3$

$$\dfrac{x^2 + 2x + 1}{x + 1} = \dfrac{(-3)^2 + 2(-3) + 1}{-3 + 1}$$
$$= \dfrac{9 - 6 + 1}{-2}$$
$$= \dfrac{4}{-2}$$
$$= -2$$

73. $\dfrac{x^2 + 2x + 1}{x + 1}$ for $x = 3$

$$\dfrac{x^2 + 2x + 1}{x + 1} = \dfrac{3^2 + 2(3) + 1}{3 + 1}$$
$$= \dfrac{9 + 6 + 1}{4}$$
$$= \dfrac{16}{4}$$
$$= 4$$

75. Let x = number of adult tickets
y = number of children's tickets
Total revenue = $4x + 0.75y$
$= 4(150) + 0.75(200)$
$= 600 + 150$
$= 750$
The revenue from the tickets was $750.

77. Let d = diameter
Circumference = $\pi d = \pi(12) \approx 37.699$
The circumference of the pool is
approximately 37.7 feet.

2.1 Experiencing Algebra the Calculator Way

1. a.
```
120→A:150→B:(A+1
5)/(B-10)
          .9642857143
```

b.
```
275→A:13→B:(A+15
)/(B-10)
          96.66666667
```

c.
```
16→A:375→B:(A+15
)/(B-10)
          .0849315068
```

2. a.
```
19→R:πR²
          1134.114948
```

b.
```
23.76→R:πR²
          1773.547177
```

c.
```
.05→R:πR²
          .0078539816
```

d.
```
7/8→R:πR²
          2.405281875
```

e.
```
3+5/9→R:πR²
          39.71593676
```

f.
```
100→R:πR²
          31415.92654
```

3. a.
```
120→A:160→B:√(A²
+B²)
                   200
```

b.
```
3→A:4→B:√(A²+B²)
                     5
```

c.
```
2.4→A:3.2→B:√(A²
+B²)
                     4
```

d.
```
1→A:1→B:√(A²+B²)
          1.414213562
```

e.
```
3/5→A:4/5→B:√(A²
+B²)
                     1
```

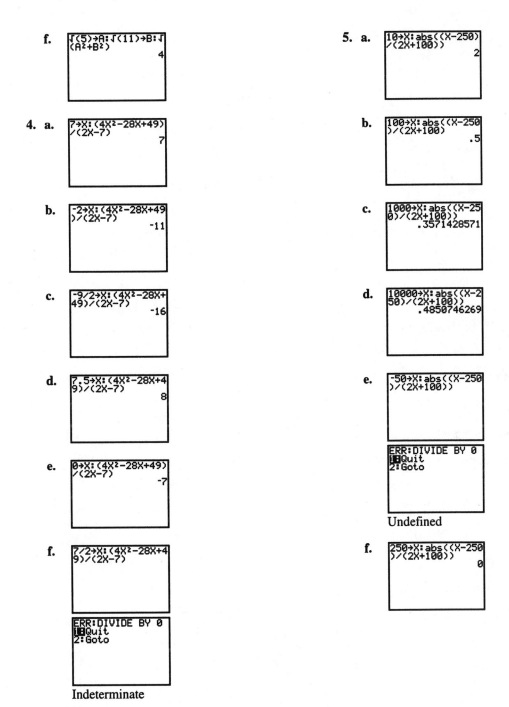

f.
```
√(5)→A:√(11)→B:√
(A²+B²)
                4
```

4. a.
```
7→X:(4X²-28X+49)
/(2X-7)
                7
```

b.
```
-2→X:(4X²-28X+49
)/(2X-7)
              -11
```

c.
```
-9/2→X:(4X²-28X+
49)/(2X-7)
              -16
```

d.
```
7.5→X:(4X²-28X+4
9)/(2X-7)
                8
```

e.
```
0→X:(4X²-28X+49)
/(2X-7)
               -7
```

f.
```
7/2→X:(4X²-28X+4
9)/(2X-7)
```
```
ERR:DIVIDE BY 0
1■Quit
2:Goto
```
Indeterminate

5. a.
```
10→X:abs((X-250)
/(2X+100))
                2
```

b.
```
100→X:abs((X-250
)/(2X+100)
               .5
```

c.
```
1000→X:abs((X-25
0)/(2X+100))
       .3571428571
```

d.
```
10000→X:abs((X-2
50)/(2X+100))
       .4850746269
```

e.
```
-50→X:abs((X-250
)/(2X+100))
```
```
ERR:DIVIDE BY 0
1■Quit
2:Goto
```
Undefined

f.
```
250→X:abs((X-250
)/(2X+100))
                0
```

2.2 Experiencing Algebra the Exercise Way

1. $12x - 11$

   a. number of terms = 2

   b. coefficient of each term: 12, –11

   c. like terms: none

3. $-15y + 12z - y + 9$

   a. number of terms = 4

   b. coefficient of each term: –15, 12, –1, 9

   c. like terms: $-15y$ and $-y$

5. $2x^2 - 6x + x + 12$

   a. number of terms = 4

   b. coefficient of each term: 2, –6, 1, 12

   c. like terms: $-6x$ and x

7. $3.4a - 11.2b - 0.3a$

   a. number of terms = 3

   b. coefficient of each term: 3.4, –11.2, –0.3

   c. like terms: $3.4a$ and $-0.3a$

9. $2m + 3(n - 5) + 6(n - 5)$

   a. number of terms = 3

   b. coefficient of each term: 2, 3, 6

   c. like terms: $3(n - 5)$ and $6(n - 5)$

11. $x^2 + 3xy - y^2 + 7$

    a. number of terms = 4

    b. coefficient of each term: 1, 3, –1, 7

    c. like terms: none

13. $27x + 44x = 71x$

15. $17a + a = 18a$

17. $5x + 9 - 13x + 17 - 12 + 9x$
    $= 5x - 13x + 9x + 9 + 17 - 12$
    $= 1x + 14$
    $= x + 14$

19. $2x^3 + 7x^2 - 2x + 8 - x^3 - 7x^2 + 3x - 2$
    $= 2x^3 - x^3 + 7x^2 - 7x^2 - 2x + 3x + 8 - 2$
    $= 1x^3 + 0x^2 + 1x + 6$
    $= x^3 + x + 6$

21. $5 - xy + 2yz + 5xz - 17 + xy - 12yz$
    $= 5 - 17 - xy + xy + 2yz - 12yz + 5xz$
    $= -12 + 0xy - 10yz + 5xz$
    $= -12 - 10yz + 5xz$

23. $3.05a + 6.29b - 1.18a + 0.49b$
    $= 3.05a - 1.18a + 6.29b + 0.49b$
    $= 1.87a + 6.78b$

25. $\frac{1}{6}x + \frac{2}{9} - \frac{2}{3}x + \frac{5}{18} = \frac{1}{6}x - \frac{2}{3}x + \frac{2}{9} + \frac{5}{18}$
    $= \frac{1}{6}x - \frac{4}{6}x + \frac{4}{18} + \frac{5}{18}$
    $= -\frac{3}{6}x + \frac{9}{18}$
    $= -\frac{1}{2}x + \frac{1}{2}$

27. $6x^3 + 3x^2y - 5xy^2 + 3y^3 - 5x^2y + xy^2 + x^3 + 6y^3$
$= 6x^3 + x^3 + 3x^2y - 5x^2y - 5xy^2 + xy^2 + 3y^3 + 6y^3$
$= 7x^3 - 2x^2y - 4xy^2 + 9y^3$

29. $\dfrac{5}{7}x + \dfrac{1}{6}y + \dfrac{5}{6}x - \left(-\dfrac{2}{7}y\right) = \dfrac{5}{7}x + \dfrac{5}{6}x + \dfrac{1}{6}y + \dfrac{2}{7}y$
$= \dfrac{30}{42}x + \dfrac{35}{42}x + \dfrac{7}{42}y + \dfrac{12}{42}y$
$= \dfrac{65}{42}x + \dfrac{19}{42}y$

31. $9.35a^2 - 4.31b^2 + 2.35ab - 1.61 + 4.39a^2 - 5.77b^2 + 10.06$
$= 9.35a^2 + 4.39a^2 - 4.31b^2 - 5.77b^2 + 2.35ab - 1.61 + 10.06$
$= 13.74a^2 - 10.08b^2 + 2.35ab + 8.45$

33. $35a^3 - 41b^3 + 5ab - 11 + 4a^2 - 7b^2 + 16$
$= 35a^3 - 41b^3 + 5ab + 4a^2 - 7b^2 - 11 + 16$
$= 35a^3 + 41b^3 + 5ab + 4a^2 - 7b^2 + 5$

35. $15a + 14b - 12 + 3a + 9 - 8b - 17a - 6b$
$= 15a + 3a - 17a + 14b - 8b - 6b - 12 + 9$
$= 1a + 0b - 3$
$= a - 3$

37. $8w + (7w - 5) = (8w + 7w) - 5$
$= 15w - 5$

39. $5.3x + 1.4 + (3.4 - 1.7x)$
$= 5.3x + 1.4 + 3.4 - 1.7x$
$= 5.3x - 1.7x + 1.4 + 3.4$
$= 3.6x + 4.8$

41. $(45x - 112) + (21x + 33)$
$= 45x - 112 + 21x + 33$
$= 45x + 21x - 112 + 33$
$= 66x - 79$

43. $-(29d - 7c) = -29d + 7c$

45. $4.9z + 1.8 - (2.6z + 0.5)$
$= 4.9z + 1.8 - 2.6z - 0.5$
$= 4.9z - 2.6z + 1.8 - 0.5$
$= 2.3z + 1.3$

47. $(235 - 12y) - (307 + 31y)$
$= 235 - 12y - 307 - 31y$
$= 235 - 307 - 12y - 31y$
$= -72 - 43y$

49. $(x + y + 4z) - (2x - 5y + z)$
$= x + y + 4z - 2x + 5y - z$
$= x - 2x + y + 5y + 4z - z$
$= -x + 6y + 3z$

51. $25z - (12z + 7) = 25z - 12z - 7 = 13z - 7$

53. $-(1.8y - 3.5z) + (4.1y - 2.7z)$
$= -1.8y + 3.5z + 4.1y - 2.7z$
$= -1.8y + 4.1y + 3.5z - 2.7z$
$= 2.3y + 0.8z$

55. $(a + b) - (a + b) + (a + b) - (a + b)$
$= a + b - a - b + a + b - a - b$
$= a - a + a - a + b - b + b - b$
$= 0a + 0b$
$= 0$

57. $(-x + y) + (x - y) = -x + y + x - y$
$= -x + x + y - y$
$= 0x + 0y$
$= 0$

59. $(a + 5b) - (-a + 5b) = a + 5b + a - 5b$
$= a + a + 5b - 5b$
$= 2a + 0b$
$= 2a$

61. $2(250 + 400) - (65 + 3) = 2(650) - 68$
$= 1300 - 68$
$= 1232$
Willy needs 1232 feet of fencing.

63. $6(7)^2 = 6(49) = 294$
Rusty needs 294 square inches of balsa wood.

65. Let x = number of square yards of carpet
Profit $= 20 + 1.55x - 0.65x - 6.5$
$= 13.5 + 0.9x$
$= 13.5 + 0.9(250)$
$= 238.5$
Carl will realize $238.50 profit on the job.

67. Let x = number of ounces
Profit $= 5 + 1.5x - 2.25$
$= 2.75 + 1.5x$
$= 2.75 + 1.5(4)$
$= 8.75$
The net profit is $8.75 for a 4-ounce letter.

69. Let x = number of discs
Cost $= 9.99x + 0.05(9.99x)$
$= 9.99x + 0.4995x$
$= 10.4895x$
Change $= 50 - 10.4895x$
Cost for 3 discs $= 10.4895(3) = 31.4685$
Change $= 50 - 31.47 = 18.53$
Laurie will pay $31.47 for the 3 CD's and get $18.53 in change.

71. Let x = number of adults
y = number of children
$ made $= 15x + 7y - 6x - 2.5y$
$= 9x + 4.5y$
$= 9(75) + 4.5(40)$
$= 675 + 180$
$= 855$
$855 was made on the dinner

2.2 Experiencing Algebra the Calculator Way

1. $12.078x + 2.093 - 17.42x - 13.9035$
$= 12.078x - 17.42x + 2.093 - 13.9035$
$= -5.342x - 11.8105$

```
12.078-17.42
           -5.342
2.093-13.9035
          -11.8105
```

2. $(2579x - 4302) - (1087x - 306)$
$= 2579x - 1087x - 4302 + 306$
$= 1492x - 3996$

```
2579-1087
            1492
-4302+306
           -3996
```

3. $\dfrac{10}{13}x - \dfrac{5}{52}y - \dfrac{17}{20}x - \dfrac{7}{13}y$
$= \dfrac{10}{13}x - \dfrac{17}{20}x - \dfrac{5}{52}y - \dfrac{7}{13}y$
$= -\dfrac{21}{260}x - \dfrac{33}{52}y$

```
10/13-17/20▶Frac
          -21/260
-5/52-7/13▶Frac
           -33/52
```

4. $\left(2\dfrac{11}{25}\right)x + 5\dfrac{17}{30} - \left(3\dfrac{13}{15}\right)x - 3\dfrac{23}{75}$
$= \left(2\dfrac{11}{25}\right)x - \left(3\dfrac{13}{15}\right)x + 5\dfrac{17}{30} - 3\dfrac{23}{75}$
$= -1\dfrac{32}{75}x + 2\dfrac{13}{50}$

```
2+11/25-3-13/15▶
Frac
          -107/75
5+17/30-3-23/75▶
Frac
           113/50
```

5. $(1.0009x + 0.0004) - (0.0909x - 1.0031)$
$= 1.0009x - 0.0909x + 0.0004 + 1.0031$
$= 0.91x + 1.0035$

```
1.0009-.0909
            .91
.0004+1.0031
          1.0035
```

6. $-(935.3376x + 701.315) - (83.027x - 581.9534)$
$= -935.3376x - 701.315 - 83.027x + 581.9534$
$= -935.3376x - 83.027x - 701.315 + 581.9534$
$= -1018.3646x - 119.3616$

```
-935.3376-83.027
       -1018.3646
-701.315+581.953
4
        -119.3616
```

7. $3.995x + 12.083 - 2.995x - 9.083$
$= 3.995x - 2.995x + 12.083 - 9.083$
$= x + 3$

```
3.995-2.995
              1
12.083-9.083
              3
```

2.3 Experiencing Algebra the Exercise Way

1. $31(2xy) = (31 \cdot 2)xy = 62xy$

3. $(-120a)(8b) = (-120 \cdot 8)ab = -960ab$

5. $-55(-4mn) = [-55 \cdot (-4)]mn = 220mn$

7. $-4.3(-0.7st) = [-4.3 \cdot (-0.7)]st = 3.01st$

9. $-\dfrac{16}{27}\left(\dfrac{15}{24}cd\right) = \left(-\dfrac{16}{27} \cdot \dfrac{15}{24}\right)cd$
$= -\dfrac{240}{648}cd$
$= -\dfrac{10}{27}cd$

11. $12(4a + 7) = 12(4a) + 12(7)$
$= (12 \cdot 4)a + 12(7)$
$= 48a + 84$

13. $-15(2x + 3) = -15(2x) - 15(3)$
$= (-15 \cdot 2)x - 15(3)$
$= -30x - 45$

15. $-4(-5z - 14) = -4(-5z) - 4(-14)$
$= 20z + 56$

17. $2.2(3.5x - 7.3) = 2.2(3.5x) + 2.2(-7.3)$
$= 7.7x - 16.06$

19. $-8(-4.2z - 5) = -8(-4.2z) - 8(-5)$
$= 33.6z + 40$

21. $\dfrac{36}{49}\left(-\dfrac{7}{6}m + \dfrac{49}{72}\right) = \dfrac{36}{49}\left(-\dfrac{7}{6}m\right) + \dfrac{36}{49}\left(\dfrac{49}{72}\right)$
$= -\dfrac{6}{7}m + \dfrac{1}{2}$

23. $-4\left(-\dfrac{5}{12}b + \dfrac{3}{16}\right) = -4\left(-\dfrac{5}{12}b\right) - 4\left(\dfrac{3}{16}\right)$

$ = \dfrac{5}{3}b - \dfrac{3}{4}$

25. $-\dfrac{2}{3}\left(-\dfrac{5}{8}d - \dfrac{9}{16}\right) = -\dfrac{2}{3}\left(-\dfrac{5}{8}d\right) - \dfrac{2}{3}\left(-\dfrac{9}{16}\right)$

$\phantom{-\dfrac{2}{3}} = \dfrac{5}{12}d + \dfrac{3}{8}$

27. $3x(5x + 4y) = 3x(5x) + 3x(4y)$

$ = (3 \cdot 5)x \cdot x + (3 \cdot 4)xy$

$ = 15x^2 + 12xy$

29. $-12d(-3c - 7d) = -12d(-3c) - 12d(-7d)$

$ = 36cd + 84d^2$

31. $-0.9m(-2.2m + 3.6n)$

$ = -0.9m(-2.2m) - 0.9m(3.6n)$

$ = 1.98m^2 - 3.24mn$

33. $-\dfrac{3}{5}a\left(\dfrac{15}{17}a + b\right) = -\dfrac{3}{5}a\left(\dfrac{15}{17}a\right) - \dfrac{3}{5}a(b)$

$\phantom{-\dfrac{3}{5}a} = -\dfrac{9}{17}a^2 - \dfrac{3}{5}ab$

35. $\dfrac{3}{8}m(m - 4n) = \dfrac{3}{8}m(m) + \dfrac{3}{8}m(-4n)$

$\phantom{\dfrac{3}{8}m} = \dfrac{3}{8}m^2 - \dfrac{3}{2}mn$

37. $\dfrac{108p}{9} = 108p \cdot \dfrac{1}{9}$

$\phantom{\dfrac{108p}{9}} = \left(108 \cdot \dfrac{1}{9}\right) \cdot p$

$\phantom{\dfrac{108p}{9}} = 12p$

39. $\dfrac{-138x}{46} = \dfrac{-138}{46} \cdot x = -3x$

41. $\dfrac{-58z}{-29} = \dfrac{-58}{-29} \cdot z = 2z$

43. $\dfrac{5.4c}{-2.7} = \dfrac{5.4}{-2.7} \cdot c = -2c$

45. $\dfrac{-\frac{1}{2}x}{2} = -\dfrac{1}{2}x \div 2$

$\phantom{\dfrac{-\frac{1}{2}x}{2}} = -\dfrac{1}{2}x \cdot \dfrac{1}{2}$

$\phantom{\dfrac{-\frac{1}{2}x}{2}} = \left(-\dfrac{1}{2} \cdot \dfrac{1}{2}\right)x$

$\phantom{\dfrac{-\frac{1}{2}x}{2}} = -\dfrac{1}{4}x$

47. $\dfrac{-\frac{2}{3}a}{-4} = -\dfrac{2}{3}a \div (-4)$

$\phantom{\dfrac{-\frac{2}{3}a}{-4}} = -\dfrac{2}{3}a \cdot \left(-\dfrac{1}{4}\right)$

$\phantom{\dfrac{-\frac{2}{3}a}{-4}} = -\dfrac{2}{3}\left(-\dfrac{1}{4}\right)a$

$\phantom{\dfrac{-\frac{2}{3}a}{-4}} = \dfrac{1}{6}a$

49. $\dfrac{120xy}{24y} = \dfrac{120}{24} \cdot x \cdot \dfrac{y}{y} = 5 \cdot x \cdot 1 = 5x$

51. $\dfrac{-1.9rs}{9.5r} = \dfrac{-1.9}{9.5} \cdot \dfrac{r}{r} \cdot s = -0.2 \cdot 1 \cdot s = -0.2s$

53. $\dfrac{12.321pqr}{-1.11pr} = \dfrac{12.321}{-1.11} \cdot \dfrac{p}{p} \cdot q \cdot \dfrac{r}{r}$

$\phantom{\dfrac{12.321pqr}{-1.11pr}} = -11.1 \cdot 1 \cdot q \cdot 1$

$\phantom{\dfrac{12.321pqr}{-1.11pr}} = -11.1q$

55. $\dfrac{-\frac{5}{9}mn}{5n} = -\dfrac{5}{9} \cdot \dfrac{1}{5} \cdot m \cdot \dfrac{n}{n} = -\dfrac{1}{9} \cdot m \cdot 1 = -\dfrac{1}{9}m$

57. $\dfrac{36x + 60}{12} = \dfrac{36x}{12} + \dfrac{60}{12} = 3x + 5$

59. $\dfrac{20.4b - 3.4c}{-6.8} = \dfrac{20.4b}{-6.8} - \dfrac{3.4c}{-6.8}$

$\phantom{\dfrac{20.4b - 3.4c}{-6.8}} = -3b + \dfrac{1}{2}c$

61. $\dfrac{96a + 24b - 120c}{8} = \dfrac{96a}{8} + \dfrac{24b}{8} - \dfrac{120c}{8}$

$= 12a + 3b - 15c$

63. $11(-3a + 2b - 4c) - 8(5a - 7b + 2c)$
$= -33a + 22b - 44c - 40a + 56b - 16c$
$= -73a + 78b - 60c$

65. $-4.6(2x - 5y) + 9.9(5x - 3y)$
$= -9.2x + 23y + 49.5x - 29.7y$
$= 40.3x - 6.7y$

67. $\dfrac{3}{8}\left(-\dfrac{4}{9}p - \dfrac{2}{9}q\right) + \dfrac{2}{3}\left(\dfrac{7}{8}p - \dfrac{6}{7}q\right)$
$= -\dfrac{1}{6}p - \dfrac{1}{12}q + \dfrac{7}{12}p - \dfrac{4}{7}q$
$= -\dfrac{2}{12}p + \dfrac{7}{12}p - \dfrac{7}{84}q - \dfrac{48}{84}q$
$= \dfrac{5}{12}p - \dfrac{55}{84}q$

69. $[15 - 2(3x + 6y - 10) + 4x] + [6x + 2(8y - 12)]$
$= (15 - 6x - 12y + 20 + 4x) + (6x + 16y - 24)$
$= 35 - 2x - 12y + 6x + 6y - 24$
$= 11 + 4x + 4y$

71. $2[-5a + 3(2b - 4c) + 15] - [7(2a + 6b - c) + 12]$
$= 2(-5a + 6b - 12c + 15) - (14a + 42b - 7c + 12)$
$= -10a + 12b - 24c + 30 - 14a - 42b + 7c - 12$
$= -24a - 30b - 17c + 18$

73. $6\{2[x + 2(3y - 4z)] - [x - y + 3(y + 2z)]\}$
$= 6[2(x + 6y - 8z) - (x - y + 3y + 6z)]$
$= 6(2x + 12y - 16z - x + y - 3y - 6z)$
$= 6(x + 10y - 22z)$
$= 6x + 60y - 132z$

75. $\dfrac{8(5a + 7c) - 6(2a + 4c)}{4} = \dfrac{40a + 56c - 12a - 24c}{4}$
$= \dfrac{40a}{4} + \dfrac{56c}{4} - \dfrac{12a}{4} - \dfrac{24c}{4}$
$= 10a + 14c - 3a - 6c$
$= 7a + 8c$

77. $\dfrac{2.6m + 3(1.2m - 2.6n) - (4.8m + 7.4n) - 1.6n}{3m + 2(-2m + 1) + m}$
$= \dfrac{2.6m + 3.6m - 7.8n - 4.8m - 7.4n - 1.6n}{3m - 4m + 2 + m}$
$= \dfrac{1.4m - 16.8n}{2}$
$= \dfrac{1.4m}{2} - \dfrac{16.8n}{2}$
$= 0.7m - 8.4n$

79. Let w = width
$2w$ = length
w = height
Surface area = $2[w(2w)] + 2[2w(w)] + 2[w \cdot w]$
$= 2(2w^2) + 2(2w^2) + 2w^2$
$= 4w^2 + 4w^2 + 2w^2$
$= 10w^2$
$= 10(2)^2$
$= 10(4)$
$= 40$
The surface area is 40 square feet.

81. Let x = number of days
y = miles driven
Expense = $35x + 0.2y + 2(120x) + 2(50)$
$= 275x + 0.2y + 100$
$= 275(4) + 0.2(625) + 100$
$= 1325$
$1325 is the total trip expense

83. Let x = number of days
y = number of miles
Expense $= 30x + 0.22y + 450 + 85(2x) + 2(50)$
$= 200x + 0.22y + 550$
$= 200(2) + 0.22(375) + 550$
$= 1032.5$
$1032.50 is the total expense

85. Let x = number of hours
Cost = $2(38x) + 2(22) + 3(16x)$
$= 124x + 44$
$= 124(7) + 44$
$= 912$
The total cost is $912.

2.3 Experiencing Algebra the Calculator Way

1. $-679(138x - 349y) + 903(287x + 423y)$
$= -679(138x) - 679(-349y) + 903(287x) + 903(423y)$
$= 165,459x + 618,940y$

2. $\dfrac{1173.04a + 2147.893b}{23.65} = \dfrac{1173.04a}{23.65} + \dfrac{2147.893b}{23.65}$
$= 49.6a + 90.82b$

3. $27.12[39.675(x - 41.56) - 21.876(y - 22.7)] + 109.35[30.4(x + 33.9) + 43.76(y - 52.7)]$
$= 27.12(39.675x - 1648.893 - 21.876y + 496.5852) + 109.35(30.4x + 1030.56 + 43.76y - 2306.152)$
$= 27.12(39.675x - 21.876y - 1152.3078) + 109.35(30.4x + 43.76y - 1275.592)$
$= 1075.986x - 593.27712y - 31,250.58754 + 3324.24x + 4785.156y - 139,485.9852$
$= 4400.226x + 4191.87888y - 170,736.5727$

4. $\left(3\frac{5}{12}\right)\left[\left(4\frac{1}{2}\right)x - \left(3\frac{3}{5}\right)y\right]$

$= \frac{41}{12}\left(\frac{9}{2}\right)x - \frac{41}{12}\left(\frac{18}{5}\right)y$

$= \frac{123}{8}x - \frac{123}{10}y$

5. $\dfrac{12.0832abc}{2.36b} = 5.12ac$

6. $\dfrac{-2102.17yz}{41.3yz} = -50.9$

7. $\dfrac{-34.155xyz}{-2.07xz} = 16.5y$

8. $\dfrac{\frac{5}{12}y}{-\frac{15}{16}} = \frac{5}{12}y \cdot \left(-\frac{16}{15}\right) = -\frac{4}{9}y$

9. $\dfrac{-\frac{2}{5}ab}{\frac{3}{5}ab} = -\frac{2ab}{5} \cdot \left(\frac{5}{3ab}\right) = -\frac{2}{3}$

10. $\dfrac{\frac{3}{5}xy}{-\frac{2}{5}x} = \frac{3xy}{5} \cdot \left(-\frac{5}{2x}\right) = -\frac{3}{2}y$

11. $\dfrac{-5bc}{-\frac{5}{7}b} = -5bc \cdot \left(-\frac{7}{5b}\right) = 7c$

12. $\dfrac{-\frac{5}{21}x + \frac{15}{49}y}{-\frac{5}{7}} = \left(-\frac{5x}{21}\right)\left(-\frac{7}{5}\right) + \left(\frac{15y}{49}\right)\left(-\frac{7}{5}\right)$

$= \frac{1}{3}x - \frac{3}{7}y$

2.4 Experiencing Algebra the Exercise Way

1. equation; equality

3. expression; no equality

5. equation; equality

7. equation; equality

9. expression; no equality

11. $7 + 15 = 2 \cdot 11$

$$22 \mid 22$$

Since $22 = 22$, the equation is true.

13. $4 \cdot 7 = 5 + 24$

$$28 \mid 29$$

Since $28 \neq 29$, the equation is false.

15. $\dfrac{1}{7} + \dfrac{2}{3} = \dfrac{6}{7} - \dfrac{1}{3}$

$$\dfrac{3}{21} + \dfrac{14}{21} \ \bigg| \ \dfrac{18}{21} - \dfrac{7}{21}$$

$$\dfrac{17}{21} \ \bigg| \ \dfrac{11}{21}$$

Since $\dfrac{17}{21} \neq \dfrac{11}{21}$, the equation is false.

17. $\dfrac{1}{2} + \dfrac{1}{3} = \dfrac{2}{3} + \dfrac{1}{6}$

$$\dfrac{3}{6} + \dfrac{2}{6} \ \bigg| \ \dfrac{4}{6} + \dfrac{1}{6}$$

$$\dfrac{5}{6} \ \bigg| \ \dfrac{5}{6}$$

Since $\dfrac{5}{6} = \dfrac{5}{6}$, the equation is true.

19. $3.7 - 4.8 = 1.7 - 0.6$

$$-1.1 \mid 1.1$$

Since $-1.1 \neq 1.1$, the equation is false.

21. $4.9 + 1.2 = 8.4 - 2.2$

$$6.1 \mid 6.2$$

Since $6.1 \neq 6.2$, the equation is false.

23. $2(16 - 5) - 3(19 - 12)$

$$2(11) \mid 3(7)$$

$$22 \mid 21$$

Since $22 \neq 21$, the equation is false.

25.
$$\frac{2}{3} + \frac{3}{2} = 2$$

| $\frac{4}{6} + \frac{9}{6}$ | 2 |
| $\frac{13}{6}$ | |

Since $\frac{13}{6} \neq 2$, the equation is false.

27. $2x + 4 = 10$

$2(3) + 4$	10
$6 + 4$	
10	

Since $10 = 10$, 3 is a solution.

29. $5y - 7 = 9$

$5(3) - 7$	9
$15 - 7$	
8	

Since $8 \neq 9$, 3 is not a solution.

31. $6a + 5 = 3a + 17$

$6(3) + 5$	$3(3) + 17$
$18 + 5$	$9 + 17$
23	26

Since $23 \neq 26$, 3 is not a solution.

33. $9z - 23 = 6z - 29$

$9(-2) - 23$	$6(-2) - 29$
$-18 - 23$	$-12 - 29$
-41	-41

Since $-41 = -41$, -2 is a solution.

35. $3(x - 5) + 9 = 4(6x - 5) - 7$

$3(1 - 5) + 9$	$4(6 \cdot 1 - 5) - 7$
$3(-4) + 9$	$4(1) - 7$
$-12 + 9$	$4 - 7$
-3	-3

Since $-3 = -3$, 1 is a solution.

37. $2[3(x - 4)\,1 - 6] + x = 3x$

$2[3(8 - 4) - 6] + 8$	$3(8)$
$2(3 \cdot 4 - 6) + 8$	24
$2(12 - 6) + 8$	
$2(6) + 8$	
$12 + 8$	
20	

Since $20 \neq 24$, 8 is not a solution.

39. $x^2 + 5 = 33 - 3x$

$4^2 + 5$	$33 - 3(4)$
$16 + 5$	$33 - 12$
21	21

Since $21 = 21$, 4 is a solution.

41. $7 + 6(5) = 3(7) + 4^2$

43. Let x = a number
$x + 6 = 15$

45. Let x = a number
$2x = 12$

47. Let x = a number
$x^2 - 21 = 100$

49. Let x = a number
$2(x + 5^2) = x + 100$

51. Let x = a number
$2(x + 2) = 4 + 2x$

53. Let x = a number

$$17 + \frac{x}{2} = 4 + 3x$$

55. Let p = perimeter
d = diameter
r = radius
$p = d + \pi r$

57. $I = A - P$

2.4 Experiencing Algebra the Calculator Way

1.
```
4-6*3+28/7
              -10
16/4+2( -16+9)
              -10
```

True

2.
```
14/2+6*3-2²
               21
5²-2²
               21
```

True

3.
```
4(8+7*3)-100
               16
5+3(2*6-20)
               -19
```

False

4.
```
-7( -5+4)+20/5
               11
-2(3²+1)+19
               -1
```

False

5.
```
13(1/5+3/7)▶Frac
            286/35
11(3/5+1/7)▶Frac
            286/35
```

True

6.
```
1/2(2/3+4/7)-3/7
▶Frac
              4/21
1-17/21▶Frac
              4/21
```

True

7.
```
(3/4)/(1/2)+2(3/
8)-(1+1/4)▶Frac
               1
1/2▶Frac
              1/2
```

False

8.
```
1/3-(5/7)/(5/3)+
6(1/7)-(2+1/3)▶F
rac
            -11/7
-(1+4/7)▶Frac
            -11/7
```

True

9.
```
1.2-3(4.7)+6.5/5
            -11.6
-3(3.2)+2(4.3-2.
4-2.9)
            -11.6
```

True

10.
```
-1.2+4.7-6.9/2.3
               5
3(5.6)-4(4.2-0.5
)
               2
```

False

11.

```
5.7/3+7(-1.2)
            -6.5
-2(3.3
            -6.6
```

False

12.

```
2(2.7)-5.8/2
            2.5
7.8/3-0.1
            2.5
```

True

13.

```
269.1/7.5
         35.88
13.8(2.6)
         35.88
```

True

14.

```
-6-1/3→Z:5Z+13=2
Z-6
            1
```

Yes

15.

```
-2/7→X:15X-2=X-7
            0
```

No

16.

```
27.5→X:15X-13=6(
2X+7)+X
            1
```

Yes

17.

```
-.5→Z:3Z+20=5-3(
Z-4)
            1
```

Yes

2.5 Experiencing Algebra the Exercise Way

1. $A = \dfrac{1}{2}bh = \dfrac{1}{2}(39)(24)$

   $A = 468 \text{ in}^2$

   $P = a + b + c = 39 + 40 + 25 = 104 \text{ in.}$

3. $A = LW = 6(4) = 24 \text{ cm}^2$

   $P = 2L + 2W = 2(6) + 2(4) = 12 + 8 = 20 \text{ cm}$

5. $A = s^2 = \left(2\dfrac{1}{2}\right)^2 = \left(\dfrac{5}{2}\right)^2 = \dfrac{25}{4} \text{ ft}^2$

   $P = 4s = 4\left(2\dfrac{1}{2}\right) = 4\left(\dfrac{5}{2}\right) = 10 \text{ ft}$

7. $A = bh = 68(45) = 3060 \text{ m}^2$

   $P = 2a + 2b$

   $= 2(53) + 2(68)$

   $= 106 + 136$

   $= 242 \text{ m}$

9. $A = \dfrac{1}{2}h(b + B)$

   $= \dfrac{1}{2}(8)(2 + 23)$

   $= 4(25)$

   $= 100 \text{ ft}^2$

   $P = a + b + c + B = 10 + 2 + 17 + 23 = 52 \text{ ft}$

11. $A = \pi r^2 = (\pi)\left(5\dfrac{1}{4}\right)^2 \approx 86.6 \text{ in}^2$

    $C = 2\pi r = 2(\pi)\left(5\dfrac{1}{4}\right) \approx 33.0 \text{ in.}$

13. $V = LWH = 5(3)(1) = 15 \text{ ft}^3$

    $S = 2LW + 2WH + 2LH$

    $= 2(5)(3) + 2(3)(1) + 2(5)(1)$

    $= 30 + 6 + 10$

    $= 46 \text{ ft}^2$

15. $V = s^3 = (7.5)^3 = 421.875$ in^3
$S = 6s^2 = 6(7.5)^2 = 337.5$ in^2

17. $V = \pi r^2 h = (\pi)\left(\dfrac{3}{2}\right)^2 (5) \approx 35.3$ in^3

$S = 2\pi r^2 + 2\pi rh$

$= 2(\pi)\left(\dfrac{3}{2}\right)^2 + 2(\pi)\left(\dfrac{3}{2}\right)(5)$

≈ 61.3 in^2

19. $V = \dfrac{4}{3}\pi r^3 = \dfrac{4}{3}(\pi)\left(\dfrac{20}{2}\right)^3 \approx 4189$ cm^3

$S = 4\pi r^2 = 4(\pi)(10)^2 \approx 1257$ cm^2

21. $V = \dfrac{1}{3}\pi r^2 h$

$= \dfrac{1}{3}(\pi)(0.25)^2 (0.75)$

≈ 0.049 ft^3

$S = \pi r\sqrt{r^2 + h^2} + \pi r^2$

$= \pi(0.25)\sqrt{(0.25)^2 + (0.75)^2} + \pi(0.25)^2$

≈ 0.817 ft^2

23. $V = \dfrac{1}{3}Bh = \dfrac{1}{3}(8)^2 (15) = 320$ cm^3

25. $90° - 65° = 25°$

27. $180° - 65° = 115°$

29. $180° - (33° + 68°) = 79°$

31. $c = \sqrt{a^2 + b^2} = \sqrt{(40)^2 + (42)^2} = 58$ in.

33. $c = \sqrt{a^2 + b^2} = \sqrt{(6)^2 + (2.5)^2} = 6.5$ cm

35. $c = \sqrt{a^2 + b^2}$

$= \sqrt{(4.81)^2 + (6)^2}$

$= 7.69$ mm

37. $A = \dfrac{1}{2}h(b + B)$

$= \dfrac{1}{2}(14)(12 + 15)$

$= 7(27)$

$= 189$ ft^2

John will need to cover 189 ft^2.

39. $A = s^2 = (52)^2 = 2704$ in^2
$P = 4s = 4(52) = 208$ in.

The area of coverage is 2704 in^2. Gretchen will need 208 in. of fringe material.

41. $A = LW = 12(8.5) = 102$ ft^2
$P = 2L + 2W = 2(12) + 2(8.5) = 41$ ft

Linda had 102 ft^2 of sod removed. She needed 41 ft of border.

43. $A = \pi r^2 = \pi(5)^2 \approx 78.5$ ft^2
$C = 2\pi r = 2\pi(5) \approx 31.4$ ft

The square footage of the area is 78.5 ft^2. The border will be 31.4 ft.

45. $A = bh = 220(250) = 55,000$ ft^2
The lot is 55,000 ft^2.

47. $V = LWH = 4(2)(2) = 16$ ft^3
$S = 2LW + 2WH + 2LH$
$= 2(4)(2) + 2(2)(2) + 2(4)(2)$
$= 16 + 8 + 16$
$= 40$ ft^2
$40 \div 20 = 2$

The box will hold 16 ft^3 of toys. Jim painted 40 ft^2 of surface area, using 2 pints of paint.

49. $V = \pi r^2 h = \pi(1)^2 (4) \approx 12.6$ ft^3
$S = 2\pi r^2 + 2\pi rh$
$= 2\pi(1)^2 + 2\pi(1)(4)$
≈ 31.4 ft^2

The barrel contains 12.6 ft^3 of space. 31.4 ft^2 of surface area will need to be painted.

51. $V = \frac{4}{3}\pi r^3 = \frac{4}{3}\pi(10)^3 \approx 4188.8$ in^3

$S = 4\pi r^2 = 4\pi(10)^2 \approx 1256.6$ in^2

The tank will hold 4188.8 in^3 of gas. The surface area is 1256.6 in^2.

53. $V = s^3 = (18)^3 = 5832$ in^3

$S = 6s^2 = 6(18)^2 = 1944$ in^2

The case contains 5832 in^3, with a surface area of 1944 in^2.

55. $V = \frac{1}{3}Bh = \frac{1}{3}(3.5)^2(7) = 28\frac{7}{12}$ in^3

The volume of the paperweight is

$28\frac{7}{12}$ in^3.

57. $180° - (30° + 65°) = 85°$

The third angle is 85°.

59. $90° - 50° = 40°$

The pitch of the roof is 40°.

61. $180° - 45° = 135°$

The other angle is 135°.

63. $c = \sqrt{a^2 + b^2} = \sqrt{6^2 + 8^2} = 10$ ft

The diagonal must measure 10 ft.

2.5 Experiencing Algebra the Calculator Way

Students should develop other programs for various formulas.

2.6 Experiencing Algebra the Exercise Way

1. $I = Prt = 2500(0.065)(1) \approx 162.50$

JoAnne paid $162.50 in interest.

3. $A = P(1+r)^t$

$= 2500(1 + 0.006)^{12}$

≈ 2686.06

$I = A - P = 2686.06 - 2500 \approx 186.06$

Simple interest is better.

5. $A = Pe^{nr} = 2500(e)^{1(0.065)} \approx 2667.90$

$I = A - P = 2667.90 - 2500 = 167.90$

The option in exercise 1 is the best choice.

7. $I = Prt = 500(0.07)\left(\frac{1}{12}\right) = 2.91\overline{6}$

$2.92 of interest is earned.

9. $I = Prt = 1200(0.08)\left(\frac{1}{2}\right) = 48$

$48 in interest is paid.

11. $A = P(1+r)^t = 2000(1 + 0.09)^2 = 2376.2$

$I = A - P = 2376.2 - 2000 = 376.2$

Steve earned $376.20 in interest.

13. $A = P(1+r)^t$

$= 4000(1 + 0.07)^5$

≈ 5610.2069

$I = A - P$

$= 5610.2069 - 4000$

$= 1610.2069$

Chauncie had $1610.21 in interest.

15. $A = P(1+r)^t$

$= 3000(1 + 0.04)^3$

$= 3374.592$

$I = 3374.592 - 3000 = 374.592$

The interest earned will be $374.59.

17. $A = Pe^{nr}$

$= 1800e^{2(0.05)}$

≈ 1989.307653

$I = A - P \approx 1989.307653 - 1800$

≈ 189.3076525

The interest is $189.31.

19. $A = Pe^{nr} = 3000e^{.06(4.5)} \approx 3929.893352$

$I = A - P \approx 3929.893352 - 3000$

≈ 929.8933522

The interest is $929.89.

21. $F = \frac{9}{5}C + 32 = \frac{9}{5}(25) + 32 = 77$

The temperature is 77°F.

23. $C = \dfrac{5}{9}(F - 32) = \dfrac{5}{9}(92 - 32) = 33\dfrac{1}{3}$

The temperature is $33\dfrac{1}{3}°C$.

25. $F = \dfrac{9}{5}C + 32 = \dfrac{9}{5}(100) + 32 = 212$

The temperature is $212°F$.

27. $F = \dfrac{9}{5}C + 32 = \dfrac{9}{5}(95) + 32 = 203$

The temperature is $203°F$.

29. $C = \dfrac{5}{9}(F - 32)$

$= \dfrac{5}{9}(-37.97 - 32)$

$= -38.87\overline{2}$

The temperature is $-38.87°C$.

31. $C = \dfrac{5}{9}(F - 32) = \dfrac{5}{9}(84 - 32) = 28.\overline{8}$

The temperature is $28.9°C$.

33. $d = rt = 55\left(9\dfrac{1}{2}\right) = 522.5$

The distance covered was 522.5 miles.

35. $d = rt = 74.602(3) = 223.806$

He drove 223.806 miles.

37. $s = -16t^2 + v_0 t + s_0$

$= -16(3)^2 + 0(3) + 200$

$= 56$

The pie is 56 feet high.

39. $s = -16t^2 + v_0 t + s_0$

$= -16(3)^2 + 25(3) + 200$

$= 131$

The baseball is 131 feet high.

41. $T = 2\pi\sqrt{\dfrac{L}{32}} = 2\pi\sqrt{\dfrac{3}{32}} \approx 1.924$

It will take 1.924 seconds.

43. $T = 2\pi\sqrt{\dfrac{L}{32}} = 2\pi\sqrt{\dfrac{2.25}{32}} \approx 1.666$

The period is 1.666 seconds.

45. $T = 2\pi\sqrt{\dfrac{L}{32}} = 2\pi\sqrt{\dfrac{0.75}{32}} \approx 0.962$

The period is 0.962 second.

47. $T = 2\pi\sqrt{\dfrac{L}{32}} = 2\pi\sqrt{\dfrac{2.5}{32}} \approx 1.756$

The leg takes 1.756 seconds.

$r = \dfrac{d}{t} = \dfrac{4}{1.756} = 2.28$

Eydie's speed is 2.28 feet per second.

2.6 Experiencing Algebra the Calculator Way

$A = Pe^{nr}$

$A = 80,000e^{n(0.085)}$

Number of Years (n)	Amount earned (A)	Interest $= A - 80,000$
5	$122,367	$42,367
10	$187,172	$107,172
15	$286,296	$206,296
20	$437,916	$357,916
25	$669,832	$589,832
30	$1,024,568	$944,568

Chapter 2 Review

Reflections

Answers will vary.

Exercises

1. Let $x = $ a number
   $55x + 4$

2. Let $x = $ a number
   $55(x + 4)$

3. Let $x = $ a number
   $\dfrac{3}{4}(x + 35)$

4. Let $x = $ a number
   $\dfrac{3}{4}x + 35$

5. Let $x = $ a number
   $2x - 20$

6. Let $x = $ a number
   $20 - 2x$

7. $2500 + 275n$

8. $\dfrac{650}{h}$

9. $200 + 5.5k$

For problems 10–18, answers will vary. One possibility is given.

10. $5 + x$
    The sum of 5 and a number

11. $8 - 6n$
    The difference of 8 and 6 times some number

12. $9x \div 6$
    The product of 9 and a number, divided by 6

13. $\dfrac{2}{3}(x - 75)$
    Two-thirds of the difference of a number and 75

14. $x^2 + y$
    The square of one number added to another number

15. xyz
    The product of three different numbers

16. $5x - 25 = 5(8) - 25 = 40 - 25 = 15$

17. $5x - 25 = 5(-4.6) - 25 = -23 - 25 = -48$

18. $\sqrt{6x + 10} = \sqrt{6(9) + 10}$
    $= \sqrt{54 + 10}$
    $= \sqrt{64}$
    $= 8$

19. $\sqrt{6x + 10} = \sqrt{6(-1) + 10}$
    $= \sqrt{-6 + 10}$
    $= \sqrt{4}$
    $= 2$

20. $\dfrac{12y - 84}{6y + 36} = \dfrac{12(9) - 84}{6(9) + 36} = \dfrac{24}{90} = \dfrac{4}{15}$

21. $\dfrac{12y - 84}{6y + 36} = \dfrac{12(-6) - 84}{6(-6) + 36} = \dfrac{-156}{0} = $
    undefined

22. $\left|2x^2 - 45x - 75\right| = \left|2(0.1)^2 - 45(0.1) - 75\right|$
    $= |0.02 - 4.5 - 75|$
    $= |-79.48|$
    $= 79.48$

23. $\left|2x^2 - 45x - 75\right| = \left|2(10)^2 - 45(10) - 75\right|$
    $= |200 - 450 - 75|$
    $= |-325|$
    $= 325$

24. $x^2 = (-15)^2 = 225$

25. $-x^2 = -(-15)^2 = -225$

26. $(-x)^2 = [-(-15)]^2 = (15)^2 = 225$

27. Let m = number of months
$1200 + 75m = 1200 + 75(12) = 2100$
Sandi will have $2100 after 12 months.
Yes, she will have enough money.

	Algebraic Expression	# of terms	terms
28.	$3x^2 - 16x + 35$	3	$3x^2, -16x, 35$
29.	$5x + 12 - 3y - 16$	4	$5x, 12, -3y, -16$
30.	$2(x + 1) - 4(y + 2) - 4$	3	$2(x + 1), -4(y + 2), -4$
31.	$a^2 + 2ab + b^2$	3	$a^2, 2ab, b^2$

	Algebraic Expression	Coefficients of terms	like terms
32.	$3x - 2y + 4x + 9y$	3, –2, 4, 9	$3x$ and $4x$, $-2y$ and $9y$
33.	$2a^2 - a + 3a^2 - 5a^3$	2, –1, 3, –5	$2a^2$ and $3a^2$
34.	$2.4x + 5.1 + 6.2x$	2.4, 5.1, 6.2	$2.4x$ and $6.2x$
35.	$4(a + b) - 2(a + b) + 2a$	4, –2, 2	$4(a + b)$ and $-2(a + b)$

36. $6c - 17c = (6 - 17)c = -11c$

37. $2.4z + 1.7z - 3.9z = (2.4 + 1.7 - 3.9)z = 0.2z$

38. $17x + 51 + 26x - 86 - 19x - 7$
$= 17x + 26x - 19x + 51 - 86 - 7$
$= 24x - 42$

39. $\dfrac{3}{4}x + \dfrac{5}{8}y - \dfrac{2}{3}x + \dfrac{1}{4}y = \dfrac{3}{4}x - \dfrac{2}{3}x + \dfrac{5}{8}y + \dfrac{1}{4}y$
$= \dfrac{9}{12}x - \dfrac{8}{12}x + \dfrac{5}{8}y + \dfrac{2}{8}y$
$= \dfrac{1}{12}x + \dfrac{7}{8}y$

40. $15x^2 - 14xy + 12y^2 - 23 + 42xy - 7y^2 + 21 - 6x^2$
$= 15x^2 - 6x^2 - 14xy + 42xy + 12y^2 - 7y^2 - 23 + 21$
$= 9x^2 + 28xy + 5y^2 - 2$

41. $-(19p - 21) = -19p + 21$

42. $4a + 6 - (a - 1) = 4a + 6 - a + 1$
$= 4a - a + 6 + 1$
$= 3a + 7$

43. $(3a + 4b) - (-2a - 6b) + (a - b) - (-a + b)$
$= 3a + 4b + 2a + 6b + a - b + a - b$
$= 3a + 2a + a + a + 4b + 6b - b - b$
$= 7a + 8b$

44. Let x = number of cans of peanuts
y = number of cans of bridge mix
$2.75(x + y) - (1.25x + 1.5y)$
$= 2.75x + 2.75y - 1.25x - 1.5y$
$= 2.75x - 1.25x + 2.75y - 1.5y$
$= 1.5x + 1.25y$
$= 1.5(220) + 1.25(480)$
$= 330 + 600 = 930$
The net profit is \$930.

45. Let x = number of pins
$20 - [5x + 0.06(5x)] = 20 - (5x + 0.3x)$
$= 20 - 5.3x$
$= 20 - 5.3(3)$
$= 20 - 15.9$
$= 4.1$
Katie received \$4.10 in change.

46. Let x = number of frames
$-3.25x + 12x - 52.65 = 8.75x - 52.65$
$= 8.75(32) - 52.65$
$= 280 - 52.65$
$= 227.35$
Margaret's net profit is \$227.35.

47. $14(5ab) = (14 \cdot 5)ab = 70ab$

48. $-2(-4.1m) = [-2 \cdot (-4.1)]m = 8.2m$

49. $-2a(3.8a - 4.7b) = -2a(3.8a) - 2a(-4.7b)$
$= -7.6a + 9.4ab$

50. $-\dfrac{5}{12}\left(\dfrac{6}{7}x - \dfrac{4}{15}y\right) = -\dfrac{5}{12}\left(\dfrac{6}{7}x\right) - \dfrac{5}{12}\left(-\dfrac{4}{15}y\right)$
$= -\dfrac{5}{14}x + \dfrac{1}{9}y$

51. $2x(3x - 17y) = 2x(3x) + 2x(-17y)$
$= 6x^2 - 34xy$

52. $\dfrac{-123m}{-3} = -123m \div (-3)$
$= -123m \cdot \left(-\dfrac{1}{3}\right)$
$= -123\left(-\dfrac{1}{3}\right)m$
$= 41m$

53. $\dfrac{\frac{15}{16}z}{-5} = \dfrac{15}{16}z \div (-5)$
$= \dfrac{15}{16}z \cdot \left(-\dfrac{1}{5}\right)$
$= \dfrac{15}{16}\left(-\dfrac{1}{5}\right) \cdot z$
$= -\dfrac{3}{16}z$

54. $\dfrac{36.6ab}{2} = \dfrac{36.6}{2} \cdot ab = 18.3ab$

55. $\dfrac{-308.88ab}{4b} = \dfrac{-308.88}{4} \cdot a \cdot \dfrac{b}{b}$
$= -77.22 \cdot a \cdot 1$
$= -77.22a$

56. $\dfrac{27a - 36b + 18c}{9} = \dfrac{27a}{9} - \dfrac{36b}{9} + \dfrac{18c}{9}$
$= 3a - 4b + 2c$

57. $12(7x + 9y) + 15(3x - 7y)$
$= 84x + 108y + 45x - 105y$
$= 129x + 3y$

58. $3[-2(x + 3y) - 5] - [3(2x + y) + 16]$
$= 3(-2x - 6y - 5) - (6x + 3y + 16)$
$= -6x - 18y - 15 - 6x - 3y - 16$
$= -12x - 21y - 31$

59. $\dfrac{25(2a - 6b) + 5(4b - 3c) + 75}{4a - 2(2a - 3) - 1}$
$= \dfrac{50a - 150b + 20b - 15c + 75}{4a - 4a + 6 - 1}$
$= \dfrac{50a - 130b - 15c + 75}{5}$
$= \dfrac{50a}{5} - \dfrac{130b}{5} - \dfrac{15c}{5} + \dfrac{75}{5}$
$= 10a - 26b - 3c + 15$

60. Tom: x pushups
Charles: $2x - 5$ pushups
Jim: $x + 7$ pushups
All three: $x + 2x - 5 + x + 7 = 4x + 2$
Total $= 4(45) + 2 = 182$
The three students can do a total of
182 pushups.

61. Let x = number of days
y = number of miles
Total cost $= 45x + 0.2y + 2(125x) + 2(20x)$
$= 45x + 0.2y + 250x + 40x$
$= 335x + 0.2y$
$= 335(4) + 0.2(345)$
$= 1409$
The total cost is \$1409.

62. expression; no equality

63. equation; equality

64. equation; equality

65. expression; no equality

66.
$$17 - 3 \cdot 4 = 20 \cdot 2 - 6^2$$

$17 - 12$	$40 - 36$
5	4

Since $5 \neq 4$, the equation is false.

67.
$$2(3.5) - 3(0.7) = 2.5 + 2(1.2)$$

$7 - 2.1$	$2.5 + 2.4$
4.9	4.9

Since $4.9 = 4.9$, the equation is true.

68.
$$3 + 5(9 - 8) + 6 \cdot 5 = 4 + 6(12 - 8) + 45 \div 9 + 5$$

$3 + 5(1) + 30$	$4 + 6(4) + 5 + 5$
38	38

Since $38 = 38$, the equation is true.

69.
$$\frac{2}{3} + 5\left(\frac{1}{6} - \frac{2}{9}\right) = \frac{4}{9} - 3\left(\frac{2}{3} - \frac{17}{18}\right)$$

$\frac{2}{3} + 5\left(\frac{3}{18} - \frac{4}{18}\right)$	$\frac{4}{9} - 3\left(\frac{12}{18} - \frac{17}{18}\right)$
$\frac{2}{3} + 5\left(-\frac{1}{18}\right)$	$\frac{4}{9} - 3\left(-\frac{5}{18}\right)$
$\frac{12}{18} - \frac{5}{18}$	$\frac{8}{18} + \frac{15}{18}$
$\frac{7}{18}$	$\frac{23}{18}$

Since $\dfrac{7}{18} \neq \dfrac{23}{18}$, the equation is false.

70.
$$15y - 35 = 12$$

$15(3) - 35$	12
$45 - 35$	
10	

Since $10 \neq 12$, 3 is not a solution.

71.
$$8a + 2(3a - 7) = 11a - 32$$

$8(-6) + 23(-6) - 7]$	$11(-6) - 32$
$-48 + 2(-18 - 7)$	$-66 - 32$
$-48 + 2(-25)$	-98
-98	

Since $-98 = -98$, -6 is a solution.

72.
$$x^3 - 25x = 2x^2 - 32 - 35$$

$5^3 - 25(5)$	$2(5)^2 - 3(5) - 35$
$125 = 125$	$50 - 15 - 35$
0	0

Since $0 = 0$, 5 is a solution.

73.

$2(x - 3.4) = x - 5$	
$2(1.8 - 3.4)$	$1.8 - 5$
$2(-1.6)$	-3.2
-3.2	

Since $-3.2 = -3.2$, 1.8 is a solution.

74.

$x - \dfrac{2}{3} = -\dfrac{1}{9}$	
$\dfrac{4}{9} - \dfrac{2}{3}$	$-\dfrac{1}{9}$
$\dfrac{4}{9} - \dfrac{6}{9}$	
$-\dfrac{2}{9}$	

Since $-\dfrac{2}{9} \neq -\dfrac{1}{9}$, $\dfrac{4}{9}$ is not a solution.

75. $3 + 2(27 - 15) = 3^3$

76. Let x = a number

$$5 + 4x = 65 + \frac{x}{4}$$

77. $A = \dfrac{1}{2}bh = \dfrac{1}{2}(26)(16) = 208 \text{ m}^2$

$P = a + b + c = 22 + 26 + 20 = 68 \text{ m}$

78. $A = LW = 44(20) = 880 \text{ in}^2$

$P = 2L + 2W = 2(44) + 2(20) = 88 + 40$
$= 128 \text{ in.}$

79. $A = s^2 = (15)^2 = 225 \text{ cm}^2$

$P = 4s = 4(15) = 60 \text{ cm}$

80. $A = bh = 10.0(6.5) = 65 \text{ m}^2$

$P = 2a + 2b = 2(7.0) + 2(10.0) = 34 \text{ m}$

81. $A = \dfrac{1}{2}h(b + B)$

$= \dfrac{1}{2}(70)(110 + 160)$

$= 9450 \text{ yd}^2$

$P = a + b + c + B$
$= 75 + 110 + 80 + 160$
$= 425 \text{ yd}$

82. $A = \pi r^2 = \pi(12.2)^2 \approx 467.59 \text{ ft}^2$

$C = 2\pi r = 2\pi(12.2) \approx 76.65 \text{ ft}$

83. $V = LWH = 35(9)(21) = 6615 \text{ in}^3$

$S = 2LW + 2WH + 2LH$
$= 2(35)(9) + 2(9)(21) + 2(35)(21)$
$= 630 + 378 + 1470$
$= 2478 \text{ in}^2$

84. $V = s^3 = (14.6)^3 = 3112.136 \text{ mm}^3$

$S = 6s^2 = 6(14.6)^2 = 1278.96 \text{ mm}^2$

85. $V = \pi r^2 h = \pi(18)^2(54) \approx 54,965.3 \text{ in}^3$

$S = 2\pi r^2 + 2\pi rh$
$= 2\pi(18)^2 + 2\pi(18)(54)$
$\approx 8143.01 \text{ in}^2$

86. $V = \dfrac{4}{3}\pi r^3 = \dfrac{4}{3}\pi(32.6)^3 \approx 145,124.7 \text{ cm}^3$

$S = 4\pi r^2 = 4\pi(32.6)^2 \approx 13,355.04 \text{ cm}^2$

87. $V = \dfrac{1}{3}\pi r^2 h = \dfrac{1}{3}\pi(4)^2(7) \approx 117.3 \text{ cm}^3$

$S = \pi r\sqrt{r^2 + h^2} + \pi r^2$

$= \pi(4)\sqrt{4^2 + 7^2} + \pi(4)^2$

$\approx 151.6 \text{ cm}^2$

88. $V = \dfrac{1}{3}Bh = \dfrac{1}{3}(5)^2(8) = 66\dfrac{2}{3} \text{ in}^3$

89. $180° - 58° = 122°$
The other angle is 122°.

90. $90° - 58° = 32°$
The other angle is 32°.

91. $180° - (67° + 88°) = 25°$
The third angle is 25°.

92. $c = \sqrt{a^2 + b^2} = \sqrt{(20)^2 + (21)^2} = 29$ in.
The hypotenuse is 29 inches.

93. $c = \sqrt{a^2 + b^2}$
$= \sqrt{(5.39)^2 + (29.4)^2}$
$= 29.89$ cm
The hypotenuse is 29.89 cm.

94. $A = LW = 85(60) = 5100 \text{ ft}^2$
$P = 2L + 2W = 2(85) + 2(60) = 290$
fence $= 290 - 3 - 35 = 252$ ft
Dan should order 5100 ft^2 of sod and 252 feet of fencing.

95. $A = \pi r^2 - \pi\left(\dfrac{2}{2}\right)^2 = \pi(8)^2 - \pi(1)^2$
$\approx 197.9 \text{ ft}^2$
The garden will have about 197.9 ft^2.

96. Area of 2 10" pizzas $= 2(\pi r^2)$
$= 2(\pi)(5)^2$
$= 50\pi \text{ in}^2$
Area of one 14" pizza $= \pi r^2$
$= \pi(7)^2$
$= 49\pi \text{ in}^2$
The 2 10-inch pizzas are the better deal because 50π is greater than 49π.

97. $V = LWH = 8(5)(2) = 80 \text{ ft}^3$
The truck will hold 80 ft^3.

98. $S = 4\pi r^2 = 4\pi\left(\dfrac{6}{2}\right)^2 \approx 113.1 \text{ ft}^2$
The surface area is approximately 113.1 ft^2.

99. $90° - 40° = 50°$
The other angle is 50°.

100. $c = \sqrt{a^2 + b^2} = \sqrt{(18)^2 + (13.5)^2} = 22.5$ ft
The diagonal must be 22.5 feet.

101. $I = Prt = 850(0.125)(1) = 106.25$
$106.25 interest
$A = P + I = 850 + 106.25 = \$956.25$
The total amount of the loan is $956.25.

102. $A = P(1 + r)^t$
$= 15,000(1 + 0.085)^6$
$\approx 24{,}472.013$
$I = A - P = 24{,}472.013 - 15{,}000 = 9472.013$
The interest is $9472.01.

103. $A = Pe^{nr} = 10{,}000e^{12(0.055)} \approx 19{,}347{,}923$
$I = A - P = 19{,}347.923 - 10{,}000 = 9347.923$
The amount of interest is $9347.92.

104. $F = \dfrac{9}{5}C + 32 = \dfrac{9}{5}(50) + 32 = 122$
The temperature is 122°F.

105. $C = \dfrac{5}{9}(F - 32) = \dfrac{5}{9}(80 - 32) = 26\dfrac{2}{3}$
The temperature is $26\dfrac{2}{3}°$C.

106. $d = rt = 62\left(5\dfrac{3}{4}\right) = 356.5$
LuAnn traveled 356.5 miles.

107. $s = -16t^2 + v_0 t + s_0$
$= -16(3)^2 + 0(3) + 180$
$= 36$
The wedding bouquet is 36 feet above the ground.

108. $s = -16t^2 + v_0 t + s_0$
$= -16(2)^2 + 96(2) + 0$
$= 128$
The height is 128 feet after 2 seconds.
$-16(3)^2 + 96(3) + 0 = 144$
The height is 144 feet after 3 seconds.
$-16(4)^2 + 96(4) + 0 = 128$
The height is 128 feet after 4 seconds.
$-16(5)^2 + 96(5) + 0 = 80$
The height is 80 feet after 5 seconds.

109. $T = 2\pi\sqrt{\dfrac{L}{32}} = 2\pi\sqrt{\dfrac{6}{32}} \approx 2.72$
The period is 2.72 seconds.

110. $I = Prt = 1000(0.08)(2) = 160$
$160 simple interest
$A = P(1+r)^t = 1000(1+0.07)^2 = 1144.9$
$I = 1144.9 - 1000 = 144.9$
$144.90 compounded interest
The first option is the better choice.

Chapter 2 Mixed Review

	Algebraic Expression	# of terms	terms
1.	$12 + y - z + 23$	4	$12x, y, -z, 23$
2.	$3(a-2) + 5(b-4) + 75$	3	$3(a-2), 5(b-4), 75$
3.	$12 - 7x + 14x - 18 + x$	5	$12, -7x, 14x, -18, x$

	Algebraic Expression	Coefficients of terms	like terms
4.	$x - 2y - 5 + 4x - y + 6$	$1, -2, -5, 4, -1, 6$	x and $4x$, $-2y$ and $-y$, -5 and 6
5.	$b^2 + 2b - 3b^2 + 6b + b^3$	$1, 2, -3, 6, 1$	b^2 and $-3b^2$, $2b$ and $6b$

6. $3x = 3(-18) = -54$

7. $-3x = -3(-18) = 54$

8. $-(-3x) = -[-3(-18)] = -54$

9. $-(-(-3x)) = -(-54) = 54$

10. $x^2 = (-18)^2 = 324$

11. $-x^2 = -(-18)^2 = -324$

12. $(-x)^2 = [-(-18)]^2 = 324$

13. $-(-x)^2 = -324$

14. $\sqrt{12y+20} = \sqrt{12(8)+20} = \sqrt{116}$
≈ 10.77033

15. $\sqrt{12y+20} = \sqrt{12\left(-\dfrac{1}{3}\right)+20}$
$= \sqrt{-4+20}$
$= \sqrt{16}$
$= 4$

16. $\sqrt{12y+20} = \sqrt{12(-5)+20} = \sqrt{-40}$
Not a real number

17. $\dfrac{7x+84}{2x-3} = \dfrac{7(-12)+84}{2(-12)-3} = \dfrac{0}{-27} = 0$

18. $\dfrac{7x+84}{2x-3} = \dfrac{7(3)+84}{2(3)-3} = \dfrac{105}{3} = 35$

19. $\dfrac{7x+84}{2x-3} = \dfrac{7(1.5)+84}{2(1.5)-3} = \dfrac{94.5}{0}$
    Undefined

20. $3 + 5 \cdot 12 = 7(3 + 6)$

$3 + 60$	$7(9)$
63	63

    Since $63 = 63$, the equation is true.

21. $38 - 2 \cdot 11 = (1+3)^2$

$38 - 22$	4^2
16	16

    Since $16 = 16$, the equation is true.

22. $5.9 + 3.6(8.7 - 3.1) + 7(6.3) = 100 - 3(8.52) - 4.3$

$5.9 + 3.6(5.6) + 44.1$	$100 - 25.56 - 4.3$
70.16	70.14

    Since $70.16 \neq 70.14$, the equation is false.

23. $\dfrac{25 + 5(14 - 2 \cdot 9) - 33(85 - 5 \cdot 7)}{100 - 5 \cdot 13} = -5 \cdot 9 - 2(50 - 7^2)$

$\dfrac{25 + 5(-4) - 33(50)}{100 - 65}$	$-45 - 2(1)$
$\dfrac{-1645}{35}$	$-45 - 2$
-47	-47

    Since $-47 = -47$, the equation is true.

24. $3x - 7 = x + 1$

$3(4) - 7$	$4 + 1$
$12 - 7$	5
5	

    Since $5 = 5$, 4 is a solution.

25.
$$5x + 17 = 10x + 5$$

$5(3) + 17$	$10(3) + 5$
$15 + 17$	$30 + 5$
32	35

Since $32 \neq 35$, 3 is not a solution.

26.
$$3x + 17 = 2(x - 5)$$

$3(-27) + 17$	$2(-27 - 5)$
$-81 + 17$	$2(-32)$
-64	-64

Since $-64 = -64$, -27 is a solution.

27.
$$2.1x - 1.9 = 0.6x - 4.6$$

$2.1(-1.8) - 1.9$	$0.6(-1.8) - 4.6$
$-3.78 - 1.9$	$-1.08 - 4.6$
-5.68	-5.68

Since $-5.68 = -5.68$, -1.8 is a solution.

28.
$$\frac{3}{4}\left(x - \frac{8}{9}\right) = \frac{1}{2}\left(x + \frac{2}{3}\right)$$

$\frac{3}{4}\left(5\frac{1}{3} - \frac{8}{9}\right)$	$\frac{1}{2}\left(5\frac{1}{3} + \frac{2}{3}\right)$
$\frac{3}{4}\left(4\frac{4}{9}\right)$	$\frac{1}{2}(6)$
$3\frac{1}{3}$	3

Since $3\frac{1}{3} \neq 3$, $5\frac{1}{3}$ is not a solution.

29.
$$3x^2 - 6x - 10 = x^2 + x - 5$$

$3(-5)^2 - 6(-5) - 10$	$(-5)^2 - 5 - 5$
$75 + 30 - 10$	$25 - 5 - 5$
95	15

Since $95 \neq 15$, -5 is not a solution.

30. $12h + 9h - 4h = (12 + 9 - 4)h = 17h$

31. $6m + 22 - m - 12 + 3m$
$= 6m - m + 3m + 22 - 12$
$= 8m + 10$

32. $3x - 35 + 4y - 5x - 6y + 7x + 27 + 17y + 22x$
$= 3x - 5x + 7x + 22x - 35 + 27 + 4y - 6y + 17y$
$= 27x - 8 + 15y$

33. $3x^4 + 5x - 7x^2 + 12x^4 - 17x - 34x + x^3 - 1$
$= 3x^4 + 12x^4 + x^3 - 7x^2 + 5x - 17x - 34x - 1$
$= 15x^4 + x^3 - 7x^2 - 46x - 1$

34. $(6.2a + 5.3b) + (4.7a - 1.9b)$
$= 6.2a + 5.3b + 4.7a - 1.9b$
$= 6.2a + 4.7a + 5.3b - 1.9b$
$= 10.9a + 3.4b$

35. $-(27y - 15) = -27y + 15$

36. $5g + 8 - (g + 4) = 5g + 8 - g - 4$
$= 5g - g + 8 - 4$
$= 4g + 4$

37. $(-2x + 4y - 7z) - (-x + 6y + 8z)$
$= -2x + 4y - 7z + x - 6y - 8z$
$= -2x + x + 4y - 6y - 7z - 8z$
$= -x - 2y - 15z$

38. $-14x(3y) = (-14)(3)xy = -42xy$

39. $\frac{15}{22}\left(\frac{11}{25}a + \frac{33}{50}b\right) = \frac{165}{550}a + \frac{495}{1100}b$
$= \frac{3}{10}a + \frac{9}{20}b$

40. $\frac{-115.388z}{-4.55} = 25.36z$

41. $\frac{104x - 156y + 221z}{13} = \frac{104x}{13} - \frac{156y}{13} + \frac{221z}{13}$
$= 8x - 12y + 17z$

42. $\frac{\frac{21}{25}uv}{-3u} = \frac{21uv}{25} \cdot \left(-\frac{1}{3u}\right) = -\frac{21uv}{75u} = -\frac{7}{25}v$

43. $\frac{-18x + 24y - 36z}{-6} = \frac{-18x}{-6} + \frac{24y}{-6} - \frac{36z}{-6}$
$= 3x - 4y + 6z$

44. $4(3.9x - 11.1y) + 7(2.9x - 0.7y)$
$= 4(3.9x) + 4(-11.1y) + 7(2.9x) + 7(-0.7y)$
$= 15.6x - 44.4y + 20.3x - 4.9y$
$= 35.9x - 49.3y$

45. $12[-3(2a - 5b) + 9] + 8[-9(a + 13) - 6(b - 12)]$
$= 12(-6a + 15b + 9) + 8(-9a - 117 - 6b + 72)$
$= -72a + 180b + 108 - 72a - 936 - 48b + 576$
$= -144a + 132b - 252$

46. $\dfrac{14.4(2x + 5) - 21.6(5x - 2) + 7.2(x - 1)}{3x - 4(x + 8) + x + 34.4}$

$= \dfrac{28.8x + 72 - 108x + 43.2 + 7.2x - 7.2}{3x - 4x - 32 + x + 34.4}$

$= \dfrac{-72x + 108}{2.4}$

$= -30x + 45$

47. Let x = a number
$2(x + 50) - \dfrac{1}{2}x$

48. Let x = a number
$11 + \dfrac{1}{4}x$

49. Let x = a number
$x^2 + 2x = x + 306$

50. Let x = number of hours
$225 + 45x = 225 + 45(120) = 5625$
Lakeetha will make $5625 in earnings.

51. Let n = number of weeks
$500 + 145n - 15n = 500 + 130n$
$= 500 + 130(15) = 2450$
Carmen has $2450 deposited after 15 weeks.
$2450 - 625 = 1825$
She has a $1825 balance after the withdrawal.

52. a. Beatrice: x appliances
Marie: $2x - 5$ appliances

Ann: $\dfrac{1}{2}x$ appliances

Magdalene: x appliances

b. Beatrice: $100 + 25x$ dollars

Marie: $100 + 25(2x - 5) = 100 + 50x - 125 = 50x - 25$ dollars

Ann: $100 + 25\left(\dfrac{1}{2}x\right) = 100 + \dfrac{25}{2}x$ dollars

Magdalene: $100 + 25x$ dollars

c. $100 + 25x + 100 + 25(2x - 5) + 100 + 25\left(\dfrac{1}{2}x\right) + 100 + 25x$

d. $100 + 25x + 50x - 25 + 100 + 12.5x + 100 + 25x = 275 + 112.5x$

e. $275 + 112.5(20) = 2525$

A total of $2525 is earned.

53. $A = \dfrac{1}{2}h(b + B)$

$= \dfrac{1}{2}(100)(120 + 200)$

$= 50(320)$

$= 16,000$

Fuad must buy 16,000 ft^2 of sod.

54. $V = \pi r^2 h = \pi(10)^2(4.5) \approx 1413.7$

Chum's pool will hold 1413.7 ft^3 of water.

55. $C = \dfrac{5}{9}(F - 32) = \dfrac{5}{9}(96 - 32) = 35\dfrac{5}{9}$

The temperature is $35\dfrac{5}{9}°C$.

56. $A = P(1 + r)^t$

$= 18,500(1 + 0.055)^{10}$

$\approx 31,600.67$

You will have $31,600.67 over 10 years.

57. $d = rt = 13(1.25) = 16.25$

Randy bicycled 16.25 miles.

58. $c = \sqrt{a^2 + b^2} = \sqrt{(95)^2 + (168)^2} = 193$

The hypotenuse is 193 mm.

59. $V = \pi r^2 h = \pi(1.625)^2(1.5) \approx 12.44$

The can's volume is about 12.44 in^3.

$S = 2\pi r^2 + 2\pi rh$

$= 2\pi(1.625)^2 + 2\pi(1.625)(1.5)$

≈ 31.91

The can's surface area is about 31.91 in^2.

60. $s = -16t^2 + v_0 t + s_0$

$= -16(1)^2 + 80(1) + 6$

$= 70$ feet

The ball is 70 ft high after 1 second.

$s = -16(2)^2 + 80(2) + 6 = 102$ feet

The ball is 102 ft high after 2 seconds.

$s = -16(3)^2 + 80(3) + 6 = 102$ feet

The ball is 102 ft high after 3 seconds.

$s = -16(4)^2 + 80(4) + 6 = 70$ feet

The ball is 70 ft high after 4 seconds.

$s = -16(5)^2 + 80(5) + 6 = 6$ feet

The ball is 6 ft high after 5 seconds.

61. Answer will vary.

Possible answer: The sum of $\dfrac{2}{3}$ times a number and 25

62. Answer will vary.

Possible answer:

Seven times the difference of a number and 45, added to 15

63. equation: equality

64. expression: no equality

65. expression: no equality

66. equation: equality

67. $90° - 85° = 5°$
The complementary angle is 5°.

68. $180° - 43° = 137°$
The supplementary angle is 137°.

69. $180° - (31° + 58°) = 91°$
The third angle is 91°.

Chapter 2 Test

1. Let x = a number
$$\frac{x}{12}$$

2. Let x = a number
$$\frac{12}{x}$$

3. $x + 0.08x = 1.08x = 1.08(15) = 16.2$
$16.20

4. Answer will vary.
Possible answer: b subtracted from the square of a

5. $\sqrt{4x^2 - 20x + 25} = \sqrt{4(2)^2 - 20(2) + 25}$
$= \sqrt{1}$
$= 1$

6. $\dfrac{-18x + 54}{x - 8} = \dfrac{-18(-1) + 54}{-1 - 8}$
$= \dfrac{18 + 54}{-9}$
$= \dfrac{72}{-9}$
$= -8$

7. $x^2 = (-6)^2 = 36$

8. $-x^2 = -(-6)^2 = -36$

9. $(-x)^2 = [-(-6)]^2 = 6^2 = 36$

10. 8 terms

11. $y^3, -5y^2, 15y, 7y^2, 4y, 6y^3$

12. $-3, -12$

13. $1, -5, 15, -3, 7, -12, 4, 6$

14. y^3 and $6y^3$, $-5y^2$ and $7y^2$, $15y$ and $4y$, -3 and -12

15. $\dfrac{2}{3}x + \dfrac{5}{6}y - \dfrac{8}{9} + \dfrac{1}{6}x + \dfrac{7}{9}y + \dfrac{1}{3}$
$= \dfrac{2}{3}x + \dfrac{1}{6}x + \dfrac{5}{6}y + \dfrac{7}{9}y - \dfrac{8}{9} + \dfrac{1}{3}$
$= \dfrac{4}{6}x + \dfrac{1}{6}x + \dfrac{15}{18}y + \dfrac{14}{18}y - \dfrac{8}{9} + \dfrac{3}{9}$
$= \dfrac{5}{6}x + \dfrac{29}{18}y - \dfrac{5}{9}$

16. $-(5p + 2q) - (-9p + q) + (p + q) - (-p - q)$
$= -5p - 2q + 9p - q + p + q + p + q$
$= -5p + 9p + p + p - 2q - q + q + q$
$= 6p - q$

17. $\dfrac{25x - 45}{-5} = \dfrac{25x}{-5} - \dfrac{45}{-5} = -5x + 9$

18. $2a(3a + 6b - 8c)$
$= 2a(3a) + 2a(6b) + 2a(-8c)$
$= 6a^2 + 12ab - 16ac$

19.

$-7x - 4 = 6x + 9$	
$-7(-2) - 4$	$6(-2) + 9$
$14 - 4$	$-12 + 9$
10	-3

Since $10 \neq -3$, -2 is not a solution.

20.

$$8x^2 + 40x + 45 = 2x^2 - 2x - 27$$

$8(-4)^2 + 40(-4) + 45$	$2(-4)^2 - 2(-4) - 27$
$8(16) - 160 + 45$	$2(16) + 8 - 27$
$128 - 160 + 45$	$32 + 8 - 27$
13	13

Since $13 = 13$, -4 is a solution.

21. $V = LWH = 4(2)(1.5) = 12 \text{ ft}^3$
The tool box's volume is 12 ft^3.
$S = 2LW + 2WH + 2LH$
$= 2(4)(2) + 2(2)(1.5) + 2(4)(1.5)$
$= 34 \text{ ft}^2$
The outside surface of the tool box is
34 ft^2.

22. $A = Pe^{nr} = 2000e^{40(0.055)} \approx 18{,}050.027$
Tracy's account will have \$18,050.03 in
40 years.

23. $180° - 78° = 102°$
The supplementary angle is $102°$.

24. $c = \sqrt{a^2 + b^2} = \sqrt{(20)^2 + (48)^2} = 52$
The hypotenuse is 52 inches.

25. $F = \dfrac{9}{5}C + 32 = \dfrac{9}{5}(25) + 32 = 77$
The temperature is $77°F$.

26. Answers will vary.
Possible answer: The area is the space
contained within the boundaries, whereas the
perimeter is the distance around the
boundaries.

Chapter 3

3.1 Experiencing Algebra the Exercise Way

1.

x	$y = 5x + 4$	y
–2	$y = 5(-2) + 4$ $y = -6$	–6
–1	$y = 5(-1) + 4$ $y = -1$	–1
0	$y = 5(0) + 4$ $y = 4$	4
1	$y = 5(1) + 4$ $y = 9$	9
2	$y = 5(2) + 4$ $y = 14$	14
3	$y = 5(3) + 4$ $y = 19$	19

3.

x	$y = \dfrac{3}{5}x - 2$	y
–15	$y = \dfrac{3}{5}(-15) - 2$ $y = -11$	–11
–10	$y = \dfrac{3}{5}(-10) - 2$ $y = -8$	–8
–5	$y = \dfrac{3}{5}(-5) - 2$ $y = -5$	–5
0	$y = \dfrac{3}{5}(0) - 2$ $y = -2$	–2
5	$y = \dfrac{3}{5}(5) - 2$ $y = 1$	1
10	$y = \dfrac{3}{5}(10) - 2$ $y = 4$	4
15	$y = \dfrac{3}{5}(15) - 2$ $y = 7$	7

5.

x	$y = 2.3x + 1.6$	y
–2	$y = 2.3(-2) + 1.6$ $y = -3$	–3
–1	$y = 2.3(-1) + 1.6$ $y = -0.7$	–0.7
0	$y = 2.3(0) + 1.6$ $y = 1.6$	1.6
1	$y = 2.3(1) + 1.6$ $y = 3.9$	3.9
2	$y = 2.3(2) + 1.6$ $y = 6.2$	6.2

7.

x	$y = \dfrac{1}{3}(x + 7)$	y
–1	$y = \dfrac{1}{3}(-1 + 7)$ $y = 2$	2
–4	$y = \dfrac{1}{3}(-4 + 7)$ $y = 1$	1
–7	$y = \dfrac{1}{3}(-7 + 7)$ $y = 0$	0
–10	$y = \dfrac{1}{3}(-10 + 7)$ $y = -1$	–1
–13	$y = \dfrac{1}{3}(-13 + 7)$ $y = -2$	–2

For exercises 9–15, answers will vary. One possibility is given.

9.

x	$y = 6x - 8$	y
–2	$y = 6(-2) - 8$ $y = -20$	–20
–1	$y = 6(-1) - 8$ $y = -14$	–14
0	$y = 6(0) - 8$ $y = -8$	–8
1	$y = 6(1) - 8$ $y = -2$	–2
2	$y = 6(2) - 8$ $y = 4$	4

11.

x	$y = \dfrac{2}{7}x - 2$	y
–14	$y = \dfrac{2}{7}(-14) - 2$ $y = -6$	–6
–7	$y = \dfrac{2}{7}(-7) - 2$ $y = -4$	–4
0	$y = \dfrac{2}{7}(0) - 2$ $y = -2$	–2
7	$y = \dfrac{2}{7}(7) - 2$ $y = 0$	0
14	$y = \dfrac{2}{7}(14) - 2$ $y = 2$	2

13.

x	$y = -4.6x + 2.1$	y
-2	$y = -4.6(-2) + 2.1$ $y = 11.3$	11.3
-1	$y = -4.6(-1) + 2.1$ $y = 6.7$	6.7
0	$y = -4.6(0) + 2.1$ $y = 2.1$	2.1
1	$y = -4.6(1) + 2.1$ $y = -2.5$	-2.5
2	$y = -4.6(2) + 2.1$ $y = -7.1$	-7.1

15.

x	$y = \dfrac{1}{4}(3x - 2)$	y
-6	$y = \dfrac{1}{4}[3(-6) - 2]$ $y = -5$	-5
-2	$y = \dfrac{1}{4}[3(-2) - 2]$ $y = -2$	-2
0	$y = \dfrac{1}{4}(3 \cdot 0 - 2)$ $y = -\dfrac{1}{2}$	$-\dfrac{1}{2}$
2	$y = \dfrac{1}{4}(3 \cdot 2 - 2)$ $y = 1$	1
6	$y = \dfrac{1}{4}(3 \cdot 6 - 2)$ $y = 4$	4

17.

x	$y = 12x - 13$	y
-2	$y = 12(-2) - 13$ $y = -37$	-37
0	$y = 12(0) - 13$ $y = -13$	-13
2	$y = 12(2) - 13$ $y = 11$	11

19.

y	$z = \dfrac{1}{3}y + 5$	z
-3	$z = \dfrac{1}{3}(-3) + 5$ $z = 4$	4
0	$z = \dfrac{1}{3}(0) + 5$ $z = 5$	5
3	$z = \dfrac{1}{3}(3) + 5$ $z = 6$	6

21.

b	$a = 14.2b + 5.7$	a
-1	$a = 14.2(-1) + 5.7$ $a = -8.5$	-8.5
0	$a = 14.2(0) + 5.7$ $a = 5.7$	5.7
1	$a = 14.2(1) + 5.7$ $a = 19.9$	19.9

23.

x	$y = 2x^2 + 3x + 1$	y
-3	$y = 2(-3)^2 + 3(-3) + 1$ $y = 10$	10
-2	$y = 2(-2)^2 + 3(-2) + 1$ $y = 3$	3
-1	$y = 2(-1)^2 + 3(-1) + 1$ $y = 0$	0
0	$y = 2(0)^2 + 3(0) + 1$ $y = 1$	1
1	$y = 2(1)^2 + 3(1) + 1$ $y = 6$	6
2	$y = 2(2)^2 + 3(2) + 1$ $y = 15$	15
3	$y = 2(3)^2 + 3(3) + 1$ $y = 28$	28

25.

x	$y = (2x - 3)(3x + 4)$	y
-1	$y = [2(-1) - 3][3(-1) + 4]$ $y = -5$	-5
0	$y = (2 \cdot 0 - 3)(3 \cdot 0 + 4)$ $y = -12$	-12
1	$y = (2 \cdot 1 - 3)(3 \cdot 1 + 4)$ $y = -7$	-7

27.

x	$y = \dfrac{3x + 7}{x - 1}$	y
-1	$y = \dfrac{3(-1) + 7}{-1 - 1}$ $y = -2$	-2
1	$y = \dfrac{3(1) + 7}{1 - 1}$ $y = \dfrac{10}{0}$	undefined

29.

H	$V = 4(2)H$ $V = 8H$	V
1	$V = 8(1)$ $V = 8$	8
3	$V = 8(3)$ $V = 24$	24
5	$V = 8(5)$ $V = 40$	40
7	$V = 8(7)$ $V = 56$	56
9	$V = 8(9)$ $V = 72$	72

31.

t	$I = 5000(0.045)t$ $I = 225t$	I
1	$I = 225(1)$ $I = 225$	225
2	$I = 225(2)$ $I = 450$	450
3	$I = 225(3)$ $I = 675$	675
4	$I = 225(4)$ $I = 900$	900
5	$I = 225(5)$ $I = 1125$	1125
6	$I = 225(6)$ $I = 1350$	1350
7	$I = 225(7)$ $I = 1575$	1575
8	$I = 225(8)$ $I = 1800$	1800
9	$I = 225(9)$ $I = 2025$	2025
10	$I = 225(10)$ $I = 2250$	2250
11	$I = 225(11)$ $I = 2475$	2475
12	$I = 225(12)$ $I = 2700$	2700

33.

R	$I = \dfrac{9}{R}$	I
1	$I = \dfrac{9}{1}$ $I = 9$	9
2	$I = \dfrac{9}{2}$ $I = 4.5$	4.5
3	$I = \dfrac{9}{3}$ $I = 3$	3
4	$I = \dfrac{9}{4}$ $I = 2.25$	2.25
5	$I = \dfrac{9}{5}$ $I = 1.8$	1.8
6	$I = \dfrac{9}{6}$ $I = 1.5$	1.5
7	$I = \dfrac{9}{7}$ $I \approx 1.2857$	$\dfrac{9}{7}$
8	$I = \dfrac{9}{8}$ $I = 1.125$	1.125
9	$I = \dfrac{9}{9}$ $I = 1$	1

35. $y = 1.2x + 4$
$y = 1.2(-2) + 4$
$y = 1.6$
$y = 1.2(-1) + 4$
$y = 2.8$
$y = 1.2(0) + 4$
$y = 4$
$y = 1.2(1) + 4$
$y = 5.2$
$y = 1.2(2) + 4$
$y = 6.4$
$(-2, 1.6), (-1, 2.8), (0, 4), (1, 5.2), (2, 6.4)$

37. $q = 1 - p$
$q = 1 - \dfrac{1}{6}$
$q = \dfrac{5}{6}$
$q = 1 - \dfrac{1}{5}$
$q = \dfrac{4}{5}$
$q = 1 - \dfrac{1}{4}$
$q = \dfrac{3}{4}$
$q = 1 - \dfrac{1}{3}$
$q = \dfrac{2}{3}$
$q = 1 - \dfrac{1}{2}$
$q = \dfrac{1}{2}$
$\left(\dfrac{1}{6}, \dfrac{5}{6}\right), \left(\dfrac{1}{5}, \dfrac{4}{5}\right), \left(\dfrac{1}{4}, \dfrac{3}{4}\right), \left(\dfrac{1}{3}, \dfrac{2}{3}\right), \left(\dfrac{1}{2}, \dfrac{1}{2}\right)$

39. $P = 4s$
$P = 4(2)$
$P = 8$
$P = 4(4)$
$P = 16$
$P = 4(6)$
$P = 24$
$P = 4(8)$
$P = 32$
$P = 4(10)$
$P = 40$
$(2, 8), (4, 16), (6, 24), (8, 32), (10, 40)$

41. $R = \{(3, 15.8), (5, 17.8), (7, 19.8), (9, 21.8)\}$
domain: $\{3, 5, 7, 9\}$
range: $\{15.8, 17.8, 19.8, 21.8\}$

43. $S = \{(4, -3), (4, -1), (4, 1), (4, 3), \ldots\}$
domain: $\{4\}$
range: $\{-3, -1, 1, 3, \ldots\}$

45. $T = \{\ldots, (2, -2), (2, -1), (2, 0), (2, 1), (2, 2),$
$\ldots\}$
domain: $\{2\}$
range: $\{\ldots, -2, -1, 0, 1, 2, \ldots\}$

47. $y = 4x - 5$ for $x = \{2, 4, 6\}$
$y = 4(2) - 5$
$y = 3$
$y = 4(4) - 5$
$y = 11$
$y = 4(6) - 5$
$y = 19$
domain: $\{2, 4, 6\}$
range: $\{3, 11, 19\}$

49. $y = 6 - x$
domain: all real numbers
range: all real numbers

51. $y = x^2 + 1$
The x-value can be any real number,
whereas x^2 will always be zero or positive.
domain: all real numbers
range: all real numbers ≥ 1

53. $y = \sqrt{x-2}$

$x - 2$ must be positive or zero, so $x \geq 2$, which makes y zero or a positive number.
domain: all real numbers ≥ 2
range: all real numbers ≥ 0

55. $y = \dfrac{6}{x-2}$

$x - 2$ must not equal zero, so $x \neq 2$, which means $y \neq 0$.
domain: all real numbers $x \neq 2$
range: all real numbers $y \neq 0$

57. $V = s^3$

s must be positive, which makes V positive.
domain: all real numbers $s > 0$
range: all real numbers $V > 0$

59. $s = -16t^2 + v_0 t + s_0$

$s = -16t^2 + 0t + 50$
$s = -16t^2 + 50$
His position will be s feet high t seconds after the jump.

61. $d = rt$
$d = 65t$

t	$d = 65t$	d
1	$d = 65(1)$	65
	$d = 65$	
2	$d = 65(2)$	130
	$d = 130$	
3	$d = 65(3)$	195
	$d = 195$	
4	$d = 65(4)$	260
	$d = 260$	

$(1, 65), (2, 130), (3, 195), (4, 260)$
The distance traveled (in miles) after 1, 2, 3, and 4 hours is 65, 130, 195, and 260 respectively.

3.1 Experiencing Algebra the Calculator Way

1. $y = 2.75x - 15.8$
$Y1 = 2.75x - 15.8$

X	Y1
-250	-703.3
-200	-565.8
-150	-428.3
-100	-290.8
-50	-153.3
0	-15.8
50	121.7

Y1◻2.75X-15.8

X	Y1
100	259.2
150	396.7
200	534.2
250	671.7

Y1◻2.75X-15.8

2. $y = 3x^3 + 5x^2 + 7x + 9$

$Y1 = 3x^3 + 5x^2 + 7x + 9$

X	Y1
-5	-276
-3.873	-117.4
-3.142	-56.66
-3	-48
-2	-9
-1.442	.30467
0	9

Y1◻3X^3+5X²+7X+9

X	Y1
2	67
3.1416	173.36
3.1623	176

Y1◻3X^3+5X²+7X+9

3. $y = (2x+1)^4$

$Y1 = (2x+1)^4$

X	Y1
25	6.77E6
30	1.38E7
35	2.54E7
40	4.3E7
45	6.86E7
50	1.04E8

Y1◻(2X+1)^4

4. $y = (8)2\pi x + 2\pi x^2$

$Y1 = (8)2\pi x + 2\pi x^2$

X	Y1
2	125.66
2.5	164.93
3	207.35
3.5	252.9
4	301.59
4.5	353.43
5	408.41

Y1◻8*2πX+2πX²

5. $y = 12,000(1 + 0.075)^x$

$y = 12,000(1.075)^x$

$Y1 = 12000(1.075)^x$

X	Y1
1	12900
2	13868
3	14908
4	16026
5	17228
6	18520
7	19909

Y1∎12000(1.075)...

X	Y1
8	21402
9	23007
10	24732
11	26587
12	28581

Y1∎12000(1.075)...

6. $y = 12,000e^{(0.075x)}$

$Y1 = 12000e^{(0.075x)}$

X	Y1
1	12935
2	13942
3	15028
4	16198
5	17460
6	18820
7	20286

Y1∎12000e^(0.07...

X	Y1
8	21865
9	23568
10	25404
11	27383
12	29515

Y1∎12000e^(0.07...

3.2 Experiencing Algebra the Exercise Way

1. $A(-7, -5)$, $B(-5, -7)$

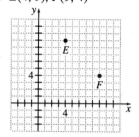

3. $E(4, 9)$, $F(9, 4)$

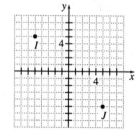

5. $I(-5, 5)$, $J(5, -5)$

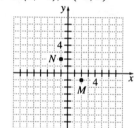

7. $M(2, -1)$, $N(-1, 2)$

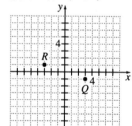

9. $Q(3, -1)$, $R(-3, 1)$

11. $U(-8, -2)$, $V(8, 2)$

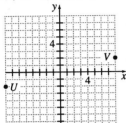

13. $A(1.2, 2.4)$

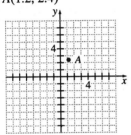

15. $C(-4.5, -2.6)$

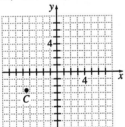

17. $E(-2.4, 2.1)$

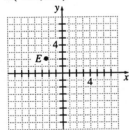

19. $G(1.8, -2.7)$

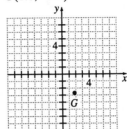

21. $(-21, 35)$
Coordinates are $(-, +)$
quadrant II

23. $(4, 96)$
Coordinates are $(+, +)$
quadrant I

25. $(-3, -19)$
Coordinates are $(-, -)$
quadrant III

27. $(0, -31)$
x-coordinate is 0
y-axis

29. $(90, -100)$
Coordinates are $(+, -)$
quadrant IV

31. $(24, 0)$
y-coordinate is 0
x-axis

33. $(-19, 0)$
y-coordinate is zero
x-axis

35. $(0.05, 1.003)$
Coordinates are $(+, +)$
quadrant I

37. $(0, 3.7)$
x-coordinate is 0
y-axis

39. $\left(\dfrac{13}{27}, -\dfrac{11}{19} \right)$
Coordinates are $(+, -)$
quadrant IV

41. $\left(-\dfrac{53}{100}, -\dfrac{39}{100}\right)$

Coordinates are $(-, -)$
quadrant III

43. $\left(0, \dfrac{28}{51}\right)$

x-coordinate is zero
y-axis

45. $A(8, 2)$, $B(-9, 7)$, $C(0, 0)$, $D(-2, -3)$, $E(0, 4)$, $F(5, -6)$, $G(-5, 0)$

47. $A = \{(-3, -3), (-2, -2), (-1, -1), (0, 0),$
$(1, 1), (2, 2), (3, 3)\}$

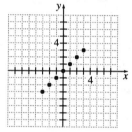

49. $C = \{(-4, 4), (-2, 2), (0, 0), (2, -2), (4, -4)\}$

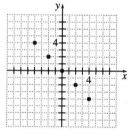

51. $E = \{(5, -3), (5, -1), (5, 1), (5, 3)\}$

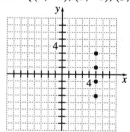

53. $G = \{\dots, (-2, -1), (-1, 0), (0, 1), (1, 2),$
$(2, 3), \dots\}$

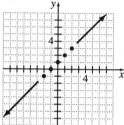

55. $I = \{\dots, (-2, -6), (-1, -5), (0, -4), (1, -3),$
$(2, -2), (3, -1), \dots\}$

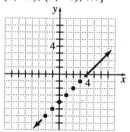

57. $y = 12x - 15$
Possible points:

x	y
$-\dfrac{1}{2}$	-21
0	-15
$\dfrac{1}{2}$	-9

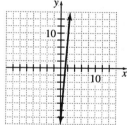

59. $y = -10x + 9$
Possible points:

x	y
-1	19
0	9
1	-1

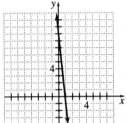

61. $y = \dfrac{1}{2}x + 3$
Possible points:

x	y
-2	2
0	3
2	4

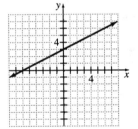

63. $y = -\dfrac{2}{3}x - 2$
Possible points:

x	y
-3	0
0	-2
3	-4

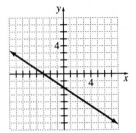

65. $y = 2x^2 - 5$
Possible points:

x	y
-2	3
0	-5
2	3

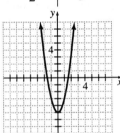

67. $y = |2x|$
Possible points:

x	y
-2	4
0	0
2	4

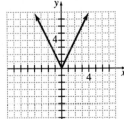

69. domain: all real numbers
range: all real numbers ≤ -1

71. domain: all real numbers
range: all real numbers ≥ 3

73. domain: 1960 through 1992
range: approximately 45% to 62%
The domain represents the time period
between the years 1960 and 1992 during
which the data was collected.
The range represents the percent of the adult
population who identify themselves as
Democrats for each year.

75. domain: all real numbers $0 \leq x \leq 2.5$
range: all real numbers $0 \leq y \leq 100$
The domain represents the time that has
passed from 0 seconds to 2.5 seconds. The
range represents the height from 100 feet to
0 feet.

3.2 Experiencing Algebra the Calculator Way

Part A

1. Integer setting

2. Decimal setting

3. Standard setting

Part B

1. $Y1 = 0.6x - 1.2$

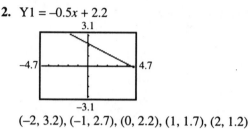

$(-2, -2.4), (-1, -1.8), (0, -1.2), (1, -0.6),$
$(2, 0)$

2. $Y1 = -0.5x + 2.2$

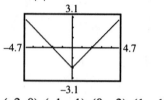

$(-2, 3.2), (-1, 2.7), (0, 2.2), (1, 1.7), (2, 1.2)$

3. $Y1 = |x| - 2$

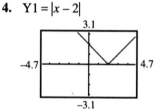

$(-2, 0), (-1, -1), (0, -2), (1, -1), (2, 0)$

4. $Y1 = |x - 2|$

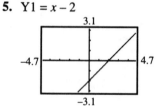

$(-2, 4), (-1, 3), (0, 2), (1, 1), (2, 0)$

5. $Y1 = x - 2$

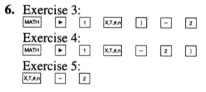

$(-2, -4), (-1, -3), (0, -2), (1, -1), (2, 0)$

6. Exercise 3:
$\boxed{\text{MATH}}$ $\boxed{\blacktriangleright}$ $\boxed{1}$ $\boxed{\text{X,T,}\theta\text{,n}}$ $\boxed{)}$ $\boxed{-}$ $\boxed{2}$
Exercise 4:
$\boxed{\text{MATH}}$ $\boxed{\blacktriangleright}$ $\boxed{1}$ $\boxed{\text{X,T,}\theta\text{,n}}$ $\boxed{-}$ $\boxed{2}$ $\boxed{)}$
Exercise 5:
$\boxed{\text{X,T,}\theta\text{,n}}$ $\boxed{-}$ $\boxed{2}$

3.3 Experiencing Algebra the Exercise Way

1. *A* is not a function because two elements in
the domain, −2 and 2, each correspond to
two elements in the range, 1 and 3, and −3
and −1, respectively.

3. *C* is a function because every element in the domain corresponds to only one element in the range.

5. *E* is not a function because an element in the domain, 6, corresponds to 4 elements in the range, $-1, -3, -5,$ and -7.

7. *G* is a function because every element in the domain corresponds to only one element in the range.

9. $y = 22x - 11$ is a function. Each element in the domain corresponds to a single element in the range.

11. $y = 13x^2 + 4$ is a function. Each element in the domain corresponds to a single element in the range.

13. $y^2 = 16x + 25$ is not a function. Each element in the domain, except -1.5625, corresponds to two elements in the range.

15. $y = 25x + 125$ is a function. Each element in the domain corresponds to a single element in the range.

17. $y = 6x^3 + 4x^2 + 2x$ is a function. Each element in the domain corresponds to a single element in the range.

19. $y^2 = 25 - x^2$ is not a function. Each element in the domain, except -5 and 5, corresponds to two elements in the range.

21. $y = |-3x - 18|$ is a function. Each element in the domain corresponds to a single element in the range.

23. $y = \sqrt{2x^2 + 4x + 1}$ is a function. Each element in the domain corresponds to a single element in the range.

25. This graph represents a function. All possible vertical lines cross the graph a maximum of one time.

27. This graph does not represent a function. Several vertical lines can be drawn to pass through more than one point.

29. This graph represents a function. All possible vertical lines cross the graph a maximum of one time.

31. $f(x) = 20x + 12$
$f(5) = 20(5) + 12$
$f(5) = 100 + 12$
$f(5) = 112$

33. $f(x) = 20x + 12$
$f(-7) = 20(-7) + 12$
$f(-7) = -140 + 12$
$f(-7) = -128$

35. $f(x) = 20x + 12$
$f(2.4) = 20(2.4) + 12$
$f(2.4) = 48 + 12$
$f(2.4) = 60$

37. $f(x) = 20x + 12$
$f\left(-\dfrac{1}{4}\right) = 20\left(-\dfrac{1}{4}\right) + 12$
$f\left(-\dfrac{1}{4}\right) = -5 + 12$
$f\left(-\dfrac{1}{4}\right) = 7$

39. $f(x) = 20x + 12$
$f(a) = 20a + 12$

41. $f(x) = 20x + 12$
$f(a + 2) = 20(a + 2) + 12$
$f(a + 2) = 20a + 40 + 12$
$f(a + 2) = 20a + 52$

43. $f(x) = 20x + 12$
$f(a - 4) = 20(a - 4) + 12$
$f(a - 4) = 20a - 80 + 12$
$f(a - 4) = 20a - 68$

45. $f(x) = 20x + 12$
$f(a + b) = 20(a + b) + 12$
$f(a + b) = 20a + 20b + 12$

47. $h(x) = 2x^2 - 4x + 5$

$h(7) = 2(7)^2 - 4(7) + 5$

$h(7) = 98 - 28 + 5$

$h(7) = 75$

49. $h(x) = 2x^2 - 4x + 5$

$h(-4) = 2(-4)^2 - 4(-4) + 5$

$h(-4) = 32 + 16 + 5$

$h(-4) = 53$

51. $h(x) = 2x^2 - 4x + 5$

$h(-1.1) = 2(-1.1)^2 - 4(-1.1) + 5$

$h(-1.1) = 2.42 + 4.4 + 5$

$h(-1.1) = 11.82$

53. $h(x) = 2x^2 - 4x + 5$

$h\left(-\dfrac{2}{5}\right) = 2\left(-\dfrac{2}{5}\right)^2 - 4\left(-\dfrac{2}{5}\right) + 5$

$h\left(-\dfrac{2}{5}\right) = \dfrac{8}{25} + \dfrac{40}{25} + \dfrac{125}{25}$

$h\left(-\dfrac{2}{5}\right) = \dfrac{173}{25}$

55. $h(x) = 2x^2 - 4x + 5$

$h\left(2\dfrac{3}{5}\right) = 2\left(2\dfrac{3}{5}\right)^2 - 4\left(2\dfrac{3}{5}\right) + 5$

$h\left(2\dfrac{3}{5}\right) = 2\left(\dfrac{13}{5}\right)^2 - 4\left(\dfrac{13}{5}\right) + 5$

$h\left(2\dfrac{3}{5}\right) = \dfrac{338}{25} - \dfrac{260}{25} + \dfrac{125}{25}$

$h\left(2\dfrac{3}{5}\right) = \dfrac{203}{25}$

57. $h(x) = 2x^2 - 4x + 5$

$h(b) = 2b^2 - 4b + 5$

59. $g(x) = |-3x + 9|$

$g(5) = |-3(5) + 9|$

$g(5) = |-15 + 9|$

$g(5) = |-6|$

$g(5) = 6$

61. $g(x) = |-3x + 9|$

$g(-5) = |-3(-5) + 9|$

$g(-5) = |15 + 9|$

$g(-5) = |24|$

$g(-5) = 24$

63. $g(x) = |-3x + 9|$

$g(4.5) = |-3(4.5) + 9|$

$g(4.5) = |-13.5 + 9|$

$g(4.5) = |-4.5|$

$g(4.5) = 4.5$

65. $g(x) = |-3x + 9|$

$g(-4.5) = |-3(-4.5) + 9|$

$g(-4.5) = |13.5 + 9|$

$g(-4.5) = |22.5|$

$g(-4.5) = 22.5$

67. $g(x) = |-3x + 9|$

$g\left(\dfrac{2}{3}\right) = \left|-3\left(\dfrac{2}{3}\right) + 9\right|$

$g\left(\dfrac{2}{3}\right) = |-2 + 9|$

$g\left(\dfrac{2}{3}\right) = |7|$

$g\left(\dfrac{2}{3}\right) = 7$

69. $g(x) = |-3x + 9|$

$g\left(-4\dfrac{2}{3}\right) = \left|-3\left(-4\dfrac{2}{3}\right) + 9\right|$

$g\left(-4\dfrac{2}{3}\right) = \left|-3\left(-\dfrac{14}{3}\right) + 9\right|$

$g\left(-4\dfrac{2}{3}\right) = |14 + 9|$

$g\left(-4\dfrac{2}{3}\right) = |23|$

$g\left(-4\dfrac{2}{3}\right) = 23$

71. $F(x) = \sqrt{x+15}$

$F(85) = \sqrt{85+15}$

$F(85) = \sqrt{100}$

$F(85) = 10$

73. $F(x) = \sqrt{x+15}$

$F(-6) = \sqrt{-6+15}$

$F(-6) = \sqrt{9}$

$F(-6) = 3$

75. $F(x) = \sqrt{x+15}$

$F(-25) = \sqrt{-25+15}$

$F(-25) = \sqrt{-10}$, which is not a real number.

77. $F(x) = \sqrt{x+15}$

$F(5.25) = \sqrt{5.25+15}$

$F(5.25) = \sqrt{20.25}$

$F(5.25) = 4.5$

79. $F(x) = \sqrt{x+15}$

$F\left(-2\frac{3}{4}\right) = \sqrt{-2\frac{3}{4}+15}$

$F\left(-2\frac{3}{4}\right) = \sqrt{-\frac{11}{4}+\frac{60}{4}}$

$F\left(-2\frac{3}{4}\right) = \sqrt{\frac{49}{4}}$

$F\left(-2\frac{3}{4}\right) = \frac{7}{2}$

81. Let x = the number of televisions.

$f(x) = 1500 + 35x$

$f(400) = 1500 + 35(400)$

$f(400) = 1500 + 14,000$

$f(400) = 15,500$

The production run costs $15,500.

83. Let x = the number of players.

$f(x) = 125x - 470$

$f(400) = 125(400) - 470$

$f(400) = 50,000 - 470$

$f(400) = 49,530$

The net revenue is $49,530.

85. Let x = the number of seats sold.
The cost of each seat = $175 - 3x$.

$f(x) = 175x - 3x^2$

$f(22) = 175(22) - 3(22)^2$

$f(22) = 3850 - 1452$

$f(22) = 2398$

If 22 customers make reservations for the tour, the company will make $2398.

87. Let x = the number of CD's, that is, 5 or more.

$f(x) = 25 + 4(x - 5)$

$f(x) = 25 + 4x - 20$

$f(x) = 5 + 4x$

$f(12) = 5 + 4(12)$

$f(12) = 5 + 48$

$f(12) = 53$

The cost of 12 used CD's will be $53.

89. Let x = the number of days.

$f(x) = 39 + 25x$

$f(3) = 39 + 25(3)$

$f(3) = 39 + 75$

$f(3) = 114$

The charge for renting a truck for 3 days will be $114.

91. Let x = the number of half-hour increments.

$f(x) = 2.5 + x$

$f(7) = 2.5 + 7$

$f(7) = 9.5$

The charge is $9.50.

3.3 Experiencing Algebra the Calculator Way

1–6.

X	Y1
65	278916
-83	-5.6E5
3.1416	45.017
1.4142	7.2426
5634	1.8E11
-3.142	-23.28

Y1☐X^3+X²+X+1

7–12.

X	Y1
-8	4
0	4
-4	0
.8	4.8
-6.3	2.3
.75	4.75

Y1☐√(X²+8X+16)

13–18.

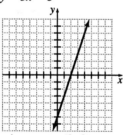

3.4 Experiencing Algebra the Exercise Way

1. **a.** $(-2, 0), (6, 0)$

   b. $(0, 3)$

   c. $x < 2$

   d. $x > 2$

   e.   4

   f.   none

3. **a.** $(-5, 0), (-1, 0), (4, 0)$

   b. $(0, -1)$

   c. $x < -3, x > 2$

   d. $-3 < x < 2$

   e.   2

   f. -3

5. $y = 3x - 6$

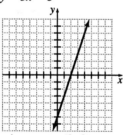

x-intercept: $(2, 0)$
y-intercept: $(0, -6)$

7. $y = \dfrac{1}{2}x + 1$

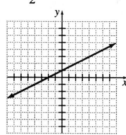

x-intercept: $(-2, 0)$
y-intercept: $(0, 1)$

9. $y = 1.2x - 6$

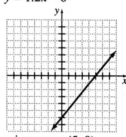

x-intercept: $(5, 0)$
y-intercept: $(0, -6)$

11. $f(x) = -12x + 24$

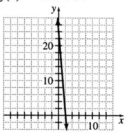

x-intercept: $(2, 0)$
y-intercept: $(0, 24)$

13. $f(x) = 9x + 15$

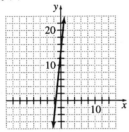

x-intercept: $\left(-\dfrac{5}{3}, 0\right)$

y-intercept: $(0, 15)$

15. $y = x^2 - 9$

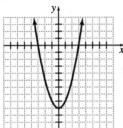

x-intercepts: $(3, 0)$, $(-3, 0)$

y-intercept: $(0, -9)$

17. $y = x^2 + 6x + 9$

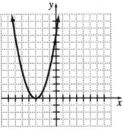

x-intercept: $(-3, 0)$

y-intercept: $(0, 9)$

19. $y = 4x^2 + 4x + 1$

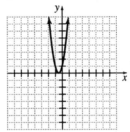

x-intercept: $\left(-\dfrac{1}{2}, 0\right)$

y-intercept: $(0, 1)$

21. $g(x) = x^2 + 10x - 3$

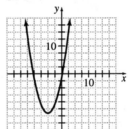

x-intercepts: $(-10.29, 0)$, $(0.29, 0)$

y-intercept: $(0, -3)$

23. $H(x) = x^2 - 5x - 24$

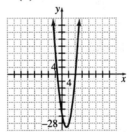

x-intercepts: $(-3, 0)$, $(8, 0)$

y-intercept: $(0, -24)$

25. $y = x^3 + x^2 - 2x$

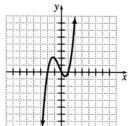

x-intercepts: $(-2, 0)$, $(0, 0)$, $(1, 0)$
y-intercept: $(0, 0)$

27. $f(x) = x^3 + 2x^2 - x - 2$

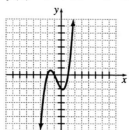

x-intercepts: $(-2, 0)$, $(-1, 0)$, $(1, 0)$
y-intercept: $(0, -2)$

29. $h(x) = |x| - 6$

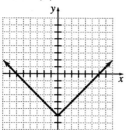

x-intercepts: $(-6, 0)$, $(6, 0)$
y-intercept: $(0, -6)$

31. $y = |2x - 3| - 1$

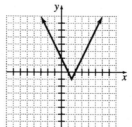

x-intercepts: $(1, 0)$, $(2, 0)$
y-intercept: $(0, 2)$

33. $y = |x^2 - 2| - 1$

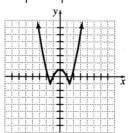

x-intercepts: $(-1.73, 0)$, $(-1, 0)$, $(1, 0)$,
$(1.73, 0)$
y-intercept: $(0, 1)$

35. $y = 2x + 8$

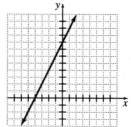

increasing for all x-values

37. $f(x) = 3 - 2x$

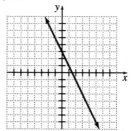

decreasing for all x-values

39. $y = 1 - x^2$

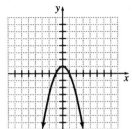

increasing for $x < 0$
decreasing for $x > 0$
relative maximum is 1

41. $g(x) = x^2 + 4x + 3$

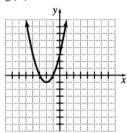

increasing for $x > -2$
decreasing for $x < -2$
relative minimum is -1

43. $y = |x + 3|$

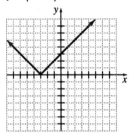

increasing for $x > -3$
decreasing for $x < -3$
relative minimum is 0

45. $f(x) = x^3 + 2x^2 - x - 2$

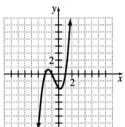

increasing for $x < -1.55$ and $x > 0.22$
decreasing for $-1.55 < x < 0.22$
relative maximum ≈ 0.63
relative minimum ≈ -2.11

47. $y = 3x - 5$ and $y = -2x + 15$

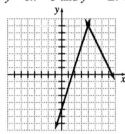

intersect at (4, 7)

49. $f(x) = 2x + 7$ and $g(x) = -x + 1$

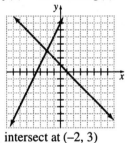

intersect at $(-2, 3)$

51. $y = -5x + 2$ and $y = 3x + 8$

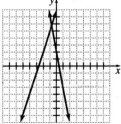

intersect at $(-0.75, 5.75)$

53. $r(x) = 5x - 7$ and $c(x) = 12$

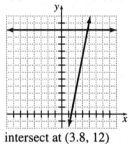

intersect at (3.8, 12)

55. $y = 3$ and $y = -x^2 + 4$

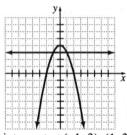

intersect at (−1, 3), (1, 3)

57. $f(x) = 2x^2 - 4x + 5$ and $g(x) = 4x - 1$

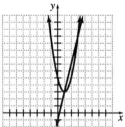

intersect at (1, 3), (3, 11)

59. $y = \frac{1}{4}x^2 - 2$ and $y = \frac{1}{2}x$

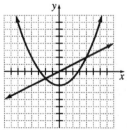

intersect at (−2, −1), (4, 2)

61. $y = |x| - 5$ and $y = 2$

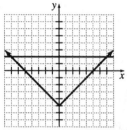

intersect at (−7, 2), (7, 2)

63. Let x = the number of crafts.
Y1 = $5x - 50$

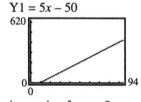

increasing for $x > 0$

65. Let x = the number of pieces of equipment.
Y1 = $400x - 10x^2$ for $0 < x \le 25$

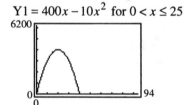

increasing for $0 < x < 20$
dereasing for $20 < x < 25$
relative maximum at $x = 20$

67. a. Let x = number of containers made in
one run.
$f(x) = 4x + 50$

b. $g(x) = 10x$

c. Y1 = $4x + 50$
Y2 = $10x$

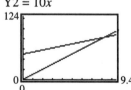

intersection at $(8.\overline{3}, 83.\overline{3})$

d. At $x = 8.3$ containers, the cost for producing and the cost for selling is the same at $83.33. This means that 9 containers must be sold to cover the costs of production.

69. a. $f(x) = 200 + 50x$
$g(x) = 75x$

b. $Y1 = 200 + 50x$
$Y2 = 75x$

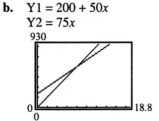

intersect at (8, 600)

c. At 8 credit hours, her pay will be the same, at $600.

3.4 Experiencing Algebra the Calculator Way

For exercises 1–3, graphs will vary.

1. $g(x) = \dfrac{1}{3}x + 3$ and $f(x) = \dfrac{1}{4}x^2 - 4$
intersect at $(-4.\overline{6}, 1.\overline{4})$ and (6, 5)

2. $y = |x| - 6$ and $y = -|x| + 4$
intersect at (−5, −1) and (5, −1)

3. $y = x^2 - 18$ and $y = -x^2 + 54$
intersect at (−6, 18) and (6, 18)

Chapter 3 Review

Reflections

Answers will vary.

Exercises

1.

a	$b = -2a + 7$	b
−3	$b = -2(-3) + 7$ $b = 13$	13
−2	$b = -2(-2) + 7$ $b = 11$	11
−1	$b = -2(-1) + 7$ $b = 9$	9
0	$b = -2(0) + 7$ $b = 7$	7
1	$b = -2(1) + 7$ $b = 5$	5
2	$b = -2(2) + 7$ $b = 3$	3
3	$b = -2(3) + 7$ $b = 1$	1

2.

x	$y = \dfrac{2}{3}x + 4$	y
9	$y = \dfrac{2}{3}(9) + 4$ $y = 10$	10
6	$y = \dfrac{2}{3}(6) + 4$ $y = 8$	8
3	$y = \dfrac{2}{3}(3) + 4$ $y = 6$	6
0	$y = \dfrac{2}{3}(0) + 4$ $y = 4$	4
–3	$y = \dfrac{2}{3}(-3) + 4$ $y = 2$	2
–6	$y = \dfrac{2}{3}(-6) + 4$ $y = 0$	0
–9	$y = \dfrac{2}{3}(-9) + 4$ $y = -2$	–2

3.

x	$y = 0.4x - 1.2$	y
–3	$y = 0.4(-3) - 1.2$ $y = -2.4$	–2.4
–2	$y = 0.4(-2) - 1.2$ $y = -2$	–2
–1	$y = 0.4(-1) - 1.2$ $y = -1.6$	–1.6
0	$y = 0.4(0) - 1.2$ $y = -1.2$	–1.2
1	$y = 0.4(1) - 1.2$ $y = -0.8$	–0.8
2	$y = 0.4(2) - 1.2$ $y = -0.4$	–0.4
3	$y = 0.4(3) - 1.2$ $y = 0$	0

4.

b	$a = 3b^2 - 2b + 5$	a
–2	$a = 3(-2)^2 - 2(-2) + 5$ $a = 21$	21
–1	$a = 3(-1)^2 - 2(-1) + 5$ $a = 10$	10
1	$a = 3(1)^2 - 2(1) + 5$ $a = 6$	6
2	$a = 3(2)^2 - 2(2) + 5$ $a = 13$	13

5.

x	$y = 5x^3 - 3x^2 + 2x - 22$	y
–18	$y = 5(-18)^3 - 3(-18)^2 + 2(-18) - 22$ $y = -30{,}190$	–30,190
–7	$y = 5(-7)^3 - 3(-7)^2 + 2(-7) - 22$ $y = -1898$	–1898
0	$y = 5(0)^3 - 3(0)^2 + 2(0) - 22$ $y = -22$	–22
6	$y = 5(6)^3 - 3(6)^2 + 2(6) - 22$ $y = 962$	962
21	$y = 5(21)^3 - 3(21)^2 + 2(21) - 22$ $y = 45{,}002$	45,002
22.5	$y = 5(22.5)^3 - 3(22.5)^2 + 2(22.5) - 22$ $y = 55{,}457.375$	55,457.375

6.

x	$y = \left| x^2 - 6x + 5 \right|$	y
–2	$y = \left| (-2)^2 - 6(-2) + 5 \right|$ $y = 21$	21
–1	$y = \left| (-1)^2 - 6(-1) + 5 \right|$ $y = 12$	12
0	$y = \left| 0^2 - 6 \cdot 0 + 5 \right|$ $y = 5$	5
1	$y = \left| 1^2 - 6 \cdot 1 + 5 \right|$ $y = 0$	0
2	$y = \left| 2^2 - 6 \cdot 2 + 5 \right|$ $y = 3$	3
3	$y = \left| 3^2 - 6 \cdot 3 + 5 \right|$ $y = 4$	4

7.

x	$y = 3.6x^2 + 1.5x - 14.2$	y
–2.7	$y = 3.6(-2.7)^2 + 1.5(-2.7) - 14.2$ $y = 7.994$	7.994
–1.9	$y = 3.6(-1.9)^2 + 1.5(-1.9) - 14.2$ $y = -4.054$	–4.054
–0.6	$y = 3.6(-0.6)^2 + 1.5(-0.6) - 14.2$ $y = -13.804$	–13.804
0	$y = 3.6(0)^2 + 1.5(0) - 14.2$ $y = -14.2$	–14.2
0.8	$y = 3.6(0.8)^2 + 1.5(0.8) - 14.2$ $y = -10.696$	–10.696
1.5	$y = 3.6(1.5)^2 + 1.5(1.5) - 14.2$ $y = -3.85$	–3.85
2.4	$y = 3.6(2.4)^2 + 1.5(2.4) - 14.2$ $y = 10.136$	10.136

8. Answers will vary. Possible answer:

x	$y = 12x - 21$	y
–1	$y = 12(-1) - 21$ $y = -33$	–33
0	$y = 12(0) - 21$ $y = -21$	–21
1	$y = 12(1) - 21$ $y = -9$	–9
2	$y = 12(2) - 21$ $y = 3$	3
3	$y = 12(3) - 21$ $y = 15$	15

9. Answers will vary. Possible answer:

x	$y = -1.6x + 4.5$	y
–2	$y = -1.6(-2) + 4.5$ $y = 7.7$	7.7
–1	$y = -1.6(-1) + 4.5$ $y = 6.1$	6.1
0	$y = -1.6(0) + 4.5$ $y = 4.5$	4.5
1	$y = -1.6(1) + 4.5$ $y = 2.9$	2.9
2	$y = -1.6(2) + 4.5$ $y = 1.3$	1.3

10. Answers will vary. Possible answer:

x	$y = \dfrac{3}{2}x - 6$	y
–4	$y = \dfrac{3}{2}(-4) - 6$ $y = -12$	–12
–2	$y = \dfrac{3}{2}(-2) - 6$ $y = -9$	–9
0	$y = \dfrac{3}{2}(0) - 6$ $y = -6$	–6
2	$y = \dfrac{3}{2}(2) - 6$ $y = -3$	–3
4	$y = \dfrac{3}{2}(4) - 6$ $y = 0$	0

11. Answers will vary. Possible answer:

x	$y = |2x - 9|$	y
–4	$y = |2(-4) - 9|$ $y = 17$	17
–2	$y = |2(-2) - 9|$ $y = 13$	13
0	$y = |2 \cdot 0 - 9|$ $y = 9$	9
2	$y = |2 \cdot 2 - 9|$ $y = 5$	5
4	$y = |2 \cdot 4 - 9|$ $y = 1$	1

12.

x	$y = (5x + 2)(x - 4)$	y
–3	$y = [5(-3) + 2](-3 - 4)$ $y = 91$	91
–1	$y = [5(-1) + 2](-1 - 4)$ $y = 15$	15
1	$y = (5 \cdot 1 + 2)(1 - 4)$ $y = -21$	–21
3	$y = (5 \cdot 3 + 2)(3 - 4)$ $y = -17$	–17

13.

x	$y = \dfrac{3}{5}x + 8$	y
–15	$y = \dfrac{3}{5}(-15) + 8$ $y = -1$	–1
–10	$y = \dfrac{3}{5}(-10) + 8$ $y = 2$	2
–5	$y = \dfrac{3}{5}(-5) + 8$ $y = 5$	5
0	$y = \dfrac{3}{5}(0) + 8$ $y = 8$	8
5	$y = \dfrac{3}{5}(5) + 8$ $y = 11$	11
10	$y = \dfrac{3}{5}(10) + 8$ $y = 14$	14
15	$y = \dfrac{3}{5}(15) + 8$ $y = 17$	17

14.

x	$y = 17.1x - 12.9$	y
–3	$y = 17.1(-3) - 12.9$ $y = -64.2$	–64.2
–2	$y = 17.1(-2) - 12.9$ $y = -47.1$	–47.1
–1	$y = 17.1(-1) - 12.9$ $y = -30$	–30
0	$y = 17.1(0) - 12.9$ $y = -12.9$	–12.9
1	$y = 17.1(1) - 12.9$ $y = 4.2$	4.2
2	$y = 17.1(2) - 12.9$ $y = 21.3$	21.3
3	$y = 17.1(3) - 12.9$ $y = 38.4$	38.4

15.

x	$y = 3x^2 - 5x + 17$	y
–6	$y = 3(-6)^2 - 5(-6) + 17$ $y = 155$	155
–2	$y = 3(-2)^2 - 5(-2) + 17$ $y = 39$	39
0	$y = 3(0)^2 - 5(0) + 17$ $y = 17$	17
3	$y = 3(3)^2 - 5(3) + 17$ $y = 29$	29
8	$y = 3(8)^2 - 5(8) + 17$ $y = 169$	169
11	$y = 3(11)^2 - 5(11) + 17$ $y = 325$	325

16.

r	$A = \pi r^2$	A
4	$A = \pi(4)^2$ $A \approx 50.265$	50.265
6	$A = \pi(6)^2$ $A \approx 113.097$	113.097
8	$A = \pi(8)^2$ $A \approx 201.062$	201.062
10	$A = \pi(10)^2$ $A \approx 314.159$	314.159

17.

r	$V = \pi(6)^2 h$ $V = 36\pi h$	V
3	$V = 36\pi(3)$ $V \approx 339.3$	339.3
6	$V = 36\pi(6)$ $V \approx 678.6$	678.6
9	$V = 36\pi(9)$ $V \approx 1017.9$	1017.9
12	$V = 36\pi(12)$ $V \approx 1357.2$	1357.2

18.

a	$b = 90 - a$	b
10	$b = 90 - 10$ $b = 80$	80
20	$b = 90 - 20$ $b = 70$	70
30	$b = 90 - 30$ $b = 60$	60
40	$b = 90 - 40$ $b = 50$	50
45	$b = 90 - 45$ $b = 45$	45

19.

t	$A = 2000(1.06)^t$	A
2	$A = 2000(1.06)^2$ $A = 2247.2$	2247.2
3	$A = 2000(1.06)^3$ $A \approx 2382.03$	2382.03
4	$A = 2000(1.06)^4$ $A \approx 2524.95$	2524.95

20.

F	$C = \dfrac{5}{9}(F - 32)$	C
–23	$C = \dfrac{5}{9}(-23 - 32)$ $C \approx -30.6$	–30.6
–14	$C = \dfrac{5}{9}(-14 - 32)$ $C \approx -25.6$	–25.6
0	$C = \dfrac{5}{9}(0 - 32)$ $C \approx -17.8$	–17.8
41	$C = \dfrac{5}{9}(41 - 32)$ $C = 5$	5
50	$C = \dfrac{5}{9}(50 - 32)$ $C = 10$	10
59	$C = \dfrac{5}{9}(59 - 32)$ $C = 15$	15
100	$C = \dfrac{5}{9}(100 - 32)$ $C \approx 37.8$	37.8

21.

t	$d = 65t$	d
2	$d = 65(2)$ $d = 130$	130
3	$d = 65(3)$ $d = 195$	195
4	$d = 65(4)$ $d = 260$	260
5	$d = 65(5)$ $d = 325$	325
6	$d = 65(6)$ $d = 390$	390
7	$d = 65(7)$ $d = 455$	455

22.

x	$y = 7 - 3x$	y
-10	$y = 7 - 3(-10)$ $y = 37$	37
-5	$y = 7 - 3(-5)$ $y = 22$	22
0	$y = 7 - 3(0)$ $y = 7$	7
5	$y = 7 - 3(5)$ $y = -8$	-8
10	$y = 7 - 3(10)$ $y = -23$	-23

ordered pairs: (-10, 37), (-5, 22), (0, 7), (5, -8), (10, -23)

23.

x	$y = \sqrt{x + 8}$	y
-8	$y = \sqrt{-8 + 8}$ $y = 0$	0
-7	$y = \sqrt{-7 + 8}$ $y = 1$	1
-4	$y = \sqrt{-4 + 8}$ $y = 2$	2
1	$y = \sqrt{1 + 8}$ $y = 3$	3
8	$y = \sqrt{8 + 8}$ $y = 4$	4

ordered pairs: (-8, 0), (-7, 1), (-4, 2), (1, 3), (8, 4)

24.

c	$d = \dfrac{2}{3}c - 1$	d
-6	$d = \dfrac{2}{3}(-6) - 1$ $d = -5$	-5
-3	$d = \dfrac{2}{3}(-3) - 1$ $d = -3$	-3
0	$d = \dfrac{2}{3}(0) - 1$ $d = -1$	-1
3	$d = \dfrac{2}{3}(3) - 1$ $d = 1$	1
6	$d = \dfrac{2}{3}(6) - 1$ $d = 3$	3

ordered pairs: (-6, -5), (-3, -3), (0, -1), (3, 1), (6, 3)

25.

d	$r = \dfrac{d}{2}$	r
2	$r = \dfrac{2}{2}$ $r = 1$	1
4	$r = \dfrac{4}{2}$ $r = 2$	2
6	$r = \dfrac{6}{2}$ $r = 3$	3
8	$r = \dfrac{8}{2}$ $r = 4$	4
10	$r = \dfrac{10}{2}$ $r = 5$	5

ordered pairs: (2, 1), (4, 2), (6, 3), (8, 4), (10, 5)

26. domain: {1, 3, 5, 7, 9}
range: {2, 6, 10, 14, 18}

27. domain: {…, −6, −4, −2, 0, 2, 4, 6, …}
range: {…, 6, 4, 2, 0, −2, −4, −6, …}

28. $y = 4(x + 5) - 1$
$y = 4(-5 + 5) - 1$
$y = 0 - 1 = -1$
$y = 4(-4 + 5) - 1$
$y = 4 - 1 = 3$
$y = 4(-3 + 5) - 1$
$y = 8 - 1 = 7$
$y = 4(-2 + 5) - 1$
$y = 12 - 1 = 11$
$y = 4(-1 + 5) - 1$
$y = 16 - 1 = 15$
domain: {−5, −4, −3, −2, −1}
range: {−1, 3, 7, 11, 15}

29. domain: all real numbers
range: all real numbers ≥ 2.5

30. $5 + x \geq 0$
$x \geq -5$
domain: all real numbers ≥ -5
range: all real numbers ≥ 0

31. $1 - x \neq 0$
$x \neq 1$
domain: all real numbers $x \neq 1$
range: all real numbers $y \neq 0$

32. $y = 5 + 4x$
domain: all real numbers
range: all real numbers

33. $\dfrac{L}{32} \geq 0$
$L \geq 0$
domain: all real numbers ≥ 0
range: all real numbers ≥ 0

34. $A(3, 2)$

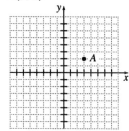

35. $B(4, -3)$

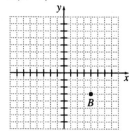

36. $C(-3, 2)$

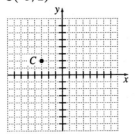

37. $D(-4, -3)$

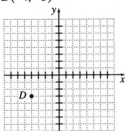

38. $E(0, 5)$

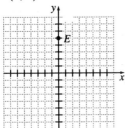

39. $F(-5, 0)$

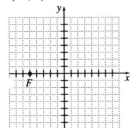

40. $A(5, 3)$

41. $B(-2, -5)$

42. $C(2, -2)$

43. $D(-3, 5)$

44. $E(5, 0)$

45. $F(0, -4)$

46. $G(0, 0)$

47. quadrant I

48. quadrant III

49. quadrant IV

50. quadrant II

51. x-axis

52. y-axis

53. origin

54. $S = \{(0, -4), (1, -3), (2, -2), (3, -1), (4, 0), (5, 1)\}$

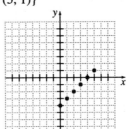

55.

x	$y = 2x - 3$	y
-1	$y = 2(-1) - 3$	-5
	$y = -5$	
0	$y = 2(0) - 3$	-3
	$y = -3$	
1	$y = 2(1) - 3$	-1
	$y = -1$	
2	$y = 2(2) - 3$	1
	$y = 1$	
3	$y = 2(3) - 3$	3
	$y = 3$	
4	$y = 2(4) - 3$	5
	$y = 5$	

$(-1, -5), (0, -3), (1, -1), (2, 1), (3, 3), (4, 5)$

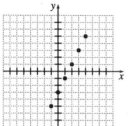

56. $y = 3 - 2x$

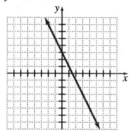

57. $T = \{(-5, 3), (-3, 3), (-1, 3), (1, 3), (3, 3), (5, 3)\}$

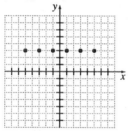

58. domain: all real numbers
range: all real numbers ≤ 2

59. domain: all real numbers
range: $\{-3\}$

60. $y = 6x - 5$
domain: all real numbers
range: all real numbers

61. $y = \sqrt{8 - 4x}$
$8 - 4x \geq 0$
$4x \leq 8$
$x \leq 2$
domain: all real numbers ≤ 2
range: all real numbers ≥ 0

62. $y = x^2 + 6x + 9$

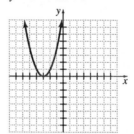

domain: all real numbers
range: all real numbers ≥ 0

63. $y = -x^2$

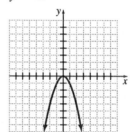

domain: all real numbers
range: all real numbers ≤ 0

64. domain: years 1970 to 1993
range: all real numbers $65 \leq y \leq 85$
For a year (independent variable) the energy consumption (dependent variable) is graphed

65. domain: integers 1 to 7
range: $\{40, 50, 55, 60\}$
The domain represents the number of children from one family enrolled in the child care center.
The range represents the charge in dollars for a week's care for the children from a family.

66. domain: all real numbers > 0
range: all real numbers > 0
The length of a side (independent variable) and its corresponding area (dependent variable) are graphed.

67. S is not a function. Two elements in the domain, -2 and -1, each correspond to 2 elements in the range, 3 and 11, and 5 and 9, respectively.

68. T is a function. Every element in the domain corresponds to only one element in the range.

69. $y = x^2 - 10$ is a function. Every element in the domain corresponds to only one element in the range.

70. $y^2 = x - 10$ is not a function. There are two elements in the range for each element in the domain (except $x = 10$).

71. $y = 4$ for all x is a function. Each element in the domain corresponds to only one element in the range.

72. $x = 2$ for all y is not a function. Each element in the domain corresponds to the same element in the domain.

73. No; does not pass the vertical line test.

74. Yes; does pass the vertical line test.

75. $f(x) = -4x + 13$
$f(13) = -4(13) + 13$
$f(13) = -39$

76. $f(x) = -4x + 13$
$f(-21) = -4(-21) + 13$
$f(-21) = 97$

77. $f(x) = -4x + 13$
$f(2.5) = -4(2.5) + 13$
$f(2.5) = 3$

78. $f(x) = -4x + 13$
$f(-3.7) = -4(-3.7) + 13$
$f(-3.7) = 27.8$

79. $f(x) = -4x + 13$
$f(a + 3) = -4(a + 3) + 13$
$f(a + 3) = -4a - 12 + 13$
$f(a + 3) = -4a + 1$

80. $f(x) = -4x + 13$
$f(-b) = -4(-b) + 13$
$f(-b) = 4b + 13$

81. $g(x) = 5x^2 + x - 4$
$g(3) = 5(3)^2 + 3 - 4$
$g(3) = 45 + 3 - 4$
$g(3) = 44$

82. $g(x) = 5x^2 + x - 4$
$g(-2) = 5(-2)^2 + (-2) - 4$
$g(-2) = 20 - 2 - 4$
$g(-2) = 14$

83. $g(x) = 5x^2 + x - 4$
$g(0.5) = 5(0.5)^2 + (0.5) - 4$
$g(0.5) = 1.25 + 0.5 - 4$
$g(0.5) = -2.25$

84. $g(x) = 5x^2 + x - 4$
$g(a) = 5a^2 + a - 4$

85. $g(x) = 5x^2 + x - 4$
$g(-a) = 5(-a)^2 + (-a) - 4$
$g(-a) = 5a^2 - a - 4$

86. $g(x) = 5x^2 + x - 4$
$g\left(-\dfrac{1}{4}\right) = 5\left(-\dfrac{1}{4}\right)^2 + \left(-\dfrac{1}{4}\right) - 4$
$g\left(-\dfrac{1}{4}\right) = \dfrac{5}{16} - \dfrac{1}{4} - 4$
$g\left(-\dfrac{1}{4}\right) = \dfrac{5}{16} - \dfrac{4}{16} - \dfrac{64}{16}$
$g\left(-\dfrac{1}{4}\right) = -\dfrac{63}{16}$

87. $S(x) = \sqrt{2x + 3}$
$S(3) = \sqrt{2 \cdot 3 + 3}$
$S(3) = \sqrt{9}$
$S(3) = 3$

88. $S(x) = \sqrt{2x + 3}$
$S(11) = \sqrt{2 \cdot 11 + 3}$
$S(11) = \sqrt{25}$
$S(11) = 5$

89. $S(x) = \sqrt{2x + 3}$

$S(59) = \sqrt{2 \cdot 59 + 3}$

$S(59) = \sqrt{121}$

$S(59) = 11$

90. Let x = the number of widgets.

$f(x) = 4500 + 17x$

$f(1200) = 4500 + 17(1200)$

$f(1200) = 4500 + 20{,}400$

$f(1200) = 24{,}900$

The cost of producing 1200 widgets is $24,900.

91. Let x = the number of days.

$f(x) = 50 + 25x$

$f(4) = 50 + 25(4)$

$f(4) = 50 + 100$

$f(4) = 150$

The cost of a 4-day rental is $150.

92. Let x = the number of employees.

$f(x) = 1500 + 125x$

$f(20) = 1500 + 125(20)$

$f(20) = 1500 + 2500$

$f(20) = 4000$

The charge for a training session for 20 employees is $4,000.

93. Let x = the number of faces.

$f(x) = 1.5x - 15$

$f(135) = 1.5(135) - 15$

$f(135) = 202.5 - 15$

$f(135) = 187.5$

He will make a profit of $187.50.

94. a. $(-5, 0), (1, 0), (6, 0)$

    b. $(0, 1)$

    c. $x < -3, x > 3$

    d. $-3 < x < 3$

    e. 3

    f. -3

95. $y = 3x + 9$

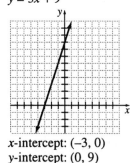

x-intercept: $(-3, 0)$

y-intercept: $(0, 9)$

96. $y = 6 - x$

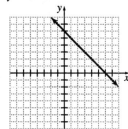

x-intercept: $(6, 0)$

y-intercept: $(0, 6)$

97. $y = \dfrac{3}{4}x - 9$

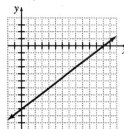

x-intercept: $(12, 0)$

y-intercept: $(0, -9)$

98. $y = x^2 - 0.36$

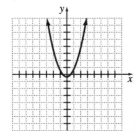

x-intercepts: $(-0.6, 0)$, $(0.6, 0)$
y-intercept: $(0, -0.36)$

99. $y = |x| - 4$

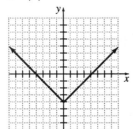

x-intercepts: $(-4, 0)$, $(4, 0)$
y-intercept: $(0, -4)$

100. $y = 4x - 3$

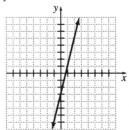

increasing for all values of x

101. $h(x) = 6 - 2x$

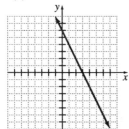

decreasing for all values of x

102. $y = 3 - x^2$

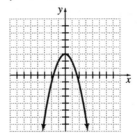

increasing for $x < 0$
decreasing for $x > 0$
relative maximum is 3

103. $y = |x| + 2$

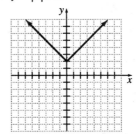

increasing for $x > 0$
decreasing for $x < 0$
relative minimum is 2

104. $y = \left| x^2 - 1 \right|$

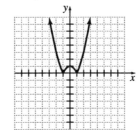

Increasing for $-1 < x < 0$, $x > 1$
decreasing for $x < -1$, $0 < x < 1$
relative maximum is 1
relative minima are 0, 0

105. $y = 2x - 2$ and $y = -\dfrac{1}{3}x + 5$

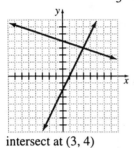

intersect at (3, 4)

106. $y = x^2 - 6$ and $y = x$

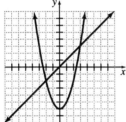

intersect at (–2, –2), (3, 3)

107. $f(x) = |x + 5|$ and $g(x) = 2$

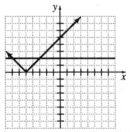

intersect at (–7, 2), (–3, 2)

108. $f(x) = 10.45x$

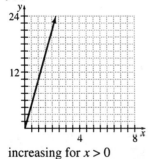

increasing for $x > 0$

109. $f(x) = 216 - 4.5x$

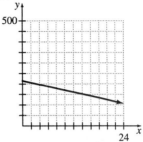

decreasing for $x > 0$

110. Let x = the number of additional desks
Y1 = $(325 - 15x)(x + 1)$

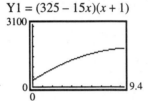

increasing for $0 < x < 7$

111. a. Let x = credit hours.
$f(x) = 400 + 65x$
$g(x) = 100x$

    b. Y1 = $400 + 65x$
Y2 = $100x$

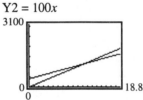

intersect at approximately (11.4, 1143)

    c. At about 11.4 credit hours the stipend is the same, at about $1143, for both options.

112. a. Let x = the number of items.
$f(x) = 500 + 12x$
$g(x) = 25x$

b. $Y1 = 500 + 12x$
$Y2 = 25x$

intersect at approximately (38.5, 961.5)

c. At about 38.5 items, the cost to produce and the revenue are equal at about $961.50 each. This means 39 items must be sold to break even.

Chapter 3 Mixed Review

1. $f(x) = -x + 9$
$f(9) = -9 + 9$
$f(9) = 0$

2. $f(x) = -x + 9$
$f(-9) = -(-9) + 9$
$f(-9) = 9 + 9$
$f(-9) = 18$

3. $f(x) = -x + 9$
$f(1.8) = -1.8 + 9$
$f(1.8) = 7.2$

4. $f(x) = -x + 9$
$f(-2.7) = -(-2.7) + 9$
$f(-2.7) = 2.7 + 9$
$f(-2.7) = 11.7$

5. $f(x) = -x + 9$
$f(-b) = -(-b) + 9$
$f(-b) = b + 9$

6. $f(x) = -x + 9$
$f(h + 1) = -(h + 1) + 9$
$f(h + 1) = -h - 1 + 9$
$f(h + 1) = -h + 8$

7. $g(x) = x^2 - 3x - 4$
$g(4) = 4^2 - 3 \cdot 4 - 4$
$g(4) = 16 - 12 - 4$
$g(4) = 0$

8. $g(x) = x^2 - 3x - 4$
$g(-1) = (-1)^2 - 3(-1) - 4$
$g(-1) = 1 + 3 - 4$
$g(-1) = 0$

9. $g(x) = x^2 - 3x - 4$
$g(1.5) = (1.5)^2 - 3(1.5) - 4$
$g(1.5) = 2.25 - 4.5 - 4$
$g(1.5) = -6.25$

10. $g(x) = x^2 - 3x - 4$
$g(v) = v^2 - 3v - 4$

11. $g(x) = x^2 - 3x - 4$
$g(-v) = (-v)^2 - 3(-v) - 4$
$g(-v) = v^2 + 3v - 4$

12. $g(x) = x^2 - 3x - 4$
$g\left(-\dfrac{2}{3}\right) = \left(-\dfrac{2}{3}\right)^2 - 3\left(-\dfrac{2}{3}\right) - 4$
$g\left(-\dfrac{2}{3}\right) = \dfrac{4}{9} + \dfrac{6}{3} - 4$
$g\left(-\dfrac{2}{3}\right) = \dfrac{4}{9} + \dfrac{18}{9} - \dfrac{36}{9}$
$g\left(-\dfrac{2}{3}\right) = -\dfrac{14}{9}$

13. $S(x) = \sqrt{6x - 8}$
$S(4) = \sqrt{6 \cdot 4 - 8}$
$S(4) = \sqrt{16}$
$S(4) = 4$

14. $S(x) = \sqrt{6x - 8}$
$S(12) = \sqrt{6(12) - 8}$
$S(12) = \sqrt{64}$
$S(12) = 8$

15. $S(x) = \sqrt{6x - 8}$
$S(44) = \sqrt{6(44) - 8}$
$S(44) = \sqrt{256}$
$S(44) = 16$

16. Yes; each element in the domain corresponds to a single element in the range.

17. No; an element in the domain, 2, corresponds to 5 elements in the range, –3, –2, –1, 0, and 1.

18. $y = x^2 + 5$

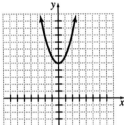

Yes; the graph passes the vertical line test.

19. $y^2 = x + 5$

No; for each element in the domain, except –5, there are two corresponding elements in the range.

20. $y = -4$ for all x.
Yes; each element in the domain corresponds to only one element in the range.

21. $y = 5x - 10$

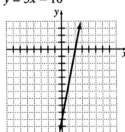

x-intercept at (2, 0)
y-intercept at (0, –10)
increasing for all x-values

22. $y = 8 - 2x$

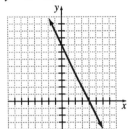

x-intercept at (0, 8)
y-intercept at (4, 0)
decreasing for all x-values

23. $y = 4.8x - 1.2$

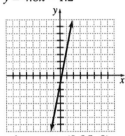

x-intercept: (0.25, 0)
y-intercept: (0, –1.2)
increasing for all x-values

24. $y = \dfrac{2}{5}x + 4$

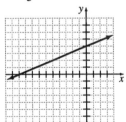

x-intercept at (–10, 0)
y-intercept at (0, 4)
increasing for all x-values

25. $y = x^2 - 1.21$

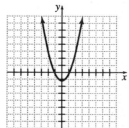

x-intercepts at (–1.1, 0) and (1.1, 0)
y-intercept at (0, –1.21)
increasing for $x > 0$
decreasing for $x < 0$
relative minimum is –1.21

26. $y = 2 - |x|$

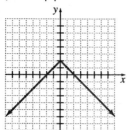

x-intercepts at (–2, 0), (2, 0)
y-intercept at (0, 2)
increasing for $x < 0$
decreasing for $x > 0$
relative maximum is 2

27. $y = 2x + 2$ and $y = -2x - 10$

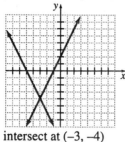

intersect at (–3, –4)

28. $y = x^2$ and $y = 3x$

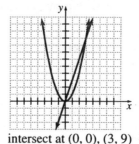

intersect at (0, 0), (3, 9)

29. $f(x) = |2x|$ and $g(x) = x + 3$

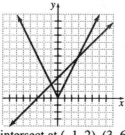

intersect at (–1, 2), (3, 6)

30. domain: {2, 4, 6, 8, 10}
range: {1, 2, 3, 4, 5}

31. domain: {..., –6, –4, –2, 0, 2, 4, 6, ...}
range: {3}

32. domain: {–5, –4, –3, –2, –1}
range: {25, 16, 9, 4, 1}

33. domain: all real numbers
range: all real numbers ≥ -1.5

34. $x - 8 \geq 0$
$x \geq 8$
domain: all real numbers ≥ 8
range: all real numbers ≥ 0

35. $x^2 \neq 0$
$x \neq 0$
domain: all real numbers $\neq 0$
range: all real numbers > 0

36. domain: all real numbers
range: all real numbers

136

37. $6 - 2x \geq 0$
$6 \geq 2x$
$x \leq 3$
domain: all real numbers ≤ 3
range: all real numbers ≥ 0

38. $y = x^2 - 2x + 1$

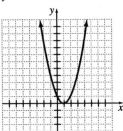

domain: all real numbers
range: all real numbers ≥ 0

39. $y = -x^2 + 3$

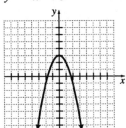

domain: all real numbers
range: all real numbers ≤ 3

40.

x	$y = 12 - 8x$	y
–6	$y = 12 - 8(-6)$	60
	$y = 60$	
–3	$y = 12 - 8(-3)$	36
	$y = 36$	
0	$y = 12 - 8(0)$	12
	$y = 12$	
3	$y = 12 - 8(3)$	–12
	$y = -12$	
6	$y = 12 - 8(6)$	–36
	$y = -36$	

ordered pairs: (–6, 60), (–3, 36), (0, 12), (3, –12), (6, –36)

41.

x	$y = \sqrt{10 - 3x}$	y
3	$y = \sqrt{10 - 3 \cdot 3}$	1
	$y = 1$	
2	$y = \sqrt{10 - 3 \cdot 2}$	2
	$y = 2$	
1	$y = \sqrt{10 - 3 \cdot 1}$	$\sqrt{7}$
	$y = \sqrt{7}$	
0	$y = \sqrt{10 - 3 \cdot 0}$	$\sqrt{10}$
	$y = \sqrt{10}$	
–1	$y = \sqrt{10 - 3(-1)}$	$\sqrt{13}$
	$y = \sqrt{13}$	
–2	$y = \sqrt{10 - 3(-2)}$	4
	$y = 4$	

ordered pairs: (3, 1), (2, 2), $\left(1, \sqrt{7}\right)$, $\left(0, \sqrt{10}\right)$, $\left(-1, \sqrt{13}\right)$, (–2, 4)

42.

s	$t = \dfrac{4}{7}s + 5$	t
–7	$t = \dfrac{4}{7}(-7) + 5$ $t = 1$	1
0	$t = \dfrac{4}{7}(0) + 5$ $t = 5$	5
7	$t = \dfrac{4}{7}(7) + 5$ $t = 9$	9
14	$t = \dfrac{4}{7}(14) + 5$ $t = 13$	13
21	$t = \dfrac{4}{7}(21) + 5$ $t = 17$	17

ordered pairs: (–7, 1), (0, 5), (7, 9), (14, 13), (21, 17)

43.

x	$y = (3x - 5)(2x + 1)$	y
–2	$y = [3(-2) - 5][2(-2)+1]$ $y = 33$	33
0	$y = (3 \cdot 0 - 5)(2 \cdot 0 + 1)$ $y = -5$	–5
2	$y = (3 \cdot 2 - 5)(2 \cdot 2 + 1)$ $y = 5$	5

44.

x	$y = \dfrac{3}{4}x - 5$	y
–8	$y = \dfrac{3}{4}(-8) - 5$ $y = -11$	–11
–4	$y = \dfrac{3}{4}(-4) - 5$ $y = -8$	–8
0	$y = \dfrac{3}{4} \cdot 0 - 5$ $y = -5$	–5
4	$y = \dfrac{3}{4} \cdot 4 - 5$ $y = -2$	–2
8	$y = \dfrac{3}{4} \cdot 8 - 5$ $y = 1$	1

45.

x	$y = 15.8 - 4.7x$	y
–2	$y = 15.8 - 4.7(-2)$ $y = 25.2$	25.2
–1	$y = 15.8 - 4.7(-1)$ $y = 20.5$	20.5
0	$y = 15.8 - 4.7(0)$ $y = 15.8$	15.8
1	$y = 15.8 - 4.7(1)$ $y = 11.1$	11.1
2	$y = 15.8 - 4.7(2)$ $y = 6.4$	6.4

46.

x	$y = 4x^2 - 17x - 15$	y
-2	$y = 4(-2)^2 - 17(-2) - 15$ $y = 35$	35
$-\dfrac{3}{4}$	$y = 4\left(-\dfrac{3}{4}\right)^2 - 17\left(-\dfrac{3}{4}\right) - 15$ $y = 0$	0
0	$y = 4 \cdot 0^2 - 17 \cdot 0 - 15$ $y = -15$	-15
$\dfrac{3}{4}$	$y = 4\left(\dfrac{3}{4}\right)^2 - 17\left(\dfrac{3}{4}\right) - 15$ $y = -25.5$	-25.5
5	$y = 4(5)^2 - 17(5) - 15$ $y = 0$	0

47.

x	$y = x^3 + x^2 + x + 1$	y
-15	$y = (-15)^3 + (-15)^2 + (-15) + 1$ $y = -3164$	-3164
-5	$y = (-5)^3 + (-5)^2 + (-5) + 1$ $y = -104$	-104
0	$y = 0^3 + 0^2 + 0 + 1$ $y = 1$	1
5	$y = 5^3 + 5^2 + 5 + 1$ $y = 156$	156
15	$y = (15)^3 + (15)^2 + 15 + 1$ $y = 3616$	3616
25	$y = (25)^3 + (25)^2 + 25 + 1$ $y = 16{,}276$	$16{,}276$

48.

x	$y = \left|1 - 2x - 3x^2\right|$	y
-6	$y = \left|1 - 2(-6) - 3(-6)^2\right|$ $y = 95$	95
-3	$y = \left|1 - 2(-3) - 3(-3)^2\right|$ $y = 20$	20
0	$y = \left|1 - 2(0) - 3(0)^2\right|$ $y = 1$	1
3	$y = \left|1 - 2 \cdot 3 - 3(3)^2\right|$ $y = 32$	32
6	$y = \left|1 - 2(6) - 3(6)^2\right|$ $y = 119$	119
9	$y = \left|1 - 2(9) - 3(9)^2\right|$ $y = 260$	260

49.

x	$y = 4.6x^2 + 2.8x + 10.4$	y
−3.7	$y = 4.6(-3.7)^2 + 2.8(-3.7) + 10.4$ $y = 63.014$	63.014
−2.2	$y = 4.6(-2.2)^2 + 2.8(-2.2) + 10.4$ $y = 26.504$	26.504
−0.7	$y = 4.6(-0.7)^2 + 2.8(-0.7) + 10.4$ $y = 10.694$	10.694
0	$y = 4.6(0)^2 + 2.8(0) + 10.4$ $y = 10.4$	10.4
0.8	$y = 4.6(0.8)^2 + 2.8(0.8) + 10.4$ $y = 15.584$	15.584
2.3	$y = 4.6(2.3)^2 + 2.8(2.3) + 10.4$ $y = 41.174$	41.174
3.8	$y = 4.6(3.8)^2 + 2.8(3.8) + 10.4$ $y = 87.464$	87.464

50. Answers will vary. Possible answer:

x	$y = 17 - 5x$	y
−1	$y = 17 - 5(-1)$ $y = 22$	22
0	$y = 17 - 5(0)$ $y = 17$	17
1	$y = 17 - 5(1)$ $y = 12$	12

51. Answers will vary. Possible answer:

x	$y = 4.5x - 1.6$	y
−1	$y = 4.5(-1) - 1.6$ $y = -6.1$	−6.1
0	$y = 4.5(0) - 1.6$ $y = -1.6$	−1.6
1	$y = 4.5(1) - 1.6$ $y = 2.9$	2.9

52. Answers will vary. Possible answer:

x	$y = \frac{1}{4}x + 3$	y
-4	$y = \frac{1}{4}(-4) + 3$ $y = 2$	2
0	$y = \frac{1}{4}(0) + 3$ $y = 3$	3
4	$y = \frac{1}{4}(4) + 3$ $y = 4$	4

53. Answers will vary. Possible answer:

x	$y = |3x - 10|$	y
-3	$y = |3(-3) - 10|$ $y = 19$	19
0	$y = |3(0) - 10|$ $y = 10$	10
3	$y = |3(3) - 10|$ $y = 1$	1

54.

r	$C = 2\pi r$	C
$\frac{1}{4}$	$C = 2\pi\left(\frac{1}{4}\right)$ $C \approx 1.57$	1.57
$\frac{1}{2}$	$C = 2\pi\left(\frac{1}{2}\right)$ $C \approx 3.14$	3.14
1	$C = 2\pi(1)$ $C \approx 6.28$	6.28
$\frac{3}{2}$	$C = 2\pi\left(\frac{3}{2}\right)$ $C \approx 9.42$	9.42
2	$C = 2\pi(2)$ $C \approx 12.57$	12.57

ordered pairs: $\left(\frac{1}{4}, 1.5708\right)$, $\left(\frac{1}{2}, 3.1416\right)$, $(1, 6.2832)$, $\left(\frac{3}{2}, 9.4248\right)$, $(2, 12.566)$

55.

s	$A = s^2$	A
3	$A = 3^2$ $A = 9$	9
5	$A = 5^2$ $A = 25$	25

56.

H	$V = 2H$	V
1	$V = 2(1)$ $V = 2$	2
1.25	$V = 2(1.25)$ $V = 2.5$	2.5
1.5	$V = 2(1.5)$ $V = 3$	3
1.75	$V = 2(1.75)$ $V = 3.5$	3.5
2	$V = 2(2)$ $V = 4$	4

57.

a	$b = 180 - a$	b
30	$b = 180 - 30$ $b = 150$	150
60	$b = 180 - 60$ $b = 120$	120
90	$b = 180 - 90$ $b = 90$	90
120	$b = 180 - 120$ $b = 60$	60
150	$b = 180 - 150$ $b = 30$	30

58.

t	$I = 2000(0.06)t$ $I = 120t$	I
2	$I = 120(2)$ $I = 240$	240
3	$I = 120(3)$ $I = 360$	360
4	$I = 120(4)$ $I = 480$	480

59.

C	$F = \dfrac{9}{5}C + 32$	F
-10	$F = \dfrac{9}{5}(-10) + 32$ $F = 14$	14
-5	$F = \dfrac{9}{5}(-5) + 32$ $F = 23$	23
0	$F = \dfrac{9}{5}(0) + 32$ $F = 32$	32
5	$F = \dfrac{9}{5}(5) + 32$ $F = 41$	41
10	$F = \dfrac{9}{5}(10) + 32$ $F = 50$	50
15	$F = \dfrac{9}{5}(15) + 32$ $F = 59$	59
20	$F = \dfrac{9}{5}(20) + 32$ $F = 68$	68
25	$F = \dfrac{9}{5}(25) + 32$ $F = 77$	77

60. Let x = the number of items.
$f(x) = 2500 + 12x$
$f(1650) = 2500 + 12(1650)$
$f(1650) = 2500 + 19{,}800$
$f(1650) = 22{,}300$
The production run will cost $22,300.

61. Let x = the number of hours of rental.
$f(x) = 15 + 2x$
$f(10) = 15 + 2(10)$
$f(10) = 15 + 20$
$f(10) = 35$
It will cost $35 to rent the grinder for 10 hours.

62. Let x = the number of employees.
$f(x) = 275 + 9.5x$
$f(135) = 275 + 9.5(135)$
$f(135) = 275 + 1282.5$
$f(135) = 1557.5$
A luncheon for 135 employees will cost $1557.50.

63. Let x = the number of admissions.
$f(x) = -185 + 4x$
$f(310) = -185 + 4(310)$
$f(310) = -185 + 1240$
$f(310) = 1055$
The net profit for the game will be $1055.

64. $f(x) = 250 - 3.5x$
$Y1 = 250 - 3.5x$

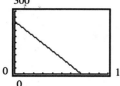

decreasing for $0 < x < 71.43$

65. $f(x) = 1000 + 50x$
$Y1 = 1000 + 50x$

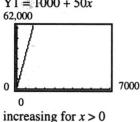

increasing for $x > 0$

66. $A = LW$
$f(x) = x(100 - x)$
$Y1 = x(100 - x)$

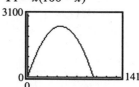

increasing for $0 < x < 50$
decreasing for $50 < x < 100$

67. a. $f(x) = 25,000 + 5000x$
$g(x) = 6000x$

b. $Y1 = 25,000 + 5000x$
$Y2 = 6000x$

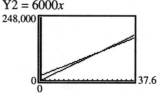

intersect at (25, 150,000)

c. At 25 years, the money received is the same at $150,000.

68. a. $f(x) = 22x + 600$
$g(x) = 75x$

b. $Y1 = 22x + 600$
$Y2 = 75x$

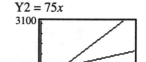

intersect at approximately (11.32, 849.06)

c. At about 11.32 appliances the total acquisition cost and total revenue are equal at about $849.06. This means 12 appliances must be sold to break even.

Chapter 3 Test

1.

x	$y = 2x^2 + 17x - 9$	y
–9	$y = 2(-9)^2 + 17(-9) - 9$ $y = 0$	0
0	$y = 2(0)^2 + 17(0) - 9$ $y = -9$	–9
3	$y = 2(3)^2 + 17(3) - 9$ $y = 60$	60

2. $y = |2x|$

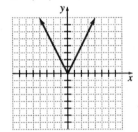

3. domain: all real numbers

4. range: all real numbers ≥ 0

5. Yes; all possible vertical lines cross the graph a maximum of one time.

6. increasing for $x > 0$

7. decreasing for $x < 0$

8. none

9. relative minimum is 0

10. x-intercept at $(0, 0)$

11. y-intercept at $(0, 0)$

12. $A(1, 2)$, $B(-2, -4)$, $C(-5, 3)$, $D(2, -5)$, $E(0, -2)$

13. quadrant IV

14. $f(x) = \dfrac{1}{2}x + 6$

$f(4) = \dfrac{1}{2}(4) + 6$
$f(4) = 2 + 6$
$f(4) = 8$

15. $f(x) = \dfrac{1}{2}x + 6$

$f(-6) = \dfrac{1}{2}(-6) + 6$
$f(-6) = -3 + 6$
$f(-6) = 3$

16. $f(x) = \dfrac{1}{2}x + 6$

$f(2a - 2) = \dfrac{1}{2}(2a - 2) + 6$
$f(2a - 2) = a - 1 + 6$
$f(2a - 2) = a + 5$

17. $f(x) = \dfrac{1}{2}x + 6$

$f(-b) = \dfrac{1}{2}(-b) + 6$
$f(-b) = -\dfrac{1}{2}b + 6$

18. $f(x) = 450 + 21.5x$

19. $f(250) = 450 + 21.5(250)$
$f(250) = 450 + 5375$
$f(250) = 5825$
The cost is $5825.

20. $y = 3x - 10$ and $y = -x - 2$

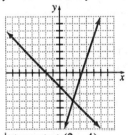

intersect at $(2, -4)$

21. Answers will vary. Possible answer: Each point in the plane corresponds to an ordered pair. The first number denotes the distance and direction along the x-axis. The second number denotes the distance and direction along the y-axis.

Chapters 1–3 Cumulative Review

1. 12 is the only whole number.

2. 0 and 12 are integers.

3. $-\dfrac{2}{3}$, 0, 12, $1\dfrac{4}{5}$, and -0.33 are rational numbers.

4. $\sqrt{7}$ is an irrational number.

5. $\dfrac{3}{8} > \dfrac{1}{3}$

6. $\dfrac{2}{3} > 0.66$

7. $-2.8 < -1.6$

8.

9. $-28 + 13 = -15$

10. $4.8 - 7.36 = 4.8 + (-7.36) = -2.56$

11. $-87 \div (-29) = 3$

12. $-\dfrac{5}{8} - \dfrac{2}{3} = -\dfrac{15}{24} - \dfrac{16}{24} = \dfrac{-15-16}{24} = -\dfrac{31}{24}$

13. $-2\dfrac{3}{4} \div 1\dfrac{3}{7} = -\dfrac{11}{4} \div \dfrac{10}{7} = -\dfrac{11}{4} \cdot \dfrac{7}{10}$

$= \dfrac{-11 \cdot 7}{4 \cdot 10} = -\dfrac{77}{40} = -1\dfrac{37}{40}$

14. $\left(-\dfrac{2}{3}\right)\left(\dfrac{3}{8}\right)\left(-\dfrac{7}{16}\right)\left(\dfrac{9}{10}\right) = \dfrac{(-2)(3)(-7)(9)}{(3)(8)(16)(10)}$

$= \dfrac{(2)(7)(9)}{(8)(16)(10)} = \dfrac{(7)(9)}{(4)(16)(10)} = \dfrac{63}{640}$

15. $(12.96)(-4.8) = -62.208$

16. $(14)(0)(5)(-6) = 0$

17. $(-12)(16) \div 4(-2)$
$= (-192) \div 4(-2)$
$= (-48)(-2)$
$= 96$

18. $14 + (-7) + 22 - 16 - (-18)$
$= 7 + 22 - 16 - (-18)$
$= 29 - 16 - (-18)$
$= 13 - (-18)$
$= 13 + 18 = 31$

19. $-[3.8 - (-2.4)] = -[3.8 + 2.4] = -[6.2] = -6.2$

20. $\dfrac{2(3^2 + 7) - 2^5}{3 \cdot 18} = \dfrac{2(9 + 7) - 32}{54}$

$= \dfrac{2(16) - 32}{54} = \dfrac{32 - 32}{54} = \dfrac{0}{54} = 0$

21. $-|12 - 20| = -|-8| = -(8) = -8$

22. $\sqrt{\dfrac{16}{25}} = \dfrac{\sqrt{16}}{\sqrt{25}} = \dfrac{4}{5}$

23. $-\sqrt{1.2} = -\sqrt{4(0.3)} = -2\sqrt{0.3} \approx -1.095$

24. $\sqrt{-16}$ is not a real number.

25. $\sqrt[3]{1\dfrac{13}{81}} = \sqrt[3]{\dfrac{94}{81}} = \sqrt[3]{\dfrac{94}{27 \cdot 3}} = \dfrac{1}{3}\sqrt[3]{\dfrac{94}{3}}$

$= \dfrac{1}{3}\sqrt[3]{\dfrac{94 \cdot 3^2}{3 \cdot 3^2}} = \dfrac{1}{3}\left(\dfrac{1}{3}\right)\sqrt[3]{94 \cdot 9}$

$= \dfrac{\sqrt[3]{846}}{9} \approx 1.051$

26. $(14)^0 = 1$

27. $1^{12} = 1$

28. 0^0 = indeterminate

29. $-8^4 = -4096$

30. $(-8)^4 = 4096$

31. $\left(\dfrac{3}{4}\right)^{-2} = \left(\dfrac{4}{3}\right)^2 = \dfrac{4^2}{3^2} = \dfrac{16}{9}$

32. $0.00000305 = 3.05 \times 10^{-6}$

33. $-4,235,600 = -4.2356 \times 10^6$

34. 3.56 E–2 = 0.0356

35. 6.78 E8 = 678,000,000

36. $\sqrt[4]{3125} = 3125^{\frac{1}{4}} \approx 7.4767$

37. $\sqrt{-3^2 + 5(3) - 2} = \sqrt{-9 + 15 - 2} = \sqrt{4} = 2$

38. a. There are six terms in the expression.

   b. The variable terms are
   a^3, $-2a^2$, a, $-2a^3$, and $7a$.

   c. The only constant term is –5.

39. $-(3y + 2z) + (4y - 2z) - (-3y - 5z)$
$= (-3y - 2z) + (4y - 2z) + (3y + 5z)$
$= -3y + 4y + 3y - 2z - 2z + 5z$
$= 4y + z$

40. $\dfrac{3}{4}x + \dfrac{5}{8}y - \dfrac{1}{16} - \dfrac{3}{4}x + \dfrac{1}{8}y - \dfrac{5}{6}$

$= \left(\dfrac{3}{4} - \dfrac{3}{4}\right)x + \left(\dfrac{5}{8} + \dfrac{1}{8}\right)y - \dfrac{1}{16} - \dfrac{5}{6}$

$= 0x + \dfrac{6}{8}y - \dfrac{3}{48} - \dfrac{40}{48}$

$= \dfrac{3}{4}y - \dfrac{43}{48}$

41. $-(-2)^2 + 3(-2) + 8 \overset{?}{=} -3(-2)$
$-4 + (-6) + 8 \overset{?}{=} 6$
$-10 + 8 \overset{?}{=} 6$
$-2 \neq 6$
$x = -2$ is not a solution of the equation.

42.

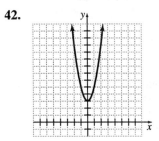

43. The domain of the function is all real numbers. The range of the function is all $y \geq 3$.

44. Since no vertical line will cross the graph more than once, the relation is a function.

45. The relation is increasing for all $x > 0$.

46. The relative minimum value of the relation is $y = 3$, which occurs when $x = 0$.

47. $f(x) = \dfrac{1}{3}x - 5$

   a. $f(9) = \dfrac{1}{3}(9) - 5$
   $f(9) = 3 - 5$
   $f(9) = -2$

   b. $f(a + h) = \dfrac{1}{3}(a + h) - 5$
   $f(a + h) = \dfrac{1}{3}a + \dfrac{1}{3}h - 5$

48. Volume is the product of length, width, and height.
$(3.5)(2.25)(1.75) = 13.78125$
The volume is 13.78125 ft^3.

49. $A = P(1+r)^t$

$A = 500(1+0.055)^4$

$A = 500(1.055)^4$

$A \approx 619.41$

The interest is the amount by which the compound amount exceed the original balance.

$619.41 - 500 = 119.41$

Kelsie's interest is $119.41.

50. The cost for one production run is the sum of the setup cost and the cost per ornament.

$c(x) = 35 + 2.80x$

$c(150) = 35 + 2.80(150)$

$c(150) = 35 + 420$

$c(150) = 455$

The cost of producing 150 ornaments in one production run is $455.

Chapter 4

4.1 Experiencing Algebra the Exercise Way

1. $6x - 55 = x + 72$ is linear because it is in the form $ax + b = ax + b$, where $a \neq 0$ in at least one of the expressions and the coefficient of a is not the same in each expression.

3. $4x^2 + 5 = 2x - 6$ is nonlinear because x has an exponent of 2.

5. $\dfrac{7}{9}z - \dfrac{2}{3} = 0$ is linear because it is in the form $ax + b = 0$.

7. $\sqrt[3]{4x + 16} = 27$ is nonlinear because the radical expression has a variable in the radicand.

9. $3(2x - 5) = x + 3(x - 9)$ is linear because it simplifies to $6x - 15 = 4x - 27$.

11. A sample table is shown below.

x	$2x - 7$	$35 - x$	
13	19	22	$19 < 22$
14	21	21	$21 = 21$
15	23	20	$23 > 20$

The solution is 14.

13. A sample table is shown below.

x	$3(2x + 11)$	$3(5 + x)$	
−5	3	0	$3 > 0$
−6	−3	−3	$-3 = -3$
−7	−9	−6	$-9 < -6$

The solution is −6.

15. A sample table is shown below.

a	$6.8a + 4.3$	$2.6a + 33.7$	
6	45.1	49.3	$45.1 < 49.3$
7	51.9	51.9	$51.9 = 51.9$
8	58.7	54.5	$58.7 > 54.5$

The solution is 7.

17. A sample table is shown below.

x	$7(x + 10) + 15$	$6(x + 15) + (x - 5)$	
0	85	85	$85 = 85$
1	92	92	$92 = 92$
2	99	99	$99 = 99$

The expressions are always equal. The equation is an identity. The solution is all real numbers.

19. A sample table is shown below.

a	$(a-4)-(a+4)$	$(a+3)-(a-2)$	
1	-8	5	$-8 \neq 5$
2	-8	5	$-8 \neq 5$
3	-8	5	$-8 \neq 5$

The expressions always equal the same value. The equation is a contradiction. There is no solution.

21. A sample table is shown below.

z	$3.5(z-1)$	$7(0.5z+0.6)+2$	
0	-3.5	6.2	$6.2-(-3.5)=9.7$
1	0	9.7	$9.7-0=9.7$
2	3.5	13.2	$13.2-3.5=9.7$

The values of the left and right expressions are never equal and their differences remain the same. The equation is a contradiction. There is no solution.

23. A sample table is shown below.

x	$\frac{4}{5}(x-1)$	$6\left(\frac{1}{15}x-\frac{1}{10}\right)$	
0	$-\frac{4}{5}$	$-\frac{3}{5}$	$-\frac{4}{5}<-\frac{3}{5}$
1	0	$-\frac{1}{5}$	$0>-\frac{1}{5}$

The solution is a noninteger between 0 and 1.

25. Y1 $= x+6$
Y2 $= 9+2x$

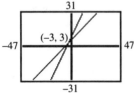

The intersection is $(-3, 3)$.
The solution is -3.

27. Y1 $= (x+4)+(x+2)$
Y2 $= (x-1)+(x-3)$

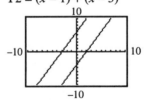

The lines are parallel.
There is no solution.

29. $Y1 = 2(x + 3$
$Y2 = 3(x - 1) - (x - 9)$

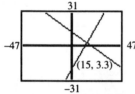

The lines are the same.
The solution is all real numbers.

31. $Y1 = 1.7x - 22.2$
$Y2 = 13.8 - 0.7x$

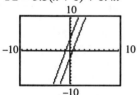

The intersection is $(15, 3.3)$.
The solution is 15.

33. $Y1 = 2.2(x - 1) + 1.7x$
$Y2 = 3.5(x + 1) + 0.4x$

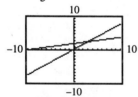

The lines are parallel.
There is no solution.

35. $Y1 = \dfrac{4}{5}x + \dfrac{1}{5}$

$Y2 = \dfrac{1}{5}x + 2$

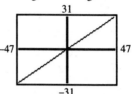

The intersection is $(3, 2.6)$.
The solution is 3.

37. $Y1 = \dfrac{2}{3}(x + 1) - \dfrac{1}{3}$

$Y2 = \dfrac{1}{3}(x + 1) + \dfrac{1}{3}x$

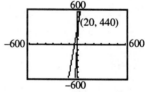

The lines are the same.
The solution is all real numbers.

39. Let x = number of miles
$49.95 = 29.95 + 0.25x$

x	49.95	$29.95 + 0.25x$	
79	49.95	49.7	$49.95 > 49.7$
80	49.95	49.95	$49.95 = 49.95$
81	49.95	50.2	$49.95 < 50.2$

The number of miles is 80.

41. Let x = number of pairs of shoes
$280 + 8x = 22x$
$Y1 = 280 + 8x$
$Y2 = 22x$

The intersection is $(20, 440)$.
The factory should produce 20 pairs of shoes.

43. Let $x =$ day 5 expenses

$$\frac{28 + 19 + 22 + 27 + x}{5} = 25$$

x	25	$\frac{96+x}{5}$	
28	25	24.8	$25 > 24.8$
29	25	25	$25 = 25$
30	25	25.2	$25 < 25.2$

She can spend $29.

45. Let $x =$ width,
then $x + 3 =$ length
$2x + 2(x + 3) = 26$
$Y1 = 2x + 2(x + 3)$
$Y2 = 26$

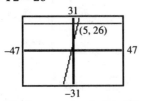

The intersection is (5, 26) so $x = 5$,
$x + 3 = 8$.
The dimensions are 5 ft by 8 ft.

47. Let $x =$ number of rolls
$25 + 9x + 5x = 25 + 14x$

x	$25 + 9x + 5x$	$25 + 14x$	
1	39	39	$39 = 39$
2	53	53	$53 = 53$
3	67	67	$67 = 67$

The values are always equal.
Any number of rolls will have the same
charge for both.

4.1 Experiencing Algebra the Calculator Way

A.

1. Graphs will vary.
The intersection is (22, 64).
The solution is 22.

2. Graphs will vary.
The intersection is (–12, –54.6).
The solution is –12.

3. Graphs will vary.
The intersection is (250, 2200).
The solution is 250.

4. Graphs will vary.
The intersection is (–500, –5200).
The solution is –500

B. Students should experiment with some
exercises.

4.2 Experiencing Algebra the Exercise Way

1. $x + 33 = 51$
$x + 33 - 33 = 51 - 33$
$x = 18$
The solution is 18.

3. $75 = a - 41$
$75 + 41 = a - 41 + 41$
$116 = a$
The solution is 116.

5. $-4.91 = y + 3.07$
$-4.91 - 3.07 = y + 3.07 - 3.07$
$-7.98 = y$
The solution is –7.98.

7. $a - \dfrac{13}{18} = \dfrac{5}{6}$

$a - \dfrac{13}{18} + \dfrac{13}{18} = \dfrac{15}{18} + \dfrac{13}{18}$

$a = \dfrac{28}{18} = \dfrac{14}{9}$

The solution is $\dfrac{14}{9}$.

9. $27 + (x - 13) = 11$
$27 + x - 13 = 11$
$14 + x = 11$
$14 + x - 14 = 11 - 14$
$x = -3$
The solution is –3.

11. $(13.9 + x) + 0.88 = -2.07$
$13.9 + x + 0.88 = -2.07$
$x + 14.78 = -2.07$
$x + 14.78 - 14.78 = -2.07 - 14.78$
$x = -16.85$
The solution is -16.85.

13. $\left(x - \dfrac{3}{10}\right) - \dfrac{2}{5} = -3\dfrac{1}{2}$

$x - \dfrac{3}{10} - \dfrac{4}{10} = -\dfrac{7}{2}$

$x - \dfrac{7}{10} = -\dfrac{35}{10}$

$x - \dfrac{7}{10} + \dfrac{7}{10} = -\dfrac{35}{10} + \dfrac{7}{10}$

$x = -\dfrac{28}{10} = -\dfrac{14}{5}$

The solution is $-\dfrac{14}{5}$.

15. $(5x - 2) - (4x + 7) = 27$
$5x - 2 - 4x - 7 = 27$
$x - 9 = 27$
$x - 9 + 9 = 27 + 9$
$x = 36$
The solution is 36.

17. $\left(\dfrac{1}{3}x + \dfrac{1}{8}\right) + \left(\dfrac{3}{4} + \dfrac{2}{3}x\right) = -\dfrac{3}{16}$

$\dfrac{1}{3}x + \dfrac{1}{8} + \dfrac{6}{8} + \dfrac{2}{3}x = -\dfrac{3}{16}$

$\dfrac{3}{3}x + \dfrac{7}{8} = -\dfrac{3}{16}$

$x + \dfrac{14}{16} - \dfrac{14}{16} = -\dfrac{3}{16} - \dfrac{14}{16}$

$x = -\dfrac{17}{16}$

The solution is $-\dfrac{17}{16}$.

19. $(3x + 76) - (2x - 45) = 31$
$3x + 76 - 2x + 45 = 31$
$x + 121 = 31$
$x + 121 - 121 = 31 - 121$
$x = -90$
The solution is -90.

21. $-324 = -4y$
$\dfrac{-324}{-4} = \dfrac{-4y}{-4}$
$81 = y$
The solution is 81.

23. $-5.1x = 0.102$
$\dfrac{-5.1x}{-5.1} = \dfrac{0.102}{-5.1}$
$x = -0.02$
The solution is -0.02.

25. $-3\dfrac{1}{3}x = -1\dfrac{1}{3}$

$-\dfrac{10}{3}x = -\dfrac{4}{3}$

$-\dfrac{3}{10}\left(-\dfrac{10}{3}x\right) = -\dfrac{3}{10}\left(-\dfrac{4}{3}\right)$

$x = \dfrac{2}{5}$

The solution is $\dfrac{2}{5}$.

27. $\dfrac{x}{4} = 1.22$

$4\left(\dfrac{x}{4}\right) = 4(1.22)$

$x = 4.88$
The solution is 4.88.

29. $-x = 57$
$-1(-x) = -1(57)$
$x = -57$
The solution is -57.

31. $57 = 2x + 17x$
$57 = 19x$
$\dfrac{57}{19} = \dfrac{19x}{19}$
$3 = x$
The solution is 3.

33. $18.22x - 12.9x = -12.76$
$5.32x = -12.76$
$$\frac{5.32x}{5.32} = \frac{-12.76}{5.32}$$
$x \approx -2.398$
The solution is about -2.398.

35. $\dfrac{5}{14} = \dfrac{9}{14}a + \dfrac{3}{7}a$
$\dfrac{5}{14} = \dfrac{9}{14}a + \dfrac{6}{14}a$
$\dfrac{5}{14} = \dfrac{15}{14}a$
$\dfrac{14}{15}\left(\dfrac{5}{14}\right) = \dfrac{14}{15}\left(\dfrac{15}{14}a\right)$
$\dfrac{1}{3} = a$

The solution is $\dfrac{1}{3}$.

37. $2(3x + 6) + 3(x - 4) = 126$
$6x + 12 + 3x - 12 = 126$
$9x = 126$
$$\frac{9x}{9} = \frac{126}{9}$$
$x = 14$
The solution is 14.

39. $2.2(x + 3.7) + 7.4(x - 1.1) = 60.48$
$2.2x + 8.14 + 7.4x - 8.14 = 60.48$
$9.6x = 60.48$
$$\frac{9.6x}{9.6} = \frac{60.48}{9.6}$$
$x = 6.3$
The solution is 6.3.

41. $3\left(\dfrac{1}{2}x - \dfrac{3}{4}\right) - 18\left(x - \dfrac{1}{8}\right) = 0$
$\dfrac{3}{2}x - \dfrac{9}{4} - 18x + \dfrac{18}{8} = 0$
$\dfrac{3}{2}x - \dfrac{36}{2}x - \dfrac{9}{4} + \dfrac{9}{4} = 0$
$-\dfrac{33}{2}x = 0$
$-\dfrac{2}{33}\left(-\dfrac{33}{2}x\right) = -\dfrac{2}{33}(0)$
$x = 0$
The solution is 0.

43. Let x = number of remaining servings
$4 + x = 10$
$4 + x - 4 = 10 - 4$
$x = 6$
6 servings remain in the box.

45. Let x = gross pay
$x - 567.32 = 1784.26$
$x - 567.32 + 567.32 = 1784.26 + 567.32$
$x = 2351.58$
His gross pay was $2351.58.

47. Let x = amount to borrow
$3\dfrac{3}{4} + x = 5\dfrac{1}{2}$
$\dfrac{15}{4} + x = \dfrac{11}{2}$
$\dfrac{15}{4} + x - \dfrac{15}{4} = \dfrac{22}{4} - \dfrac{15}{4}$
$x = \dfrac{7}{4} = 1\dfrac{3}{4}$

She should borrow $1\dfrac{3}{4}$ cups.

49. Let x = sales tax
$49.95 + x = 54.32$
$49.95 + x - 49.95 = 54.32 - 49.95$
$x = 4.37$
There was $4.37 sales tax.

51. Let x = number of feet of additional wallpaper
$35 + x = 2(18 + 22)$
$35 + x = 2(40)$
$35 + x = 80$
$35 + x - 35 = 80 - 35$
$x = 45$
She must buy 45 feet.
$2(20) = 40$ feet
$45 - 40 = 5$
No, she will not have enough.

53. Let x = total number of paid admissions

$0.55x = 264$

$$\dfrac{0.55x}{0.55} = \dfrac{264}{0.55}$$

$x = 480$

There were 480 paid admissions.

55. Let x = number of packets they must sell

$2.5x = 1450$

$$\dfrac{2.5x}{2.5} = \dfrac{1450}{2.5}$$

$x = 580$

They must sell 580 packets.

57. Let x = amount of estate

$\dfrac{3}{5}x = 45,240$

$$\dfrac{5}{3}\left(\dfrac{3}{5}x\right) = \dfrac{5}{3}(45,240)$$

$x = 75,400$

The estate was worth $75,400.

59. Let h = height

$20h = 350$

$$\dfrac{20h}{20} = \dfrac{350}{20}$$

$h = 17.5$

The height is 17.5 feet.

61. Let h = height

Volume $= \pi r^2 h$

$300 = \pi(4)^2 h$

$300 = 16\pi h$

$$\dfrac{300}{16\pi} = \dfrac{16\pi h}{16\pi}$$

$5.97 \approx h$

The height must be about 6 feet.

63. Let P = amount of principal

$864 = P(0.045)3$

$864 = 0.135P$

$$\dfrac{864}{0.135} = \dfrac{0.135P}{0.135}$$

$6400 = P$

You must place $6400 into savings.

65. Let x = amount of quarterly profits

$\dfrac{x}{5} = 12,730$

$$5\left(\dfrac{x}{5}\right) = 5(12,730)$$

$x = 63,650$

The quarterly profits were $63,650.

67. Let x = amount of sales

$0.03x = 40.5$

$$\dfrac{0.03x}{0.03} = \dfrac{40.5}{0.03}$$

$x = 1350$

Her sales were $1350.

4.2 Experiencing Algebra the Calculator Way

1. Let x = linear feet of fencing

$x + 2(5) = 2\pi(75)$

$x + 10 = 150\pi$

$x + 10 - 10 = 150\pi - 10$

$x \approx 461.2$

It will require about 461.2 feet.

2. Let A = total amount paid

$A = 8000(1 + 0.04)^5$

$A \approx 9733.22$

Let I = interest

$I = A - 8000$

$I = 1733.22$

She will pay $1733.22.

3. Let d = difference

$d + 25.8 = 25.8\pi$

$d + 25.8 - 25.8 = 25.8\pi - 25.8$

$d \approx 55.3$

The difference is about 55.3 inches.

4. Let P = amount of principal

$7325 = Pe^{3(0.045)}$

$$\dfrac{7325}{e^{0.135}} = \dfrac{Pe^{0.135}}{e^{0.135}}$$

$6400 \approx P$

About $6400 was invested.

5. Let x = amount of increase per share

$$250x = 531\frac{1}{4}$$

$$\frac{250x}{250} = \frac{531.25}{250}$$

$$x = 2.125 = 2\frac{1}{8}$$

The increase was $2\frac{1}{8}$ per share.

6. Let x = number of pieces

$$5\frac{3}{8}x = 34\frac{1}{2}$$

$$\frac{43}{8}x = \frac{69}{2}$$

$$\frac{8}{43}\left(\frac{43}{8}x\right) = \frac{8}{43}\left(\frac{69}{2}\right)$$

$$x = \frac{276}{43} \approx 6.4$$

You can cut 6 pieces.

7. Let x = gallons of fuel

$$18.5x = 220$$

$$\frac{18.5x}{18.5} = \frac{220}{18.5}$$

$$x \approx 11.89$$

The car will need about 12 gallons.

8. Let x = number of square feet of wall

$$6.5x = 800$$

$$\frac{6.5x}{6.5} = \frac{800}{6.5}$$

$$x \approx 123.1$$

About 123 square feet of wall can be constructed.

9. Let x = daily recommended amount of fat

$$0.04x = 2.5$$

$$\frac{0.04x}{0.04} = \frac{2.5}{0.04}$$

$$x = 62.5$$

The recommended amount of fat is 62.5 grams.

10. Let x = daily recommended amount of fat

$$0.05x = 3$$

$$\frac{0.05x}{0.05} = \frac{3}{0.05}$$

$$x = 60$$

The recommended amount is 60 grams. This does not agree with problem 9. The difference may be round-off error.

11. Let r = speed

$$250 = r(15)$$

$$\frac{250}{15} = \frac{15r}{15}$$

$$16.7 \approx r$$

The speed was about 16.7 feet per second.

4.3 Experiencing Algebra the Exercise Way

1. $4x + 8 = 0$

$4x + 8 - 8 = 0 - 8$

$4x = -8$

$$\frac{4x}{4} = \frac{-8}{4}$$

$x = -2$

The solution is –2.

3. $-3x + 7 = 7$

$-3x + 7 - 7 = 7 - 7$

$-3x = 0$

$$\frac{-3x}{-3} = \frac{0}{-3}$$

$x = 0$

The solution is 0.

5. $15.17 = 5.9x - 4.3$

$15.17 + 4.3 = 5.9x - 4.3 + 4.3$

$19.47 = 5.9x$

$$\frac{19.47}{5.9} = \frac{5.9x}{5.9}$$

$3.3 = x$

The solution is 3.3.

7. $6.1 = -0.55a + 6.1$

$6.1 - 6.1 = -0.55a + 6.1 - 6.1$

$0 = -0.55a$

$$\frac{0}{-0.55} = \frac{-0.55a}{-0.55}$$

$0 = a$

The solution is 0.

9. $-9.2x - 4.3 = -70.6$
$-9.2x - 4.3 + 4.3 = -70.6 + 4.3$
$-9.2x = -66.3$
$\dfrac{-9.2x}{-9.2} = \dfrac{-66.3}{-9.2}$
$x \approx 7.21$
The solution is about 7.21.

11. $-\dfrac{5}{9}b + \dfrac{11}{12} = \dfrac{23}{36}$
$36\left(-\dfrac{5}{9}b\right) + 36\left(\dfrac{11}{12}\right) = 36\left(\dfrac{23}{36}\right)$
$-20b + 33 = 23$
$-20b + 33 - 33 = 23 - 33$
$-20b = -10$
$\dfrac{-20b}{-20} = \dfrac{-10}{-20}$
$b = \dfrac{1}{2}$

The solution is $\dfrac{1}{2}$.

13. $-2\dfrac{2}{3}z - 3\dfrac{1}{2} = -8\dfrac{5}{6}$
$-\dfrac{8}{3}z - \dfrac{7}{2} = -\dfrac{53}{6}$
$6\left(-\dfrac{8}{3}z\right) - 6\left(\dfrac{7}{2}\right) = 6\left(-\dfrac{53}{6}\right)$
$-16z - 21 = -53$
$-16z - 21 + 21 = -53 + 21$
$-16z = -32$
$\dfrac{-16z}{-16} = \dfrac{-32}{-16}$
$z = 2$
The solution is 2.

15. $5x + 6 = x + 126$
$5x + 6 - 6 = x + 126 - 6$
$5x = x + 120$
$5x - x = x + 120 - x$
$4x = 120$
$\dfrac{4x}{4} = \dfrac{120}{4}$
$x = 30$
The solution is 30.

17. $27x - 49 = -12x - 10$
$27x - 49 + 49 = -12x - 10 + 49$
$27x = -12x + 39$
$27x + 12x = -12x + 39 + 12x$
$39x = 39$
$\dfrac{39x}{39} = \dfrac{39}{39}$
$x = 1$
The solution is 1.

19. $156z - 210 = 47z + 662$
$156z - 210 + 210 = 47z + 662 + 210$
$156z = 47z + 872$
$156z - 47z = 47z + 872 - 47z$
$109z = 872$
$\dfrac{109z}{109} = \dfrac{872}{109}$
$z = 8$
The solution is 8.

21. $4x - (3x + 5) = x - 5$
$4x - 3x - 5 = x - 5$
$x - 5 = x - 5$
$x - 5 - x = x - 5 - x$
$-5 = -5$
This is a true equation for all real numbers.

23. $6x - (x + 1) = 5x + 7$
$6x - x - 1 = 5x + 7$
$5x - 1 = 5x + 7$
$5x - 1 - 5x = 5x + 7 - 5x$
$-1 = 7$
This is a false equation.
There is no solution.

25. $5(0.3x + 8.7) = 1.5x + 43.5$
$1.5x + 43.5 = 1.5x + 43.5$
$1.5x + 43.5 - 1.5x = 1.5x + 43.5 - 1.5x$
$43.5 = 43.5$
This is a true equation for all real numbers.

27. $5.5x = 1.2x + 3.3(x - 2)$
$5.5x = 1.2x + 3.3x - 6.6$
$5.5x = 4.5x - 6.6$
$5.5x - 4.5x = 4.5x - 6.6 - 4.5x$
$x = -6.6$
The solution is -6.6.

29. $\dfrac{3}{4}x + 6 = \dfrac{1}{2}x + \dfrac{1}{4}x$

$\dfrac{3}{4}x + 6 = \dfrac{2}{4}x + \dfrac{1}{4}x$

$\dfrac{3}{4}x + 6 = \dfrac{3}{4}x$

$\dfrac{3}{4}x + 6 - \dfrac{3}{4}x = \dfrac{3}{4}x - \dfrac{3}{4}x$

$6 = 0$

This is a false equation.
There is no solution.

31. $3x - \dfrac{1}{4} = \left(x + \dfrac{1}{2}\right) + \left(2x + \dfrac{1}{3}\right)$

$3x - \dfrac{1}{4} = x + \dfrac{3}{6} + 2x + \dfrac{2}{6}$

$3x - \dfrac{1}{4} = 3x + \dfrac{5}{6}$

$3x - \dfrac{1}{4} - 3x = 3x + \dfrac{5}{6} - 3x$

$-\dfrac{1}{4} = \dfrac{5}{6}$

This is a false equation.
There is no solution.

33. $11x - 12 = 7(3x - 6) - 2(x + 9)$

$11x - 12 = 21x - 42 - 2x - 18$

$11x - 12 = 19x - 60$

$11x - 12 + 12 = 19x - 60 + 12$

$11x = 19x - 48$

$11x - 19x = 19x - 48 - 19x$

$-8x = -48$

$\dfrac{-8x}{-8} = \dfrac{-48}{-8}$

$x = 6$

The solution is 6.

35. $7x - 5(3x + 9) = -2(4x + 35)$

$7x - 15x - 45 = -8x - 70$

$-8x - 45 = -8x - 70$

$-8x - 45 + 8x = -8x - 70 + 8x$

$-45 = -70$

This is a false equation.
There is no solution.

37. $\dfrac{1}{4}x + \dfrac{5}{9} = \dfrac{5}{6}$

$36\left(\dfrac{1}{4}x\right) + 36\left(\dfrac{5}{9}\right) = 36\left(\dfrac{5}{6}\right)$

$9x + 20 = 30$

$9x + 20 - 20 = 30 - 20$

$9x = 10$

$\dfrac{9x}{9} = \dfrac{10}{9}$

$x = \dfrac{10}{9}$

The solution is $\dfrac{10}{9}$.

39. $3x + \dfrac{1}{4} = 2x + \dfrac{7}{36}$

$36(3x) + 36\left(\dfrac{1}{4}\right) = 36(2x) + 36\left(\dfrac{7}{36}\right)$

$108x + 9 = 72x + 7$

$108x + 9 - 9 = 72x + 7 - 9$

$108x = 72x - 2$

$108x - 72x = 72x - 2 - 72x$

$36x = -2$

$\dfrac{36x}{36} = \dfrac{-2}{36}$

$x = -\dfrac{1}{18}$

The solution is $-\dfrac{1}{18}$.

41. $\dfrac{2}{5}b - 12 = \dfrac{2}{3}b + 20$

$15\left(\dfrac{2}{5}b\right) - 15(12) = 15\left(\dfrac{2}{3}b\right) + 15(20)$

$6b - 180 = 10b + 300$

$6b - 180 + 180 = 10b + 300 + 180$

$6b = 10b + 480$

$6b - 10b = 10b + 480 - 10b$

$-4b = 480$

$\dfrac{-4b}{-4} = \dfrac{480}{-4}$

$b = -120$

The solution is -120.

43. $\frac{3}{4}\left(x+\frac{4}{5}\right) \pm -\frac{7}{8}x - \frac{2}{5}$

$\frac{3}{4}x + \frac{3}{5} = -\frac{7}{8}x - \frac{2}{5}$

$40\left(\frac{3}{4}x\right) + 40\left(\frac{3}{5}\right) = 40\left(-\frac{7}{8}x\right) - 40\left(\frac{2}{5}\right)$

$30x + 24 = -35x - 16$

$30x + 24 - 24 = -35x - 16 - 24$

$30x = -35x - 40$

$30x + 35x = -35x - 40 + 35x$

$65x = -40$

$\frac{65x}{65} = \frac{-40}{65}$

$x = -\frac{8}{13}$

The solution is $-\frac{8}{13}$.

45. $0.05x + 10.5 = 0.15x - 0.25$

$100(0.05x) + 100(10.5)$
$\qquad = 100(0.15x) - 100(0.25)$

$5x + 1050 = 15x - 25$

$5x + 1050 - 1050 = 15x - 25 - 1050$

$5x = 15x - 1075$

$5x - 15x = 15x - 1075 - 15x$

$-10x = -1075$

$\frac{-10x}{-10} = \frac{-1075}{-10}$

$x = \frac{1075}{10} = 107.5$

The solution is 107.5.

47. $21.1x + 0.46 = 10.9x + 0.46$

$100(21.1x) + 100(0.46)$
$\qquad = 100(10.9x) + 100(0.46)$

$2110x + 46 = 1090x + 46$

$2110x + 46 - 46 = 1090x + 46 - 46$

$2110x = 1090x$

$2110x - 1090x = 1090x - 1090x$

$1020x = 0$

$\frac{1020x}{1020} = \frac{0}{1020}$

$x = 0$

The solution is 0.

49. $15.2y - 175.43 = -2.4y - 176.31$

$100(15.2y) - 100(175.43)$
$\qquad = 100(-2.4y) - 100(176.31)$

$1520y - 17,543 = -240y - 17,631$

$1520y - 17,543 + 17,543$
$\qquad = -240y - 17,631 + 17,543$

$1520y = -240y - 88$

$1520y + 240y = -240y - 88 + 240y$

$1760y = -88$

$\frac{1760y}{1760} = \frac{-88}{1760}$

$y = -0.05$

The solution is -0.05.

51. Let $x =$ number of miles

$49.95 + 0.12x = 200$

$49.95 + 0.12x - 49.95 = 200 - 49.95$

$0.12x = 150.05$

$\frac{0.12x}{0.12} = \frac{150.05}{0.12}$

$x \approx 1250.4$

You could drive about 1250 miles.

53. Let $x =$ monthly payment

$24x = 2252 - 200$

$24x = 2052$

$\frac{24x}{24} = \frac{2052}{24}$

$x = 85.5$

The monthly payments would be \$85.50.

55. Let $x =$ number of liters of 30% solution

$4(0.10) + x(0.30) = 0.25(4 + x)$

$0.4 + 0.3x = 1 + 0.25x$

$100(0.4) + 100(0.3x) = 100(1) + 100(0.25)$

$40 + 30x = 100 + 25x$

$40 + 30x - 40 = 100 + 25x - 40$

$30x = 60 + 25x$

$30x - 25x = 60 + 25x - 25x$

$5x = 60$

$\frac{5x}{5} = \frac{60}{5}$

$x = 12$

There were 12 liters of 30% solution.

57. Let x = brother's earnings
$25 + 2x = 730.10$
$25 + 2x - 25 = 730.1 - 25$
$2x = 705.1$
$\dfrac{2x}{2} = \dfrac{705.1}{2}$
$x = 352.55$
Her brother's average weekly earnings are $352.55.

59. Let x = amount of sales
$150 + 0.10x - 0.02x = 200 + 0.06x + 0.02x$
$150 + 0.8x = 200 + 0.08x$
$150 + 0.08x - 0.08x = 200 + 0.08x - 0.08x$
$150 = 200$
This is a false equation. There is no solution. There is no value of sales for which the two are equal.

61. Let x = total sales
$300 + 0.04x = 700 + 0.04(x - 10,000)$
$300 + 0.04x = 700 + 0.04x - 400$
$300 + 0.04x = 300 + 0.04x$
$300 + 0.04x - 0.04x = 300 + 0.04x - 0.04x$
$300 = 300$
This is a true equation. The solution is all real numbers. The plans are the same for any amount of total sales that exceeds $10,000.

4.3 Experiencing Algebra the Calculator Way

1. Let x = total sales
$0.0375x + 25 = 49.48$
$0.0375x + 25 - 25 = 49.48 - 25$
$0.0375x = 24.48$
$\dfrac{0.0375x}{0.0375} = \dfrac{24.48}{0.0375}$
$x = 652.8$
The sales were $652.80.

2. Let x = number of 120-grain tablets
$200 + 120x = 620$
$200 + 120x - 200 = 620 - 200$
$120x = 420$
$\dfrac{120x}{120} = \dfrac{420}{120}$
$x = 3.5$
She should administer 3.5 tablets.

3. Let x = number of sheets of wood
$2 + \dfrac{3}{4}x = 4\dfrac{1}{4}$
$2 + \dfrac{3}{4}x - 2 = 4\dfrac{1}{4} - 2$
$\dfrac{3}{4}x = 2\dfrac{1}{4}$
$\dfrac{3}{4}x = \dfrac{9}{4}$
$\dfrac{4}{3}\left(\dfrac{3}{4}x\right) = \dfrac{4}{3}\left(\dfrac{9}{4}\right)$
$x = 3$
He needs 3 sheets.

4.4 Experiencing Algebra the Exercise Way

1. $P = 4s$, for s
$\dfrac{P}{4} = \dfrac{4s}{4}$
$\dfrac{P}{4} = s$ or $s = \dfrac{P}{4}$

3. $C = \pi d$, for d
$\dfrac{C}{\pi} = \dfrac{\pi d}{\pi}$
$\dfrac{C}{\pi} = d$ or $d = \dfrac{C}{\pi}$

5. $V = LWH$, for L
$\dfrac{V}{WH} = \dfrac{LWH}{WH}$
$\dfrac{V}{WH} = L$ or $L = \dfrac{V}{WH}$

7. $S = 2LW + 2LH + 2WH$, for L
$S - 2WH = 2LW + 2LH + 2WH - 2WH$
$S - 2WH = 2LW + 2LH$
$S - 2WH = L(2W + 2H)$
$\dfrac{S - 2WH}{2W + 2H} = \dfrac{L(2W + 2H)}{2W + 2H}$
$\dfrac{S - 2WH}{2W + 2H} = L$ or $L = \dfrac{S - 2WH}{2W + 2H}$

9. $V = \pi r^2 h$, for h

$$\frac{V}{\pi r^2} = \frac{\pi r^2 h}{\pi r^2}$$

$$\frac{V}{\pi r^2} = h \text{ or } h = \frac{V}{\pi r^2}$$

11. $F = \frac{9}{5}C + 32$, for C

$$F - 32 = \frac{9}{5}C + 32 - 32$$

$$F - 32 = \frac{9}{5}C$$

$$\frac{5}{9}(F - 32) = \frac{5}{9}\left(\frac{9}{5}C\right)$$

$$\frac{5}{9}(F - 32) = C \text{ or } C = \frac{5}{9}(F - 32)$$

13. $I = Prt$, for P

$$\frac{I}{rt} = \frac{Prt}{rt}$$

$$\frac{I}{rt} = P \text{ or } P = \frac{I}{rt}$$

15. $A = P(1 + i)^t$, for P

$$\frac{A}{(1+i)^t} = \frac{P(1+i)^t}{(1+i)^t}$$

$$\frac{A}{(1+i)^t} = P \text{ or } P = \frac{A}{(1+i)^t}$$

17. $v = gt$, for g

$$\frac{v}{t} = \frac{gt}{t}$$

$$\frac{v}{t} = g \text{ or } g = \frac{v}{t}$$

19. $I = \frac{V}{R}$, for R

$$RI = R\left(\frac{V}{R}\right)$$

$$RI = V$$

$$\frac{RI}{I} = \frac{V}{I}$$

$$R = \frac{V}{I}$$

21. $z = \frac{x - m}{s}$, for m

$$zs = s\left(\frac{x - m}{s}\right)$$

$$zs = x - m$$

$$zs - x = x - m - x$$

$$zs - x = -m$$

$$\frac{zs - x}{-1} = \frac{-m}{-1}$$

$$-zs + x = m \text{ or } m = x - zs$$

23. $4x + 3y = 0$

$$4x + 3y - 4x = 0 - 4x$$

$$3y = -4x$$

$$\frac{3y}{3} = \frac{-4x}{3}$$

$$y = -\frac{4}{3}x$$

25. $-5x + 10y = 0$

$$-5x + 10y + 5x = 0 + 5x$$

$$10y = 5x$$

$$\frac{10y}{10} = \frac{5x}{10}$$

$$y = \frac{1}{2}x$$

27. $-x - y = 0$

$$-x - y + x = 0 + x$$

$$-y = x$$

$$\frac{-y}{-1} = \frac{x}{-1}$$

$$y = -x$$

29. $5x + 4y = 20$

$$5x + 4y - 5x = 20 - 5x$$

$$4y = 20 - 5x$$

$$\frac{4y}{4} = \frac{20 - 5x}{4}$$

$$y = \frac{20}{4} - \frac{5}{4}x$$

$$y = 5 - \frac{5}{4}x$$

$$y = -\frac{5}{4}x + 5$$

31. $-x - y = 7$
$-x - y + x = 7 + x$
$-y = 7 + x$
$\dfrac{-y}{-1} = \dfrac{7 + x}{-1}$
$y = -7 - x$
$y = -x - 7$

33. $7x - 14y = -28$
$7x - 14y - 7x = -28 - 7x$
$-14y = -28 - 7x$
$\dfrac{-14y}{-14} = \dfrac{-28 - 7x}{-14}$
$y = \dfrac{-28}{-14} - \dfrac{7x}{-14}$
$y = 2 + \dfrac{1}{2}x$
$y = \dfrac{1}{2}x + 2$

35. $-x + y = -1$
$-x + y + x = -1 + x$
$y = -1 + x$
$y = x - 1$

37. $y - 5 = 4(x - 6)$
$y - 5 = 4x - 24$
$y - 5 + 5 = 4x - 24 + 5$
$y = 4x - 19$

39. $y + 6 = -2(x - 7)$
$y + 6 = -2x + 14$
$y + 6 - 6 = -2x + 14 - 6$
$y = -2x + 8$

41. $y + 2 = -1(x + 4)$
$y + 2 = -x - 4$
$y + 2 - 2 = -x - 4 - 2$
$y = -x - 6$

43. $y - 4 = \dfrac{2}{3}(x + 9)$
$y - 4 = \dfrac{2}{3}x + 6$
$y - 4 + 4 = \dfrac{2}{3}x + 6 + 4$
$y = \dfrac{2}{3}x + 10$

45. $y + \dfrac{5}{9} = -\dfrac{2}{3}\left(x - \dfrac{1}{3}\right)$
$y + \dfrac{5}{9} = -\dfrac{2}{3}x + \dfrac{2}{9}$
$y + \dfrac{5}{9} - \dfrac{5}{9} = -\dfrac{2}{3}x + \dfrac{2}{9} - \dfrac{5}{9}$
$y = -\dfrac{2}{3}x - \dfrac{3}{9}$
$y = -\dfrac{2}{3}x - \dfrac{1}{3}$

47. $y = 5x + 15$
$y - 15 = 5x + 15 - 15$
$y - 15 = 5x$
$\dfrac{y - 15}{5} = \dfrac{5x}{5}$
$\dfrac{y}{5} - \dfrac{15}{5} = x$
$\dfrac{y}{5} - 3 = x \text{ or } x = \dfrac{1}{5}y - 3$

49. $y = -2x + 8$
$y - 8 = -2x + 8 - 8$
$y - 8 = -2x$
$\dfrac{y - 8}{-2} = \dfrac{-2x}{-2}$
$\dfrac{y}{-2} - \dfrac{8}{-2} = x$
$-\dfrac{1}{2}y + 4 = x \text{ or } x = -\dfrac{1}{2}y + 4$

51. $y = -\dfrac{6}{7}x + 8$
$y - 8 = -\dfrac{6}{7}x + 8 - 8$
$y - 8 = -\dfrac{6}{7}x$
$-\dfrac{7}{6}(y - 8) = -\dfrac{7}{6}\left(-\dfrac{6}{7}x\right)$
$-\dfrac{7}{6}y + \dfrac{28}{3} = x \text{ or } x = -\dfrac{7}{6}y + \dfrac{28}{3}$

53. $y = \dfrac{5}{9}x - \dfrac{2}{3}$

$y + \dfrac{2}{3} = \dfrac{5}{9}x - \dfrac{2}{3} + \dfrac{2}{3}$

$y + \dfrac{2}{3} = \dfrac{5}{9}x$

$\dfrac{9}{5}\left(y + \dfrac{2}{3}\right) = \dfrac{9}{5}\left(\dfrac{5}{9}x\right)$

$\dfrac{9}{5}y + \dfrac{6}{5} = x$ or $x = \dfrac{9}{5}y + \dfrac{6}{5}$

55. $y = mx + b$

$y - b = mx + b - b$

$y - b = mx$

$\dfrac{y - b}{m} = \dfrac{mx}{m}$

$\dfrac{y}{m} - \dfrac{b}{m} = x$

$\dfrac{1}{m}y - \dfrac{b}{m} = x$ or $x = \dfrac{1}{m}y - \dfrac{b}{m}$

57. $P = 200 + 85m$

$P - 200 = 200 + 85m - 200$

$P - 200 = 85m$

$\dfrac{P - 200}{85} = \dfrac{85m}{85}$

$\dfrac{1}{85}P - \dfrac{40}{17} = m$ or $m = \dfrac{1}{85}P - \dfrac{40}{17}$

$m = \dfrac{1}{85}(2240) - \dfrac{40}{17}$

$m = 24$

It will take 24 months to pay off \$2240.

$m = \dfrac{1}{85}(1200) - \dfrac{40}{17}$

$m \approx 11.76$

It will take 12 months to pay off \$1200.

59. $T = 22c + 12.50$

$T - 12.50 = 22c + 12.50 - 12.50$

$T - 12.50 = 22c$

$\dfrac{T - 12.50}{22} = \dfrac{22c}{22}$

$\dfrac{1}{22}T - \dfrac{12.5}{22} = c$ or $c = \dfrac{1}{22}T - \dfrac{25}{44}$

$c = \dfrac{1}{22}(35) - \dfrac{25}{44}$

$c \approx 1.02$

She can spend about \$1.02 on each student for a \$35 party.

$c = \dfrac{1}{22}(50) - \dfrac{25}{44}$

$c \approx 1.70$

She can spend about \$1.70 on each student for a \$50 party.

61. $B = 75 + 3T$

$B - 75 = 75 + 3T - 75$

$B - 75 = 3T$

$\dfrac{B - 75}{3} = \dfrac{3T}{3}$

$\dfrac{1}{3}B - 25 = T$ or $T = \dfrac{1}{3}B - 25$

$T = \dfrac{1}{3}(725) - 25$

$T \approx 216.67$

Ted's weekly earnings are about \$216.67 when his boss averages \$725 per week.

$T = \dfrac{1}{3}(1275) - 25$

$T = 400$

Ted's weekly earnings are \$400 when his boss averages \$1275 per week.

63. $C = 75 + 65d$
$C - 75 = 75 + 65d - 75$
$C - 75 = 65d$
$$\frac{C-75}{65} = \frac{65d}{65}$$
$\frac{1}{65}C - \frac{15}{13} = d$ or $d = \frac{1}{65}C - \frac{15}{13}$
$d = \frac{1}{65}(250) - \frac{15}{13}$
$d \approx 2.69$
The equipment could be rented for 2 days for under $250.
$d = \frac{1}{65}(400) - \frac{15}{13}$
$d = 5$
The equipment could be rented for 5 days for $400.

65. $V = 3(5)h$
$V = 15h$
$$\frac{V}{15} = \frac{15h}{15}$$
$\frac{V}{15} = h$ or $h = \frac{V}{15}$
$h = \frac{60}{15}$
$h = 4$
The height should be 4 feet for a volume of 60 cubic feet.
$h = \frac{100}{15}$
$h = 6\frac{2}{3}$

The height should be $6\frac{2}{3}$ feet for a volume of 100 cubic feet.

4.4 Experiencing Algebra the Calculator Way

$A = Pe^{rt}$
$$\frac{A}{e^{rt}} = \frac{Pe^{rt}}{e^{rt}}$$
$\frac{A}{e^{rt}} = P$ or $P = \frac{A}{e^{rt}}$
$P = \frac{10,000}{e^{5(0.045)}}$
$P \approx 7985.16$
The amount to invest is about $7985.16 if you want $10,000 at the end of 5 years at 4.5% interest.

A	t	r	P
$10,000	5	4.5%	$7985.16
$10,000	7	4.5%	$7297.89
$10,000	10	4.5%	$6376.28
$10,000	12	4.5%	$5827.48
$25,000	5	7%	$17,617.20
$25,000	7	7%	$15,315.66
$25,000	10	7%	$12,414.63
$25,000	12	7%	$10,792.76

4.5 Experiencing Algebra the Exercise Way

Method of solution may vary.

1. Let x = number of grains in first dosage
$x + 2$ = number of grains in second dosage
$x + 4$ = number of grains in third dosage
$x + x + 2 + x + 4 = 24$
$Y1 = x + x + 2 + x + 4$
$Y2 = 24$

X	Y₁	Y₂
1	9	24
2	12	24
3	15	24
4	18	24
5	21	24
6	24	24
7	27	24

X=7

The solution is 6.
$x = 6$
$x + 2 = 8$
$x + 4 = 10$
The doses are 6 grains, 8 grains, and 10 grains.

3. Let x = number of prizes in first stage
$x + 2$ = number of prizes in second stage
$x + 4$ = number of prizes in third stage
$x + 6$ = number of prizes in fourth stage
$x + x + 2 + x + 4 + x + 6 = 24$
$Y1 = x + x + 2 + x + 4 + x + 6$
$Y2 = 24$

X	Y1	Y2
1	16	24
2	20	24
3	24	24
4	28	24
5	32	24
6	36	24
7	40	24

X=1

$x = 3$
$x + 2 = 5$
$x + 4 = 7$
$x + 6 = 9$
The number awarded at each stage is
3 prizes, 5 prizes, 7 prizes, and 9 prizes.

5. Let x = lowest grade
$x + 1$ = second lowest grade
$\vdots$
$x + 7$ = highest grade
$x + x + 1 + x + 2 + x + 3 + x + 4 + x + 5 + x$
$\qquad + 6 + x + 7 = 676$
$8x + 28 = 676$
$8x + 28 - 28 = 676 - 28$
$8x = 648$
$\dfrac{8x}{8} = \dfrac{648}{8}$
$x = 81$
$x + 7 = 88$
The lowest grade was 81 points and the highest grade was 88 points.

7. Let x = first angle measurement
$x + 10$ = second angle measurement
$x + 20$ = third angle measurement
$x + x + 10 + x + 20 = 180$
$3x + 30 = 180$
$3x + 30 - 30 = 180 - 30$
$3x = 150$
$\dfrac{3x}{3} = \dfrac{150}{3}$
$x = 50$
$x + 10 = 60$
$x + 20 = 70$
The angles measure 50°, 60°, and 70°.

9. Let x = measure of each side
$x + x + x = 29\dfrac{1}{4}$
$Y1 = x + x + x$
$Y2 = 29\dfrac{1}{4}$

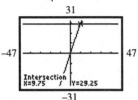

The intersection is (9.75, 29.25).
The solution is 9.75 or $9\dfrac{3}{4}$. Each side
measures $9\dfrac{3}{4}$ inches.

11. Let x = length of the first side
x = length of the second side
$\dfrac{2}{3}x$ = length of the third side

$x + x + \dfrac{2}{3}x = 16$

$Y1 = \dfrac{8}{3}x$
$Y2 = 16$

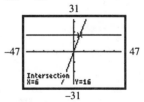

$x = 6$
$\dfrac{2}{3}x = 4$
The sides measure 6 feet, 6 feet, and 4 feet.

13. Let x = width
$x + 30$ = length
$2x + 2(x + 30) = 400$
$2x + 2x + 60 = 400$
$4x + 60 = 400$
$4x + 60 - 60 = 400 - 60$
$4x = 340$
$\dfrac{4x}{4} = \dfrac{340}{4}$
$x = 85$
$x + 30 = 115$
The dimensions are 85 yards by 115 yards.

15. Let x = length
$0.55x$ = width
$2x + 2(0.55x) = 294.5$
$2x + 1.1x = 294.5$
$3.1x = 294.5$
$\dfrac{3.1x}{3.1} = \dfrac{294.5}{3.1}$
$x = 95$
$0.55x = 52.25$
The dimensions are 95 cm by 52.25 cm.

17. Let x = width
$5x$ = length
$2x + 2(5x) = 96$
$2x + 10x = 96$
$12x = 96$
$\dfrac{12x}{12} = \dfrac{96}{12}$
$x = 8$
$5x = 40$
The dimensions should be 8 feet by 40 feet.
Area = $8(40) = 320$
It will cover 320 square feet of yard.

19. Let x = width
$5x$ = length
$2x + 5x = 96$
$7x = 96$
$\dfrac{7x}{7} = \dfrac{96}{7}$
$x \approx 13.7$
$5x \approx 68.5$
The dimensions are about 13.7 feet by 68.6 feet.
Area $\approx (13.7)(68.6) \approx 938$ ft^2
The area of this run is larger than the run in exercise 17.

21. Let $3x$ = first angle measurement
x = second angle measurement
$x - 5$ = third angle measurement
$3x + x + x - 5 = 180$
$5x - 5 = 180$
$5x - 5 + 5 = 180 + 5$
$5x = 185$
$\dfrac{5x}{5} = \dfrac{185}{5}$
$x = 37$
$3x = 111$
$x - 5 = 32$
The angles measure 37°, 111°, and 32°.

23. $I = PRT$
$I = P(0.125)(1)$
$I = 0.125P$
$P + 0.125P = 4500$
$1.125P = 4500$
$\dfrac{1.125P}{1.125} = \dfrac{4500}{1.125}$
$P = 4000$
$0.125P = 500$
The amount borrowed was $4000.
The interest was $500.

25. $I = PRT$
$I = P(0.09)(1)$
$I = 0.09P$
$P + 0.09P = 5000$
$1.09P = 5000$
$\dfrac{1.09P}{1.09} = \dfrac{5000}{1.09}$
$P = 4587.156$
You must invest about $4587.16.

27. $I = PRT$
$I = P(0.10)(1)$
$I = 0.1P$
$P + 0.1P = 500,000$
$1.1P = 500,000$
$\dfrac{1.1P}{1.1} = \dfrac{500,000}{1.1}$
$P \approx 454,545$
He should invest about \$454,545.

29. Let x = amount invested at 8%
$15,000 - x$ = amount invested at 6.5%
$0.08x + 0.065(15,000 - x) = 1117.50$
$0.08x + 975 - 0.065x = 1117.5$
$975 + 0.015x = 1117.5$
$975 + 0.015x - 975 = 1117.5 - 975$
$0.015x = 142.5$
$\dfrac{0.015x}{0.015} = \dfrac{142.5}{0.015}$
$x = 9500$
$15,000 - x = 5500$
\$9500 was invested at 8%.
\$5500 was invested at 6.5%.

31. $A = P + 0.07P$
$A = 1.07P$
$\dfrac{A}{1.07} = \dfrac{1.07P}{1.07}$
$\dfrac{A}{1.07} = P$ or $P = \dfrac{A}{1.07}$
$P = \dfrac{1350}{1.07} \approx 1261.68$
About \$1261.68 should be invested to have \$1350.
$P = \dfrac{2500}{1.07} \approx 2336.45$
About \$2336.45 should be invested to have \$2500.

33. Let x = original price
$x - 0.20x = 68$
$0.8x = 68$
$\dfrac{0.8x}{0.8} = \dfrac{68}{0.8}$
$x = 85$
The original price was \$85.

35. Let x = original cost
$x + 0.6x = 19.95$
$1.6x = 19.95$
$\dfrac{1.6x}{1.6} = \dfrac{19.95}{1.6}$
$x \approx 12.47$
The original cost was about \$12.47.

37. Let x = markup percentage
$x = \dfrac{17.10 - 9.50}{9.50}$
$x = \dfrac{7.6}{9.5}$
$x = 0.8$ or 80%
The markup percentage was 80%.

39. Let x = regular price
$x - 0.25x = 195$
$0.75x = 195$
$\dfrac{0.75x}{0.75} = \dfrac{195}{0.75}$
$x = 260$
The regular price is \$260.

41. Let x = SRP
$x - 0.35x = 12,500$
$0.65x = 12,500$
$\dfrac{0.65x}{0.65} = \dfrac{12,500}{0.65}$
$x \approx 19,230.77$
The SRP should be about \$19,230.77.

43. $y = x - 0.1x$
$y = 0.9x$
$\dfrac{y}{0.9} = \dfrac{0.9x}{0.9}$
$\dfrac{y}{0.9} = x$ or $x = \dfrac{y}{0.9}$
$x = \dfrac{53.96}{0.9} \approx 59.96$
The original price is about \$59.96 if the sale price is \$53.96.
$x = \dfrac{98.95}{0.9} \approx 109.94$
The original price is about \$109.94 if the sale price is \$98.95.

45. Let x = hourly wage before increase

$x + 0.022x = 13.75$

$1.022x = 13.75$

$\dfrac{1.022x}{1.022} = \dfrac{13.75}{1.022}$

$x \approx 13.45$

His hourly wage before the increase was about $13.45.

47. Let x = amount of bill before gratuity

$x + 0.15x = 143.24$

$1.15x = 143.24$

$\dfrac{1.15x}{1.15} = \dfrac{143.24}{1.15}$

$x \approx 124.56$

The bill before the gratuity was added was about $124.56.

4.5 Experiencing Algebra the Calculator Way

1.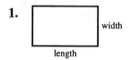

2. Let x = length in inches

3.

4. $P = 2L + 2W$

$323.61 = 2x + 2Gx$

5. $323.61 = x(2 + 2G)$

$\dfrac{323.61}{2 + 2G} = \dfrac{x(2 + 2G)}{2 + 2G}$

$\dfrac{323.61}{2 + 2G} = x$

```
(√(5)-1)/2→G:323
.61/(2+2G)
        100.0009895
```

$x \approx 100$

The length is about 100 inches.

6.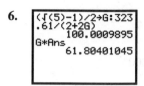
```
(√(5)-1)/2→G:323
.61/(2+2G)
        100.0009895
G*Ans
        61.80401045
```

$Gx \approx 61.804$

The width is about 61.804 inches.

7. $P = 2L + 2W$

323.61	$2L + 2W$
323.61	$2(100) + 2(61.804)$
323.61	323.608

The perimeter is 323.61 inches (after rounding).

Chapter 4 Review

Reflections

1–8. Answers will vary.

Exercises

1. $4x^2 - 2x + 1 = 0$ is nonlinear because x has an exponent of 2.

2. $5x + 3 = x - 4$ is linear because it is in the form $ax + b = ax + b$.

3. $5.7x - 8.2(x + 46) = 0$ is linear because it simplifies to $-2.5x - 37.72 = 0$ which is in the form $ax + b = c$.

4. $\sqrt{x} + 2 = 3x - 6$ is nonlinear because there is a radical expression with a variable in its radicand.

5. $\dfrac{1}{8}x + \dfrac{3}{4} = \dfrac{11}{16}$ is linear because it is in the form $ax + b = c$.

6. $\dfrac{3}{x} + 5 = 12x$ is nonlinear because there is a variable in the denominator of a term.

7.

x	$4x + 7$	$2x - 5$	
-7	-21	-19	$-21 < -19$
-6	-17	-17	$-17 = -17$
-5	-13	-15	$-13 > -15$

The solution is -6.

8.

x	$2.4x - 9.6$	4.8	
5	2.4	4.8	$2.4 < 4.8$
6	4.8	4.8	$4.8 = 4.8$
7	7.2	4.8	$7.2 > 4.8$

The solution is 6.

9.

x	$\frac{3}{5}x - \frac{7}{10}$	$\frac{1}{5}x + \frac{1}{2}$	
2	0.5	0.9	$0.5 < 0.9$
3	1.1	1.1	$1.1 = 1.1$
4	1.7	1.3	$1.7 > 1.3$

The solution is 3.

10.

x	$14x + 12$	$11(x - 5) + 60$	
-3	-30	-28	$-30 < -28$
-2	-16	-17	$-16 > -17$

The solution is a noninteger between -3 and -2.

11.

x	$3(x - 2) - 1$	$4(x - 1) - (x + 3)$	
0	-7	-7	$-7 = -7$
1	-4	-4	$-4 = -4$
2	-1	-1	$-1 = -1$

The values of the expressions are always equal. The equation is an identity. The solution is all real numbers.

12.

x	$2(2x + 1) + x$	$3(2x - 1) - (x + 1)$	
0	2	-4	$2 - (-4) = 6$
1	7	1	$7 - 1 = 6$
2	12	6	$12 - 6 = 6$

The difference in the values of the expressions is a constant, 6. There is no solution.

13. $2x - 2 = -x + 4$
$Y1 = 2x - 2$
$Y2 = -x + 4$

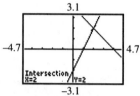

The graphs intersect at (2, 2).
The solution is 2.

14. $\frac{1}{2}x - 2 = -\frac{1}{3}x - \frac{11}{3}$

$Y1 = \frac{1}{2}x - 2$

$Y2 = -\frac{1}{3}x - \frac{11}{3}$

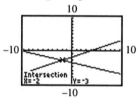

The graphs intersect at (−2, −3).
The solution is −2.

15. $(x + 3) + (x + 1) = 3(x + 1) - (x - 1)$
$Y1 = (x + 3) + (x + 1)$
$Y2 = 3(x + 1) - (x - 1)$

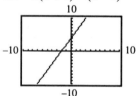

The graphs are the same.
The solution is all real numbers.

16. $(x + 6) - 3(x + 1) = (2x + 5) - 2(2x + 3)$
$Y1 = (x + 6) - 3(x + 1)$
$Y2 = (2x + 5) - 2(2x + 3)$

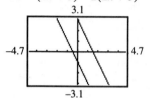

The graphs are parallel. There is no solution.

17. $1.2x + 0.72 = -2.1x + 8.64$
$Y1 = 1.2x + 0.72$
$Y2 = -2.1x + 8.64$

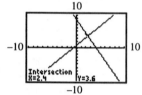

The graphs intersect at (2.4, 3.6). The
solution is 2.4.

18. Let $x =$ the number of hours worked
$90 = 25 + 6.5x$
$Y1 = 90$
$Y2 = 25 + 6.5x$

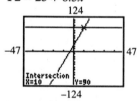

The graphs intersect at (10, 90).
The solution is 10.
The two offers are equivalent at 10 hours.

19. Let x = the amount of the fourth week
donation

$$\frac{2200 + 1750 + 1885 + x}{4} = 2000$$

$$Y1 = \frac{2200 + 1750 + 1885 + x}{4}$$

$Y2 = 2000$

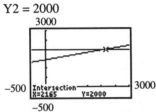

The graphs intersect at (2165, 2000). The
solution is 2165. The fourth week donation
should be $2165.

20. Let x = length of first side
x = length of second side
$3 + x$ = length of third side
$x + x + 3 + x = 26.25$
$Y1 = x + x + 3 + x$
$Y2 = 26.25$

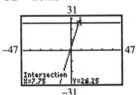

The graphs intersect at (7.75, 26.25). The
solution is 7.75, so $3 + x = 10.75$. The sides
measure 7.75 feet, 7.75 feet, and 10.75 feet.

21. $41 + x = 67$
$41 + x - 41 = 67 - 41$
$x = 26$
The solution is 26.

22. $y - \dfrac{7}{13} = \dfrac{11}{39}$

$y - \dfrac{7}{13} + \dfrac{7}{13} = \dfrac{11}{39} + \dfrac{7}{13}$

$y = \dfrac{11}{39} + \dfrac{21}{39}$

$y = \dfrac{32}{39}$

The solution is $\dfrac{32}{39}$.

23. $5 - (2 - x) = 1$
$5 - 2 + x = 1$
$3 + x = 1$
$3 + x - 3 = 1 - 3$
$x = -2$
The solution is -2.

24. $0.59(z - 1) + 0.41(z + 2) = 3.163$
$0.59z - 0.59 + 0.41z + 0.82 = 3.163$
$1z + 0.23 = 3.163$
$z + 0.23 - 0.23 = 3.163 - 0.23$
$z = 2.933$
The solution is 2.933.

25. $-4x = 272$

$\dfrac{-4x}{-4} = \dfrac{272}{-4}$

$x = -68$
The solution is -68.

26. $45.86z = -1765.61$

$\dfrac{45.86z}{45.86} = \dfrac{-1765.61}{45.86}$

$z = -38.5$
The solution is -38.5.

27. $\dfrac{a}{7} = 15$

$7\left(\dfrac{a}{7}\right) = 7(15)$

$a = 105$
The solution is 105.

28. $\dfrac{4}{5}x = \dfrac{64}{125}$

$\dfrac{5}{4}\left(\dfrac{4}{5}x\right) = \dfrac{5}{4}\left(\dfrac{64}{125}\right)$

$x = \dfrac{16}{25}$

The solution is $\dfrac{16}{25}$.

29. $15x - 16x = -12$
$-x = -12$

$\dfrac{-x}{-1} = \dfrac{-12}{-1}$

$x = 12$
The solution is 12.

30. $-y = 2.98$

$\dfrac{-y}{-1} = \dfrac{2.98}{-1}$

$y = -2.98$

The solution is -2.98.

31. $4(2x - 6) + 6(x + 4) = 49$

$8x - 24 + 6x + 24 = 49$

$14x = 49$

$\dfrac{14x}{14} = \dfrac{49}{14}$

$x = 3.5$

The solution is 3.5.

32. Let x = the number of passes given

$189 + x = 247$

$189 + x - 189 = 247 - 189$

$x = 58$

There were 58 passes given.

33. Let x = total graduates

$0.55x = 154$

$\dfrac{0.55x}{0.55} = \dfrac{154}{0.55}$

$x = 280$

There were 280 graduates.

34. Let x = the number of books they must sell

$15x = 80,000$

$\dfrac{15x}{15} = \dfrac{80,000}{15}$

$x = 5333\dfrac{1}{3}$

They must sell 5334 books.

35. Let x = total proceeds

$\dfrac{3}{7}x = 18,270$

$\dfrac{7}{3}\left(\dfrac{3}{7}x\right) = \dfrac{7}{3}(18,270)$

$x = 42,630$

The total proceeds were \$42,630.

36. $3x + 7 = 4x + 21$

$3x + 7 - 7 = 4x + 21 - 7$

$3x = 4x + 14$

$3x - 4x = 4x + 14 - 4x$

$-x = 14$

$\dfrac{-x}{-1} = \dfrac{14}{-1}$

$x = -14$

The solution is -14.

37. $14 - 2x = 5x$

$14 - 2x + 2x = 5x + 2x$

$14 = 7x$

$\dfrac{14}{7} = \dfrac{7x}{7}$

$2 = x$

The solution is 2.

38. $8.7x + 4.33 = -2.4x - 33.41$

$8.7x + 4.33 - 4.33 = -2.4x - 33.41 - 4.33$

$8.7x = -2.4x - 37.74$

$8.7x + 2.4x = -2.4x - 37.74 + 2.4x$

$11.1x = -37.74$

$\dfrac{11.1x}{11.1} = \dfrac{-37.74}{11.1}$

$x = -3.4$

The solution is -3.4.

39. $6.8z - 9.52 = 0$

$6.8z - 9.52 + 9.52 = 0 + 9.52$

$6.8z = 9.52$

$\dfrac{6.8z}{6.8} = \dfrac{9.52}{6.8}$

$z = 1.4$

The solution is 1.4.

40. $\dfrac{3}{8}x + \dfrac{1}{2} = \dfrac{3}{4}$

$8\left(\dfrac{3}{8}x\right) + 8\left(\dfrac{1}{2}\right) = 8\left(\dfrac{3}{4}\right)$

$3x + 4 = 6$

$3x + 4 - 4 = 6 - 4$

$3x = 2$

$\dfrac{3x}{3} = \dfrac{2}{3}$

$x = \dfrac{2}{3}$

The solution is $\dfrac{2}{3}$.

41. $\dfrac{5}{7}a + \dfrac{11}{14} = \dfrac{2}{7}a$

$14\left(\dfrac{5}{7}a\right) + 14\left(\dfrac{11}{14}\right) = 14\left(\dfrac{2}{7}a\right)$

$10a + 11 = 4a$

$10a - 4a + 11 = 4a - 4a$

$6a + 11 = 0$

$6a + 11 - 11 = 0 - 11$

$6a = -11$

$\dfrac{6a}{6} = \dfrac{-11}{6}$

$a = -\dfrac{11}{6}$

The solution is $-\dfrac{11}{6}$.

42. $2(x + 5) - (x + 6) = 2(x + 2) - x$

$2x + 10 - x - 6 = 2x + 4 - x$

$x + 4 = x + 4$

$x + 4 - x = x + 4 - x$

$4 = 4$

This is a true equation.

The solution is all real numbers.

43. $3(x - 4) + 2(x + 1) = 5x + 10$

$3x - 12 + 2x + 2 = 5x + 10$

$5x - 10 = 5x + 10$

$5x - 10 - 5x = 5x + 10 - 5x$

$-10 = 10$

This is a false equation. There is no solution.

44. Let x = the number of miles.

$49.95 + 0.22x = 250$

$49.95 + 0.22x - 49.95 = 250 - 49.95$

$0.22x = 200.05$

$\dfrac{0.22x}{0.22} = \dfrac{200.05}{0.22}$

$x \approx 909.3$

You can drive 909 miles.

45. Let x = annual depreciation.

$7x + 25{,}000 = 150{,}000$

$7x + 25{,}000 - 25{,}000 = 150{,}000 - 25{,}000$

$7x = 125{,}000$

$\dfrac{7x}{7} = \dfrac{125{,}000}{7}$

$x \approx 17{,}857.14$

The annual depreciation is about \$17,857.

46. Let x = number of hours.

$175 + 35x = 100 + 40x$

$175 + 35x - 175 = 100 + 40x - 175$

$35x = -75 + 40x$

$35x - 40x = -75 + 40x - 40x$

$-5x = -75$

$\dfrac{-5x}{-5} = \dfrac{-75}{-5}$

$x = 15$

The offers are the same at 15 hours.

47. $A = \dfrac{1}{2}h(b + B)$ for h

$2A = 2\left[\dfrac{1}{2}h(b + B)\right]$

$2A = h(b + B)$

$\dfrac{2A}{b + B} = \dfrac{h(b + B)}{b + B}$

$\dfrac{2A}{b + B} = h$ or $h = \dfrac{2A}{b + B}$

48. $S = 2LW + 2WH + 2LH$ for W

$S - 2LH = 2LW + 2WH + 2LH - 2LH$

$S - 2LH = 2LW + 2WH$

$S - 2LH = W(2L + 2H)$

$\dfrac{S - 2LH}{2L + 2H} = \dfrac{W(2L + 2H)}{2L + 2H}$

$\dfrac{S - 2LH}{2L + 2H} = W$ or $W = \dfrac{S - 2LH}{2L + 2H}$

49. $6x + 7y = 42$ for x

$6x + 7y - 7y = 42 - 7y$

$6x = 42 - 7y$

$\dfrac{6x}{6} = \dfrac{42 - 7y}{6}$

$x = \dfrac{42}{6} - \dfrac{7y}{6}$

$x = 7 - \dfrac{7}{6}y$ or $x = -\dfrac{7}{6}y + 7$

50. $\frac{3}{4}x - \frac{5}{8}y = \frac{11}{12}$ for y

$24\left(\frac{3}{4}x\right) - 24\left(\frac{5}{8}y\right) = 24\left(\frac{11}{12}\right)$

$18x - 15y = 22$

$18x - 15y - 18x = 22 - 18x$

$-15y = 22 - 18x$

$\frac{-15y}{-15} = \frac{22 - 18x}{-15}$

$y = \frac{22}{-15} - \frac{18x}{-15}$

$y = -\frac{22}{15} + \frac{6}{5}x$ or $y = \frac{6}{5}x - \frac{22}{15}$

51. $A = 6000 + 8000n$

$A - 6000 = 6000 + 8000n - 6000$

$\frac{A - 6000}{8000} = \frac{8000n}{8000}$

$\frac{A - 6000}{8000} = n$ or $n = \frac{A - 6000}{8000}$

$n = \frac{78,000 - 6000}{8000}$

$n = 9$

It will last 9 years if the amount is $78,000.

$n = \frac{126,000 - 6000}{8000}$

$n = 15$

It will last 15 years if the amount is $126,000.

52. Let x = length of first piece

$x + 2$ = length of second piece

$x + 4$ = length of third piece

$x + x + 2 + x + 4 = 3(12)$

$3x + 6 = 36$

$3x + 6 - 6 = 36 - 6$

$3x = 30$

$\frac{3x}{3} = \frac{30}{3}$

$x = 10$

$x + 2 = 12$

$x + 4 = 14$

The lengths should be 10 in., 12 in., and 14 in.

53. Let x = length of first piece

$x + 1$ = length of second piece

$x + 2$ = length of third piece

$x + x + 1 + x + 2 = 3(12)$

$3x + 3 = 36$

$3x + 3 - 3 = 36 - 3$

$3x = 33$

$\frac{3x}{3} = \frac{33}{3}$

$x = 11$

$x + 1 = 12$

$x + 2 = 13$

The lengths should be 11 in., 12 in., and 13 in.

54. Let x = width

$3x$ = length

$2x + 2(3x) - 4 - 6 = 230$

$8x - 10 = 230$

$8x - 10 + 10 = 230 + 10$

$8x = 240$

$\frac{8x}{8} = \frac{240}{8}$

$x = 30$

$3x = 90$

The dimensions are 30 ft by 90 ft.

55. $I = PRT$

$2250 = 5000(0.075)T$

$2250 = 375T$

$\frac{2250}{375} = \frac{375T}{375}$

$6 = T$

It should be invested for 6 years.

56. $I = PRT$

$2250 = P(0.075)(3)$

$2250 = 0.225P$

$\frac{2250}{0.225} = \frac{0.225P}{0.225}$

$10,000 = P$

You should invest $10,000.

57. Let x = the amount that the retail price
should be
$x - 0.2x = 210$
$x(1 - 0.2) = 210$
$0.8x = 210$
$$\frac{0.8x}{0.8} = \frac{210}{0.8}$$
$x = 262.5$
No, the store was not being honest. The
suit's retail price should have been $262.50.

58. Let x = the artist's price
$x + 0.3x = 32.50$
$x(1 + 0.3) = 32.50$
$1.3x = 32.5$
$$\frac{1.3x}{1.3} = \frac{32.5}{1.3}$$
$x = 25$
The artist was paid $25.

Chapter 4 Mixed Review

1. $3(x + 2) - 2(x - 1) = (2x + 5) - (x - 3)$
Y1 = $3(x + 2) - 2(x - 1)$
Y2 = $(2x + 5) - (x - 3)$

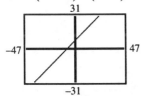

The graphs are the same.
The solution is all real numbers.

2. $(2x + 1) - (3x - 7) = (x + 5) - 2(x - 2)$
Y1 = $(2x + 1) - (3x - 7)$
Y2 = $(x + 5) - 2(x - 2)$

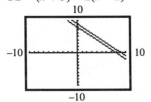

The graphs are parallel.
There is no solution.

3. $x + 1 = 2x + 5$
Y1 = $x + 1$
Y2 = $2x + 5$

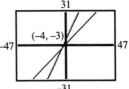

The graphs intersect at $(-4, -3)$.
The solution is -4.

4. $\frac{1}{3}x + 3 = 6 - \frac{2}{3}x$

Y1 = $\frac{1}{3}x + 3$

Y2 = $6 - \frac{2}{3}x$

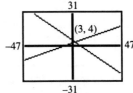

The graphs intersect at $(3, 4)$.
The solution is 3.

5. $1.2x - 6.12 = -2.2x + 4.42$
Y1 = $1.2x - 6.12$
Y2 = $-2.2x + 4.42$

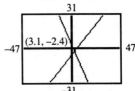

The graphs intersect at $(3.1, -2.4)$.
The solution is 3.1.

6.

x	$4x - 5$	$7 - 2x$	
1	−1	5	−1 < 5
2	3	3	3 = 3
3	7	1	7 > 1

The solution is 2.

7.

x	$14(x + 6) - 17$	$12(x + 5)$	
−4	11	12	11 < 12
−3	25	24	25 > 24

The solution is a noninteger between −4 and −3.

8.

x	$1.5x + 5.5$	$-2.4x - 6.2$	
−4	−0.5	3.4	−0.5 < 3.4
−3	1	1	1 = 1
−2	2.5	−1.4	2.5 > −1.4

The solution is −3.

9.

x	$2(2x - 3) + 3(x + 1)$	$6(x - 1) + (x + 3)$	
0	−3	−3	−3 = −3
1	4	4	4 = 4
2	11	11	11 = 11

The values of the expressions are always equal. The solution is all real numbers.

10.

x	$\frac{1}{3}x + \frac{14}{3}$	$\frac{5}{2} - \frac{3}{4}x$	
−3	3.67	4.75	3.67 < 4.75
−2	4	4	4 = 4
−1	4.33	3.25	4.33 > 3.25

The solution is −2.

11.

x	$4(x - 2) - (x - 1)$	$3x + 1$	
0	−7	1	−7 − 1 = −8
1	−4	4	−4 − 4 = −8
2	−1	7	−1 − 7 = −8

The difference in the values of the expressions is a constant, −8.
There is no solution.

12. $5 - \dfrac{3}{7}x = \dfrac{2}{3}x + 2$ is linear because it is in the form $b + ax = ax + b$.

13. $2x - 4 = 1 + \dfrac{9}{x}$ is nonlinear because a variable appears in the denominator of a term.

14. $-2x^3 + 3x = x - 5$ is nonlinear because the variable x has an exponent of 3.

15. $15 = 2(x - 7) - (x - 2)$ is linear because it simplifies to $15 = x - 12$.

16. $3.9(x - 1.2) - 6.7x = 0$ is linear because it simplifies to $-2.8x - 4.68 = 0$.

17. $\sqrt[3]{x + 1} - 5 = 4x$ is nonlinear because there is a radical expression with a variable in its radicand.

18. $\dfrac{z}{29} = -12$

$29\left(\dfrac{z}{29}\right) = 29(-12)$

$z = -348$

The solution is -348.

19. $\dfrac{5}{22} = \dfrac{25}{33}y$

$\dfrac{33}{25}\left(\dfrac{5}{22}\right) = \dfrac{33}{25}\left(\dfrac{25}{33}y\right)$

$\dfrac{3}{10} = y$

The solution is $\dfrac{3}{10}$.

20. $59a = 1888$

$\dfrac{59a}{59} = \dfrac{1888}{59}$

$a = 32$

The solution is 32.

21. $-174.243 = 2.41x$

$\dfrac{-174.243}{2.41} = \dfrac{2.41x}{2.41}$

$-72.3 = x$

The solution is -72.3.

22. $2.3x - 3.3x = 14$

$-x = 14$

$\dfrac{-x}{-1} = \dfrac{14}{-1}$

$x = -14$

The solution is -14.

23. $-b = 14.59$

$\dfrac{-b}{-1} = \dfrac{14.59}{-1}$

$b = -14.59$

The solution is -14.59.

24. $3(x - 8) + 8(x + 3) = 77$

$3x - 24 + 8x + 24 = 77$

$11x = 77$

$\dfrac{11x}{11} = \dfrac{77}{11}$

$x = 7$

The solution is 7.

25. $z + 193 = -251$

$z + 193 - 193 = -251 - 193$

$z = -444$

The solution is -444.

26. $\dfrac{13}{17} + a = \dfrac{5}{51}$

$\dfrac{13}{17} + a - \dfrac{13}{17} = \dfrac{5}{51} - \dfrac{13}{17}$

$a = \dfrac{5}{51} - \dfrac{39}{51}$

$a = -\dfrac{34}{51} = -\dfrac{2}{3}$

The solution is $-\dfrac{2}{3}$.

27. $0.92(x - 2) + 0.08(x + 1) = 5.73$

$0.92x - 1.84 + 0.08x + 0.08 = 5.73$

$x - 1.76 = 5.73$

$x - 1.76 + 1.76 = 5.73 + 1.76$

$x = 7.49$

The solution is 7.49.

28. $6.2x + 5.67 = 4.9x + 16.98$
$6.2x + 5.67 - 5.67 = 4.9x + 16.98 - 5.67$
$6.2x = 4.9x + 11.31$
$6.2x - 4.9x = 4.9x + 11.31 - 4.9x$
$1.3x = 11.31$
$$\frac{1.3x}{1.3} = \frac{11.31}{1.3}$$
$x = 8.7$
The solution is 8.7.

29. $24.96 - 3.9a = 0$
$24.96 - 3.9a + 3.9a = 0 + 3.9a$
$24.96 = 3.9a$
$$\frac{24.96}{3.9} = \frac{3.9a}{3.9}$$
$6.4 = a$
The solution is 6.4.

30. $\dfrac{2}{3}x - \dfrac{3}{4} = -\dfrac{5}{6}$
$$12\left(\frac{2}{3}x\right) - 12\left(\frac{3}{4}\right) = 12\left(-\frac{5}{6}\right)$$
$8x - 9 = -10$
$8x - 9 + 9 = -10 + 9$
$8x = -1$
$$\frac{8x}{8} = \frac{-1}{8}$$
$$x = -\frac{1}{8}$$
The solution is $-\dfrac{1}{8}$.

31. $\dfrac{4}{9}y + \dfrac{11}{18} = \dfrac{5}{6}y$
$$18\left(\frac{4}{9}y\right) + 18\left(\frac{11}{18}\right) = 18\left(\frac{5}{6}y\right)$$
$8y + 11 = 15y$
$8y + 11 - 8y = 15y - 8y$
$11 = 7y$
$$\frac{11}{7} = \frac{7y}{7}$$
$$\frac{11}{7} = y$$
The solution is $\dfrac{11}{7}$.

32. $7x - 4 = 3x + 20$
$7x - 4 + 4 = 3x + 20 + 4$
$7x = 3x + 24$
$7x - 3x = 3x + 24 - 3x$
$4x = 24$
$$\frac{4x}{4} = \frac{24}{4}$$
$x = 6$
The solution is 6.

33. $7x = 15 - 3x$
$7x + 3x = 15 - 3x + 3x$
$10x = 15$
$$\frac{10x}{10} = \frac{15}{10}$$
$x = 1.5$
The solution is 1.5.

34. $2(x + 2) + (x + 1) = 5(x + 1) - 2x$
$2x + 4 + x + 1 = 5x + 5 - 2x$
$3x + 5 = 3x + 5$
$3x + 5 - 3x = 3x + 5 - 3x$
$5 = 5$
This is a true equation. The solution is all real numbers.

35. $4(2x - 1) = 3(2x + 1) + 2(x + 1)$
$8x - 4 = 6x + 3 + 2x + 2$
$8x - 4 = 8x + 5$
$8x - 4 - 8x = 8x + 5 - 8x$
$-4 = 5$
This is a false equation. There is no solution.

36. $1.2x - 2.4y = 0.36$ for x
$1.2x - 2.4y + 2.4y = 0.36 + 2.4y$
$1.2x = 0.36 + 2.4y$
$$\frac{1.2x}{1.2} = \frac{0.36 + 2.4y}{1.2}$$
$$x = \frac{0.36}{1.2} + \frac{2.4y}{1.2}$$
$x = 0.3 + 2y$ or $x = 2y + 0.3$

37. $\frac{1}{9}x + \frac{2}{3}y = \frac{1}{6}$ for y

$$18\left(\frac{1}{9}x\right) + 18\left(\frac{2}{3}y\right) = 18\left(\frac{1}{6}\right)$$

$2x + 12y = 3$

$2x + 12y - 2x = 3 - 2x$

$12y = 3 - 2x$

$$\frac{12y}{12} = \frac{3 - 2x}{12}$$

$$y = \frac{3}{12} - \frac{2}{12}x$$

$$y = \frac{1}{4} - \frac{1}{6}x \text{ or } y = -\frac{1}{6}x + \frac{1}{4}$$

38. $S = 2\pi r^2 + 2\pi rh$ for h

$S - 2\pi r^2 = 2\pi r^2 + 2\pi rh - 2\pi r^2$

$S - 2\pi r^2 = 2\pi rh$

$$\frac{S - 2\pi r^2}{2\pi r} = \frac{2\pi rh}{2\pi r}$$

$$\frac{S - 2\pi r^2}{2\pi r} = h \text{ or } h = \frac{S - 2\pi r^2}{2\pi r}$$

39. $I = PRT$ for P

$$\frac{I}{RT} = \frac{PRT}{RT}$$

$$\frac{I}{RT} = P \text{ or } P = \frac{I}{RT}$$

40. Let x = the subtotal

$0.0825x = 5.41$

$$\frac{0.0825x}{0.0825} = \frac{5.41}{0.0825}$$

$x \approx 65.58$

$65.58 + 5.41 = 70.99$

The subtotal was about \$65.58.
The total bill was about \$70.99.

41. Let x = the total amount he expects to receive

$$\frac{2}{5}x = 5650$$

$$\frac{5}{2}\left(\frac{2}{5}x\right) = \frac{5}{2}(5650)$$

$x = 14,125$

He expects to receive \$14,125.

42. Let x = the number of hours

$13.25x = 16,562.50$

$$\frac{13.25x}{13.25} = \frac{16,562.5}{13.25}$$

$x = 1250$

She worked 1250 hours.

43. Let x = the number of houses

$15 + 0.25x = 45$

$15 + 0.25x - 15 = 45 - 15$

$0.25x = 30$

$$\frac{0.25x}{0.25} = \frac{30}{0.25}$$

$x = 120$

An employee needs 120 houses.

44. Let x = the annual depreciation

$20x + 15,000 = 250,000$

$20x + 15,000 - 15,000 = 250,000 - 15,000$

$20x = 235,000$

$$\frac{20x}{20} = \frac{235,000}{20}$$

$x = 11,750$

The annual depreciation is \$11,750.

45. Let x = additional money

$x + 985 = 1399$

$x + 985 - 985 = 1399 - 985$

$x = 414$

She needs an additional \$414.

46. Let x = length of first piece

$x + 1$ = length of second piece

$x + 2$ = length of third piece

$x + 3$ = length of fourth piece

$x + x + 1 + x + 2 + x + 3 = 26$

$4x + 6 = 26$

$4x + 6 - 6 = 26 - 6$

$4x = 20$

$$\frac{4x}{4} = \frac{20}{4}$$

$x = 5$

$x + 1 = 6$

$x + 2 = 7$

$x + 3 = 8$

The pieces should be 5 in., 6 in., 7 in., and 8 in.

47. Let x = length of first piece
$x + 2$ = length of second piece
$x + 4$ = length of third piece
$x + 6$ = length of fourth piece
$x + x + 2 + x + 4 + x + 6 = 26 - 2$
$4x + 12 = 24$
$4x + 12 - 12 = 24 - 12$
$4x = 12$
$$\frac{4x}{4} = \frac{12}{4}$$
$x = 3$
$x + 2 = 5$
$x + 4 = 7$
$x + 6 = 9$
The pieces should be 3 in., 5 in., 7 in., and 9 in.

48. Let x = price before markdown
$x - 0.15x = 259.95$
$x(1 - 0.15) = 259.95$
$0.85x = 259.95$
$$\frac{0.85x}{0.85} = \frac{259.95}{0.85}$$
$x \approx 305.82$
The original price was about $305.82.

49. $I = PRT$
$1560 = 8000(R)(3)$
$1560 = 24,000R$
$$\frac{1560}{24,000} = \frac{24,000R}{24,000}$$
$0.065 = R$
$R = 6.5\%$
The simple interest rate should be 6.5%.

50. $I = PRT$
$562.50 = P(0.045)(1)$
$562.5 = 0.045P$
$$\frac{562.5}{0.045} = \frac{0.045P}{0.045}$$
$12,500 = P$
You should invest $12,500.

51. Let x = width
$1.5x$ = length
$2x + 2(1.5x) = 84$
$2x + 3x = 84$
$5x = 84$
$$\frac{5x}{5} = \frac{84}{5}$$
$x = 16.8$
$1.5x = 25.2$
The dimensions are 16.8 feet by 25.2 feet.

52. Let x = the price before tax
$x + 0.0875x = 325.16$
$1.0875x = 325.16$
$$\frac{1.0875x}{1.0875} = \frac{325.16}{1.0875}$$
$x \approx 299.00$
$299(0.0875) \approx 26.16$
The price before tax was about $299.
Sales tax was about $26.16.

53. $A = 40 + 22.5h$
$A - 40 = 40 + 22.5h - 40$
$A - 40 = 22.5h$
$$\frac{A - 40}{22.5} = \frac{22.5h}{22.5}$$
$$\frac{A - 40}{22.5} = h \text{ or } h = \frac{A - 40}{22.5}$$
$$h = \frac{208.75 - 40}{22.5}$$
$h = 7.5$
A job that cost $208.75 lasts 7.5 hours.
$$h = \frac{85 - 40}{22.5}$$
$h = 2$
A job that cost $85 lasts 2 hours.

54. Let x = the measure of the angle made with width
$x + 35 + 90 = 180$
$x + 125 = 180$
$x + 125 - 125 = 180 - 125$
$x = 55$
The angle made with the width of the rug is 55°.

Chapter 4 Test

1. $3.14x + 9.07 = 5.72x$ is linear because it is in the form $ax + b = ax$.

2. $5x = 12 + \dfrac{19}{x}$ is nonlinear because a variable appears in the denominator of a term.

3. $4x + 21 = 5x^2$ is nonlinear because the variable x has an exponent of 2.

4. $4(x - 6) = 3(5 - x) + 12$ is linear because it simplifies to $4x - 24 = 27 - 3x$.

5. $2(2x - 5) - 2(2 - x) = 6(x + 1) + 1$
   $4x - 10 - 4 + 2x = 6x + 6 + 1$
   $6x - 14 = 6x + 7$
   $6x - 14 - 6x = 6x + 7 - 6x$
   $-14 = 7$
   This is a false equation. There is no solution.

6. $2(x + 5) = -3(x + 1) - 2$
   $2x + 10 = -3x - 3 - 2$
   $2x + 10 = -3x - 5$
   $2x + 10 - 10 = -3x - 5 - 10$
   $2x = -3x - 15$
   $2x + 3x = -3x - 15 + 3x$
   $5x = -15$
   $\dfrac{5x}{5} = \dfrac{-15}{5}$
   $x = -3$
   The solution is -3.

7. $1.41(x + 5.08) + 1.17x + 0.00102$
   $\qquad\qquad = -3.46x - 5.39334$
   $1.41x + 7.1628 + 1.17x + 0.00102$
   $\qquad\qquad = -3.46x - 5.39334$
   $2.58x + 7.16382 = -3.46x - 5.39334$
   $2.58x + 7.16382 - 7.16382$
   $\qquad\qquad = -3.46x - 5.39334 - 7.16382$
   $2.58x = -3.46x - 12.55716$
   $2.58x + 3.46x = -3.46x - 12.55716 + 3.46x$
   $6.04x = -12.55716$
   $\dfrac{6.04x}{6.04} = \dfrac{-12.55716}{6.04}$
   $x = -2.079$
   The solution is -2.079.

8. $(x + 1) - 4(x - 1) = 3(2 - x) - 1$
   $x + 1 - 4x + 4 = 6 - 3x - 1$
   $-3x + 5 = 5 - 3x$
   $-3x + 5 + 3x = 5 - 3x + 3x$
   $5 = 5$
   This is a true equation. The solution is all real numbers.

9. $\dfrac{4}{5}x + \dfrac{31}{10} = -\dfrac{4}{3}x + \dfrac{41}{6}$
   $30\left(\dfrac{4}{5}x\right) + 30\left(\dfrac{31}{10}\right) = 30\left(-\dfrac{4}{3}x\right) + 30\left(\dfrac{41}{6}\right)$
   $24x + 93 = -40x + 205$
   $24x + 93 - 93 = -40x + 205 - 93$
   $24x = -40x + 112$
   $24x + 40x = -40x + 112 + 40x$
   $64x = 112$
   $\dfrac{64x}{64} = \dfrac{112}{64}$
   $x = 1\dfrac{3}{4}$
   The solution is $1\dfrac{3}{4}$.

10. Let x = length of first piece
    $x + 1$ = length of second piece
    $x + 2$ = length of third piece
    $x + x + 1 + x + 2 = 45$
    $3x + 3 = 45$
    $3x + 3 - 3 = 45 - 3$
    $3x = 42$
    $\dfrac{3x}{3} = \dfrac{42}{3}$
    $x = 14$
    $x + 1 = 15$
    $x + 2 = 16$
    The pieces should be cut into sections measuring 14 in., 15in., and 16 in.

11. Let x = his monthly payment
    $12x + 850 = 2470$
    $12x + 850 - 850 = 2470 - 850$
    $12x = 1620$
    $\dfrac{12x}{12} = \dfrac{1620}{12}$
    $x = 135$
    His monthly payments will be $135.

12. Let x = the original price
$x - 0.25x = 179.95$
$x(1 - 0.25) = 179.95$
$0.75x = 179.95$
$$\frac{0.75x}{0.75} = \frac{179.95}{0.75}$$
$x \approx 239.93$
The price before it went on sale was about $239.93.

13. $P = 2L + 2W$ for W
$P - 2L = 2L + 2W - 2L$
$P - 2L = 2W$
$$\frac{P - 2L}{2} = \frac{2W}{2}$$
$$\frac{P - 2L}{2} = W \text{ or } W = \frac{P - 2L}{2}$$
$$W = \frac{44.8 - 2(14.8)}{2}$$
$$W = \frac{15.2}{2}$$
$W = 7.6$
The width is 7.6 inches.

14. Let x = first angle measurement
$x + 10$ = second angle measurement
$x + x + 10$ = third angle measurement
$x + x + 10 + x + x + 10 = 180$
$4x + 20 = 180$
$4x + 20 - 20 = 180 - 20$
$4x = 160$
$$\frac{4x}{4} = \frac{160}{4}$$
$x = 40$
$x + 10 = 50$
$x + x + 10 = 90$
The measures of the angles are 40°, 50°, and 90°.

15. Let x = number of liters of 60% apple juice.
$0.60x + 0.20(500) = 0.50(x + 500)$
$0.60x + 100 = 0.50x + 250$
$0.60x + 100 - 100 = 0.50x + 250 - 100$
$0.60x = 0.50x + 150$
$0.60x - 0.50x = 0.50x + 150 - 0.50x$
$0.10x = 150$
$$\frac{0.10x}{0.10} = \frac{150}{0.10}$$
$x = 1500$
1500 liters of 60% apple juice must be added.

16. Let x = amount of sales.
$1500 = 1200 + 0.4x$
$1500 - 1200 = 1200 + 0.4x - 1200$
$300 = 0.4x$
$$\frac{300}{0.4} = \frac{0.4x}{0.4}$$
$7500 = x$
The two payment plans will be equal for $7500 of sales.

17. Let x = the number of hours to do the job.
$12x + 50 = 15x$
$12x + 50 - 12x = 15x - 12x$
$50 = 3x$
$$\frac{50}{3} = \frac{3x}{3}$$
$$x = \frac{50}{3} = 16\frac{2}{3}$$
The two plans will cost the same if the job lasts $16\frac{2}{3}$ hours.

18. Answers will vary.

Chapter 5

5.1 Experiencing Algebra the Exercise Way

1. $5x + 7y = 35$ is linear and it is written in standard form, $5x + 7y = 35$.
$a = 5, b = 7, c = 35$

3. $-4\sqrt{x} + y = 8$ is nonlinear because the variable x is in the radicand.

5. $6x^2 + 2y = 12$ is nonlinear because the variable x is squared.

7. $y = \dfrac{5}{6}x - 2$ is linear.
$$y - \frac{5}{6}x = \frac{5}{6}x - 2 - \frac{5}{6}x$$
$$y - \frac{5}{6}x = -2$$
$$-\frac{5}{6}x + y = -2 \text{ is standard form.}$$
$a = -\dfrac{5}{6}, b = 1, c = -2$

9. $x - 2y + 1 = x - 4y + 9$ is linear.
$x - 2y + 1 - 1 = x - 4y + 9 - 1$
$x - 2y = x - 4y + 8$
$x - 2y - x = x - 4y + 8 - x$
$-2y = -4y + 8$
$-2y + 4y = -4y + 8 + 4y$
$2y = 8$
$\dfrac{2y}{2} = \dfrac{8}{2}$
$y = 4$ is standard form.
$a = 0, b = 1, c = 4$

11. $2x - 5 = 0$ is linear.
$2x - 5 + 5 = 0 + 5$
$2x = 5$ is standard form.
$a = 2, b = 0, c = 5$

13. $4.2x + 0.3y = 1.4$ is linear and it is in standard form, $4.2x + 0.3y = 1.4$.
$a = 4.2, b = 0.3, c = 1.4$

15.

$2x - 3y = 11$	
$2(5) - 3(7)$	11
$10 - 21$	
-11	

Since $-11 \neq 11$, $(5, 7)$ is not a solution.

17.

$-2x + y = 2$	
$-2\left(\frac{1}{2}\right) + 3$	2
$-1 + 3$	
2	

Since $2 = 2$, $\left(\dfrac{1}{2}, 3\right)$ is a solution.

19.

$m(x) = 2.4x + 3.8$	
11	$2.4(3) + 3.8$
	$7.2 + 3.8$
	11

Since $11 = 11$, $(3, 11)$ is a solution.

21.

$y = -6.4x + 10.4$	
-4	$-6.4(4) + 10.4$
	$-25.6 + 10.4$
	-15.2

Since $-4 \neq -15.2$, $(4, -4)$ is not a solution.

23.

$y = 5$	
5	5

Since $5 = 5$, $(11, 5)$ is a solution.

25.

$-4x = 12$	
$-4(-3)$	12
12	

Since $12 = 12$, $(-3, 9)$ is a solution.

27.

$y = 2$	
-5	-2

Since $-5 \neq -2$, $(4, -5)$ is not a solution.

29.

$$x = 15$$

$$3\;\big|\;15$$

Since $3 \neq 15$, $(3, 15)$ is not a solution.

31.

$$x - 4y = 7$$

$$18.6 - 4(-2.9)\;\big|\;7$$

$$18.6 + 11.6$$

$$30.2$$

Since $30.2 \neq 7$, $(18.6, -2.9)$ is not a solution.

33. Answers will vary. Possible answer:

$$x - 6y = 12$$
$$x - 6y - x = 12 - x$$
$$-6y = 12 - x$$
$$\frac{-6y}{-6} = \frac{12 - x}{-6}$$
$$\frac{-6y}{-6} = \frac{12 - x}{-6}$$
$$y = \frac{12}{-6} - \frac{x}{-6}$$
$$y = \frac{1}{6}x - 2$$

x	$y = \frac{1}{6}x - 2$	y
-6	$y = \frac{1}{6}(-6) - 2 = -3$	-3
0	$y = \frac{1}{6}(0) - 2 = -2$	-2
6	$y = \frac{1}{6}(6) - 2 = -1$	-1

$(-6, -3)$, $(0, -2)$, and $(6, -1)$ are three possible solutions.

35. Answers will vary. Possible answer:

x	$y = -\frac{5}{8}x + 1$	y
-8	$y = -\frac{5}{8}(-8) + 1 = 6$	6
0	$y = -\frac{5}{8}(0) + 1 = 1$	1
8	$y = -\frac{5}{8}(8) + 1 = -4$	-4

$(-8, 6)$, $(0, 1)$, and $(8, -4)$ are three possible solutions.

37. Answers will vary. Possible answer:

x	$p(x) = \frac{4x+1}{3}$	$p(x)$
-1	$p(x) = \frac{4(-1)+1}{3} = -1$	-1
2	$p(x) = \frac{4(2)+1}{3} = 3$	3
5	$p(x) = \frac{4(5)+1}{3} = 7$	7

$(-1, -1)$, $(2, 3)$, and $(5, 7)$ are three possible solutions.

39. Answers will vary. Possible answer:

x	$y = 8$	y
0	$y = 8$	8
1	$y = 8$	8
2	$y = 8$	8

$(0, 8)$, $(1, 8)$, and $(2, 8)$ are three possible solutions.

41. Answers will vary. Possible answer:

x	$y = 2.6x - 4.2$	y
-1	$y = 2.6(-1) - 4.2 = -6.8$	-6.8
0	$y = 2.6(0) - 4.2 = -4.2$	-4.2
1	$y = 2.6(1) - 4.2 = -1.6$	-1.6

$(-1, -6.8)$, $(0, -4.2)$, and $(1, -1.6)$ are three possible solutions.

43. Answers will vary. Possible answer:
$$2x + y = 3$$
$$2x + y - 2x = 3 - 2x$$
$$y = 3 - 2x$$
$$y = -2x + 3$$

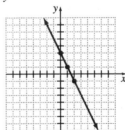

$(0, 3)$, $(1, 1)$, and $(2, -1)$ are three possible solutions.

45. $5x - 3y = 6$

$5x - 3y - 5x = 6 - 5x$

$-3y = 6 - 5x$

$\dfrac{-3y}{-3} = \dfrac{6 - 5x}{-3}$

$y = -2 + \dfrac{5}{3}x$

$y = \dfrac{5}{3}x - 2$

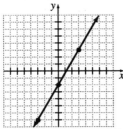

$(-3, -7)$, $(0, -2)$, and $(3, 3)$ are three possible solutions.

47. Answers will vary. Possible answer:

$r(x) = -\dfrac{3}{4}x + 4$

$y = -\dfrac{3}{4}x + 4$

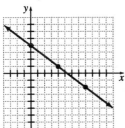

$(0, 4)$, $(4, 1)$, and $(8, -2)$ are three possible solutions.

49. Answers will vary. Possible answer:

$y = 2.8x - 1.6$

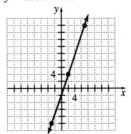

$(-3, -10)$, $(2, 4)$, and $(7, 18)$ are three possible solutions.

51. Answers will vary. Possible answer:

$y = \dfrac{3x - 5}{2}$

$y = \dfrac{3}{2}x - \dfrac{5}{2}$

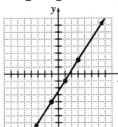

$(-1, -4)$, $(1, -1)$, and $(3, 2)$ are three possible solutions.

53. Answers will vary. Possible answer:
$$3x + y - 4 = x + 2y - 3$$
$$3x - y - 4 - 3x = x + 2y - 3 - 3x$$
$$y - 4 = -2x + 2y - 3$$
$$y - 4 + 4 = -2x + 2y - 3 + 4$$
$$y = -2x + 2y + 1$$
$$y - 2y = -2x + 2y + 1 - 2y$$
$$-y = -2x + 1$$
$$\frac{-y}{-1} = \frac{-2x + 1}{-1}$$
$$y = 2x - 1$$

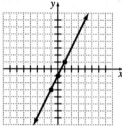

$(-1, -3)$, $(0, -1)$, and $(1, 1)$ are three possible solutions.

55. Answers will vary. Possible answer:
$$5y = -20$$
$$\frac{5y}{5} = \frac{-20}{5}$$
$$y = -4$$

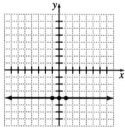

$(-1, -4)$, $(0, -4)$ and $(1, -4)$ are three possible solutions.

57. a. $(1, 0)$, $(2, 5)$, $(4, 15)$

b.

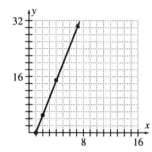

c. At $x = 5$, $y = 20$
A crew of 5 people would pack 20 boxes per minute.

d. Answers will vary. Possible answer: There is probably a maximum number of boxes that can be packed by a large number of people given the size limitations of the packing plant.

59. a. $p(x) = 8x + 25$
$p(7) = 8(7) + 25$
$p(7) = 81$
She would earn $81.

b. $p(5) = 8(5) + 25$
$p(5) = 65$
No, she should be paid $65.

c. $p(0) = 8(0) + 25$
$p(0) = 25$
She would be paid $25.

d. Answers will vary. Possible answer:
Domain: 0 to 50
Range: 25 to 425

61. a. (2, 6), (3.5, 10.5), (10.5, 31.5)

b.

c. At $x = 4$, $y = 12$
The border would be 12 inches.

d. Yes, all equilateral triangles have three sides of equal measure.

63. a. $D(x) = 3.75x + 250$
$D(20) = 3.75(20) + 250$
$D(20) = 325$
The cost would be $325.

b. $D(30) = 3.75(30) + 250$
$D(30) = 362.5$
The cost would be $362.50.

c. $D(25) = 3.75(25) + 250$
$D(25) = 343.75$
Yes, $343.75 is less than $350.

d.

x	$D(x) = 3.75x + 250$	$D(x)$
0	$D(0) = 3.75(0) + 250 = 250$	250
5	$D(5) = 3.75(5) + 250 = 268.75$	268.75
10	$D(10) = 3.75(10) + 250 = 287.5$	287.50
15	$D(15) = 3.75(15) + 250 = 306.25$	306.25
20	$D(20) = 3.75(20) + 250 = 325$	325
25	$D(25) = 3.75(25) + 250 = 343.75$	343.75
30	$D(30) = 3.75(30) + 250 = 362.5$	362.50
35	$D(35) = 3.75(35) + 250 = 381.25$	381.25
40	$D(40) = 3.75(40) + 250 = 400$	400
45	$D(45) = 3.75(45) + 250 = 418.75$	418.75
50	$D(50) = 3.75(50) + 250 = 437.5$	437.50

65. Answers will vary. Possible answer:
(5, 225), (10, 350), (15, 475)
The cost of producing 5 items, 10 items, and
15 items is $225, $350, and $475,
respectively.

5.1 Experiencing Algebra the Calculator Way

1.

$$y = 6.871x - 35.711249$$

3.287	$6.871(4.719) - 35.711249$
	-3.287

Since $3.287 \ne -3.287$, (4.719, 3.287) is not a
solution.

2.

$$1.05x + 1.45y = -23.75$$

$1.05(25) + 1.45(50)$	-23.75
98.75	

Since $98.75 \ne -23.75$, (25, 50) is not a
solution.

3.

$$132x - 297y = 1526$$

$132\left(\frac{17}{33}\right) - 297\left(-4\frac{10}{11}\right)$	1526
1526	

Since $1526 = 1526$, $\left(\frac{17}{33}, -4\frac{10}{11}\right)$ is a

solution.

4. $y = 3.721x - 1.857$

x	y
1.5	3.7245
1.6	4.0966
1.7	4.4687
1.8	4.8408
1.9	5.2129
2.0	5.585

5. $y = \dfrac{13}{28}x - 5$

x	y
0	-5
7	-1.75
14	1.5
21	4.75
28	8
35	11.25

6. $y = \sqrt{3}x - \sqrt{2}$
$Y1 = \sqrt{3}x - \sqrt{2}$

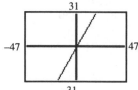

(0, −1.41), (2, 2.05), (4, 5.51), (6, 8.98),
(8, 12.44)

7. $y = \pi r^2 x$
$y = \pi(0.25)^2 x$
$y = 0.0625\pi x$
$Y1 = 0.0625\pi x$

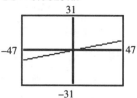

(1, 0.20), (2, 0.39), (3, 0.59), (4, 0.79),
(5, 0.98)

8. $Y1 = 9.64x$

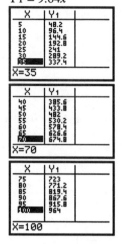

5.2 Experiencing Algebra the Exercise Way

1. x-intercept: $(-2, 0)$
y-intercept: $(0, 4)$

3. x-intercept: $(4, 0)$
y-intercept: $(0, 2)$

5. x-intercept: $(0, 0)$
y-intercept: $(0, 0)$

7. x-intercept: $(3, 0)$
y-intercept: none

9. x-intercept: none
y-intercept: $(0, 1)$

11. $Y1 = 5x - 24$

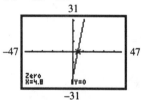

x-intercept: $(4.8, 0)$
y-intercept: $(0, -24)$

13. $Y1 = 3x + 18$

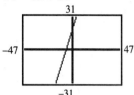

x-intercept: $(-6, 0)$
y-intercept: $(0, 18)$

15. $5x + 4y = 30$
$5x + 4y - 5x = 30 - 5x$
$4y = 30 - 5x$
$$\frac{4y}{4} = \frac{30 - 5x}{4}$$
$$y = \frac{30}{4} - \frac{5}{4}x$$
$$y = -\frac{5}{4}x + \frac{15}{2}$$
$$Y1 = -\frac{5}{4}x + \frac{15}{2}$$

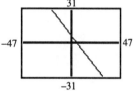

x-intercept: $(6, 0)$
y-intercept: $(0, 7.5)$

17. $5x - 7y = 28$

$5x - 7y - 5x = 28 - 5x$

$-7y = 28 - 5x$

$\dfrac{-7y}{-7} = \dfrac{28 - 5x}{-7}$

$y = -\dfrac{28}{7} + \dfrac{5}{7}x$

$y = \dfrac{5}{7}x - 4$

$Y1 = \dfrac{5}{7}x - 4$

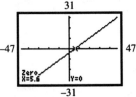

x-intercept: $(5.6, 0)$

y-intercept: $(0, -4)$

19. $-2x - 3y = 10$

$-2x - 3y + 2x = 10 + 2x$

$-3y = 10 + 2x$

$\dfrac{-3y}{-3} = \dfrac{10 + 2x}{-3}$

$y = -\dfrac{10}{3} - \dfrac{2}{3}x$

$y = -\dfrac{2}{3}x - \dfrac{10}{3}$

$Y1 = -\dfrac{2}{3}x - \dfrac{10}{3}$

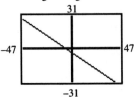

x-intercept: $(-5, 0)$

y-intercept: $(0, -3.\overline{3})$

21. $3x + 5y = 12$

To find the x-intercept, substitute 0 for y.

$3x + 5(0) = 12$

$3x = 12$

$\dfrac{3x}{3} = \dfrac{12}{3}$

$x = 4$

The x-intercept is $(4, 0)$.

To find the y-intercept, substitute 0 for x.

$3(0) + 5y = 12$

$5y = 12$

$\dfrac{5y}{5} = \dfrac{12}{5}$

$y = \dfrac{12}{5}$

The y-intercept is $\left(0, \dfrac{12}{5}\right)$.

23. $4x - 7y = 14$

Substitute 0 for y.

$4x - 7(0) = 14$

$4x = 14$

$\dfrac{4x}{4} = \dfrac{14}{4}$

$x = \dfrac{7}{2}$

The x-intercept is $\left(\dfrac{7}{2}, 0\right)$.

Substitute 0 for x.

$4(0) - 7y = 14$

$-7y = 14$

$\dfrac{-7y}{-7} = \dfrac{14}{-7}$

$y = -2$

The y-intercept is $(0, -2)$.

25. $-2x - 9y = 27$
Substitute 0 for y.
$-2x - 9(0) = 27$
$-2x = 27$
$\dfrac{-2x}{-2} = \dfrac{27}{-2}$
$x = -\dfrac{27}{2}$
The x-intercept is $\left(-\dfrac{27}{2},\, 0 \right)$.
Substitute 0 for x.
$-2(0) - 9y = 27$
$-9y = 27$
$\dfrac{-9y}{-9} = \dfrac{27}{-9}$
$y = -3$
The y-intercept is $(0, -3)$.

27. $6x + 9y - 36 = 0$
Substitute 0 for y.
$6x + 9(0) - 36 = 0$
$6x - 36 = 0$
$6x - 36 + 36 = 0 + 36$
$6x = 36$
$\dfrac{6x}{6} = \dfrac{36}{6}$
$x = 6$
The x-intercept is $(6, 0)$.
Substitute 0 for x.
$6(0) + 9y - 36 = 0$
$9y - 36 = 0$
$9y - 36 + 36 = 0 + 36$
$9y = 36$
$\dfrac{9y}{9} = \dfrac{36}{9}$
$y = 4$
The y-intercept is $(0, 4)$.

29. $3x + 7y = 0$
Substitute 0 for y.
$3x + 7(0) = 0$
$3x = 0$
$\dfrac{3x}{3} = \dfrac{0}{3}$
$x = 0$
The x-intercept is $(0, 0)$.
Substitute 0 for x.
$3(0) + 7y = 0$
$7y = 0$
$\dfrac{7y}{7} = \dfrac{0}{7}$
$y = 0$
The y-intercept is $(0, 0)$.

31. $6x - 8 = 2x + 32$
$6x - 8 - 2x = 2x + 32 - 2x$
$4x - 8 = 32$
$4x - 8 + 8 = 32 + 8$
$4x = 40$
$\dfrac{4x}{4} = \dfrac{40}{4}$
$x = 10$
The x-intercept is $(10, 0)$.
There is no y-intercept.

33. $y = 3y - 22$
$y - 3y = 3y - 22 - 3y$
$-2y = -22$
$\dfrac{-2y}{-2} = \dfrac{-22}{-2}$
$y = 11$
The y-intercept is $(0, 11)$.
There is no x-intercept.

35. $12x - y = 24$
$12x - y - 12x = 24 - 12x$
$-y = 24 - 12x$
$\dfrac{-y}{-1} = \dfrac{24 - 12x}{-1}$
$y = -24 + 12x$
$y = 12x - 24$
The y-coordinate is the constant -24.
The x-coordinate is 0. The y-intercept is $(0, -24)$.

37. $y = 5(x - 3)$
$y = 5x - 15$
The y-intercept is $(0, -15)$.

39. $5x - 15y = 0$
$5x - 15y - 5x = 0 - 5x$
$-15y = -5x$
$\dfrac{-15y}{-15} = \dfrac{-5x}{-15}$
$y = \dfrac{1}{3}x$
The constant is 0.
The y-intercept is $(0, 0)$.

41. $3y = 12y + 18$
$3y - 12y = 12y + 18 - 12y$
$-9y = 18$
$\dfrac{-9y}{-9} = \dfrac{18}{-9}$
$y = -2$
The y-intercept is $(0, -2)$.

43. $y = 0$
The y-intercept is $(0, 0)$.

45. $-17.6x + 2.2y = 19.8$
$-17.6x + 2.2y + 17.6x = 19.8 + 17.6x$
$2.2y = 19.8 + 17.6x$
$\dfrac{2.2y}{2.2} = \dfrac{19.8 + 17.6x}{2.2}$
$y = \dfrac{19.8}{2.2} + \dfrac{17.6x}{2.2}$
$y = 8x + 9$
The y-intercept is $(0, 9)$.

47. $x = 12y$
$\dfrac{x}{12} = \dfrac{12y}{12}$
$\dfrac{x}{12} = y$
$y = \dfrac{1}{12}x$
The constant is 0.
The y-intercept is $(0, 0)$.

49. $3x + 5y = 30$
Substitute 0 for y.
$3x + 5(0) = 30$
$3x = 30$
$\dfrac{3x}{3} = \dfrac{30}{3}$
$x = 10$
The x-intercept is $(10, 0)$.
Substitute 0 for x.
$3(0) + 5y = 30$
$5y = 30$
$\dfrac{5y}{5} = \dfrac{30}{5}$
$y = 6$
The y-intercept is $(0, 6)$.

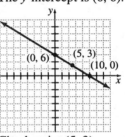

Check point $(5, 3)$.

$3x + 5y = 30$	
$3(5) + 5(3)$	30
$15 + 15$	
30	

Since $30 = 30$, $(5, 3)$ is a solution.

51. $4x - 3y = 24$
Substitute 0 for y.
$4x - 3(0) = 24$
$4x = 24$
$\dfrac{4x}{4} = \dfrac{24}{4}$
$x = 6$
The x-intercept is $(6, 0)$.
Substitute 0 for x.
$4(0) - 3y = 24$
$-3y = 24$
$\dfrac{-3y}{-3} = \dfrac{24}{-3}$
$y = -8$
The y-intercept is $(0, -8)$.

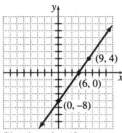

Check point (9, 4).

$$\frac{4x - 3y = 24}{}$$

$4(9) - 3(4)$	24
$36 - 12$	
24	

Since 24 = 24, (9, 4) is a solution.

53. $x - y = 9$
Substitute 0 for y.
$x - 0 = 9$
$x = 9$
The x-intercept is (9, 0).
Substitute 0 for x.
$0 - y = 9$
$\dfrac{-y}{-1} = \dfrac{9}{-1}$
$y = -9$
The y-intercept is (0, –9).

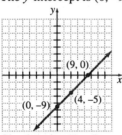

Check point (4, –5).

$$\frac{x - y = 9}{}$$

$4 - (-5)$	9
$4 + 5$	
9	

Since 9 = 9, (4, –5) is a solution.

55. $-x - y = 9$
Substitute 0 for y.
$-x - 0 = 9$
$\dfrac{-x}{-1} = \dfrac{9}{-1}$
$x = -9$
The x-intercept is (–9, 0).
Substitute 0 for x.
$0 - y = 9$
$\dfrac{-y}{-1} = \dfrac{9}{-1}$
$y = -9$
The y-intercept is (0, –9).

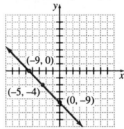

Check point (–5, –4).

$$\frac{-x - y = 9}{}$$

$-(-5) - (-4)$	9
$5 + 4$	
9	

Since 9 = 9, (–5, –4) is a solution.

57. $3x - 7y = -14$
Substitute 0 for y.
$3x - 7(0) = -14$
$3x = -14$
$\dfrac{3x}{3} = \dfrac{-14}{3}$
$x = -\dfrac{14}{3}$

The x-intercept is $\left(-\dfrac{14}{3}, 0 \right)$.

Substitute 0 for x.
$3(0) - 7y = -14$
$-7y = -14$
$\dfrac{-7y}{-7} = \dfrac{-14}{-7}$
$y = 2$
The y-intercept is (0, 2).

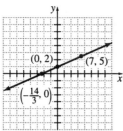

Check point (7, 5).

$$
\begin{array}{c|c}
\multicolumn{2}{c}{3x - 7y = -14} \\
\hline
3(7) - 7(5) & -14 \\
21 - 35 & \\
-14 &
\end{array}
$$

Since $-14 = -14$, (7, 5) is a solution.

59. $y = 6.1x - 23.18$
Substitute 0 for y.
$0 = 6.1x - 23.18$
$0 + 23.18 = 6.1x - 23.18 + 23.18$
$23.18 = 6.1x$
$\dfrac{23.18}{6.1} = \dfrac{6.1x}{6.1}$
$3.8 = x$
The x-intercept is (3.8, 0).
The y-intercept is (0, -23.18).

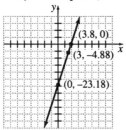

Check point (3, -4.88).

$$
\begin{array}{c|c}
\multicolumn{2}{c}{y = 6.1x - 23.18} \\
\hline
-4.88 & 6.1(3) - 23.18 \\
& 18.3 - 23.18 \\
& -4.88
\end{array}
$$

Since $-4.88 = -4.88$, (3, -4.88) is a solution.

61. $y = \dfrac{5}{3}x + 10$
Substitute 0 for y.
$0 = \dfrac{5}{3}x + 10$
$0 - 10 = \dfrac{5}{3}x + 10 - 10$
$-10 = \dfrac{5}{3}x$
$\dfrac{3}{5}(-10) = \dfrac{3}{5}\left(\dfrac{5}{3}x\right)$
$-6 = x$
The x-intercept is (-6, 0).
The y-intercept is (0, 10).

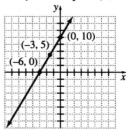

Check point (-3, 5).

$$
\begin{array}{c|c}
\multicolumn{2}{c}{y = \dfrac{5}{3}x + 10} \\
\hline
5 & \dfrac{5}{3}(-3) + 10 \\
& -5 + 10 \\
& 5
\end{array}
$$

Since $5 = 5$, (-3, 5) is a solution.

63. $y = -40x + 200$
Substitute 0 for y.
$0 = -40x + 200$
$40x + 0 = -40x + 200 + 40x$
$40x = 200$
$\dfrac{40x}{40} = \dfrac{200}{40}$
$x = 5$
Substitute 0 for x.
$y = -40(0) + 200$
$y = 200$
The x-intercept is (5, 0) which represents after 5 hours, there will be 0 miles left. The y-intercept is (0, 200) which represents after 0 hours, there are 200 miles to travel.

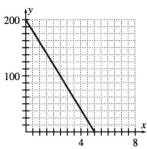

The destination is 100 miles away after 2.5 hours of travel.

65. $y = -4x + 150$
Substitute 0 for y.
$0 = -4x + 150$
$4x + 0 = -4x + 150 + 4x$
$4x = 150$
$\dfrac{4x}{4} = \dfrac{150}{4}$
$x = 37.5$
Substitute 0 for x.
$y = -4(0) + 150$
$y = 150$
The x-intercept is (37.5, 0) which represents after 37.5 minutes, the tank is empty. The y-intercept is (0, 150) which represents after 0 minutes, the tank is full.

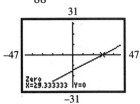

At $x = 20$, $y = 70$.
After 20 minutes there are 70 gallons remaining.

67. The x-intercept is (0, 0).
The y-intercept is (0, 0).
The cost of producing 0 units is $0.
At $x = 50$, $y = 250$.
The cost of producing 50 units is $250.
At $x = 75$, $y = 375$.
The cost of producing 75 units is $375.

5.2 Experiencing Algebra the Calculator Way

1. $Y1 = \dfrac{45}{88}x - 15$

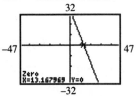

x-intercept $(29.\overline{3}, 0)$
y-intercept $(0, -15)$

2. $Y1 = -2.56x + 33.71$

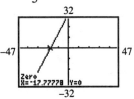

x-intercept (13.168, 0)
y-intercept (0, 33.71)

3. $F = \dfrac{9}{5}C + 32$

$Y1 = \dfrac{9}{5}x + 32$

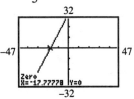

x-intercept $(-17.\overline{7}, 0)$
y-intercept (0, 32)

4. $C = \dfrac{5}{9}(F - 32)$

$Y1 = \dfrac{5}{9}(x - 32)$

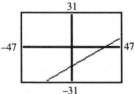

x-intercept $(32, 0)$
y-intercept $(0, -17.\overline{7})$

5. At $0°C$, the Fahrenheit temperature is $32°F$.
At $0°F$, the Celsius temperature is $-17.\overline{7}°C$.

5.3 Experiencing Algebra the Exercise Way

1.

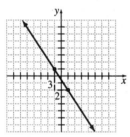

The slope is negative.
The slope is $-\dfrac{3}{2}$.

3. The slope is zero.

5.

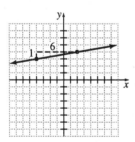

The slope is positive.
The slope is $\dfrac{1}{6}$.

7. The slope is undefined.

9. $(-7, -2)$ and $(5, 6)$

$m = \dfrac{y_2 - y_1}{x_2 - x_1}$

$m = \dfrac{6 - (-2)}{5 - (-7)}$

$m = \dfrac{8}{12}$

$m = \dfrac{2}{3}$

The slope is $\dfrac{2}{3}$.

11. $(-12, -9)$ and $(4, -9)$

$m = \dfrac{y_2 - y_1}{x_2 - x_1}$

$m = \dfrac{-9 - (-9)}{4 - (-12)}$

$m = \dfrac{0}{16}$

$m = 0$

The slope is 0.

13. $(0, 3)$ and $(0, 8)$

$m = \dfrac{y_2 - y_1}{x_2 - x_1}$

$m = \dfrac{8 - 3}{0 - 0}$

$m = \dfrac{5}{0}$

The slope is undefined.

15. $(6, -4)$ and $(7, -6)$

$m = \dfrac{y_2 - y_1}{x_2 - x_1}$

$m = \dfrac{-6 - (-4)}{7 - 6}$

$m = \dfrac{-2}{1}$

$m = -2$

The slope is -2.

17. (0, 4) and (5, 0)

$$m = \frac{y_2 - y_1}{x_2 - x_1}$$

$$m = \frac{0 - 4}{5 - 0}$$

$$m = -\frac{4}{5}$$

The slope is $-\frac{4}{5}$.

19. (11.5, –9.2) and (6.9, 18.4)

$$m = \frac{y_2 - y_1}{x_2 - x_1}$$

$$m = \frac{18.4 - (-9.2)}{6.9 - 11.5}$$

$$m = \frac{27.6}{-4.6}$$

$$m = -6$$

The slope is –6.

21. $\left(\frac{1}{2}, \frac{3}{4}\right)$ and $\left(-\frac{1}{2}, -\frac{5}{6}\right)$

$$m = \frac{y_2 - y_1}{x_2 - x_1}$$

$$m = \frac{-\frac{5}{6} - \frac{3}{4}}{-\frac{1}{2} - \frac{1}{2}}$$

$$m = \frac{-\frac{19}{12}}{-1}$$

$$m = \frac{19}{12}$$

The slope is $\frac{19}{12}$.

23. $y = 21x + 15$
$m = 21, b = 15$
The slope is 21 and the *y*-intercept is (0, 15).

25. $y = \frac{11}{15}x - \frac{21}{25}$

$$m = \frac{11}{15}, b = -\frac{21}{25}$$

The slope is $\frac{11}{15}$ and the *y*-intercept is

$\left(0, -\frac{21}{25}\right)$.

27. $4 = 5.95x - 2.01$
$m = 5.95, b = -2.01$
The slope is 5.95 and the *y*-intercept is
(0, –2.01).

29. $y = 85{,}600 - 1255x$
$y = -1255x + 85{,}600$
$m = -1255, b = 85{,}600$
The slope is –1255 and the *y*-intercept is
(0, 85,600).

31. $16x - 4y = 64$
$16x - 4y - 16x = 64 - 16x$
$-4y = 64 - 16x$

$$\frac{-4y}{-4} = \frac{64 - 16x}{-4}$$

$$y = \frac{64}{-4} - \frac{16x}{-4}$$

$y = -16 + 4x$
$y = 4x - 16$
$m = 4, b = -16$
The slope is 4 and the *y*-intercept is (0, –16).

33. $7y + 18 = 2(y + 6) - 4$
$7y + 18 = 2y + 12 - 4$
$7y + 18 = 2y + 8$
$7y + 18 - 2y = 2y + 8 - 2y$
$5y + 18 = 8$
$5y + 18 - 18 = 8 - 18$
$5y = -10$

$$\frac{5y}{5} = \frac{-10}{5}$$

$y = -2$
or $y = 0x - 2$
$m = 0, b = -2$
The slope is 0 and the *y*-intercept is (0, –2).

35. $7.83x - 2.61y = 10.44$

$7.83x - 2.61y - 7.83x = 10.44 - 7.83x$

$-2.61y = 10.44 - 7.83x$

$$\frac{-2.61y}{-2.61} = \frac{10.44 - 7.83x}{-2.61}$$

$$y = \frac{10.44}{-2.61} - \frac{7.83x}{-2.61}$$

$y = -4 + 3x$

$y = 3x - 4$

$m = 3, b = -4$

The slope is 3 and the y-intercept is $(0, -4)$.

37. $\dfrac{3}{2}x - \dfrac{3}{5}y = \dfrac{21}{10}$

$$\frac{3}{2}x - \frac{3}{5}y - \frac{3}{2}x = \frac{21}{10} - \frac{3}{2}x$$

$$-\frac{3}{5}y = \frac{21}{10} - \frac{3}{2}x$$

$$-\frac{5}{3}\left(-\frac{3}{5}y\right) = -\frac{5}{3}\left(\frac{21}{10} - \frac{3}{2}x\right)$$

$$y = -\frac{7}{2} + \frac{5}{2}x$$

$$y = \frac{5}{2}x - \frac{7}{2}$$

$$m = \frac{5}{2}, b = -\frac{7}{2}$$

The slope is $\dfrac{5}{2}$ and the y-intercept is

$\left(0, -\dfrac{7}{2}\right)$.

39. $x = -4\dfrac{7}{8}$

The slope is undefined and there is no y-intercept.

41. $(8, 3)$ and $m = \dfrac{4}{7}$

43. $(-10, 4)$ and $m = -\dfrac{5}{9}$

45. $(5, 7)$ and $m = 4$

47. $(0, 9)$ and $m = 0$

49. $(9, 0)$ and m is undefined

51. $y = \dfrac{5}{3}x - 4$

$m = \dfrac{5}{3}$, $b = -4$

The slope is $\dfrac{5}{3}$ and the y-intercept is $(0, -4)$.

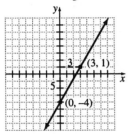

53. $16x - 8y = 40$

$16x - 8y - 16x = 40 - 16x$

$-8y = 40 - 16x$

$\dfrac{-8y}{-8} = \dfrac{40 - 16x}{-8}$

$y = \dfrac{40}{-8} - \dfrac{16x}{-8}$

$y = -5 + 2x$

$y = 2x - 5$

$m = 2$, $b = -5$

The slope is 2 and the y-intercept is $(0, -5)$.

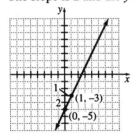

55. $7x + 2y = -16$

$7x + 2y - 7x = -16 - 7x$

$2y = -16 - 7x$

$\dfrac{2y}{2} = \dfrac{-16 - 7x}{2}$

$y = \dfrac{-16}{2} - \dfrac{7x}{2}$

$y = -\dfrac{7}{2}x - 8$

$m = -\dfrac{7}{2}$, $b = -8$

The slope is $-\dfrac{7}{2}$ and the y-intercept is $(0, -8)$.

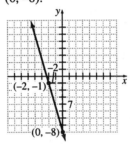

57. $14y + 21 = 6y + 5$

$14y + 21 - 6y = 6y + 5 - 6y$

$8y + 21 = 5$

$8y + 21 - 21 = 5 - 21$

$8y = -16$

$\dfrac{8y}{8} = \dfrac{-16}{8}$

$y = -2$

$m = 0$, $b = -2$

The slope is 0 and the y-intercept is $(0, -2)$.

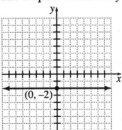

59. $5y = 150x + 350$

$$\frac{5y}{5} = \frac{150x + 350}{5}$$

$y = 30x + 70$

$m = 30, b = 70$

The slope is 30 and the y-intercept is (0, 70).

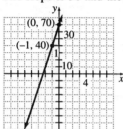

61. $f(x) = 0.3x - 1.2$

$m = 0.3, b = -1.2$

The slope is 0.3 or $\dfrac{3}{10}$ and the y-intercept is

(0, −1.2).

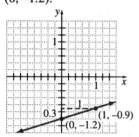

63.

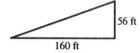

grade $= \dfrac{56}{160} \times 100\%$

$= 35\%$

The grade of the advertised terrain is 35%.

65.

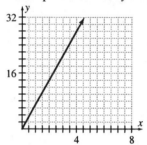

pitch $= \dfrac{3.3}{12} \times 100\% = 27.5\%$

The pitch of the roof is 27.5%.

67. $R = \dfrac{D}{T}$

$R = \dfrac{10.5}{1.5}$

$R = 7$

Their average speed was 7 miles per hour.

$y = 7x$

$m = 7, b = 0$

The slope is 7 and the y-intercept is (0, 0).

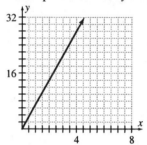

At $x = 2.5, y = 17.5$

They would travel 17.5 miles.

69. a. $\dfrac{1045 - 1624}{10 - 1} = \dfrac{-579}{9}$

b. $y = -\dfrac{579}{9}x + 1624$

c. $Y1 = -\dfrac{579}{9}x + 1624$

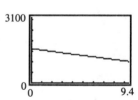

d. $Y1 = -\dfrac{579}{9}x + 1624$

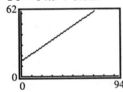

Answers will vary.

71. a. $Y1 = 14.9 + 0.66x$

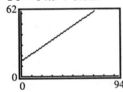

b. $y = 14.9 + 0.66(15)$
$y = 24.8$
The actual value is (15, 25) which is very close.

c.

x	actual y	predicted y	difference
27	30	32.72	–2.72
22	26	29.42	–3.42
15	25	24.8	0.2
35	42	38	4
30	38	34.7	3.3
52	40	49.22	–9.22
35	32	38	–6
55	54	51.2	2.8
40	50	41.3	8.7
40	43	41.3	1.7

73. $D(p) = 80 - \dfrac{4}{5}p$

a. $y = -\dfrac{4}{5}x + 80$

$m = -\dfrac{4}{5}, \; b = 80$

The slope is $-\dfrac{4}{5}$ and the y-intercept is (0, 80).

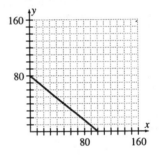

b. At $x = 10, \; y = 72$
$x = 20, \; y = 64$
$x = 40, \; y = 48$
$x = 64, \; y = 28.8$

c. Demand decreases as price increases.

d. At a price of $100 or more, demand is 0.

5.3 Experiencing Algebra the Calculator Way

1. a. $y = 4x$
$m = 4$
$b = 0$

b. $y = 2x$
$m = 2$
$b = 0$

c. $y = \left(\dfrac{1}{2}\right)x$

$m = \dfrac{1}{2}$
$b = 0$
$Y1 = 4x$
$Y2 = 2x$

$Y3 = \dfrac{1}{2}x$

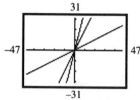

The y-intercept, $(0, 0)$, is the same.
The slopes are different, but all are positive.
The graphs rise at different rates, but all cross at $(0, 0)$.

2. **a.** $y = -4x$
$m = -4$
$b = 0$

b. $y = -2x$
$m = -2$
$b = 0$

c. $y = \left(-\dfrac{1}{2}\right)x$

$m = -\dfrac{1}{2}$
$b = 0$
$Y1 = -4x$
$Y2 = -2x$
$Y3 = -\dfrac{1}{2}x$

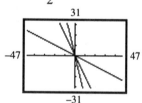

The y-intercept is the same at $(0, 0)$.
The slopes are different, but all are negative.
The graphs decline at different rates and all cross at $(0, 0)$.

3. **a.** $y = 4x + 9$
$m = 4$
$b = 9$

b. $y = 4x$
$m = 4$
$b = 0$

c. $y = 4x - 9$
$m = 4$
$b = -9$
$Y1 = 4x + 9$
$Y2 = 4x$
$Y3 = 4x - 9$

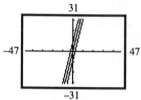

The slopes are the same. The y-intercepts are different. They are all parallel, and cross the y-axis at different points.

4. **a.** $y = -4x + 9$
$m = -4$
$b = 9$

b. $y = -4x$
$m = -4$
$b = 0$

c. $y = -4x - 9$
$m = -4$
$b = -9$
$Y1 = -4x + 9$
$Y2 = -4x$
$Y3 = -4x - 9$

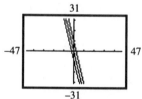

The slopes are the same. The y-intercepts are different. They are parallel, and cross the y-axis at different points.

5.4 Experiencing Algebra the Exercise Way

1. $m = \dfrac{3}{2}, b = -1$
$y = mx + b$
$y = \dfrac{3}{2}x - 1$

3. The graph is a vertical line with $c = -5$.
$$x = -5$$

5. $m = -3$, $b = 0$
$$y = mx + b$$
$$y = -3x + 0$$
$$y = -3x$$

7. The graph is a horizontal line with $b = -\dfrac{3}{2}$.
$$y = -\dfrac{3}{2}$$

9. $m = -\dfrac{2}{5}$, $b = 4$
$$y = mx + b$$
$$y = -\dfrac{2}{5}x + 4$$

11. $m = \dfrac{5}{9}$, $b = 0$
$$y = mx + b$$
$$y = \dfrac{5}{9}x + 0$$
$$y = \dfrac{5}{9}x$$

13. $m = 4$, $b = -\dfrac{3}{4}$
$$y = mx + b$$
$$y = 4x - \dfrac{3}{4}$$

15. $m = -4.1$, $b = 0.5$
$$y = mx + b$$
$$y = -4.1x + 0.5$$

17. $m = 0$, $b = -33$
$$y = mx + b$$
$$y = 0x - 33$$
$$y = -33$$

19. $m = \dfrac{2}{3}$, $(3, -3)$
$$y - (-3) = \dfrac{2}{3}(x - 3)$$
$$y + 3 = \dfrac{2}{3}x - 2$$
$$y = \dfrac{2}{3}x - 5$$

21. $m = -3$, $(0, 4)$
$$y - 4 = -3(x - 0)$$
$$y - 4 = -3x$$
$$y = -3x + 4$$

23. $m = -1.7$, $(3, -1.5)$
$$y - (-1.5) = -1.7(x - 3)$$
$$y + 1.5 = -1.7x + 5.1$$
$$y = -1.7x + 3.6$$

25. $(-1, 1)$ and $(1, -2)$
$$m = \dfrac{y_2 - y_1}{x_2 - x_1}$$
$$m = \dfrac{-2 - 1}{1 - (-1)}$$
$$m = \dfrac{-3}{2}$$
$$y - y_1 = m(x - x_1)$$
$$y - 1 = -\dfrac{3}{2}[x - (-1)]$$
$$y - 1 = -\dfrac{3}{2}(x + 1)$$
$$y - 1 = -\dfrac{3}{2}x - \dfrac{3}{2}$$
$$y - 1 + 1 = -\dfrac{3}{2}x - \dfrac{3}{2} + \dfrac{2}{2}$$
$$y = -\dfrac{3}{2}x - \dfrac{1}{2}$$

27. $(-1, -2)$ and $(-1, 5)$

$$m = \frac{y_2 - y_1}{x_2 - x_1}$$

$$m = \frac{5 - (-2)}{-1 - (-1)}$$

$$m = \frac{7}{0}$$

m is undefined.
The line is a vertical line.
The constant term is -1.
$x = -1$

29. $(-1, 1)$ and $(-2, -1)$

$$m = \frac{y_2 - y_1}{x_2 - x_1}$$

$$m = \frac{-1 - 1}{-2 - (-1)}$$

$$m = \frac{-2}{-1}$$

$m = 2$
$y - y_1 = m(x - x_1)$
$y - 1 = 2[x - (-1)]$
$y - 1 = 2(x + 1)$
$y - 1 + 1 = 2x + 2 + 1$
$y = 2x + 3$

31. $(-2, 2)$ and $(4, 2)$

$$m = \frac{y_2 - y_1}{x_2 - x_1}$$

$$m = \frac{2 - 2}{4 - (-2)}$$

$m = 0$
$y - y_1 = m(x - x_1)$
$y - 2 = 0[x - (-2)]$
$y - 2 + 2 = 0 + 2$
$y = 2$

33. $\left(4\frac{1}{2}, 5\frac{1}{4}\right)$ and $(1, 4)$

$$m = \frac{y_2 - y_1}{x_2 - x_1}$$

$$m = \frac{4 - 5\frac{1}{4}}{1 - 4\frac{1}{2}}$$

$$m = \frac{-1\frac{1}{4}}{-3\frac{1}{2}}$$

$$m = -\frac{5}{4} \cdot \left(-\frac{2}{7}\right)$$

$$m = \frac{5}{14}$$

$y - y_2 = m(x - x_2)$

$$y - 4 = \frac{5}{14}(x - 1)$$

$$y - 4 + 4 = \frac{5}{14}x - \frac{5}{14} + \frac{56}{14}$$

$$y = \frac{5}{14}x + \frac{51}{14}$$

35. $\left(-1\frac{1}{3}, 2\right)$ and $(0, 0)$

$$m = \frac{y_2 - y_1}{x_2 - x_1}$$

$$m = \frac{0 - 2}{0 - \left(-1\frac{1}{3}\right)}$$

$$m = \frac{-2}{\frac{4}{3}}$$

$$m = -2 \cdot \frac{3}{4}$$

$$m = -\frac{3}{2}$$

$y - y_2 = m(x - x_2)$

$$y - 0 = -\frac{3}{2}(x - 0)$$

$$y = -\frac{3}{2}x$$

37. $(0.5, 0)$ and $(-0.8, 4.2)$

$$m = \frac{y_2 - y_1}{x_2 - x_1}$$

$$m = \frac{4.2 - 0}{-0.8 - 0.5}$$

$$m = \frac{4.2}{-1.3}$$

$$m = -\frac{42}{13}$$

$$y - y_1 = m(x - x_1)$$

$$y - 0 = -\frac{42}{13}(x - 0.5)$$

$$y = -\frac{42}{13}x + \frac{21}{13}$$

39. $(2.4, 2.8)$ and $(-2.6, -2.2)$

$$m = \frac{y_2 - y_1}{x_2 - x_1}$$

$$m = \frac{-2.2 - 2.8}{-2.6 - 2.4}$$

$$m = \frac{5}{5}$$

$$m = 1$$

$$y - y_1 = m(x - x_1)$$

$$y - 2.8 = 1(x - 2.4)$$

$$y - 2.8 = x - 2.4$$

$$y - 2.8 + 2.8 = x - 2.4 + 2.8$$

$$y = x + 0.4$$

41. a.

$$m = \frac{y_2 - y_1}{x_2 - x_1}$$

$$m = \frac{25.5 - 42.4}{25 - 0}$$

$$m = \frac{-16.9}{25}$$

$$m = -0.676$$

b.

$$y - y_1 = m(x - x_1)$$

$$y - 42.4 = -0.676(x - 0)$$

$$y - 42.4 = -0.676x$$

$$y - 42.4 + 42.4 = -0.676x + 42.4$$

$$y = -0.676x + 42.4$$

c. In 1988, $x = 23$
At $x = 23$, $y = 26.852$
The percent is predicted to be 26.852%.
It is fairly close.

d. In 2005, $x = 40$
At $x = 40$, $y = 15.36$
The percent is predicted to be 15.36%.

e. At $x = 62$, $y \approx 0$
$1965 + 62 = 2027$
In year 2027, the predicted percentage
becomes close to zero.
Answers will vary.

43.

$$m = \frac{y_2 - y_1}{x_2 - x_1}$$

$$m = \frac{263 - 273}{-10 - 0}$$

$$m = \frac{-10}{-10}$$

$$m = 1$$

$$b = 273$$

$$y = mx + b$$

$$y = 1x + 273$$

$$y = x + 273$$

$$y = 100 + 273$$

$$y = 373$$

A Kelvin temperature of 373 corresponds to
100°C.

45. $m = \dfrac{y_2 - y_1}{x_2 - x_1}$

$m = \dfrac{16.96 - 15.92}{1.6 - 1.2}$

$m = \dfrac{1.04}{0.4}$

$m = 2.6$

$y - y_1 = m(x - x_1)$

$y - 15.92 = 2.6(x - 1.2)$

$y - 15.92 = 2.6x - 3.12$

$y - 15.92 + 15.92 = 2.6x - 3.12 + 15.92$

$y = 2.6x + 12.8$

$y = 2.6(1.2) + 12.8 = 15.92$

$y = 2.6(1.6) + 12.8 = 16.96$

$y = 2.6(2) + 12.8 = 18.00$

$y = 2.6(2.4) + 12.8 = 19.04$

$y = 2.6(2.8) + 12.8 = 20.08$

47. (2500, 40) and (3500, 35)

$m = \dfrac{y_2 - y_1}{x_2 - x_1}$

$m = \dfrac{35 - 40}{3500 - 2500}$

$m = \dfrac{-5}{1000}$

$m = -0.005$

$y - y_1 = m(x - x_1)$

$y - 40 = -0.005(x - 2500)$

$y - 40 = -0.005x + 12.5$

$y - 40 + 40 = -0.005x + 12.5 + 40$

$y = -0.005x + 52.5$

49. (6, 5000) and (15, 14,000)

$m = \dfrac{14{,}000 - 5000}{15 - 6}$

$m = \dfrac{9000}{9}$

$m = 1000$

$y - y_1 = m(x - x_1)$

$y - 5000 = 1000(x - 6)$

$y - 5000 = 1000x - 6000$

$y - 5000 + 5000 = 1000x - 6000 + 5000$

$y = 1000x - 1000$

$y = 1000(10) - 1000$

$y = 9000$

The prediction is 9000 sales.

51. $m = 120$, $b = 5470$

$y = mx + b$

$y = 120x + 5470$

$x = 1996 - 1990$

$x = 6$

$y = 120(6) + 5470$

$y = 6190$

The predicted enrollment in 1996 is 6190 students.

Yes, this is a good estimate.

$x = 2000 - 1990$

$x = 10$

$y = 120(10) + 5470$

$y = 6670$

The predicted enrollment in 2000 would be 6670 students.

5.4 Experiencing Algebra the Calculator Way

1. (12, 925) and (72, 4225)

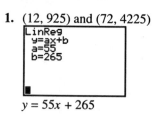

$y = 55x + 265$

2. (16, –351) and (40, 417)

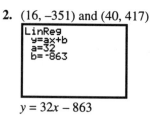

$y = 32x - 863$

3. (0, 6.4) and (36.4, 103.59)

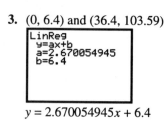

$y = 2.670054945x + 6.4$

4. $\left(1, \dfrac{17}{4}\right)$ and $\left(3, \dfrac{57}{4}\right)$

```
LinReg
 y=ax+b
 a=5
 b=-.75
```

$y = 5x - 0.75$

5. $(-5, 21)$ and $(-9, 7)$

```
LinReg
 y=ax+b
 a=3.5
 b=38.5
```

$y = 3.5x + 38.5$

6. $(16, -2)$ and $(7, 4)$

```
LinReg
 y=ax+b
 a=-.6666666667
 b=8.666666667
```

$y = -0.\overline{6}x + 8.\overline{6}$

Chapter 5 Review

Reflections

1–7. Answers will vary.

Exercises

1. $y = 0.6x + 2.3$ is linear.
$y - 2.3 - y = 0.6x + 2.3 - 2.3 - y$
$-2.3 = 0.6x - y$
$0.6x - y = -2.3$ is in standard form.

2. $y = 4x^2 - 2$ is nonlinear because the x variable is squared.

3. $3x - 6y + 12 = 0$ is linear.
$3x - 6y + 12 - 12 = 0 - 12$
$3x - 6y = -12$ is in standard form.

4. $5x - 3y + 7 = x - y + 9$ is linear
$5x - 3y + 7 - 7 = x - y + 9 - 7$
$5x - 3y = x - y + 2$
$5x - 3y - x + y = x - y + 2 - x + y$
$4x - 2y = 2$ is in standard form.

5. $3x^2 + y = 1$ is nonlinear because the x variable is squared.

6. $5y - 12 = 7 - y$ is linear
$5y - 12 + 12 = 7 - y + 12$
$5y = 19 - y$
$5y + y = 19 - y + y$
$6y = 19$ is in standard form.

7.

$$
\begin{array}{c|c}
\multicolumn{2}{c}{f(x) = 10x - 42} \\
\hline
-2 & 10(4) - 42 \\
 & 40 - 42 \\
 & -2
\end{array}
$$

Since $-2 = -2$, $(4, -2)$ is a solution.

8.

$$
\begin{array}{c|c}
\multicolumn{2}{c}{3x - 2y = 6} \\
\hline
3(1) - 2(-1) & 6 \\
3 + 2 & \\
5 &
\end{array}
$$

Since $5 \neq 6$, $(1, -1)$ is not a solution.

9.

$$
\begin{array}{c|c}
\multicolumn{2}{c}{\dfrac{3}{5}x - \dfrac{8}{9}y = 1} \\
\hline
\frac{3}{5}(15) - \frac{8}{9}(18) & 1 \\
9 - 16 & \\
-7 &
\end{array}
$$

Since $-7 \neq 1$, $(15, 18)$ is not a solution.

10.

$$
\begin{array}{c|c}
\multicolumn{2}{c}{45x - 14y = 31} \\
\hline
45\left(\frac{5}{12}\right) - 14\left(-\frac{7}{8}\right) & 31 \\
\frac{225}{12} + \frac{98}{8} & \\
31 &
\end{array}
$$

Since $31 = 31$, $\left(\dfrac{5}{12}, -\dfrac{7}{8}\right)$ is a solution.

11. Answers will vary. Possible answer:
$12x + 6y = 48$
$6y = -12x + 48$
$y = -2x + 8$

x	$y = -2x + 8$	y
-1	$y = -2(-1) + 8 = 10$	10
0	$y = -2(0) + 8 = 8$	8
1	$y = -2(1) + 8 = 6$	6

$(-1, 10)$, $(0, 8)$, and $(1, 6)$ are three possible solutions.

12. Answers will vary. Possible answer:

x	$y = \frac{8}{13}x - 7$	y
-13	$y = \frac{8}{13}(-13) - 7 = -15$	-15
0	$y = \frac{8}{13}(0) - 7 = -7$	-7
13	$y = \frac{8}{13}(13) - 7 = 1$	1

$(-13, -15)$, $(0, -7)$, and $(13, 1)$ are three possible solutions.

13. Answers will vary. Possible answer:

x	$y = -9$	y
0	$y = -9$	-9
1	$y = -9$	-9
2	$y = -9$	-9

$(0, -9)$, $(1, -9)$, and $(2, -9)$ are three possible solutions.

14. Answers will vary. Possible answer:
$9x - y = 12$
$9x - y + y = 12 + y$
$9x = 12 + y$
$9x - 12 = 12 + y - 12$
$9x - 12 = y$
$y = 9x - 12$

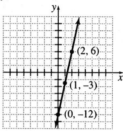

$(0, -12)$, $(1, -3)$, and $(2, 6)$ are three possible solutions.

15. Answers will vary. Possible answer:
$$y = -\frac{4}{3}x + 2$$

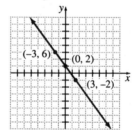

$(-3, 6)$, $(0, 2)$, and $(3, -2)$ are three possible solutions.

16. Answers will vary. Possible answer:
$$5y - 2 = y - 10$$
$$5y - 2 - y = y - 10 - y$$
$$4y - 2 = -10$$
$$4y - 2 + 2 = -10 + 2$$
$$4y = -8$$
$$\frac{4y}{4} = \frac{-8}{4}$$
$$y = -2$$

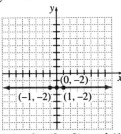

$(-1, -2)$, $(0, -2)$, and $(1, -2)$ are three possible solutions.

17. Answers will vary. Possible answer:
$$2x + 16 = x + 18$$
$$2x + 16 - x = x + 18 - x$$
$$x + 16 = 18$$
$$x + 16 - 16 = 18 - 16$$
$$x = 2$$

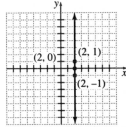

$(2, -1)$, $(2, 0)$, and $(2, 1)$ are three possible solutions.

18. a. $(10, 85)$ and $(20, 160)$

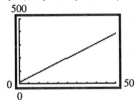

b. At $x = 30$, $y = 235$
She will receive \$235 for a job that is 30 pages long.

19. a. $y = 20 + 9x$
$$y = 20 + 9(10)$$
$$y = 110$$
He would earn \$110.

b. $y = 20 + 9(15)$
$$y = 155$$
No, he should be paid \$155.

c. $y = 20 + 9(0)$
$$y = 20$$
He would be paid \$20.

20. $8x + 12y = -24$
Substitute 0 for y.
$$8x + 12(0) = -24$$
$$8x = -24$$
$$\frac{8x}{8} = \frac{-24}{8}$$
$$x = -3$$
The x-intercept is $(-3, 0)$.
Substitute 0 for x.
$$8(0) + 12y = -24$$
$$12y = -24$$
$$\frac{12y}{12} = \frac{-24}{12}$$
$$y = -2$$
The y-intercept is $(0, -2)$.

21. $y = \dfrac{2x + 20}{5}$
Substitute 0 for y.
$$0 = \frac{2x + 20}{5}$$
$$5(0) = 5\left(\frac{2x + 20}{5}\right)$$
$$0 = 2x + 20$$
$$0 - 20 = 2x + 20 - 20$$
$$-20 = 2x$$
$$\frac{-20}{2} = \frac{2x}{2}$$
$$-10 = x$$

The x-intercept is $(-10, 0)$.
Substitute 0 for x.
$$y = \frac{2(0) + 20}{5}$$
$$y = \frac{20}{5}$$
$$y = 4$$
The y-intercept is $(0, 4)$.

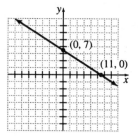

22. $9x - 3y = -12$
$9x - 3y - 9x = -12 - 9x$
$-3y = -12 - 9x$
$$\frac{-3y}{-3} = \frac{-12 - 9x}{-3}$$
$y = 4 + 3x$
$y = 3x + 4$
The y-intercept is $(0, 4)$.

23. $6x + 2y = x$
$6x + 2y - 6x = x - 6x$
$2y = -5x$
$$\frac{2y}{2} = \frac{-5x}{2}$$
$$y = -\frac{5}{2}x$$
The y-intercept is $(0, 0)$.

24. $7x + 11y = 77$
Substitute 0 for y.
$7x + 11(0) = 77$
$7x = 77$
$$\frac{7x}{7} = \frac{77}{7}$$
$x = 11$
The x-intercept is $(11, 0)$.
Substitute 0 for x.
$7(0) + 11y = 77$
$11y = 77$
$$\frac{11y}{11} = \frac{77}{11}$$
$y = 7$
The y-intercept is $(0, 7)$.

25. $-x - y = 2$
Substitute 0 for y.
$-x - 0 = 2$
$$\frac{-x}{-1} = \frac{2}{-1}$$
$x = -2$
The x-intercept is $(-2, 0)$.
Substitute 0 for x.
$-0 - y = 2$
$$\frac{-y}{-1} = \frac{2}{-1}$$
$y = -2$
The y-intercept is $(0, -2)$.

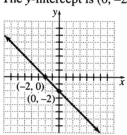

26. $7.4x + 14.8y = 29.6$
Substitute 0 for y.
$7.4x + 14.8(0) = 29.6$
$7.4x = 29.6$
$$\frac{7.4x}{7.4} = \frac{29.6}{7.4}$$
$x = 4$
The x-intercept is $(4, 0)$.
Substitute 0 for x.
$7.4(0) + 14.8y = 29.6$
$$\frac{14.8y}{14.8} = \frac{29.6}{14.8}$$
$y = 2$
The x-intercept is $(0, 2)$.

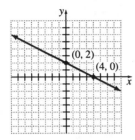

27. a. $y = 5200 - 100x$

b. Substitute 0 for y.
$0 = 5200 - 100x$
$$\frac{-5200}{-100} = \frac{-100x}{-100}$$
$52 = x$
The x-intercept is (52, 0). At 52 weeks the balance is $0.
Substitute 0 for x.
$y = 5200 - 100(0)$
$y = 5200$
The y-intercept is (0, 5200). At 0 weeks, the balance is $5200.

c.

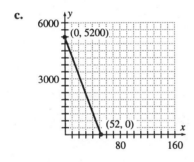

d. At $x = 32$, $y = 2000$
There is $2000 in the account.

28. a. $y = 50 - 0.5x$

b. Substitute 0 for y.
$0 = 50 - 0.5x$
$$\frac{-50}{-0.5} = \frac{-0.5x}{-0.5}$$
$100 = x$
The x-intercept is (100, 0). After 100 plays, Bryon has $0 left.
Substitute 0 for x.
$y = 50 - 0.5(0)$
$y = 50$
The y-intercept is (0, 50). After 0 plays, Bryon has $50.

c.

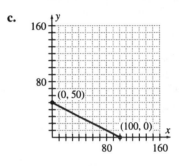

d. At $x = 60$, $y = 20$
He has $20 left.

29. The slope is 0 because the line is horizontal.

30. The slope is $\dfrac{9}{2}$.

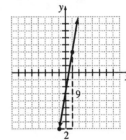

211

31. The slope is $-\dfrac{8}{5}$.

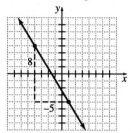

32. The slope is undefined because the line is vertical.

33. $(-5, -3)$ and $(1, 0)$

$$m = \frac{y_2 - y_1}{x_2 - x_1}$$

$$m = \frac{0 - (-3)}{1 - (-5)}$$

$$m = \frac{3}{6}$$

$$m = \frac{1}{2}$$

The slope is $\dfrac{1}{2}$.

34. $(-7, 7)$ and $(-4, 2)$

$$m = \frac{y_2 - y_1}{x_2 - x_1}$$

$$m = \frac{2 - 7}{-4 - (-7)}$$

$$m = \frac{-5}{3}$$

The slope is $-\dfrac{5}{3}$.

35. $(4, -3)$ and $(10, -3)$

$$m = \frac{y_2 - y_1}{x_2 - x_1}$$

$$m = \frac{-3 - (-3)}{10 - 4}$$

$$m = \frac{0}{6}$$

The slope is 0.

36. $(4, -3)$ and $(4, 3)$

$$m = \frac{y_2 - y_1}{x_2 - x_1}$$

$$m = \frac{3 - (-3)}{4 - 4}$$

$$m = \frac{6}{0}$$

The slope is undefined.

37. $y = 23x - 51$
$m = 23$, $b = -51$
The slope is 23 and the y-intercept is $(0, -51)$.

38. $y = -3.05x + 2.97$
$m = -3.05$, $b = 2.97$
The slope is -3.05 and the y-intercept is $(0, 2.97)$.

39. $6x + 5y = 12$
$6x + 5y - 6x = 12 - 6x$
$5y = -6x + 12$

$$\frac{5y}{5} = \frac{-6x + 12}{5}$$

$$y = -\frac{6}{5}x + \frac{12}{5}$$

$$m = -\frac{6}{5}, \ b = \frac{12}{5}$$

The slope is $-\dfrac{6}{5}$ and the y-intercept is $\left(0, \dfrac{12}{5}\right)$.

40. $4(y - 2) = 3(2y - 1) + 4$

$4y - 8 = 6y - 3 + 4$

$4y - 8 = 6y + 1$

$4y - 8 - 6y = 6y + 1 - 6y$

$-2y - 8 = 1$

$-2y - 8 + 8 = 1 + 8$

$-2y = 9$

$$\frac{-2y}{-2} = \frac{9}{-2}$$

$$y = -\frac{9}{2}$$

$m = 0$, $b = -\dfrac{9}{2}$

The slope is 0 and the *y*-intercept is

$\left(0, -\dfrac{9}{2}\right).$

41. $(-2, 3)$ and $m = \dfrac{5}{9}$

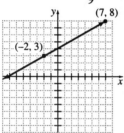

42. $(3, -2)$ and $m = 3$

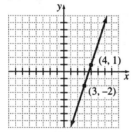

43. $(-2, -2)$ and $m = 0$

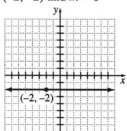

44. $(-2, -2)$ and $m =$ undefined

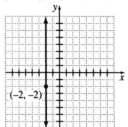

45. $y = -\dfrac{5}{3}x + 2$

$m = -\dfrac{5}{3}$, $b = 2$

The slope is $-\dfrac{5}{3}$ and the *y*-intercept is

$(0, 2)$.

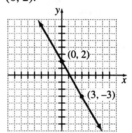

46. $3x + 2y = 12$

$3x + 2y - 3x = 12 - 3x$

$2y = -3x + 12$

$$\frac{2y}{2} = \frac{-3x + 12}{2}$$

$$y = -\frac{3}{2}x + 6$$

The slope is $-\dfrac{3}{2}$ and the *y*-intercept is

$(0, 6)$.

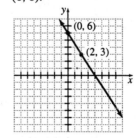

47. $m(x) = \dfrac{5}{8}x$

The slope is $\dfrac{5}{8}$ and the y-intercept is $(0, 0)$.

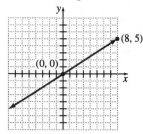

48. $y = -4$
The slope is 0 and the y-intercept is $(0, -4)$.

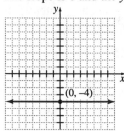

49. a. $(0, 62)$ $(4, 100)$

$m = \dfrac{y_2 - y_1}{x_2 - x_1}$

$m = \dfrac{100 - 62}{4 - 0} = \dfrac{38}{4} = 9.5$

$m = 9.5$
The slope is 9.5.

b. $y - y_1 = m(x - x_1)$
$y - 100 = 9.5(x - 4)$
$y - 100 = 9.5x - 38$
$y - 100 + 100 = 9.5x - 38 + 100$
$y = 9.5x + 62$

c.

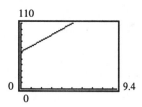

At $x = 3$, $y = 90.5$
The prediction is not very close to 98.
At $x = 2$, $y = 81$
The prediction is fairly close to 80 and 82.

d. At $x = 1.5$, $y = 76.25$
The predicted grade would be about 76 points.

50. $m = -\dfrac{1}{4}$, $b = 1$

$y = -\dfrac{1}{4}x + 1$

51. $m = -2$, $b = 3$
$y = -2x + 3$

52. $m = \dfrac{3}{5}$, $b = -2$

$y = \dfrac{3}{5}x - 2$

53. $m = 0$, $b = -3.5$
$y = 0x - 3.5$
$y = -3.5$

54. Vertical line, $c = 2.6$
$x = 2.6$

55. $m = -5$, $(2, -3)$
$y + 3 = -5(x - 2)$
$y + 3 = -5x + 10$
$y = -5x + 7$

56. $(9, 5)$ and $(-2, 2)$

$m = \dfrac{y_2 - y_1}{x_2 - x_1}$

$m = \dfrac{2 - 5}{-2 - 9}$

$m = \dfrac{-3}{-11}$

$m = \dfrac{3}{11}$

$$y - y_1 = m(x - x_1)$$

$$y - 5 = \frac{3}{11}(x - 9)$$

$$y - 5 = \frac{3}{11}x - \frac{27}{11}$$

$$y = \frac{3}{11}x + \frac{28}{11}$$

57. $y - y_1 = m(x - x_1)$

$$y - 6 = \frac{1}{3}(x - 4)$$

$$y - 6 = \frac{1}{3}x - \frac{4}{3}$$

$$y = \frac{1}{3}x + \frac{14}{3}$$

58. a. (2, 15) and (8, 30)

$$m = \frac{y_2 - y_1}{x_2 - x_1}$$

$$m = \frac{30 - 15}{8 - 2}$$

$$m = \frac{15}{6}$$

$$m = \frac{5}{2}$$

$$y - y_1 = m(x - x_1)$$

$$y - 15 = \frac{5}{2}(x - 2)$$

$$y - 15 = \frac{5}{2}x - 5$$

$$y = \frac{5}{2}x + 10$$

b. $y = \frac{5}{2}(5) + 10$

$y = 22.5$

The prediction is 22.5 thousands of records sold. This is not very close to the actual sales.

Chapter 5 Mixed Review

1. Answers will vary. Possible answer:

$3(y - 5) = y + 7$
$3y - 15 = y + 7$
$3y - 15 - y + 15 = y + 7 - y + 15$
$2y = 22$
$$\frac{2y}{2} = \frac{22}{2}$$
$y = 11$

x	$y = 11$	y
–1	$y = 11$	11
0	$y = 11$	11
1	$y = 11$	11

(–1, 11), (0, 11), and (1, 11) are three possible solutions.

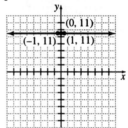

2. Answers will vary. Possible answer.

$14x + 7y = 7$
$14x + 7y - 14x = 7 - 14x$
$7y = -14x + 7$
$$\frac{7y}{7} = \frac{-14x + 7}{7}$$
$y = -2x + 1$

x	$y = -2x + 1$	y
–2	$y = -2(-2) + 1 = 5$	5
0	$y = -2(0) + 1 = 1$	1
2	$y = -2(2) + 1 = -3$	–3

(–2, 5), (0, 1), and (2, –3) are three possible solutions.

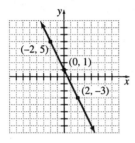

3. Answers will vary. Possible answer:

x	$y = -\frac{7}{5}x - 10$	y
-5	$y = -\frac{7}{5}(-5) - 10 = -3$	-3
0	$y = -\frac{7}{5}(0) - 10 = -10$	-10
5	$y = -\frac{7}{5}(5) - 10 = -17$	-17

$(-5, -3)$, $(0, -10)$, and $(5, -17)$ are three possible solutions.

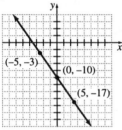

4. $(2, 6)$ and $m = 0$

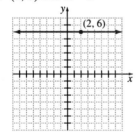

5. $(2, 6)$ and $m =$ undefined

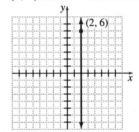

6. $(1, -4)$ and $m = -2$

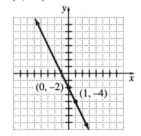

7. $(0, 3)$ and $m = \dfrac{2}{7}$

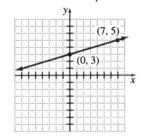

8. $y = -\dfrac{3}{8}x$

$m = -\dfrac{3}{8}$, $b = 0$

The slope is $-\dfrac{3}{8}$ and the y-intercept is $(0, 0)$.

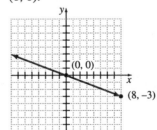

9. $y = 7$
$m = 0, b = 7$
The slope is 0 and the y-intercept is (0, 7).

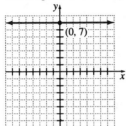

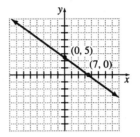

10. $5x - 4y = 12$
$5x - 4y - 5x = 12 - 5x$
$-4y = -5x + 12$
$\dfrac{-4y}{-4} = \dfrac{-5x + 12}{-4}$
$y = \dfrac{5}{4}x - 3$
$m = \dfrac{5}{4}, \ b = -3$

The slope is $\dfrac{5}{4}$ and the y-intercept is (0, –3).

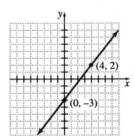

11. $15.5x + 21.7y = 108.5$
Substitute 0 for y.
$15.5x + 21.7(0) = 108.5$
$\dfrac{15.5x}{15.5} = \dfrac{108.5}{15.5}$
$x = 7$
The x-intercept is (7, 0).
Substitute 0 for x.
$15.5(0) + 21.7y = 108.5$
$\dfrac{21.7y}{21.7} = \dfrac{108.5}{21.7}$
$y = 5$
The y-intercept is (0, 5).

12. $12x - 24y = -48$
Substitute 0 for y.
$12x - 24(0) = -48$
$\dfrac{12x}{12} = \dfrac{-48}{12}$
$x = -4$
The x-intercept is (–4, 0).
Substitute 0 for x.
$12(0) - 24y = -48$
$\dfrac{-24y}{-24} = \dfrac{-48}{-24}$
$y = 2$
The y-intercept is (0, 2).

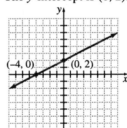

13. $-x + y = -6$
Substitute 0 for y.
$-x + 0 = -6$
$\dfrac{-x}{-1} = \dfrac{-6}{-1}$
$x = 6$
The x-intercept is (6, 0).
Substitute 0 for x.
$-0 + y = -6$
$y = -6$
The y-intercept is (0, –6).

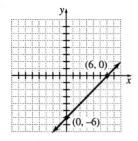

14. Answers will vary. Possible answer:

x	$y = 8$	y
-1	$y = 8$	8
0	$y = 8$	8
1	$y = 8$	8

$(-1, 8)$, $(0, 8)$, and $(1, 8)$ are three possible solutions.

15. Answers will vary. Possible answer:
$$8x - 9y = -72$$
$$8x - 9y - 8x = -72 - 8x$$
$$-9y = -8x - 72$$
$$\frac{-9y}{-9} = \frac{-8x - 72}{-9}$$
$$y = \frac{8}{9}x + 8$$

x	$y = \frac{8}{9}x + 8$	y
-9	$y = \frac{8}{9}(-9) + 8 = 0$	0
0	$y = \frac{8}{9}(0) + 8 = 8$	8
9	$y = \frac{8}{9}(9) + 8 = 16$	16

$(-9, 0)$, $(0, 8)$, and $(9, 16)$ are three possible solutions.

16. Answers will vary. Possible answer:

x	$t(x) = \frac{9}{11}x - 8$	$t(x)$
-11	$t(x) = \frac{9}{11}(-11) - 8 = -17$	-17
0	$t(x) = \frac{9}{11}(0) - 8 = -8$	-8
11	$t(x) = \frac{9}{11}(11) - 8 = 1$	1

$(-11, -17)$, $(0, -8)$, and $(11, 1)$ are three possible solutions.

17. $y = x$ is linear.
$$y - y = x - y$$
$$0 = x - y$$
$x - y = 0$ is in standard form.

18. $y = 2x^2 - 8$ is nonlinear because the variable x is squared.

19. $y = 1.3x - 0.5$ is linear.
$$y - y = 1.3x - 0.5 - y$$
$$0 = 1.3x - 0.5 - y$$
$$0.5 + 0 = 1.3x - 0.5 - y + 0.5$$
$$0.5 = 1.3x - y$$
$1.3x - y = 0.5$ is in standard form.

20. $y = 15x$ is linear.
$$y - y = 15x - y$$
$$0 = 15x - y$$
$15x - y = 0$ is in standard form.

21. $8y - 5x = 21$ is linear.
$$-5x + 8y = 21$$
$$-1(-5x) - 1(8y) = -1(21)$$
$5x - 8y = -21$ is in standard form.

22. $2x + 3y - 1 = y^2 + x$ is nonlinear because one y-variable is squared.

23.

$$\frac{2}{3}x + \frac{3}{4}y = 2$$

$\frac{2}{3}(12) + \frac{3}{4}(-8)$	2
$8 + (-6)$	
2	

Since $2 = 2$, $(12, -8)$ is a solution.

24.

$$\frac{22x - 5y = 17}{22\left(\frac{7}{12}\right) - 5\left(-\frac{5}{6}\right) \;\bigg|\; 17}$$

$$\frac{154}{12} + \frac{25}{6} \;\bigg|$$

$$17 \;\bigg|$$

Since $17 = 17$, $\left(\dfrac{7}{12}, -\dfrac{5}{6}\right)$ is a solution.

25.

$$\frac{u(x) = -8x + 2}{-14 \;\bigg|\; -8(2) + 2}$$

$$-16 + 2 \;\bigg|$$

$$-14 \;\bigg|$$

Since $-14 = -14$, $(2, -14)$ is a solution.

26.

$$\frac{7x + 8y = 8}{7(0) + 8(-3) \;\bigg|\; 8}$$

$$-24 \;\bigg|$$

Since $-24 \neq 8$, $(0, -3)$ is not a solution.

27.

$$\frac{x = -8}{-8 \;\bigg|\; -8}$$

Since $-8 = -8$, $(-8, -2)$ is a solution.

28.

$$\frac{v(x) = 12}{-2 \;\bigg|\; 12}$$

Since $-2 \neq 12$, $(12, -2)$ is not a solution.

29. $x + 4(x - 2) = 2$
$x + 4x - 8 = 2$
$5x - 8 = 2$
$5x - 8 + 8 = 2 + 8$
$5x = 10$
$\dfrac{5x}{5} = \dfrac{10}{5}$
$x = 2$
The slope is undefined.
There is no y-intercept.

30. $x - 3(y + 2) = 4(x + 1) - y$
$x - 3y - 6 = 4x + 4 - y$
$x - 3y - 6 - x + 6 = 4x + 4 - y - x + 6$
$-3y = 3x + 10 - y$
$-3y + y = 3x + 10 - y + y$
$-2y = 3x + 10$
$\dfrac{-2y}{-2} = \dfrac{3x + 10}{-2}$
$y = -\dfrac{3}{2}x - 5$
$m = -\dfrac{3}{2}, \; b = -5$

The slope is $-\dfrac{3}{2}$ and the y-intercept is $(0, -5)$.

31. $12x - 4y = 8$
$12x - 4y - 12x = 8 - 12x$
$-4y = -12x + 8$
$\dfrac{-4y}{-4} = \dfrac{-12x + 8}{-4}$
$y = 3x - 2$
$m = 3, \; b = -2$
The slope is 3 and the y-intercept is $(0, -2)$.

32. $2y - 6 = 4(y - 2) + 2$
$2y - 6 = 4y - 8 + 2$
$2y - 6 = 4y - 6$
$2y - 6 + 6 = 4y - 6 + 6$
$2y = 4y$
$2y - 4y = 4y - 4y$
$-2y = 0$
$\dfrac{-2y}{-2} = \dfrac{0}{-2}$
$y = 0$
$m = 0, \; b = 0$
The slope is 0 and the y-intercept is $(0, 0)$.

33. $y = 13x - 15$
$m = 13, \; b = -15$
The slope is 13 and the y-intercept is $(0, -15)$.

34. $y = -5.03x + 7.92$
$m = -5.03, \; b = 7.92$
The slope is -5.03 and the y-intercept is $(0, 7.92)$.

35. (5, 5) and (8, 5)

$$m = \frac{y_2 - y_1}{x_2 - x_1}$$

$$m = \frac{5 - 5}{8 - 5}$$

$$m = \frac{0}{3}$$

The slope is 0.

36. (−3, −3) and (−3, 3)

$$m = \frac{y_2 - y_1}{x_2 - x_1}$$

$$m = \frac{3 - (-3)}{-3 - (-3)}$$

$$m = \frac{6}{0}$$

The slope is undefined.

37. (−4, 4) and (−1, −5)

$$m = \frac{y_2 - y_1}{x_2 - x_1}$$

$$m = \frac{-5 - 4}{-1 - (-4)}$$

$$m = \frac{-9}{3}$$

$$m = -3$$

The slope is −3.

38. $\left(-\dfrac{2}{5}, \dfrac{4}{7}\right)$ and $\left(-\dfrac{4}{7}, \dfrac{1}{5}\right)$

$$m = \frac{y_2 - y_1}{x_2 - x_1}$$

$$m = \frac{\frac{1}{5} - \frac{4}{7}}{-\frac{4}{7} - \left(-\frac{2}{5}\right)}$$

$$m = \frac{\frac{7 - 20}{35}}{\frac{-20 + 14}{35}}$$

$$m = \frac{-13}{35} \cdot \frac{35}{-6}$$

$$m = \frac{-13}{-6}$$

$$m = \frac{13}{6}$$

The slope is $\dfrac{13}{6}$.

39. $m = \dfrac{y_2 - y_1}{x_2 - x_1}$

$$m = \frac{2 - (-2)}{5 - 4}$$

$$m = \frac{4}{1}$$

$$m = 4$$

$$y - y_1 = m(x - x_1)$$

$$y - (-2) = 4(x - 4)$$

$$y + 2 = 4x - 16$$

$$y = 4x - 18$$

40. $y - y_1 = m(x - x_1)$

$$y - (-1) = -\frac{1}{4}[x - (-1)]$$

$$y + 1 = -\frac{1}{4}x - \frac{1}{4}$$

$$y = -\frac{1}{4}x - \frac{5}{4}$$

41. $m = -\dfrac{2}{3}$, $b = 5$

$$y = -\frac{2}{3}x + 5$$

42. A vertical line has a constant of 4.1.

$$x = 4.1$$

43. $m = 0, b = 8$
$y = 0x + 8$
$y = 8$

44. $m = 4, (1.2, 8)$
$y - 8 = 4(x - 1.2)$
$y - 8 = 4x - 4.8$
$y = 4x + 3.2$

45. $m = 3, b = -2$
$y = 3x - 2$

46. a. $m = \dfrac{y_2 - y_1}{x_2 - x_1}$

$m = \dfrac{99 - 57}{94 - 50}$

$m = \dfrac{42}{44}$

$m = \dfrac{21}{22}$

$y - y_1 = m(x - x_1)$

$y - 57 = \dfrac{21}{22}(x - 50)$

$y - 57 = \dfrac{21}{22}x - \dfrac{525}{11}$

$y - 57 + 57 = \dfrac{21}{22}x - \dfrac{525}{11} + \dfrac{627}{11}$

$y = \dfrac{21}{22}x + \dfrac{102}{11}$

b. $y = \dfrac{21}{22}(80) + \dfrac{102}{11}$

$y \approx 85.6$
The predicted score of about 86 is not very close to the actual score of 78.

47. a. (7, 24) and (9, 28)

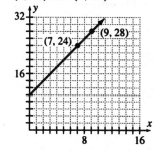

b. See the graph in part a.
At $x = 8, y = 26$
Frank would receive \$26 for 8 innings.
At $x = 10, y = 30$
Frank would receive \$30 for 10 innings.

48. a. $y = 180 + 10x - 15x$
$y = -5x + 180$

b. Substitute 0 for y.
$0 = -5x + 180$
$5x = 180$
$\dfrac{5x}{5} = \dfrac{180}{5}$
$x = 36$
The x-intercept is (36, 0). After 36 hours, the tank is empty.
Substitute 0 for x.
$y = -5(0) + 180$
$y = 180$
The y-intercept is (0, 180). After 0 hours, the tank has 180 gallons.

c.

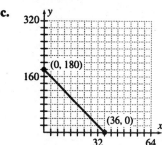

d. At $x = 16, y = 100$
After 16 hours, 100 gallons remain in the tank.

49. a. $y = 25x + 15$

b. $y = 25(2) + 15$
$y = 65$
The trainer would earn \$65.

c. $y = 25\left(1\dfrac{1}{2}\right) + 15$

$y = 52\dfrac{1}{2}$
He would earn \$52.50.

50. grade $= \dfrac{60 \text{ feet}}{125 \text{ feet}} = \dfrac{12}{25} = 48\%$

Chapter 5 Test

1.

$7x = 8(y - 3)$	
$7(8)$	$8(10 - 3)$
56	$8(7)$
	56

Since $56 = 56$, $(8, 10)$ is a solution.

2.

$x - 8y = 16$	
$-8 - 8(-2)$	16
$-8 + 16$	
8	

Since $8 \neq 16$, $(-8, -2)$ is not a solution.

3. $5x - 7y = 2(x + 1) - 3$ is linear because it simplifies to $3x - 7y = -1$.

4. $y = \dfrac{11}{12}x - 5$ is linear because it is in slope-intercept form $y = ax + b$.

5. $y^2 - 16 = 0$ is nonlinear because the variable y is squared.

6. $7x + 2y = 3x^2 + 3$ is nonlinear because one of the x-variables is squared.

7. Answers will vary. Possible answer:
$2x - 8y = 0$
$2x - 8y - 2x = 0 - 2x$
$-8y = -2x$
$\dfrac{-8y}{-8} = \dfrac{-2x}{-8}$
$y = \dfrac{1}{4}x$

x	$y = \frac{1}{4}x$	y
-4	$y = \frac{1}{4}(-4) = -1$	-1
0	$y = \frac{1}{4}(0) = 0$	0
4	$y = \frac{1}{4}(4) = 1$	1

$(-4, -1)$, $(0, 0)$, and $(4, 1)$ are three possible solutions.

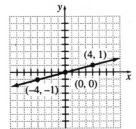

8. $12x + 15y = 60$
Substitute 0 for y.
$12x + 15(0) = 60$
$\dfrac{12x}{12} = \dfrac{60}{12}$
$x = 5$
The x-intercept is $(5, 0)$.
Substitute 0 for x.
$12(0) + 15y = 60$
$\dfrac{15y}{15} = \dfrac{60}{15}$
$y = 4$
The y-intercept is $(0, 4)$.

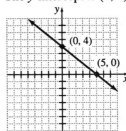

9. $g(x) = 3x - 5$
$m = 3, b = -5$
The slope is 3 and the y-intercept is $(0, -5)$.

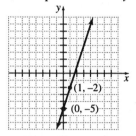

10. Graph of $y = \frac{1}{2}x - \frac{7}{2}$

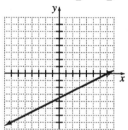

11. Graph of $x = 1$

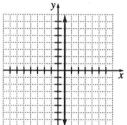

12. Graph of $y = -2$

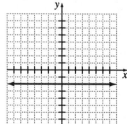

13. Graph of $y = \frac{2}{5}x - 2$

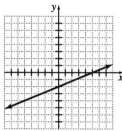

14. $(0, -1)$ and $(-1, -1)$

$m = \dfrac{y_2 - y_1}{x_2 - x_1}$

$m = \dfrac{-1 - (-1)}{-1 - 0}$

$m = \dfrac{0}{-1}$

The slope is 0.

15. $(4.8, 0.6)$ and $(-1.5, 2.4)$

$m = \dfrac{y_2 - y_1}{x_2 - x_1}$

$m = \dfrac{2.4 - 0.6}{-1.5 - 4.8}$

$m = \dfrac{1.8}{-6.3}$

$m = -\dfrac{18}{63}$

$m = -\dfrac{2}{7}$

The slope is $-\dfrac{2}{7}$.

16. $(-3, -4)$ and $(1, 0)$

$m = \dfrac{y_2 - y_1}{x_2 - x_1}$

$m = \dfrac{0 - (-4)}{1 - (-3)}$

$m = \dfrac{4}{4}$

$m = 1$

The slope is 1.

17. $(-2, 2)$ and $(-2, 8)$

$$m = \frac{y_2 - x_1}{x_2 - x_1}$$

$$m = \frac{8 - 2}{-2 - (-2)}$$

$$m = \frac{6}{0}$$

The slope is undefined.

18. $4y = 3x + 12$

$$\frac{4y}{4} = \frac{3x + 12}{4}$$

$$y = \frac{3}{4}x + 3$$

$$y = \frac{3}{4},\ b = 3$$

The slope is $\frac{3}{4}$ and the y-intercept is $(0, 3)$.

19. $7x + 7y = 21$
$7x + 7y - 7x = 21 - 7x$
$7y = -7x + 21$

$$\frac{7y}{7} = \frac{-7x + 21}{7}$$

$y = -x + 3$
$m = -1,\ b = 3$
The slope is -1 and the y-intercept is $(0, 3)$.

20. $m = 9,\ b = 7$
$y = 9x + 7$

21. $y - 2 = 6[x - (-1)]$
$y - 2 = 6(x + 1)$
$y - 2 = 6x + 6$
$y = 6x + 8$

22. $(2, 3)\ \ (-4, 1)$

$$m = \frac{1 - 3}{-4 - 2} = \frac{-2}{-6} = \frac{1}{3}$$

$$y - 3 = \frac{1}{3}(x - 2)$$

$$y - 3 = \frac{1}{3}x - \frac{2}{3}$$

$$y = \frac{1}{3}x + \frac{7}{3}$$

23. $y = 1450 - 150x$

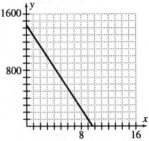

At $x = 4$, $y = 850$
After 4 months, his remaining loan is \$850.

24. Answers will vary.

Chapter 6

6.1 Experiencing Algebra the Exercise Way

1. $3x - 2y = 5(y + 7)$
$3x - 2y = 5y + 35$
$3x - 2y - 3x = 5y + 35 - 3x$
$-2y = 5y + 35 - 3x$
$-2y - 5y = 5y + 35 - 3x - 5y$
$-7y = -3x + 35$
$\dfrac{-7y}{-7} = \dfrac{-3x + 35}{-7}$
$y = \dfrac{3}{7}x - 5 \quad m = \dfrac{3}{7}, b = -5$
$7x = 3(1 - y)$
$7x = 3 - 3y$
$7x - 3 = 3 - 3y - 3$
$7x - 3 = -3y$
$\dfrac{7x - 3}{-3} = \dfrac{-3y}{-3}$
$-\dfrac{7}{3}x + 1 = y$
$y = -\dfrac{7}{3}x + 1 \quad m = -\dfrac{7}{3}, b = 1$
The lines are intersecting and perpendicular because the slopes (m) are opposite reciprocals of each other.

3. $x = 4(y - 3)$
$x = 4y - 12$
$x + 12 = 4y - 12 + 12$
$x + 12 = 4y$
$\dfrac{x + 12}{4} = \dfrac{4y}{4}$
$y = \dfrac{1}{4}x + 3 \quad m = \dfrac{1}{4}, b = 3$
$x = 4(y + 5)$
$x = 4y + 20$
$x - 20 = 4y + 20 - 20$
$x - 20 = 4y$
$\dfrac{x - 20}{4} = \dfrac{4y}{4}$
$y = \dfrac{1}{4}x - 5 \quad m = \dfrac{1}{4}, b = -5$
The lines are parallel because the slopes (m) are equal and the y-intercepts (b) are not equal.

5. $4x - y = 6$
$y = 4x - 6 \quad m = 4, b = -6$
$2x - y + 3 = 0$
$y = 2x + 3 \quad m = 2, b = 3$
The lines are only intersecting because the slopes (m) are not equal and their product is not -1.

7. $x = 2(y - 7)$
$x = 2y - 14$
$2y = x + 14$
$y = \dfrac{1}{2}x + 7 \quad m = \dfrac{1}{2}, b = 7$
$y = \dfrac{1}{2}x + 7 \quad m = \dfrac{1}{2}, b = 7$
The lines are coinciding because the slopes (m) are equal and the y-intercepts (b) are equal.

9. $5x + y = -6$
$y = -5x - 6 \quad m = -5, b = -6$
$3x + y = 0$
$y = -3x \quad m = -3, b = 0$
The lines are only intersecting because the slopes (m) are not equal and their product is not -1.

11. $4x + y = 8$
$y = -4x + 8 \quad m = -4, b = 8$
$4x + y + 2 = 0$
$y = -4x - 2 \quad m = -4, b = -2$
The lines are parallel because the slopes (m) are equal and the y-intercepts are not equal.

13. $y - 5 = 0$
$y = 5$
$2x + 6 = x + 9$
$x = 5$
The lines are intersecting and perpendicular because one is vertical and the other is horizontal.

15. $2y - 3 = 13$
$2y = 16$
$y = 8$ $m = 0, b = 8$
$y + 1 = 4$
$y = 3$ $m = 0, b = 3$
The lines are parallel because both lines are horizontal with different y-intercepts.

17. $x + 3 = 0$
$x = -3$
$x - 5 = 0$
$x = 5$
The lines are parallel because they are both vertical with different constants.

19. $2x - 9 = 0$
$x = \dfrac{9}{2}$
$x - 4 = 5 - x$
$2x = 9$
$x = \dfrac{9}{2}$
The lines are coinciding because both are vertical with the same constant.

21. $3(y - 3) = 1$
$3y - 9 = 1$
$3y = 10$
$y = \dfrac{10}{3}$ $m = 0, b = \dfrac{10}{3}$

$5y = 10 + 2y$
$3y = 10$
$y = \dfrac{10}{3}$ $m = 0, b = \dfrac{10}{3}$
The lines are coinciding because they are both horizontal lines with the same constant.

23. $x = 2$ $m =$ undefined
$y = 2x - 1$ $m = 2, b = -1$
The lines are only intersecting because one is a vertical line and the other is not a vertical or a horizontal line.

25. $y - 3 = 0$
$y = 3$ $m = 0, b = 3$

$2x + 3y = 0$
$3y = -2x$

$y = -\dfrac{2}{3}x$ $m = -\dfrac{2}{3}, b = 0$
The lines are only intersecting because the slopes are not equal and their product is not -1.

27. a. $y = 0.15x + 2.5 + 0.1x + 1$
$y = 0.25x + 3.5$

b. $y = 0.25x$

c. The lines are parallel. Since they never intersect, there is no break-even point.

d. $y = \dfrac{1}{3}x$

e. $0.25x + 3.5 = \dfrac{1}{3}x$
$\dfrac{1}{4}x + 3.5 = \dfrac{1}{3}x$
$3.5 = \dfrac{1}{12}x$
$x = 42$
At 42 candy bars, Brook will break even.

f. total cost $y = 0.15x + 1$
total revenue $y = 0.25x$
$0.15x + 1 = 0.25x$
$0.15x + 1 - 0.15x = 0.25x - 0.15x$
$1 = 0.1x$
$x = 10$
At 10 candy bars, Brook will break even.

g. $0.15x + 1 = \dfrac{1}{3}x$
$\dfrac{9}{60}x + 1 = \dfrac{20}{60}x$
$\dfrac{9}{60}x + 1 - \dfrac{9}{60}x = \dfrac{20}{60}x - \dfrac{9}{60}x$
$1 = \dfrac{11}{60}x$
$x \approx 5.45$
At about 5.45 candy bars, Brook will break even. At 6 candy bars, Brook will start making a profit.

h. Answers will vary.

29. $y_1 = 35 + 0.25x$ $m = 0.25, b = 35$
$y_2 = 60$ $m = 0, b = 60$
Their graphs will intersect because the slopes are not equal.

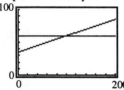

The intersection is (100, 60). At 100 miles, the prices are equal at $60 per day.

31. $y_1 = 50 + 2x$ $m = 2, b = 50$
$y_2 = 75$ $m = 0, b = 75$
$x = 8$ $m =$ undefined, $b =$ none
Yes, y_1 intersects with $x = 8$ and y_2 intersects with $x = 8$ because each has a different slope. $x = 8$ is perpendicular to y_2 because one is vertical and the other is horizontal. The intersection of y_1 and $x = 8$ is (8, 66). At 8 years, she will receive $66. The intersection of y_2 and $x = 8$ is (8, 75). At 8 years, she will receive $75. The intersection of y_1 and y_2 is (12.5, 75). At $12\frac{1}{2}$ years she will receive the same either way, $75.

33. $y_1 = 10x$ $m = 10, b = 0$
$y_2 = 15(x - 4)$ $m = 15, b = -60$
$y_3 = 150$ $m = 0, b = 150$
y_1 and y_2 intersect at (12, 120). At 12 seconds, Speedie will catch Archie at a distance of 120 feet or before the end zone 150 feet away.

6.1 Experiencing Algebra the Calculator Way

1. Xmin = –10
Xmax = 10
X3cl = 1
Ymin = –10
Ymax = 10
Yscl = 1

2. Xmin = –47
Xmax = 47
Xscl = 10
Ymin = –31
Ymax = 31
Yscl = 10

3. Xmin = 0
Xmax = 94
Xscl = 10
Ymin = 0
Ymax = 62
Yscl = 10

4. Xmin = –470
Xmax = 470
Xscl = 100
Ymin = –310
Ymax = 310
Yscl = 100

6.2 Experiencing Algebra the Exercise Way

1.

$y = x + 1$	
4	3 + 1
	4

$2y = x + 5$	
2(4)	3 + 5
8	8

Both equations are true. Therefore, (3, 4) is a solution of the system.

3. $p(x) = \frac{1}{4}x + 1$

$\frac{6}{5}$	$\frac{1}{4}\left(\frac{4}{5}\right) + 1$
	$\frac{1}{5} + \frac{5}{5}$
	$\frac{6}{5}$

$$q(x) = \frac{1}{2}x + \frac{4}{5}$$

$\frac{6}{5}$	$\frac{1}{2}\left(\frac{4}{5}\right) + \frac{4}{5}$
	$\frac{2}{5} + \frac{4}{5}$
	$\frac{6}{5}$

Both equations are true. Therefore, $\left(\frac{4}{5}, \frac{6}{5}\right)$ is a solution of the system.

5.

$3y = x + 6$	
$3(2.1)$	$0.3 + 6$
6.3	6.3

$10y = 25$	
$10(2.1)$	25
21	

The second equation is false. Therefore $(0.3, 2.1)$ is not a solution.

7.

$7y + 2 = 7$	
$7\left(\frac{5}{7}\right) + 2$	7
$5 + 2$	
7	

$5y + 3 = 0$	
$5\left(\frac{5}{7}\right) + 3$	0
$\frac{25}{7} + \frac{21}{7}$	
$\frac{46}{7}$	

The second equation is false. Therefore $\left(\frac{2}{3}, \frac{5}{7}\right)$ is not a solution.

9.

$2y = -x$	
$2(-2)$	-5
-4	

$x = 5$	
5	5

The first equation is false. Therefore, $(5, -2)$ is not a solution.

11. $2x + 3y = -6$
$x - 4y = 8$

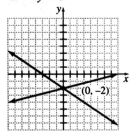

The solution is $(0, -2)$.

13. $a + b = 4$
$a - 2b = 7$

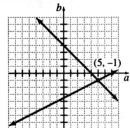

The solution is $(5, -1)$.

15. $x + 2y = 6$
$x + 2y = 2$

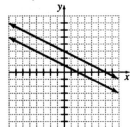

The lines are parallel. There is no solution.

17. $x - y = -1$
$3x - 3y = -3$

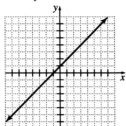

The graphs are coinciding. There is an
infinite number of solutions. **The solutions
are all ordered pairs** (x, y) **that satisfy**
$x - y = -1$.

19. $y = \dfrac{1}{2}x + 3$

$y = -\dfrac{3}{2}x - 1$

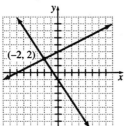

The solution is $(-2, 2)$.

21. $f(x) = 3x - 1$
$g(x) = -2x + 4$

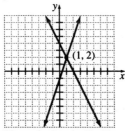

The solution is $(1, 2)$.

23. $F(x) = 2x + 2$
$G(x) = 2x - 3$

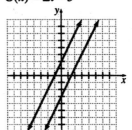

The graphs are parallel. There is no solution.

25. $y = \dfrac{1}{2}x + 2$

$6y = 3x + 12$

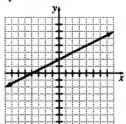

The graphs are coinciding. The solutions are
all ordered pairs (x, y) that satisfy

$y = \dfrac{1}{2}x + 2$.

27. $3x + 2y = 12$
$y - 2 = 1$

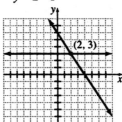

The solution is $(2, 3)$.

229

29. $3y - 2 = 2y + 2$
    $x - 1 = 0$

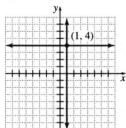

The solution is $(1, 4)$.

31. $y - 1 = 0$
    $y + 3 = 0$

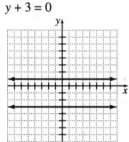

The graphs are parallel. There is no solution.

33. $y = 3x - 5$
    $x = 4 - 2y$

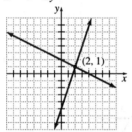

The solution is $(2, 1)$.

35. $y = x - 7$
    $x = y + 7$

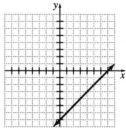

The graphs are coinciding. The solutions are all ordered pairs (x, y) that satisfy $y = x - 7$.

37. $a(x) = 1 + 4x$
    $b(x) = 4x - 2$

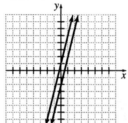

The graphs are parallel. There is no solution.

39. a. Turtle Rental: $y = 39.95 + 0.25(x - 150)$
    Snail Rental: $y = 79.95 + 0.15(x - 200)$

b.

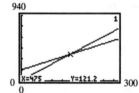

The solution is $(475, 121.2)$.

c. If Ahmet drives less than 475 miles, Turtle Rental is a better deal. If Ahmet drives more than 475 miles, Snail Rental is a better deal.

d. Turtle Rental

e. Snail Rental

41. a. Bulk-rate: $y = 75 + 0.226x$
    First-class: $y = 0.32x$

b.

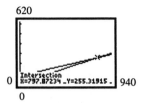

The solution is approximately (798, 255).

c. When there are more than 798 pieces of mail, it is more cost effective to use bulk-rate. The graph representing bulk-rate is lower than that representing first-class for values of x greater than 798.

d. When there are fewer than 798 pieces of mail, it is more cost effective to use first-class. The graph representing first-class is lower than that representing bulk-rate for values of x less than 798.

43. Let x = car payment
y = rent payment
$x + y = 500$
$y = 3x$

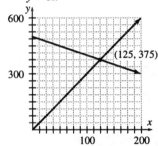

The solution is (125, 375). The car payment is \$125 and the rent payment is \$375.

45. Let x = number of hours at \$7.00 per hour
y = number of hours at \$8.20 per hour
$x + y = 40$
$7x + 8.2y = 298$

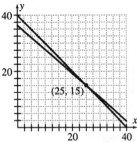

The solution is (25, 15). She should work 15 hours at the job that pays \$8.20 per hour and 25 hours at the job paying \$7 per hour.

47. Let W = width
L = length
$L = 2W - 25$
$2L + 2W = 550$

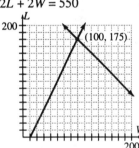

The solution is (100, 175). The width is 100 feet and the length is 175 feet.

49. Let x = number of Shania's records
y = number of Reba's records
$y = 2x + 1$
$x + y = 22$

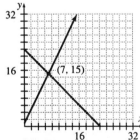

The solution is (7, 15). Shania had 7 records and Reba had 15 records.

6.2 Experiencing Algebra the Calculator Way

1. $y = 2.4x + 1.2$
$y = -1.3x - 2.7$

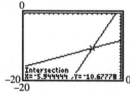

The solution is approximately
$(-1.05, -1.33)$.

2. $y = 0.4x - 8.3$
$y = 2.2x + 2.4$

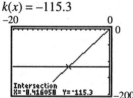

The solution is approximately
$(-5.94, -10.68)$.

3. $h(x) = 13.7x$
$k(x) = -115.3$

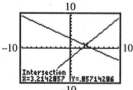

The solution is approximately
$(-8.42, -115.3)$.

4. $y = \dfrac{36 - 8x}{12}$

$y = \dfrac{18x - 45}{15}$

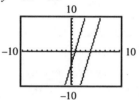

The solution is approximately $(3.21, 0.86)$.

5. $y = \dfrac{15x - 12}{3}$
$y = 5x - 20$

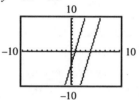

The graphs are parallel. There is no solution.

6. $y = \dfrac{49 - 7x}{21}$

$y = \dfrac{-x + 7}{3}$

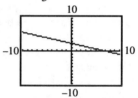

The graphs are coinciding. The solution is all ordered pairs (x, y) that satisfy
$y = \dfrac{-x + 7}{3}$.

7. $F(x) = 45x + 705$
$G(x) = -35x + 155$

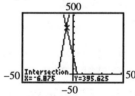

The solution is approximately
$(-6.88, 395.63)$.

8. $y = 35x + 25$
$y = 575$

The solution is approximately $(15.71, 575)$.

9. $y = 100x + 6000$
$y = 300x$

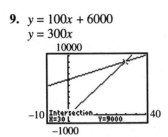

The solution is (30, 9000).

6.3 Experiencing Algebra the Exercise Way

1. $x - 2y = 26$
$5x + 10y = -10$
Solve the first equation for x: $x = 2y + 26$.
Substitute $2y + 26$ for x in the second equation.
$5(2y + 26) + 10y = -10$
$10y + 130 + 10y = -10$
$20y + 130 = -10$
$20y + 130 - 130 = -10 - 130$
$20y = -140$
$\dfrac{20y}{20} = \dfrac{-140}{20}$
$y = -7$
Substitute -7 for y in the first equation.
$x - 2(-7) = 26$
$x + 14 = 26$
$x + 14 - 14 = 26 - 14$
$x = 12$
The solution is (12, -7).

3. $y = 3x - 15$
$x - 5 = 0$
Solve the second equation for x: $x = 5$.
Substitute 5 for x in the first equation.
$y = 3(5) - 15$
$y = 15 - 15$
$y = 0$
The solution is (5, 0).

5. $x - 5y = 20$
$2y + 3 = -7$
Solve the second equation for y: $y = -5$.
Substitute -5 for y in the first equation.
$x - 5(-5) = 20$
$x + 25 = 20$
$x + 25 - 25 = 20 - 25$
$x = -5$
The solution is (-5, -5).

7. $3y = x + 6$
$x - 3y = 9$
Solve the second equation for x: $x = 3y + 9$.
Substitute $3y + 9$ for x in the first equation.
$3y = 3y + 9 + 6$
$3y = 3y + 15$
$3y - 3y = 3y + 15 - 3y$
$0 = 15$
This is a contradiction. There is no solution.

9. $y = x - 3$
$2x - y = 19$
Substitute $x - 3$ for y in the second equation.
$2x - (x - 3) = 19$
$2x - x + 3 = 19$
$x + 3 = 19$
$x + 3 - 3 = 19 - 3$
$x = 16$
Substitute 16 for x in the first equation.
$y = 16 - 3$
$y = 13$
The solution is (16, 13).

11. $y = x + 1$
$5x - 10y = 3$
Substitute $x + 1$ for y in the second equation.
$5x - 10(x + 1) = 3$
$5x - 10x - 10 = 3$
$-5x - 10 = 3$
$-5x - 10 + 10 = 3 + 10$
$-5x = 13$
$\dfrac{-5x}{-5} = \dfrac{13}{-5}$
$x = -\dfrac{13}{5}$

Substitute $-\dfrac{13}{5}$ for x in the first equation.

$y = -\dfrac{13}{5} + 1$

$y = -\dfrac{13}{5} + \dfrac{5}{5}$

$y = -\dfrac{8}{5}$

The solution is $\left(-\dfrac{13}{5}, -\dfrac{8}{5} \right)$.

13. $4y = x + 4$
$5y = 2x + 11$
Solve the first equation for x: $x = 4y - 4$.
Substitute $4y - 4$ for x in the second equation.
$5y = 2(4y - 4) + 11$
$5y = 8y - 8 + 11$
$5y = 8y + 3$
$5y - 8y = 8y + 3 - 8y$
$-3y = 3$
$\dfrac{-3y}{-3} = \dfrac{3}{-3}$
$y = -1$
Substitute -1 for y in the first equation.
$4(-1) = x + 4$
$-4 = x + 4$
$-4 - 4 = x + 4 - 4$
$-8 = x$
The solution is $(-8, -1)$.

15. $3y = x + 6$
$10y = 25$

Solve the second equation for y: $y = \dfrac{5}{2}$.

Substitute $\dfrac{5}{2}$ for y in the first equation.

$3\left(\dfrac{5}{2}\right) = x + 6$

$\dfrac{15}{2} = x + 6$

$\dfrac{15}{2} - 6 = x + 6 - 6$

$\dfrac{15}{2} - \dfrac{12}{2} = x$

$\dfrac{3}{2} = x$

The solution is $\left(\dfrac{3}{2}, \dfrac{5}{2}\right)$.

17. $y = -3x + 2$
$2y + 2x = 2 + y - x$
Substitute $-3x + 2$ for y in the second equation.
$2(-3x + 2) + 2x = 2 + (-3x + 2) - x$
$-6x + 4 + 2x = 2 - 3x + 2 - x$
$-4x + 4 = 4 - 4x$
$-4x + 4 + 4x = 4 - 4x + 4x$
$4 = 4$
This is an identity. The solutions are all ordered pairs (x, y) that satisfy $y = -3x + 2$.

19. $2x + y = -3$
$y = -0.5x + 3$
Substitute $-0.5x + 3$ for y in the first equation.
$2x + (-0.5x + 3) = -3$
$1.5x + 3 = -3$
$1.5x + 3 - 3 = -3 - 3$
$1.5x = -6$
$\dfrac{1.5x}{1.5} = \dfrac{-6}{1.5}$
$x = -4$
Substitute -4 for x in the second equation.
$y = -0.5(-4) + 3$
$y = 5$
The solution is $(-4, 5)$.

21. $5x - 3y = -13$
$4x + y = 27$
Solve the second equation for y: $y = 27 - 4x$.
Substitute $27 - 4x$ for y in the first equation.
$5x - 3(27 - 4x) = -13$
$5x - 81 + 12x = -13$
$17x - 81 = -13$
$17x - 81 + 81 = -13 + 81$
$17x = 68$
$\dfrac{17x}{17} = \dfrac{68}{17}$
$x = 4$
Substitute 4 for x in the second equation.
$4(4) + y = 27$
$16 + y - 16 = 27 - 16$
$y = 11$
The solution is $(4, 11)$.

23. $x + 2y = -28$
$3x + y = -9$
Solve the second equation for y: $y = -3x - 9$.
Substitute $-3x - 9$ for y in the first equation.
$x + 2(-3x - 9) = -28$
$x - 6x - 18 = -28$
$-5x - 18 = -28$
$-5x - 18 + 18 = -28 + 18$
$-5x = -10$
$\dfrac{-5x}{-5} = \dfrac{-10}{-5}$
$x = 2$
Substitute 2 for x in the second equation:
$3(2) + y = -9$
$6 + y - 6 = -9 - 6$
$y = -15$
The solution is $(2, -15)$.

25. $x - y = -1$
$3x - 3y = -3$
Solve the first equation for x: $x = y - 1$.
Substitute $y - 1$ for x in the second equation.
$3(y - 1) - 3y = -3$
$3y - 3 - 3y = -3$
$-3 = -3$
This is an identity. The solutions are all ordered pairs (x, y) that satisfy $x - y = -1$.

27. $y = 3x - 1$
$2x + y = 4$
Substitute $3x - 1$ for y in the second equation.
$2x + 3x - 1 = 4$
$5x - 1 = 4$
$5x - 1 + 1 = 4 + 1$
$5x = 5$
$\dfrac{5x}{5} = \dfrac{5}{5}$
$x = 1$
Substitute 1 for x in the first equation.
$y = 3(1) - 1$
$y = 2$
The solution is $(1, 2)$.

29. $y = \dfrac{1}{2}x + 3$
$3x + 2y = -2$
Substitute $\dfrac{1}{2}x + 3$ for y in the second equation.
$3x + 2\left(\dfrac{1}{2}x + 3\right) = -2$
$3x + x + 6 = -2$
$4x + 6 - 6 = -2 - 6$
$4x = -8$
$\dfrac{4x}{4} = \dfrac{-8}{4}$
$x = -2$
Substitute -2 for x in the first equation.
$y = \dfrac{1}{2}(-2) + 3$
$y = 2$
The solution is $(-2, 2)$.

31. $y = \dfrac{1}{2}x + 2$
$6y = 3x + 12$
Substitute $\dfrac{1}{2}x + 2$ for y in the second equation.
$6\left(\dfrac{1}{2}x + 2\right) = 3x + 12$
$3x + 12 = 3x + 12$
$3x + 12 - 3x = 3x + 12 - 3x$
$12 = 12$
This is an identity. The solutions are all ordered pairs (x, y) that satisfy $y = \dfrac{1}{2}x + 2$.

33. $y = 2x + 2$
$2x - y = 3$
Substitute $2x + 2$ for y in the second equation.
$2x - (2x + 2) = 3$
$2x - 2x - 2 = 3$
$-2 = 3$
This is a contradiction. There is no solution.

35. $3x + 2y = 12$
$y - 2 = 1$
Solve the second equation for y: $y = 3$.
Substitute 3 for y in the first equation.
$3x + 2(3) = 12$
$3x + 6 - 6 = 12 - 6$
$3x = 6$
$\dfrac{3x}{3} = \dfrac{6}{3}$
$x = 2$
The solution is $(2, 3)$.

37. $3x - y = 5$
$x + 2y = 4$
Solve the second equation for x: $x = 4 - 2y$.
Substitute $4 - 2y$ for x in the first equation.
$3(4 - 2y) - y = 5$
$12 - 6y - y = 5$
$12 - 7y = 5$
$12 - 7y - 12 = 5 - 12$
$-7y = -7$
$\dfrac{-7y}{-7} = \dfrac{-7}{-7}$
$y = 1$
Substitute 1 for y in the second equation.
$x + 2(1) = 4$
$x + 2 - 2 = 4 - 2$
$x = 2$
The solution is $(2, 1)$.

39. $y = x - 7$
$x = y + 7$
Substitute $x - 7$ for y in the second equation.
$x = x - 7 + 7$
$x = x$
This is an identity. The solutions are all ordered pairs (x, y) that satisfy $y = x - 7$.

41. $y = 1 + 4x$
$4x - y = 2$
Substitute $1 + 4x$ for y in the second equation.
$4x - (1 + 4x) = 2$
$4x - 1 - 4x = 2$
$-1 = 2$
This is a contradiction. There is no solution.

43. $5x + 7y = 35$
$2x - 5y = 53$
Solve the first equation for x,
$x = 7 - \dfrac{7}{5}y$.
Substitute $7 - \dfrac{7}{5}y$ for x in the second
equation.
$2\left(7 - \dfrac{7}{5}y\right) - 5y = 53$
$14 - \dfrac{14}{5}y - 5y = 53$
$14 - \dfrac{14}{5}y - \dfrac{25}{5}y = 53$
$14 - \dfrac{39}{5}y = 53$
$14 - \dfrac{39}{5}y - 14 = 53 - 14$
$-\dfrac{39}{5}y = 39$
$\left(-\dfrac{5}{39}\right)\left(-\dfrac{39}{5}y\right) = -\dfrac{5}{39}(39)$
$y = -5$
Substitute -5 for y in the second equation.
$2x - 5(-5) = 53$
$2x + 25 = 53$
$2x + 25 - 25 = 53 - 25$
$2x = 28$
$\dfrac{2x}{2} = \dfrac{28}{2}$
$x = 14$
The solution is $(14, -5)$.

45. Let W = width
  L = length
$2W + 2L = 146$
$L = 2W - 11$
Substitute $2W - 11$ for L in the first
equation.
$2W + 2(2W - 11) = 146$
$2W + 4W - 22 = 146$
$6W - 22 + 22 = 146 + 22$
$6W = 168$
$\dfrac{6W}{6} = \dfrac{168}{6}$
$W = 28$

$L = 2(28) - 11 = 45$
The width is 28 feet and the length is
45 feet.

47. Let x = measure of small angle
  y = measure of large angle
$x + y = 90$
$y = 4x - 10$
Substitute $4x - 10$ for y in the first equation.
$x + 4x - 10 = 90$
$5x - 10 + 10 = 90 + 10$
$5x = 100$
$\dfrac{5x}{5} = \dfrac{100}{5}$
$x = 20$
$y = 4(20) - 10 = 70$
The angles measure 20° and 70°.

49. Let x = measure of the two equal angles
  y = measure of third angle
$x + x + y = 180$
$y = x + x + 20$
Substitute $x + x + 20$ for y in the first equation.
$x + x + x + x + 20 = 180$
$4x + 20 - 20 = 180 - 20$
$4x = 160$
$\dfrac{4x}{4} = \dfrac{160}{4}$
$x = 40$
$y = 40 + 40 + 20 = 100$
The angles measure 40°, 40°, and 100°.

51. Let r = radius of small circle
  R = radius of large circle
$R = 2r + 5$
$2\pi R = 283$
Substitute $2r + 5$ for R in the second equation.
$2\pi(2r + 5) = 283$
$4\pi r + 10\pi = 283$
$4\pi r + 10\pi - 10\pi = 283 - 10\pi$
$4\pi r = 283 - 10\pi$
$\dfrac{4\pi r}{4\pi} = \dfrac{283 - 10\pi}{4\pi}$
$r = \dfrac{283 - 10\pi}{4\pi}$
$r \approx 20$
$R \approx 2(20) + 5 = 45$
The radius of each measures 20 inches and 45 inches.

53. Let x = amount borrowed at 7%
  y = amount borrowed at 6%
$x + y = 100{,}000$
$0.07x + 0.06y = 6450$
Solve the first equation for x.
$x = 100{,}000 - y$
Substitute $100{,}000 - y$ for x in the second equation.
$0.07(100{,}000 - y) + 0.06y = 6450$
$7000 - 0.07y + 0.06y = 6450$
$7000 - 0.01y - 7000 = 6450 - 7000$
$-0.01y = -550$
$\dfrac{-0.01y}{-0.01} = \dfrac{-550}{-0.01}$
$y = 55{,}000$
Substitute 55,000 for y in the second equation.
$x + 55{,}000 = 100{,}000$
$x + 55{,}000 - 55{,}000 = 100{,}000 - 55{,}000$
$x = 45{,}000$
She borrowed \$45,000 at 7% simple interest and \$55,000 at 6% simple interest.

55. Let x = number of acres of soybeans
  y = number of acres of corn
$x + y = 15$
$y = 25 - x$
Substitute $25 - x$ for y in the first equation.
$x + 25 - x = 15$
$25 = 15$
This is a contradiction. There is no solution.

57. Let x = amount from individuals
  y = amount from other sources
$x + y = 1.4155 \times 10^{12}$
$y = x + 1.687 \times 10^{11}$
Substitute $x + 1.687 \times 10^{11}$ for y in the first equation.
$x + x + 1.687 \times 10^{11} = 1.4155 \times 10^{12}$
$2x = 1.2468 \times 10^{12}$
$x = 6.234 \times 10^{11}$
$y = 6.234 \times 10^{11} + 1.687 \times 10^{11}$
$y = 7.921 \times 10^{11}$
The amounts are as follows:
\$$6.234 \times 10^{11}$ from individual income taxes and \$$7.921 \times 10^{11}$ from other sources.

6.3 Experiencing Algebra the Calculator Way

1. $15.8x + y = 2655.10$
   $18.4x + 73.2y = 19361.22$
   Solve the first equation for y.
   $y = 2655.1 - 15.8x$
   Substitute $2655.1 - 15.8x$ for y in the second equation.

   $18.4x + 73.2(2655.1 - 15.8x) = 19361.22$
   $18.4x + 194353.32 - 1156.56x = 19361.22$
   $-1138.16x = -174992.1$
   $x = 153.75$
   $y = 2655.1 - 15.8(153.75) = 225.85$
   The solution is $(153.75, 225.85)$.

2. $x + \dfrac{9}{17}y = \dfrac{137}{170}$
   $\dfrac{3}{8}x + \dfrac{7}{16}y = \dfrac{2303}{5280}$
   Solve the first equation for x.
   $x = \dfrac{137}{170} - \dfrac{9}{17}y$
   Substitute $\dfrac{137}{170} - \dfrac{9}{17}y$ in the second equation for x.
   $\dfrac{3}{8}\left(\dfrac{137}{170} - \dfrac{9}{17}y\right) + \dfrac{7}{16}y = \dfrac{2303}{5280}$
   $\dfrac{411}{1360} - \dfrac{27}{136}y + \dfrac{7}{16}y = \dfrac{2303}{5280}$
   $\left(\dfrac{7}{16} - \dfrac{27}{136}\right)y = \dfrac{2303}{5280} - \dfrac{411}{1360}$
   $y = \dfrac{37}{66}$
   $x = \dfrac{137}{170} - \dfrac{9}{17}\left(\dfrac{37}{66}\right) = \dfrac{28}{55}$
   The solution is $\left(\dfrac{28}{55}, \dfrac{37}{66}\right)$.

3.
   ```
   .0983→X: -1.0768→
   Y:4.055X-8.752Y=
   9.8277601
                    0
   ```

   The first equation is not true. Therefore, $(0.0983, -1.0768)$ is not a solution.

4.
   ```
   -2897.65→A:1725.
   85→B:.408A+.526B
   =-274.4441
                     1
   .005A-.802B=-139
   8.720
                    0
   ```

   The second equation is not true. Therefore, $(-2897.65, 1725.85)$ is not a solution.

5.
   ```
   (13/35)→M:(-17/5
   5)→N:(7/39)M+(5/
   34)N=7/330
                     1
   (5/42)M-(11/85)N
   =169/7350
                    0
   ```

   The second equation is not true. Therefore $\left(\dfrac{13}{35}, -\dfrac{17}{55}\right)$ is not a solution.

6.
   ```
   (-78/85)→P:(28/5
   1)→Q:(34/65)P+(5
   1/56)Q=-16/325
                    0
   ```

   The first equation is not true. Therefore $\left(-\dfrac{78}{85}, \dfrac{28}{51}\right)$ is not a solution.

6.4 Experiencing Algebra the Exercise Way

1. $2x - y = -6$
   $\underline{5x + y = -8}$
   $7x = -14$
   $\dfrac{7x}{7} = \dfrac{-14}{7}$
   $x = -2$
   Substitute -2 for x in the second equation.
   $5(-2) + y = -8$
   $-10 + y + 10 = -8 + 10$
   $y = 2$
   The solution is $(-2, 2)$.

3. $x + 7y = 19$
   $2y = x - 1$
   Write the second equation in standard from and add to the first equation.

$$x + 7y = 19$$
$$-x + 2y = -1$$
$$\overline{9y = 18}$$
$$\frac{9y}{9} = \frac{18}{9}$$
$$y = 2$$

Substitute 2 for y in the first equation.
$$x + 7(2) = 19$$
$$x + 14 - 14 = 19 - 14$$
$$x = 5$$
The solution is (5, 2).

5. $5x + y = -24$
   $3x - 2y = 9$

Multiply the first equation by 2 and add to the second equation.
$$10x + 2y = -48$$
$$3x - 2y = 9$$
$$\overline{13x = -39}$$
$$\frac{13x}{13} = -\frac{39}{13}$$
$$x = -3$$

Substitute –3 for x in the first equation.
$$5(-3) + y = -24$$
$$-15 + y + 15 = -24 + 15$$
$$y = -9$$
The solution is (–3, –9).

7. $x + 3y = 2$
   $x + 5y = -2$

Multiply the second equation by –1 and add to the first equation.
$$x + 3y = 2$$
$$-x - 5y = 2$$
$$\overline{-2y = 4}$$
$$\frac{-2y}{-2} = \frac{4}{-2}$$
$$y = -2$$

Substitute –2 for y in the first equation.
$$x + 3(-2) = 2$$
$$x - 6 + 6 = 2 + 6$$
$$x = 8$$
The solution is (8, –2).

9. $2x + 7y = 29$
   $4x + 3y = 25$

Multiply the first equation by –2 and add to the second equation.

$$-4x - 14y = -58$$
$$4x + 3y = 25$$
$$\overline{-11y = -33}$$
$$\frac{-11y}{-11} = \frac{-33}{-11}$$
$$y = 3$$

Substitute 3 for y in the first equation.
$$2x + 7(3) = 29$$
$$2x + 21 - 21 = 29 - 21$$
$$2x = 8$$
$$\frac{2x}{2} = \frac{8}{2}$$
$$x = 4$$
The solution is (4, 3).

11. $3x - 5y = 66$
   $4x + 3y = 1$

Multiply the first equation by 3 and the second equation by 5. Add the new equations.
$$9x - 15y = 198$$
$$20x + 15y = 5$$
$$\overline{29x = 203}$$
$$\frac{29x}{29} = \frac{203}{29}$$
$$x = 7$$

Substitute 7 for x in the first equation.
$$3(7) - 5y = 66$$
$$21 - 5y - 21 = 66 - 21$$
$$-5y = 45$$
$$\frac{-5y}{-5} = \frac{45}{-5}$$
$$y = -9$$
The solution is (7, –9).

13. $2x + 9y = 102$
   $5x - 11y = -147$

Multiply the first equation by –5 and the second equation by 2. Add the new equations.
$$-10x - 45y = -510$$
$$10x - 22y = -294$$
$$\overline{-67y = -804}$$
$$\frac{-67y}{-67} = \frac{-804}{-67}$$
$$y = 12$$

Substitute 12 for y in the first equation.

$2x + 9(12) = 102$

$2x + 108 - 108 = 102 - 108$

$2x = -6$

$\dfrac{2x}{2} = \dfrac{-6}{2}$

$x = -3$

The solution is $(-3, 12)$.

15. $40x = 23 + 10y$

$50x + 10y = 94$

Write the first equation in standard form and add to the second equation.

$40x - 10y = 23$

$50x + 10y = 94$

$\overline{90x = 117}$

$\dfrac{90x}{90} = \dfrac{117}{90}$

$x = \dfrac{13}{10}$

Substitute $\dfrac{13}{10}$ for x in the second equation.

$50\left(\dfrac{13}{10}\right) + 10y = 94$

$65 + 10y - 65 = 94 - 65$

$10y = 29$

$\dfrac{10y}{10} = \dfrac{29}{10}$

$y = \dfrac{29}{10}$

The solution is $\left(\dfrac{13}{10}, \dfrac{29}{10}\right)$.

17. $5x + 40y = 77$

$5x + 15y = 17$

Multiply the second equation by -1 and add to the first equation.

$5x + 40y = 77$

$-5x - 15y = -17$

$\overline{25y = 60}$

$\dfrac{25y}{25} = \dfrac{60}{25}$

$y = \dfrac{12}{5}$

Substitute $\dfrac{12}{5}$ for y in the first equation.

$5x + 40\left(\dfrac{12}{5}\right) = 77$

$5x + 96 - 96 = 77 - 96$

$5x = -19$

$\dfrac{5x}{5} = -\dfrac{19}{5}$

$x = -\dfrac{19}{5}$

The solution is $\left(-\dfrac{19}{5}, \dfrac{12}{5}\right)$.

19. $10x - 4y = 28$

$\underline{5x + 4y = 35}$

$15x = 63$

$\dfrac{15x}{15} = \dfrac{63}{15}$

$x = \dfrac{21}{5}$

Substitute $\dfrac{21}{5}$ for x in the first equation.

$10\left(\dfrac{21}{5}\right) - 4y = 28$

$42 - 4y - 42 = 28 - 42$

$-4y = -14$

$\dfrac{-4y}{-4} = \dfrac{-14}{-4}$

$y = \dfrac{7}{2}$

The solution is $\left(\dfrac{21}{5}, \dfrac{7}{2}\right)$.

21. $2x + 8y = 29$

$13y = 3x + 39$

Multiply the first equation by 3 and the second equation by 2. Write the second equation in standard form and add to the first equation.

$6x + 24y = 87$

$\underline{-6x + 26y = 78}$

$50y = 165$

$\dfrac{50y}{50} = \dfrac{165}{50}$

$y = \dfrac{33}{10}$

Substitute $\dfrac{33}{10}$ for y in the first equation.

$$2x + 8\left(\frac{33}{10}\right) = 29$$

$$2x + \frac{132}{5} - \frac{132}{5} = 29 - \frac{132}{5}$$

$$2x = \frac{145}{5} - \frac{132}{5}$$

$$2x = \frac{13}{5}$$

$$\frac{1}{2}(2x) = \frac{1}{2}\left(\frac{13}{5}\right)$$

$$x = \frac{13}{10}$$

The solution is $\left(\frac{13}{10}, \frac{33}{10}\right)$.

23. $\frac{1}{4}x + \frac{1}{3}y = \frac{5}{12}$

$\frac{1}{4}x = \frac{1}{3}y - \frac{1}{12}$

Multiply each equation by 12. Write the second equation in standard form and add to the first equation.

$$\begin{array}{r} 3x + 4y = 5 \\ 3x - 4y = -1 \\ \hline 6x = 4 \end{array}$$

$$\frac{6x}{6} = \frac{4}{6}$$

$$x = \frac{2}{3}$$

Substitute $\frac{2}{3}$ for x in the first equation.

$$\frac{1}{4}\left(\frac{2}{3}\right) + \frac{1}{3}y = \frac{5}{12}$$

$$\frac{1}{6} + \frac{1}{3}y - \frac{1}{6} = \frac{5}{12} - \frac{1}{6}$$

$$\frac{1}{3}y = \frac{1}{4}$$

$$3\left(\frac{1}{3}y\right) = 3\left(\frac{1}{4}\right)$$

$$y = \frac{3}{4}$$

The solution is $\left(\frac{2}{3}, \frac{3}{4}\right)$.

25. $\frac{1}{3}x + \frac{1}{2}y = -\frac{1}{4}$

$\frac{1}{6}x - \frac{5}{6}y = \frac{11}{16}$

Multiply the first equation by −24 and the second equation by 48. Add the new equations.

$$\begin{array}{r} -8x - 12y = 6 \\ 8x - 40y = 33 \\ \hline -52y = 39 \end{array}$$

$$\frac{-52y}{-52} = \frac{39}{-52}$$

$$y = -\frac{3}{4}$$

Substitute $-\frac{3}{4}$ for y in the first equation.

$$\frac{1}{3}x + \frac{1}{2}\left(-\frac{3}{4}\right) = -\frac{1}{4}$$

$$\frac{1}{3}x - \frac{3}{8} = -\frac{1}{4}$$

$$\frac{1}{3}x - \frac{3}{8} + \frac{3}{8} = -\frac{2}{8} + \frac{3}{8}$$

$$\frac{1}{3}x = \frac{1}{8}$$

$$3\left(\frac{1}{3}x\right) = 3\left(\frac{1}{8}\right)$$

$$x = \frac{3}{8}$$

The solution is $\left(\frac{3}{8}, -\frac{3}{4}\right)$.

27. $y = \frac{3}{2}x - 9$

$\frac{1}{6}x - \frac{1}{9}y = 1$

Multiply the first equation by 2 and write in standard form. Multiply the second equation by 18. Add the new equations.

$$\begin{array}{r} -3x + 2y = -18 \\ 3x - 2y = 18 \\ \hline 0 = 0 \end{array}$$

This is an identity. The solutions are all ordered pairs (x, y) that satisfy $y = \frac{3}{2}x - 9$.

29. $\dfrac{4}{5}x - \dfrac{3}{5}y = -1$

$\dfrac{3}{2}x - \dfrac{9}{8}y = 1$

Multiply the first equation by −15 and the second equation by 8. Add the new equations.

$-12x + 9y = 15$
$\underline{12x - 9y = \ 8}$
$\qquad\qquad 0 = 23$

This is a contradiction. There is no solution.

31. $x - 20y = 70$
$3x + 10y = 70$

Multiply the second equation by 2 and add to the first equation.

$x - 20y = \ 70$
$\underline{6x + 20y = 140}$
$7x \qquad\ = 210$

$\dfrac{7x}{7} = \dfrac{210}{7}$

$x = 30$

Substitute 30 for x in the first equation.
$30 - 20y = 70$
$30 - 20y - 30 = 70 - 30$

$\dfrac{-20y}{-20} = \dfrac{40}{-20}$

$y = -2$
The solution is (30, −2).

33. $3x + y = 40$
$x + y = 20$

Multiply the second equation by −1 and add to the first equation.

$3x + y = \ 40$
$\underline{-x - y = -20}$
$2x \qquad = \ 20$

$\dfrac{2x}{2} = \dfrac{20}{2}$

$x = 10$

Substitute 10 for x in the second equation.
$10 + y = 20$
$10 + y - 10 = 20 - 10$
$y = 10$
The solution is (10, 10).

35. $x - y = 300$
$2x - y = -100$

Multiply the second equation by −1 and add to the first equation.

$x - y = 300$
$\underline{-2x + y = 100}$
$-x \qquad = 400$

$\dfrac{-x}{-1} = \dfrac{400}{-1}$

$x = -400$

Substitute −400 for x in the first equation.
$-400 - y = 300$
$-400 - y + 400 = 300 + 400$
$-y = 700$

$\dfrac{-y}{-1} = \dfrac{700}{-1}$

$y = -700$
The solution is (−400, −700).

37. $10x - 10y = 22$
$y = 2x - 11$

Multiply the second equation by 5 and write in standard form. Add the result to the first equation.

$10x - 10y = \ 22$
$\underline{-10x + \ 5y = -55}$
$\qquad\quad -5y = -33$

$\dfrac{-5y}{-5} = \dfrac{-33}{-5}$

$y = \dfrac{33}{5}$

Substitute $\dfrac{33}{5}$ for y in the first equation.

$10x - 10\left(\dfrac{33}{5}\right) = 22$

$10x - 66 = 22$
$10x - 66 + 66 = 22 + 66$
$10x = 88$

$\dfrac{10x}{10} = \dfrac{88}{10}$

$x = \dfrac{44}{5}$

The solution is $\left(\dfrac{44}{5}, \dfrac{33}{5}\right)$.

39. $3.2x + 4.2y = 368$
$4.4x - 2.1y = 128$
Multiply the second equation by 2 and add to the first equation.
$3.2x + 4.2y = 368$
$\underline{8.8x - 4.2y = 256}$
$12x = 624$

$$\frac{12x}{12} = \frac{624}{12}$$
$$x = 52$$
Substitute 52 for x in the first equation.
$3.2(52) + 4.2y = 368$
$166.4 + 4.2y - 166.4 = 368 - 166.4$
$4.2y = 201.6$
$$\frac{4.2y}{4.2} = \frac{201.6}{4.2}$$
$y = 48$
The solution is (52, 48).

41. $0.05x + 0.10y = 0.75$
$x + y = 11$
Multiply the second equation by -0.05 and add to the first equation.
$0.05x + 0.10y = 0.75$
$\underline{-0.05x - 0.05y = -0.55}$
$0.05y = 0.2$

$$\frac{0.05y}{0.05} = \frac{0.2}{0.05}$$
$$y = 4$$
Substitute 4 for y in the second equation.
$x + 4 = 11$
$x + 4 - 4 = 11 - 4$
$x = 7$
The solution is (7, 4).

43. $0.3x + 0.45y = 43.5$
$x + y = 110$
Multiply the second equation by -0.3 and add to the first equation.
$0.3x + 0.45y = 43.5$
$\underline{-0.3x - 0.3y = -33}$
$0.15y = 10.5$

$$\frac{0.15y}{0.15} = \frac{10.5}{0.15}$$
$$y = 70$$

Substitute 70 for y in the second equation.
$x + 70 = 110$
$x + 70 - 70 = 110 - 70$
$x = 40$
The solution is (40, 70).

45. $12.50x + 6.50y = 1780$
$x + y = 200$
Multiply the second equation by -6.5 and add to the first equation.
$12.5x + 6.5y = 1780$
$\underline{-6.5x - 6.5y = -1300}$
$6x = 480$

$$\frac{6x}{6} = \frac{480}{6}$$
$$x = 80$$
Substitute 80 for x in the second equation.
$80 + y = 200$
$80 + y - 80 = 200 - 80$
$y = 120$
The solution is (80, 120).

47. Let x = number of students
y = number of others
$x + y = 683$
$1.5x + 5y = 2645$
Multiply the first equation by -5 and add to the second equation.
$-5x - 5y = -3415$
$\underline{1.5x + 5y = 2645}$
$-3.5x = -770$

$$\frac{-3.5x}{-3.5} = \frac{-770}{-3.5}$$
$$x = 220$$
There were 220 students who attended the game.

49. Let x = number of cookbooks
y = number of calendars
$x - y = 50$
$8.5x + 5y = 3462.5$
Multiply the first equation by 5 and add to the second equation.
$5x - 5y = 250$
$\underline{8.5x + 5y = 3462.5}$
$13.5x = 3712.5$

$$\frac{13.5x}{13.5} = \frac{3712.5}{13.5}$$
$$x = 275$$

Substitute 275 for x in the first equation.
$275 - y = 50$
$275 - y - 275 = 50 - 275$
$-y = -225$
$\dfrac{-y}{-1} = \dfrac{-225}{-1}$
$y = 225$
They sold 275 cookbooks and 225 calendars.

51. Let x = amount invested at 5%
    y = amount invested at 7.25%
    $x + y = 16{,}500$
    $0.05x + 0.0725y = 1000$
    Multiply the first equation by −0.05 and add to the second equation.
    $\begin{array}{r} -0.05x - 0.05y = -825 \\ 0.05x + 0.0725y = 1000 \\ \hline 0.0225y = 175 \end{array}$
    $\dfrac{0.0225y}{0.0225} = \dfrac{175}{0.0225}$
    $y \approx 7777.78$
    Substitute 7777.78 for y in the first equation.
    $x + 7777.78 = 16{,}500$
    $x + 7777.78 - 7777.78 = 16{,}500 - 7777.78$
    $x = 8722.22$
    She should invest \$8722.22 at 5% and \$7777.78 at 7.25%.

53. Let x = number of adults
    y = number of children
    $x + y = 385$
    $2x + 2y = 770$
    Multiply the first equation by −2 and add to the second equation.
    $\begin{array}{r} -2x - 2y = -770 \\ 2x + 2y = 770 \\ \hline 0 = 0 \end{array}$
    This is an identity. The solutions are any number of adults, x, and any number of children, y, where $x + y = 385$.

55. Let x = width of smaller field
    y = length of smaller field
    $2x + 2y = 620$
    $2x + 2(y + 90) = 800$
    Multiply the first equation by −1 and add to the second equation in standard form.

$\begin{array}{r} -2x - 2y = -620 \\ 2x + 2y = 620 \\ \hline 0 = 0 \end{array}$
This is an identity. The solutions are any number of yards, x, and any number of yards, y, where $x + y = 310$.

57. Let x = measure of each equal angle
    y = measure of third angle
    $x + x + y = 180$
    $y = 90 - 2x$
    Multiply the second equation by −1 and write in standard form. Add the result to the first equation in standard form.
    $\begin{array}{r} 2x + y = 180 \\ -2x - y = -90 \\ \hline 0 = 90 \end{array}$

    This is a contradiction. There is no solution.

59. Let x = distance from Earth to sun
    y = distance from Mars to sun
    $y = x + 4.864 \times 10^7$
    $\dfrac{x + y}{2} = 1.1728 \times 10^8$
    Multiply the second equation by 2 and add to the first equation written in standard form.
    $\begin{array}{r} -x + y = 4.864 \times 10^7 \\ x + y = 2.3456 \times 10^8 \\ \hline 2y = 2.832 \times 10^8 \end{array}$
    $\dfrac{2y}{2} = \dfrac{2.832 \times 10^8}{2}$
    $y = 1.416 \times 10^8$
    Substitute 1.416×10^8 for y in the first equation.
    $1.416 \times 10^8 = x + 4.864 \times 10^7$
    $1.416 \times 10^8 - 4.864 \times 10^7$
    $\quad = x + 4.864 \times 10^7 - 4.864 \times 10^7$
    $9.296 \times 10^7 = x$
    The distances from the sun are 9.296×10^7 miles for Earth and 1.416×10^8 miles for Mars.

6.4 Experiencing Algebra the Calculator Way

1.
$$2578x - 4823y = 299,758$$
$$7395x + 4823y = 3,210,738$$
$$\overline{9973x = 3,510,496}$$

$$\frac{9973x}{9973} = \frac{3,510,496}{9973}$$
$$x = 352$$

Substitute 352 for x in the first equation.
$$2578(352) - 4823y = 299,758$$
$$907,456 - 4823y = 299,758$$
$$-4823y = -607,698$$
$$y = 126$$
The solution is (352, 126).

2.
$$487x + 182y = 51,567$$
$$976x + 364y = 103,280$$
Multiply the first equation by –2 and add to the second equation.
$$-974x - 364y = -103,134$$
$$\underline{976x + 364y = 103,280}$$
$$2x = 146$$
$$x = 73$$
Substitute 73 for x in the first equation.
$$487(73) + 182y = 51,567$$
$$35,551 + 182y = 51,567$$
$$182y = 16,016$$
$$y = 88$$
The solution is (73, 88).

3.
$$4.376x - 2.659y = 4.479378$$
$$-2.188x - 5.033y = 19.787286$$
Multiply the second equation by 2 and add to the first equation.
$$4.376x - 2.659y = 4.479378$$
$$\underline{-4.376x - 10.066y = 39.574572}$$
$$-12.725y = 44.05395$$
$$y = -3.462$$
Substitute –3.462 for y in the first equation.
$$4.376x - 2.659(-3.462) = 4.479378$$
$$4.376x + 9.205458 = 4.479378$$
$$4.376x = -4.72608$$
$$x = -1.08$$
The solution is (–1.08, –3.462).

4.
$$5882x + 0.473y = 2764.366$$
$$2750x - y = -4508$$
Multiply the second equation by 0.473 and add to the first equation.

$$5882x + 0.473y = 2764.366$$
$$\underline{1300.75x - 0.473y = -2132.284}$$
$$7182.75x = 632.082$$
$$x = 0.088$$
Substitute 0.088 for x in the second equation.
$$2750(0.088) - y = -4508$$
$$242 - y = -4508$$
$$-y = -4750$$
$$y = 4750$$
The solution is (0.088, 4750).

5.
$$\frac{27}{85}x + y = -\frac{37}{95}$$
$$x + \frac{57}{82}y = \frac{7}{72}$$

Multiply the second equation by $-\dfrac{27}{85}$ and add to the first equation.

$$\frac{27}{85}x + y = -\frac{37}{95}$$
$$\underline{-\frac{27}{85}x - \frac{1539}{6970}y = -\frac{189}{6120}}$$
$$\frac{5431}{6970}y = -\frac{37}{95} - \frac{189}{6120}$$
$$y = -\frac{41}{76}$$

Substitute $-\dfrac{41}{76}$ for y in the second equation.

$$x + \frac{57}{82}\left(-\frac{41}{76}\right) = \frac{7}{72}$$
$$x = \frac{17}{36}$$

The solution is $\left(\dfrac{17}{36}, -\dfrac{41}{76}\right)$.

6.
$$\frac{5}{17}x - \frac{4}{19}y = \frac{421}{1615}$$
$$\frac{8}{19}x + \frac{11}{17}y = -\frac{2}{323}$$

Multiply the first equation by $\dfrac{11}{17}$ and add to the second equation multiplied by $\dfrac{4}{19}$.

$$\frac{55}{289}x - \frac{44}{323}y = \frac{4631}{27,455}$$

$$\frac{32}{361}x + \frac{44}{323}y = -\frac{8}{6137}$$

$$\frac{29,103}{104,329}x = \frac{87,309}{521,645}$$

$$x = \frac{3}{5}$$

Substitute $\frac{3}{5}$ for x in the second equation.

$$\frac{8}{19}\left(\frac{3}{5}\right) + \frac{11}{17}y = -\frac{2}{323}$$

$$y = -\frac{2}{5}$$

The solution is $\left(\frac{3}{5}, -\frac{2}{5}\right)$.

6.5 Experiencing Algebra the Exercise Way

1. Let x = number of pounds of peanuts
      y = number of pounds of candy
   $x + y = 30$
   $0.9x + 1.5y = 1.25(30)$
   Solve the first equation for x: $x = 30 - y$.
   Substitute $30 - y$ for x in the second
   equation.
   $0.9(30 - y) + 1.5y = 1.25(30)$
   $27 - 0.9y + 1.5y = 37.5$
   $27 + 0.6y - 27 = 37.5 - 27$
   $0.6y = 10.5$
   $$\frac{0.6y}{0.6} = \frac{10.5}{0.6}$$
   $y = 17.5$
   $x = 30 - y$
   $x = 30 - 17.5$
   $x = 12.5$
   She should mix 12.5 pounds of peanuts with
   17.5 pounds of candy.

3. Let x = number of \$5 bills
      y = number of \$10 bills
   $x + y = 65$
   $5x + 10y = 365$
   Multiply the first equation by –5 and add to
   the second equation.

$$-5x - 5y = -325$$
$$5x + 10y = 365$$
$$5y = 40$$

$$\frac{5y}{5} = \frac{40}{5}$$
$$y = 8$$

Substitute 8 for y in the first equation.
$x + 8 = 65$
$x + 8 - 8 = 65 - 8$
$x = 57$
There are 57 \$5 bills and 8 \$10 bills.

5. Let x = number of gallons of concentrate
      y = number of gallons of water
   $x + y = 130$
   $0.65x + 0y = 0.35(130)$
   Solve the second equation for x: $x = 70$.
   Substitute 70 for x in the first equation.
   $70 + y = 130$
   $70 + y - 70 = 130 - 70$
   $y = 60$
   The mixture should have 70 gallons of
   concentrate and 60 gallons of water.

7. Let x = speed of plane with no wind
      y = speed of wind
   $450 = 3(x + y)$
   $450 = 5(x - y)$
   Multiply the first equation by 5 and the
   second equation by 3. Add the new
   equations.
   $$2250 = 15x + 15y$$
   $$1350 = 15x - 15y$$
   $$3600 = 30x$$

   $$\frac{3600}{30} = \frac{30x}{30}$$
   $$120 = x$$
   Substitute 120 for x in the first equation.
   $450 = 3(120 + y)$
   $450 = 360 + 3y$
   $450 - 360 = 360 + 3y - 360$
   $90 = 3y$
   $$\frac{90}{3} = \frac{3y}{3}$$
   $30 = y$
   The speed of the plane with no wind is 120
   mph and the speed of the wind is 30 mph.

9. Let x = number of children

$\quad\quad y$ = number of adults

$4.5x + 7.5y = 3937.5$

$x + y = 625$

Multiply the second equation by -4.5 and add to the first equation.

$$\begin{array}{r} 4.5x + 7.5y = 3937.5 \\ -4.5x - 4.5y = -2812.5 \\ \hline 3y = 1125 \end{array}$$

$$\frac{3y}{3} = \frac{1125}{3}$$

$$y = 375$$

Substitute 375 for y in the second equation.

$x + 375 = 625$

$x + 375 - 375 = 625 - 375$

$x = 250$

There were 250 children and 375 adults.

11. Let x = number of pints of 70% solution

$\quad\quad y$ = number of pints of 40% solution

$x + y = 5$

$0.7x + 0.4y = 0.5(5)$

Multiply the first equation by -0.4 and add to the second equation.

$$\begin{array}{r} -0.4x - 0.4y = -2 \\ 0.7x + 0.4y = 2.5 \\ \hline 0.3x = 0.5 \end{array}$$

$$\frac{0.3x}{0.3} = \frac{0.5}{0.3}$$

$$x = 1\frac{2}{3}$$

Substitute $1\frac{2}{3}$ for x in the first equation.

$$1\frac{2}{3} + y = 5$$

$$1\frac{2}{3} + y - 1\frac{2}{3} = 5 - 1\frac{2}{3}$$

$$y = 3\frac{1}{3}$$

He should mix $1\frac{2}{3}$ pints of 70% weed killer

and $3\frac{1}{3}$ pints of 40% weed killer.

13. Let x = number of pounds of French vanilla

$\quad\quad y$ = number of pounds of hazelnut

$9.5x + 7y = 8.5(20)$

$x + y = 20$

Multiply the second equation by -7 and add to the first equation.

$$\begin{array}{r} 9.5x + 7y = 170 \\ -7x - 7y = -140 \\ \hline 2.5x = 30 \end{array}$$

$$\frac{2.5x}{2.5} = \frac{30}{2.5}$$

$$x = 12$$

Substitute 12 for x in the second equation.

$12 + y = 20$

$12 + y - 12 = 20 - 12$

$y = 8$

He should use 12 pounds of French vanilla coffee and 8 pounds of hazelnut coffee.

15. Let x = amount at 8.5%

$\quad\quad y$ = amount at 7%

$x + y = 10,000$

$0.085x + 0.07y = 752.5$

Multiply the first equation by -0.07 and add to the second equation.

$$\begin{array}{r} -0.07x - 0.07y = -700 \\ 0.085x + 0.07y = 752.5 \\ \hline 0.015x = 52.5 \end{array}$$

$$\frac{0.015x}{0.015} = \frac{52.5}{0.015}$$

$$x = 3500$$

Substitute 3500 for x in the first equation.

$3500 + y = 10,000$

$3500 + y - 3500 = 10,000 - 3500$

$y = 6500$

She invested $3500 at 8.5% and $6500 at 7%.

17. Let x = time spent walking

$\quad\quad y$ = time spent riding

$x = 2y$

$11 = 3x + 60y$

Substitute $2y$ for x in the second equation.

$11 = 3(2y) + 60y$

$11 = 66y$

$$\frac{11}{66} = \frac{66y}{66}$$

$$\frac{1}{6} = y$$

$$x = 2\left(\frac{1}{6}\right) = \frac{1}{3}$$

He walked for $\frac{1}{3}$ hour and rode for $\frac{1}{6}$ hour.

19. Let x = number of azaleas
y = number of rhododendrons
$x + y = 30$
$5x + 12y = 250$
Multiply the first equation by -5 and add to the second equation.

$$-5x - 5y = -150$$
$$\underline{5x + 12y = \ 250}$$
$$7y = \ 100$$
$$\frac{7y}{7} = \frac{100}{7}$$
$$y \approx 14$$

Substitute 14 for y in the first equation.
$x + 14 = 30$
$x + 14 - 14 = 30 - 14$
$x = 16$
She can buy 16 azaleas and 14 rhododendrons.

21. Let x = wage as waitress
y = wage as cook
$15x + 20y = 320$
$18x + 24y = 384$
Multiply the first equation by 18 and the second equation by -15. Add the new equations.

$$270x + 360y = \ 5760$$
$$\underline{-270x - 360y = -5760}$$
$$0 = 0$$

This is an identity. There are an infinite number of solutions. The solutions are any wage, x, and any wage, y, where $15x + 20y = 320$.

6.5 Experiencing Algebra the Calculator Way

1. a. $R(x) = 25x$

b. $C(x) = 10x + 1500$

c.

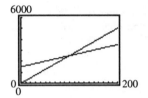

d. The graphs intersect at (125, 5625). The break-even point is at producing and selling 125 lamps.

e.

$R(x) = 25x$	
2500	25(100)
	2500

$C(x) = 10x + 1500$	
2500	10(100) + 1500
	1000 + 1500
	2500

The solution checks.

2. a. $R(x) = 45x$

b. $C(x) = 25x + 2500$

c.

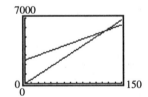

d. The graphs intersect at (125, 5625). The break-even point is at producing and selling 125 tables.

e.

$R(x) = 45x$	
5625	45(125)
	5625

$$\begin{array}{c|l} \multicolumn{2}{l}{C(x) = 25x + 2500} \\ \hline 5625 & 25(125) + 2500 \\ & 3125 + 2500 \\ & 5625 \end{array}$$

The solution checks.

3. $R(x) = 32.49x$
   $C(x) = 15.89x + 1256$

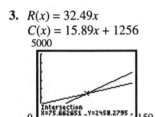

   The graphs intersect at approximately
   (75.7, 2458.3). The break-even point is at
   producing and selling 76 purses.

Chapter 6 Review

Reflections

1–9. Answers may vary.

Exercises

1. $y = 2x + 6$ $m = 2, b = 6$
   $3y - x = 15$
   $3y = x + 15$
   $y = \dfrac{1}{3}x + 5$ $m = \dfrac{1}{3}, b = 5$
   The lines are only intersecting because their
   slopes are not equal and their product is not
   –1.

2. $2y - 2x = y + 3x + 2$
   $y = 5x + 2$ $m = 5, b = 2$
   $5x - y = -2$
   $y = 5x + 2$ $m = 5, b = 2$
   The lines are coinciding because they have
   the same slopes and the same y-intercepts.

3. $4x - 20 = 0$
   $4x = 20$
   $x = 5$
   $2x + 3 = x + 4$
   $x = 1$
   The lines are parallel because they are both
   vertical with different constants.

4. $2x + y = 4$
   $y = -2x + 4$ $m = -2, b = 4$
   $y = -2(x + 2)$
   $y = -2x - 4$ $m = -2, b = -4$
   The lines are parallel because they have the
   same slopes and different y-intercepts.

5. $3(y + 2) = 2(y + 4)$
   $3y + 6 = 2y + 8$
   $y = 2$
   $2(x - 2) = 0$
   $2x - 4 = 0$
   $2x = 4$
   $x = 2$
   The lines are intersecting and perpendicular
   because one is a vertical line and the other is
   a horizontal line.

6. $5y - 4(y + 3) = -10$
   $5y - 4y - 12 = -10$
   $y - 12 = -10$
   $y = 2$ $m = 0, b = 2$

   $y - 1 = -2(x - 1)$
   $y - 1 = -2x + 2$
   $y = -2x + 3$ $m = -2, b = 3$
   The lines are only intersecting because the
   slopes are not equal and their product is not
   –1.

7. $2x - 3y = -9$
   $-3y = -2x - 9$
   $y = \dfrac{2}{3}x + 3$ $m = \dfrac{2}{3}, b = 3$

   $3x + 2y = 6$
   $2y = -3x + 6$
   $y = -\dfrac{3}{2}x + 3$ $m = -\dfrac{3}{2}, b = 3$
   The lines are intersecting and perpendicular
   because the product of the slopes equals –1.

8. $y = x + 7$ $m = 1, b = 7$
$y = -x + 7$ $m = -1, b = 7$
The lines are intersecting and perpendicular because the product of the slopes is −1.

9. a. $y = 25 + 35 + 30x + 25 + 5x$
$y = 35x + 85$

b. $y = 35x$

c.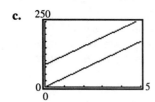

There is no break-even point.

d. $y = 60x$

e.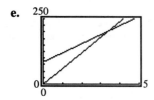

The break-even point is at (3.4, 204). He will start making a profit when he sells the fourth calculator.

f. J.R. should sell the calculators for $60 each.

10.

$3x + 2y = -2$	
$3(2) + 2(-4)$	-2
$6 - 8$	
-2	

$4x - 3y = 20$	
$4(2) - 3(-4)$	20
$8 + 12$	
20	

Both equations are true. Therefore, (2, −4) is a solution.

11.

$2x + y = 6$	
$2\left(\frac{17}{7}\right) + \frac{8}{7}$	6
$\frac{34}{7} + \frac{8}{7}$	
$\frac{42}{7}$	
6	

$-x + 3y = 1$	
$-\frac{17}{7} + 3\left(\frac{8}{7}\right)$	1
$-\frac{17}{7} + \frac{24}{7}$	
$\frac{7}{7}$	
1	

Both equations are true. Therefore, $\left(\dfrac{17}{7}, \dfrac{8}{7}\right)$ is a solution.

12.

$4x - 5y = 3$	
$4(0.25) - 5(-0.45)$	3
$1 + 2.25$	
3.25	

$8x + 5y = 0$	
$8(0.25) + 5(-0.45)$	0
$2 - 2.25$	
-0.25	

Both equations are false. Therefore, (0.25, −0.45) is not a solution.

13.

$x + 3y = 13$	
$7 + 3(-2)$	13
$7 - 6$	
1	

$x - y = 5$	
$7 - (-2)$	5
9	

Both equations are false. Therefore (7, −2) is not a solution.

14.

$$3x + 6y = -2$$

$$
\begin{array}{c|c}
3\left(\frac{2}{3}\right) + 6\left(\frac{2}{3}\right) & -2 \\
2 + 4 & \\
6 &
\end{array}
$$

$$6x - 3y = 6$$

$$
\begin{array}{c|c}
6\left(\frac{2}{3}\right) - 3\left(\frac{2}{3}\right) & 6 \\
4 - 2 & \\
2 &
\end{array}
$$

Both equations are false. Therefore, $\left(\dfrac{2}{3}, \dfrac{2}{3}\right)$ is not a solution.

15.

$$6x + 5y = -3$$

$$
\begin{array}{c|c}
6(1.5) + 5(-2.4) & -3 \\
9 - 12 & \\
-3 &
\end{array}
$$

$$2x - 10y = 27$$

$$
\begin{array}{c|c}
2(1.5) - 10(-2.4) & 27 \\
3 + 24 & \\
27 &
\end{array}
$$

Both equations are true. Therefore, $(1.5, -2.4)$ is a solution.

16.

$$2(x - 3) = 2$$

$$
\begin{array}{c|c}
2(4 - 3) & 2 \\
2(1) & \\
2 &
\end{array}
$$

$$3(y + 1) = -3$$

$$
\begin{array}{c|c}
3(2 + 1) & -3 \\
3(3) & \\
9 &
\end{array}
$$

The second equation is false. Therefore, $(4, 2)$ is not a solution.

17.

$$x + 5 = 2$$

$$
\begin{array}{c|c}
-3 + 5 & 2 \\
2 &
\end{array}
$$

$$2y - 3 = 7$$

$$
\begin{array}{c|c}
2(5) - 3 & 7 \\
10 - 3 & \\
7 &
\end{array}
$$

Both equations are true. Therefore, $(-3, 5)$ is a solution.

18. $2x + y = 17$
   $y = 3x - 18$

The solution is $(7, 3)$.

19. $3(x + 2) + 1 = -5$
   $2x - y = -10$

The solution is $(-4, 2)$.

20. $x + 6 = 4$
   $2(y + 2) = y + 3$

The solution is $(-2, -1)$.

21. $x - 2y = 11$
$2x + 3y = -6$

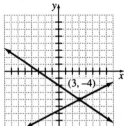

The solution is $(3, -4)$.

22. $y = 2(x + 2)$
$2x - y = 5$

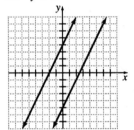

The graphs are parallel. There is no solution.

23. $y = \dfrac{3}{2}x - 6$
$3x - 2y = 12$

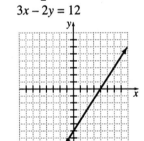

The graphs coincide. The solutions are all

ordered pairs (x, y) that satisfy $y = \dfrac{3}{2}x - 6$.

24. $y = 3x - 6$
$y = -3$

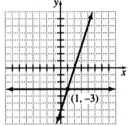

The solution is $(1, -3)$.

25. $2y - 3 = 1$
$5(y - 4) = 10$

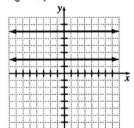

The graphs are parallel. There is no solution.

26. $2x - 11 = 3$
$14 - 2x = x - 7$

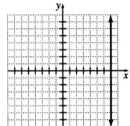

The graphs coincide. The solutions are all ordered pairs (x, y) that satisfy $2x - 11 = 3$ or $x = 7$.

27. $y - 1 = 4$
$3y - 2 = 13$

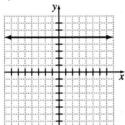

The graphs coincide. The solutions are all ordered pairs (x, y) that satisfy $y - 1 = 4$ or $y = 5$.

28. $2x + 3y = 6$
$x + y = 1$

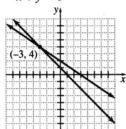

The solution is $(-3, 4)$.

29. $f(x) = 2x - 15$
$g(x) = -3x + 10$

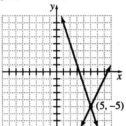

The solution is $(5, -5)$.

30. $y = 2x + 1$
$x = 2y + 7$

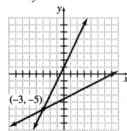

The solution is $(-3, -5)$.

31. $2y + 6 = 3y + 5$
$2(x - 3) + 1 = 5$

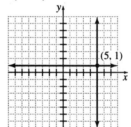

The solution is $(5, 1)$.

32. Let x = number of Democrats
y = number of Republicans
$x + y = 500 - 120$
$y = x + 40$

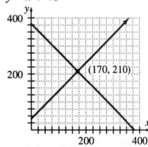

The solution is $(170, 210)$.
There were 170 Democrats.
$170 - 120 = 50$
There were 50 more Democrats than Independents.

33. Let W = width of first towel
L = length of first towel
$2W + 2L = 116.6$
$2(W + 4.7) + 2(L - 2.9) = 120.2$

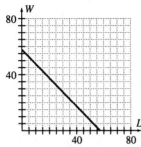

The graphs coincide. The solution is all ordered pairs (W, L) that satisfy $2W + 2L = 116.6$.

34. $8y = 5$
$4x + 8y = 2$

Solve the first equation for y: $y = \dfrac{5}{8}$.

Substitute $\dfrac{5}{8}$ for y in the second equation.

$4x + 8\left(\dfrac{5}{8}\right) = 2$
$4x + 5 = 2$
$4x + 5 - 5 = 2 - 5$
$4x = -3$
$\dfrac{4x}{4} = \dfrac{-3}{4}$
$x = -\dfrac{3}{4}$

The solution is $\left(-\dfrac{3}{4}, \dfrac{5}{8}\right)$.

35. $10x - 5y = -14$
$5x - 1 = 0$

Solve the second equation for x: $x = \dfrac{1}{5}$.

Substitute $\dfrac{1}{5}$ for x in the first equation.

$10\left(\dfrac{1}{5}\right) - 5y = -14$
$2 - 5y = -14$
$2 - 5y - 2 = -14 - 2$
$-5y = -16$
$\dfrac{-5y}{-5} = \dfrac{-16}{-5}$
$y = \dfrac{16}{5}$

The solution is $\left(\dfrac{1}{5}, \dfrac{16}{5}\right)$.

36. $4x - y = 5$
$y = 4x + 3$
Substitute $4x + 3$ for y in the first equation.
$4x - (4x + 3) = 5$
$4x - 4x - 3 = 5$
$-3 = 5$
This is a contradiction. There is no solution.

37. $x - 2y = 71$
$3x - 7y = 275$
Solve the first equation for x: $x = 2y + 71$.
Substitute $2y + 71$ for x in the second equation.
$3(2y + 71) - 7y = 275$
$6y + 213 - 7y = 275$
$213 - y = 275$
$213 - y - 213 = 275 - 213$
$-y = 62$
$\dfrac{-y}{-1} = \dfrac{62}{-1}$
$y = -62$
Substitute -62 for y in the first equation.
$x - 2(-62) = 71$
$x + 124 = 71$
$x + 124 - 124 = 71 - 124$
$x = -53$
The solution is $(-53, -62)$.

38. $2x + 3y = -69$
$2x - 4y = 218$
Solve the second equation for x.
$x = 2y + 109$
Substitute $2y + 109$ for x in the first equation.

$2(2y + 109) + 3y = -69$
$4y + 218 + 3y = -69$
$7y + 218 - 218 = -69 - 218$
$7y = -287$
$\dfrac{7y}{7} = \dfrac{-287}{7}$
$y = -41$
Substitute -41 for y in the second equation.
$2x - 4(-41) = 218$
$2x + 164 = 218$
$2x + 164 - 164 = 218 - 164$
$2x = 54$
$\dfrac{2x}{2} = \dfrac{54}{2}$
$x = 27$
The solution is $(27, -41)$.

39. $3(x + 2) - y = -1$
$y = 3x + 7$
Substitute $3x + 7$ for y in the first equation.
$3(x + 2) - (3x + 7) = -1$
$3x + 6 - 3x - 7 = -1$
$-1 = -1$
This is an identity. The solutions are all ordered pairs (x, y) that satisfy $y = 3x + 7$.

40. $x + 8y = 5$
$12x + y = 10$
Solve the first equation for x: $x = 5 - 8y$
Substitute $5 - 8y$ for x in the second equation.
$12(5 - 8y) + y = 10$
$60 - 96y + y = 10$
$60 - 95y = 10$
$60 - 95y - 60 = 10 - 60$
$-95y = -50$
$\dfrac{-95y}{-95} = \dfrac{-50}{-95}$
$y = \dfrac{10}{19}$
Substitute $\dfrac{10}{19}$ for y in the first equation.

$x + 8\left(\dfrac{10}{19}\right) = 5$
$x + \dfrac{80}{19} - \dfrac{80}{19} = 5 - \dfrac{80}{19}$
$x = \dfrac{95}{19} - \dfrac{80}{19}$
$x = \dfrac{15}{19}$
The solution is $\left(\dfrac{15}{19}, \dfrac{10}{19}\right)$.

41. $5x - 3y = 406$
$2x - y = 327$
Solve the second equation for y.
$y = 2x - 327$
Substitute $2x - 327$ for y in the first equation.
$5x - 3(2x - 327) = 406$
$5x - 6x + 981 = 406$
$-x + 981 - 981 = 406 - 981$
$-x = -575$
$\dfrac{-x}{-1} = \dfrac{-575}{-1}$
$x = 575$
Substitute 575 for x in the second equation.
$2(575) - y = 327$
$1150 - y - 1150 = 327 - 1150$
$-y = -823$
$\dfrac{-y}{-1} = \dfrac{-823}{-1}$
$y = 823$
The solution is $(575, 823)$.

42. $y = \dfrac{1}{3}x - 4$
$x - 5y = 0$
Substitute $\dfrac{1}{3}x - 4$ for y in the second equation.

$$x - 5\left(\frac{1}{3}x - 4\right) = 0$$

$$x - \frac{5}{3}x + 20 = 0$$

$$\frac{3}{3}x - \frac{5}{3}x + 20 - 20 = 0 - 20$$

$$-\frac{2}{3}x = -20$$

$$-\frac{3}{2}\left(-\frac{2}{3}x\right) = -\frac{3}{2}(-20)$$

$$x = 30$$

Substitute 30 for x in the first equation.

$$y = \frac{1}{3}(30) - 4$$

$$y = 10 - 4$$

$$y = 6$$

The solution is (30, 6).

43. $y = \frac{1}{2}x + 7$

$$y = -\frac{3}{5}x - 4$$

Substitute $\frac{1}{2}x + 7$ for y in the second

equation.

$$\frac{1}{2}x + 7 = -\frac{3}{5}x - 4$$

$$\frac{1}{2}x + 7 - 7 = -\frac{3}{5}x - 4 - 7$$

$$\frac{1}{2}x = -\frac{3}{5}x - 11$$

$$\frac{1}{2}x + \frac{3}{5}x = -\frac{3}{5}x - 11 + \frac{3}{5}x$$

$$\frac{5}{10}x + \frac{6}{10}x = -11$$

$$\frac{11}{10}x = -11$$

$$\frac{10}{11}\left(\frac{11}{10}x\right) = \frac{10}{11}(-11)$$

$$x = -10$$

Substitute −10 for x in the first equation.

$$y = \frac{1}{2}(-10) + 7$$

$$y = -5 + 7$$

$$y = 2$$

The solution is (−10, 2).

44. $3x - y = 10$

$2x - 6y = 26$

Solve the first equation for y: $y = 3x - 10$

Substitute $3x - 10$ for y in the second

equation.

$$2x - 6(3x - 10) = 26$$

$$2x - 18x + 60 = 26$$

$$-16x + 60 - 60 = 26 - 60$$

$$-16x = -34$$

$$\frac{-16x}{-16} = \frac{-34}{-16}$$

$$x = \frac{17}{8}$$

Substitute $\frac{17}{8}$ for x in the first equation.

$$3\left(\frac{17}{8}\right) - y = 10$$

$$\frac{51}{8} - y - \frac{51}{8} = 10 - \frac{51}{8}$$

$$-y = \frac{29}{8}$$

$$-1(-y) = -1\left(\frac{29}{8}\right)$$

$$y = -\frac{29}{8}$$

The solution is $\left(\frac{17}{8}, -\frac{29}{8}\right)$.

45. $x = 160y$

$3x - 440y = 30$

Substitute $160y$ for x in the second equation.

$$3(160y) - 440y = 30$$

$$480y - 440y = 30$$

$$40y = 30$$

$$\frac{40y}{40} = \frac{30}{40}$$

$$y = \frac{3}{4}$$

Substitute $\frac{3}{4}$ for y in the first equation.

$$x = 160\left(\frac{3}{4}\right)$$

$$x = 120$$

The solution is $\left(120, \frac{3}{4}\right)$.

46. Let x = measure of smaller angle
  y = measure of larger angle
$x + y = 90$
$y = 2x + 12$
Substitute $2x + 12$ for y in the first equation.
$x + 2x + 12 = 90$
$3x + 12 - 12 = 90 - 12$
$3x = 78$
$\dfrac{3x}{3} = \dfrac{78}{3}$
$x = 26$
Substitute 26 for x in the second equation.
$y = 2(26) + 12$
$y = 52 + 12$
$y = 64$
$64 - 26 = 38$
The difference in the angles is $38°$.

47. Let W = width
  L = length
$L = 3W - 2.5$
$2W + 2L = 31.8$
Substitute $3W - 2.5$ for L in the second equation.
$2W + 2(3W - 2.5) = 31.8$
$2W + 6W - 5 = 31.8$
$8W - 5 + 5 = 31.8 + 5$
$8W = 36.8$
$\dfrac{8W}{8} = \dfrac{36.8}{8}$
$W = 4.6$
Substitute 4.6 for W in the first equation.
$L = 3(4.6) - 2.5$
$L = 13.8 - 2.5$
$L = 11.3$
Area $= (11.3)(4.6) = 51.98$
The area is 51.98 cm^2.

48. Let x = amount at 4.5%
  y = amount at 6%
$x + y = 10,000,000$
$0.045x + 0.06y = 487,500$
Solve the first equation for y.
$y = 10,000,000 - x$.
Substitute $10,000,000 - x$ in the second equation.

$0.045x + 0.06(10,000,000 - x) = 487,500$
$0.045x + 600,000 - 0.06x = 487,500$
$-0.015x + 600,000 - 600,000$
  $= 487,500 - 600,000$
$-0.015x = -112,500$
$\dfrac{-0.015x}{-0.015} = \dfrac{-112,500}{-0.015}$
$x = 7,500,000$
Substitute 7,500,000 for x in the first equation.
$7,500,000 + y = 10,000,000$
$7,500,000 + y - 7,500,000$
  $= 10,000,000 - 7,500,000$
$y = 2,500,000$
She invested \$7,500,000 at 4.5% simple interest and \$2,500,000 at 6% simple interest.

49. $5x + 3y = -10$
   $\underline{5x - 3y = 80}$
   $10x \qquad = 70$
   $\dfrac{10x}{10} = \dfrac{70}{10}$
   $x = 7$
Substitute 7 for x in the first equation.
$5(7) + 3y = -10$
$35 + 3y - 35 = -10 - 35$
$3y = -45$
$\dfrac{3y}{3} = \dfrac{-45}{3}$
$y = -15$
The solution is $(7, -15)$.

50. $x = 18 - 2y$
$3x + 2y = 30$
Multiply the first equation by -1 and write in standard form. Add the result to the second equation.
   $-x - 2y = -18$
   $\underline{3x + 2y = 30}$
   $2x \qquad = 12$
   $\dfrac{2x}{2} = \dfrac{12}{2}$
   $x = 6$

Substitute 6 for x in the second equation.

$3(6) + 2y = 30$
$18 + 2y - 18 = 30 - 18$
$2y = 12$
$\dfrac{2y}{2} = \dfrac{12}{2}$
$y = 6$
The solution is (6, 6).

51. $\dfrac{1}{2}x + \dfrac{1}{3}y = 3$
$\dfrac{1}{4}x - \dfrac{2}{5}y = -7$

Multiply the first equation by 6 and the second equation by 5. Add the new equations.
$3x + 2y = 18$
$\dfrac{5}{4}x - 2y = -35$
$\dfrac{17}{4}x \quad = -17$
$\dfrac{4}{17}\left(\dfrac{17}{4}x\right) = \dfrac{4}{17}(-17)$
$x = -4$

Substitute −4 for x in the second equation.
$\dfrac{1}{4}(-4) - \dfrac{2}{5}y = -7$
$-1 - \dfrac{2}{5}y + 1 = -7 + 1$
$-\dfrac{2}{5}y = -6$
$-\dfrac{5}{2}\left(-\dfrac{2}{5}y\right) = -\dfrac{5}{2}(-6)$
$y = 15$
The solution is (−4, 15).

52. $5x + 10y = -55$
$2x - 3y = 6$
Multiply the first equation by 3 and the second equation by 10. Add the new equations.
$15x + 30y = -165$
$20x - 30y = 60$
$35x \quad = -105$
$\dfrac{35x}{35} = \dfrac{-105}{35}$
$x = -3$
Substitute −3 for x in the first equation.

$5(-3) + 10y = -55$
$-15 + 10y = -55$
$-15 + 10y + 15 = -55 + 15$
$10y = -40$
$\dfrac{10y}{10} = \dfrac{-40}{10}$
$y = -4$
The solution is (−3, −4).

53. $2x = 3y + 1$
$15y = 10x + 5$
Multiply the first equation by 5 and add both equations written in standard form.
$10x - 15y = 5$
$-10x + 15y = 5$
$0 = 10$
This is a contradiction. There is no solution.

54. $5x + 7y = 21$
$3x - 2y = 13$
Multiply the first equation by 2 and the second equation by 7. Add the new equations.
$10x + 14y = 42$
$21x - 14y = 91$
$31x \quad = 133$
$\dfrac{31x}{31} = \dfrac{133}{31}$
$x = \dfrac{133}{31}$
Substitute $\dfrac{133}{31}$ for x in the first equation.
$5\left(\dfrac{133}{31}\right) + 7y = 21$
$\dfrac{665}{31} + 7y - \dfrac{665}{31} = 21 - \dfrac{665}{31}$
$7y = \dfrac{651}{31} - \dfrac{665}{31}$
$7y = -\dfrac{14}{31}$
$\dfrac{1}{7}(7y) = \dfrac{1}{7}\left(-\dfrac{14}{31}\right)$
$y = -\dfrac{2}{31}$
The solution is $\left(\dfrac{133}{31}, -\dfrac{2}{31}\right)$.

55. $3x + y = 20$

$y = \dfrac{2}{3}x + \dfrac{16}{3}$

Multiply the second equation by -3 and write in standard form. Add the result to the first equation multiplied by 3.

$9x + 3y = 60$

$\underline{2x - 3y = -16}$

$11x \quad\;\; = 44$

$\dfrac{11x}{11} = \dfrac{44}{11}$

$x = 4$

Substitute 4 for x in the first equation.

$3(4) + y = 20$

$12 + y - 12 = 20 - 12$

$y = 8$

The solution is $(4, 8)$.

56. $7(y - 1) = 5x$

$y = \dfrac{5}{7}x + 1$

Multiply the second equation by 7. Add both equations written in standard form.

$-5x + 7y = 7$

$\underline{5x - 7y = -7}$

$0 = 0$

This is an identity. The solutions are all ordered pairs (x, y) that satisfy $y = \dfrac{5}{7}x + 1$.

57. $3x - 3y = 4$

$9x + 9y = -2$

Multiply the first equation by 3 and add to the second equation.

$9x - 9y = 12$

$\underline{9x + 9y = -2}$

$18x \quad\;\; = 10$

$\dfrac{18x}{18} = \dfrac{10}{18}$

$x = \dfrac{5}{9}$

Substitute $\dfrac{5}{9}$ for x in the second equation.

$9\left(\dfrac{5}{9}\right) + 9y = -2$

$5 + 9y = -2$

$5 + 9y - 5 = -2 - 5$

$9y = -7$

$\dfrac{9y}{9} = \dfrac{-7}{9}$

$y = -\dfrac{7}{9}$

The solution is $\left(\dfrac{5}{9}, -\dfrac{7}{9}\right)$.

58. $3x - y = 0$

$2x + 2y = 7$

Multiply the first equation by 2 and add to the second equation.

$6x - 2y = 0$

$\underline{2x + 2y = 7}$

$8x \quad\;\; = 7$

$\dfrac{8x}{8} = \dfrac{7}{8}$

$x = \dfrac{7}{8}$

Substitute $\dfrac{7}{8}$ for x in the first equation.

$3\left(\dfrac{7}{8}\right) - y = 0$

$\dfrac{21}{8} - y - \dfrac{21}{8} = 0 - \dfrac{21}{8}$

$-y = -\dfrac{21}{8}$

$-1(-y) = -1\left(-\dfrac{21}{8}\right)$

$y = \dfrac{21}{8}$

The solution is $\left(\dfrac{7}{8}, \dfrac{21}{8}\right)$.

59. $0.25x + 0.3y = 4$

$0.5x - 0.2y = 4$

Multiply the second equation by -0.5 and add to the first equation.

$0.25x + 0.3y = 4$
$\underline{-0.25x + 0.1y = -2}$
$\quad\quad\quad 0.4y = 2$
$$\frac{0.4y}{0.4} = \frac{2}{0.4}$$
$\quad\quad\quad\quad y = 5$
Substitute 5 for y in the first equation.
$0.25x + 0.3(5) = 4$
$0.25x + 1.5 - 1.5 = 4 - 1.5$
$0.25x = 2.5$
$$\frac{0.25x}{0.25} = \frac{2.5}{0.25}$$
$x = 10$
The solution is (10, 5).

60. $0.2x + 0.1y = 5$
$0.02x - 0.01y = 13.5$
Multiply the first equation by 0.1 and add to the second equation.
$0.02x + 0.01y = 0.5$
$\underline{0.02x - 0.01y = 13.5}$
$0.04x \quad\quad = 14$
$$\frac{0.04x}{0.04} = \frac{14}{0.04}$$
$\quad\quad x = 350$
Substitute 350 for x in the first equation.
$0.2(350) + 0.1y = 5$
$70 + 0.1y - 70 = 5 - 70$
$0.1y = -65$
$$\frac{0.1y}{0.1} = \frac{-65}{0.1}$$
$y = -650$
The solution is (350, –650).

61. Let x = number of pounds of 25% copper
$\quad\quad y$ = number of pounds of 30% copper
$x + y = 100$
$0.25x + 0.3y = 0.27(100)$
Multiply the first equation by –0.3 and add to the second equation.
$-0.3x - 0.3y = -30$
$\underline{0.25x + 0.3y = 27}$
$-0.05x \quad\quad = -3$
$$\frac{-0.05x}{-0.05} = \frac{-3}{-0.05}$$
$\quad\quad x = 60$
Substitute 60 for x in the first equation.
$60 + y = 100$
$60 + y - 60 = 100 - 60$
$y = 40$

The amounts are as follows:
60 pounds of 25% copper and 40 pounds of 30% copper.

62. Let x = number of small offices
$\quad\quad y$ = number of large offices
$x + y = 40$
$250x + 400y = 12{,}700$
Multiply the first equation by –250 and add to the second equation.
$-250x - 250y = -10{,}000$
$\underline{250x + 400y = 12{,}700}$
$150y = 2700$
$$\frac{150y}{150} = \frac{2700}{150}$$
$y = 18$
Substitute 18 for y in the first equation.
$x + 18 = 40$
$x + 18 - 18 = 40 - 18$
$x = 22$
She will have more of the small offices. She will have 22 small offices and 18 large offices.

63. Let x = mass of Mars
$\quad\quad y$ = mass of Earth
$y = x + 5.328 \times 10^{24}$
$$\frac{x + y}{2} = 3.306 \times 10^{24}$$
Write each equation in standard form and add them.
$-x + y = 5.328 \times 10^{24}$
$\underline{x + y = 6.612 \times 10^{24}}$
$2y = 1.194 \times 10^{25}$
$$\frac{2y}{2} = \frac{1.194 \times 10^{25}}{2}$$
$y = 5.97 \times 10^{24}$
Substitute 5.97×10^{24} for y in the first equation.
$5.97 \times 10^{24} = x + 5.328 \times 10^{24}$
$6.42 \times 10^{23} = x$
The mass of Mars is 6.42×10^{23} kg and the mass of Earth is 5.97×10^{24} kg.

64. Let x = number from A
$\quad\quad y$ = number from B
$25x + 35y = 5500$
$x + y = 200$
Multiply the second equation by -25 and add to the first equation.
$\quad 25x + 35y = 5500$
$\underline{-25x - 25y = -5000}$
$\quad\quad\quad 10y = 500$
$\quad\quad\quad \dfrac{10y}{10} = \dfrac{500}{10}$
$\quad\quad\quad\quad y = 50$
Substitute 50 for y in the second equation.
$x + 50 = 200$
$x + 50 - 50 = 200 - 50$
$x = 150$
The plant should order 150 components from Supplier A and 50 components from Supplier B.

65. Let x = number of pounds of 10% nitrogen
$\quad\quad y$ = number of pounds of 5% nitrogen
$x + y = 150$
$0.1x + 0.05y = 0.08(150)$
Multiply the first equation by -0.05 and add to the second equation.
$\quad -0.05x - 0.05y = -7.5$
$\quad\underline{0.1x + 0.05y = 12}$
$\quad 0.05x \quad\quad\quad = 4.5$
$\quad \dfrac{0.05x}{0.05} = \dfrac{4.5}{0.05}$
$\quad\quad\quad x = 90$
Substitute 90 for x in the first equation.
$90 + y = 150$
$90 + y - 90 = 150 - 90$
$y = 60$
He should use 90 pounds of 10% nitrogen with 60 pounds of 5% nitrogen.

66. Let x = number of pounds of broccoli
$\quad\quad y$ = number of pounds of cauliflower
$0.49x + 0.99y = 0.69(200)$
$x + y = 200$
Multiply the second equation by -0.49 and add to the first equation.

$\quad 0.49x + 0.99y = 138$
$\underline{-0.49x - 0.49y = -98}$
$\quad\quad\quad 0.5y = 40$
$\quad\quad\quad \dfrac{0.5y}{0.5} = \dfrac{40}{0.5}$
$\quad\quad\quad\quad y = 80$
Substitute 80 for y in the second equation.
$x + 80 = 200$
$x + 80 - 80 = 200 - 80$
$x = 120$
They should mix 120 pounds of broccoli with 80 pounds of cauliflower.

67. Let x = number of hours of interstate driving
$\quad\quad y$ = number of hours of highway driving
$x + y = 10$
$65x + 45y = 600$
Multiply the first equation by -45 and add to the first equation.
$\quad -45x - 45y = -450$
$\quad\underline{65x + 45y = 600}$
$\quad 20x \quad\quad\quad = 150$
$\quad\quad \dfrac{20x}{20} = \dfrac{150}{20}$
$\quad\quad\quad x = 7.5$
$7.5(65) = 487.5$
He drove 487.5 miles on the interstate.

68. Let x = speed of current
$\quad\quad y$ = distance of race
$y = 217.26(x + 1.603)$
$y = 217.50(x + 1.598)$
Substitute $217.26(x + 1.603)$ for y in the second equation.
$217.26(x + 1.603) = 217.50(x + 1.598)$
$217.26x + 348.26778 = 217.50x + 347.565$
$217.26x - 217.5x = 347.565 - 348.26778$
$-0.24x = -0.703$
$\dfrac{-0.24x}{-0.24} = \dfrac{-0.70278}{-0.24}$
$x = 2.92825 \approx 2.928$
(rounded to the nearest thousandth)
Substitute 2.928 for x in the first equation.
$y = 217.26(2.928 + 1.603)$
$y = 984.459$
The speed of the current was 2.928 mps and the distance of the race was 984.459 meters.

69. $C(x) = 400 + 5x$
$R(x) = 8.5x$

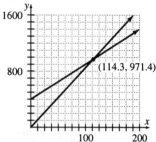

The intersection is approximately at (114.3, 971.4). The break-even point is at about 114 items.
When the company produces and sells 114 items or less, cost exceeds revenue. When the company produces and sells at least 115 items, revenue exceeds cost.

70. Let x = miles driven.
$A(x) = 150 + 0.25x$
$B(x) = 175 + 0.2x$

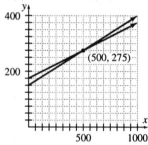

The graphs intersect at (500, 275).
He must drive more than 500 miles.

Chapter 6 Mixed Review

1. $y = \dfrac{4}{7}x + 2$

$x - 3y = 4$

Multiply the first equation by 7 and the second equation by 4. Add the new equations written in standard form.
$-4x + 7y = 14$
$\underline{4x - 12y = 16}$
$-5y = 30$
$\dfrac{-5y}{-5} = \dfrac{30}{-5}$
$y = -6$

Substitute -6 for y in the second equation.
$x - 3(-6) = 4$
$x + 18 = 4$
$x + 18 - 18 = 4 - 18$
$x = -14$
The solution is $(-14, -6)$.

2. $y = \dfrac{4}{5}x - 6$

$5y + 2 = 4(x - 7)$
Multiply the first equation by -5. Add the two equations written in standard form.
$4x - 5y = 30$
$\underline{-4x + 5y = -30}$
$0 = 0$
This is an identity. The solutions are all ordered pairs (x, y) that satisfy $y = \dfrac{4}{5}x - 6$.

3. $1.25x + 3.5y = -25$
$4.5x + 2.8y = 8$
Multiply the first equation by 2.8 and the second equation by -3.5. Add the new equations.
$3.5x + 9.8y = -70$
$\underline{-15.75x - 9.8y = -28}$
$-12.25x = -98$
$\dfrac{-12.25x}{-12.25} = \dfrac{-98}{-12.25}$
$x = 8$

Substitute 8 for x in the second equation.
$4.5(8) + 2.8y = 8$
$36 + 2.8y - 36 = 8 - 36$
$2.8y = -28$
$\dfrac{2.8y}{2.8} = \dfrac{-28}{2.8}$
$y = -10$
The solution is $(8, -10)$.

4. $0.55x - 0.68y = 48$
$\underline{-0.51x + 0.68y = 0}$
$0.04x = 48$
$\dfrac{0.04x}{0.04} = \dfrac{48}{0.04}$
$x = 1200$

Substitute 1200 for x in the first equation.

$0.55(1200) - 0.68y = 48$
$660 - 0.68y - 660 = 48 - 660$
$-0.68y = -612$
$\dfrac{-0.68y}{-0.68} = \dfrac{-612}{-0.68}$
$y = 900$
The solution is (1200, 900).

5. $8x + 2y = -6$
$\underline{6x - 2y = -78}$
$14x = -84$
$\dfrac{14x}{14} = \dfrac{-84}{14}$
$x = -6$

Substitute –6 for x in the first equation.
$8(-6) + 2y = -6$
$-48 + 2y + 48 = -6 + 48$
$2y = 42$
$\dfrac{2y}{2} = \dfrac{42}{2}$
$y = 21$
The solution is (–6, 21).

6. $2x + 8y = -12$
$y = 2x + 3$
Write the second equation in standard form
and add to the first equation.
$2x + 8y = -12$
$\underline{-2x + y = 3}$
$9y = -9$
$\dfrac{9y}{9} = \dfrac{-9}{9}$
$y = -1$

Substitute –1 for y in the first equation.
$2x + 8(-1) = -12$
$2x - 8 + 8 = -12 + 8$
$2x = -4$
$\dfrac{2x}{2} = \dfrac{-4}{2}$
$x = -2$
The solution is (–2, –1).

7. $3x + y = 4$
$12x + 4y = 9$
Multiply the first equation by –4 and add to
the second equation.

$-12x - 4y = -16$
$\underline{12x + 4y = 9}$
$0 = -7$
This is a contradiction. There is no solution.

8. $3x - 5y = 11$
$4x + 2y = 13$
Multiply the first equation by 2 and the
second equation by 5. Add the new
equations.
$6x - 10y = 22$
$\underline{20x + 10y = 65}$
$26x = 87$
$\dfrac{26x}{26} = \dfrac{87}{26}$
$x = \dfrac{87}{26}$

Substitute $\dfrac{87}{26}$ for x in the second equation.

$4\left(\dfrac{87}{26}\right) + 2y = 13$

$\dfrac{174}{13} + 2y - \dfrac{174}{13} = 13 - \dfrac{174}{13}$

$2y = \dfrac{169}{13} - \dfrac{174}{13}$

$2y = -\dfrac{5}{13}$

$\dfrac{1}{2}(2y) = \dfrac{1}{2}\left(-\dfrac{5}{13}\right)$

$y = -\dfrac{5}{26}$

The solution is $\left(\dfrac{87}{26}, -\dfrac{5}{26}\right)$.

9. $\dfrac{1}{3}x - \dfrac{2}{5}y = 10$

$\dfrac{1}{2}x + \dfrac{4}{5}y = -13$

Multiply the first equation by –15 and the
second equation by 10. Add the new
equations.

$$-5x + 6y = -150$$
$$\underline{5x + 8y = -130}$$
$$14y = -280$$
$$\frac{14y}{14} = -\frac{280}{14}$$
$$y = -20$$

Substitute –20 for y in the second equation.

$$\frac{1}{2}x + \frac{4}{5}(-20) = -13$$

$$\frac{1}{2}x - 16 + 16 = -13 + 16$$

$$\frac{1}{2}x = 3$$

$$2\left(\frac{1}{2}x\right) = 2(3)$$

$$x = 6$$

The solution is (6, –20).

10. $4x + 8y = 68$
$5x - 3y = -6$

Multiply the first equation by 3 and the second equation by 8. Add the new equations.

$$12x + 24y = 204$$
$$\underline{40x - 24y = -48}$$
$$52x = 156$$
$$\frac{52x}{52} = \frac{156}{52}$$
$$x = 3$$

Substitute 3 for x in the first equation.

$$4(3) + 8y = 68$$
$$12 + 8y - 12 = 68 - 12$$
$$8y = 56$$
$$\frac{8y}{8} = \frac{56}{8}$$
$$y = 7$$

The solution is (3, 7).

11. $2x + y = 0$
$x + 4y = 3$

Multiply the second equation by –2 and add to the first equation.

$$2x + y = 0$$
$$\underline{-2x - 8y = -6}$$
$$-7y = -6$$
$$\frac{-7y}{-7} = \frac{-6}{-7}$$
$$y = \frac{6}{7}$$

Substitute $\frac{6}{7}$ for y in the second equation.

$$x + 4\left(\frac{6}{7}\right) = 3$$

$$x + \frac{24}{7} - \frac{24}{7} = 3 - \frac{24}{7}$$

$$x = \frac{21}{7} - \frac{24}{7}$$

$$x = -\frac{3}{7}$$

The solution is $\left(-\frac{3}{7}, \frac{6}{7}\right)$.

12. $5x + 3y = 9$
$x - y = 6$

Multiply the second equation by 3 and add to the first equation.

$$5x + 3y = 9$$
$$\underline{3x - 3y = 18}$$
$$8x = 27$$
$$\frac{8x}{8} = \frac{27}{8}$$
$$x = \frac{27}{8}$$

Substitute $\frac{27}{8}$ for x in the second equation.

$$\frac{27}{8} - y = 6$$

$$\frac{27}{8} - y - \frac{27}{8} = 6 - \frac{27}{8}$$

$$-y = \frac{48}{8} - \frac{27}{8}$$

$$-1(-y) = -1\left(\frac{21}{8}\right)$$

$$y = -\frac{21}{8}$$

The solution is $\left(\frac{27}{8}, -\frac{21}{8}\right)$.

13. $2y + 2 = y$
$y = -x + 1$

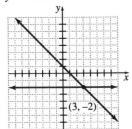

The solution is $(3, -2)$.

14. $2x + 3 = x$
$2x = 5$

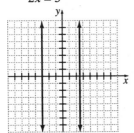

The graphs are parallel. There is no solution.

15. $y = 2x + 10$
$3x + y = -10$

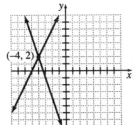

The solution is $(-4, 2)$.

16. $2(x - 1) - 4 = 0$
$2x + y = 3$

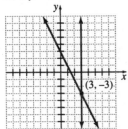

The solution is $(3, -3)$.

17. $x + y = 9 - x$
$2x + y = 5$

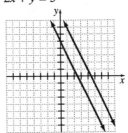

The graphs are parallel. There is no solution.

18. $y = -4x - 3$
$4x + 2y = y - 3$

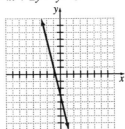

The graphs are coinciding. The solutions are all ordered pairs (x, y) satisfying $y = -4x - 3$.

19. $3x + 4 = 2x$
$x + 7 = 3$

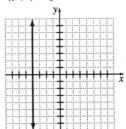

The graphs are coinciding. The solutions are all ordered pairs (x, y) that satisfy $x + 7 = 3$ or $x = -4$.

20. $3y + 7 = 2y + 5$
$\quad y + 9 = 7$

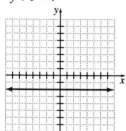

The graphs are coinciding. The solutions are all ordered pairs (x, y) that satisfy $y + 9 = 7$ or $y = -2$.

21. $3(x - 3) - 1 = 8$
$\quad\quad 3y - 10 = 15 - 2y$

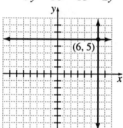

The solution is $(6, 5)$.

22. $y = 3x + 7$
$\quad x - y = -1$

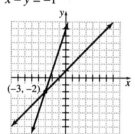

The solution is $(-3, -2)$.

23. $x - 5y = 15$
$\quad x + 5y = -5$

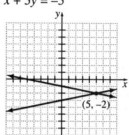

The solution is $(5, -2)$.

24. $f(x) = -4x - 6$
$\quad g(x) = 3x + 8$

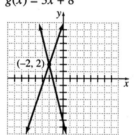

The solution is $(-2, 2)$.

25. $y = x$
$\quad x = 2 - y$

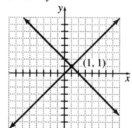

The solution is $(1, 1)$.

26. $x + 9 = 3$
$\quad 3y = 2y - 1$

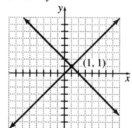

The solution is $(-6, -1)$.

27. $y = \dfrac{5}{11}x - 43$

$x + 2y = -2$

Substitute $\dfrac{5}{11}x - 43$ for y in the second

equation.

$x + 2\left(\dfrac{5}{11}x - 43\right) = -2$

$x + \dfrac{10}{11}x - 86 = -2$

$\dfrac{21}{11}x - 86 + 86 = -2 + 86$

$\dfrac{21}{11}x = 84$

$\dfrac{11}{21}\left(\dfrac{21}{11}x\right) = \dfrac{11}{21}(84)$

$x = 44$

Substitute 44 for x in the first equation.

$y = \dfrac{5}{11}(44) - 43$

$y = -23$

The solution is $(44, -23)$.

28. $y = \dfrac{3}{4}x + 14$

$y = -\dfrac{2}{3}x - 3$

Substitute $\dfrac{3}{4}x + 14$ for y in the second

equation.

$\dfrac{3}{4}x + 14 = -\dfrac{2}{3}x - 3$

$\dfrac{3}{4}x + \dfrac{2}{3}x = -3 - 14$

$\dfrac{17}{12}x = -17$

$\dfrac{12}{17}\left(\dfrac{17}{12}x\right) = \dfrac{12}{17}(-17)$

$x = -12$

Substitute -12 for x in the first equation.

$y = \dfrac{3}{4}(-12) + 14$

$y = 5$

The solution is $(-12, 5)$.

29. $2x + 3y = 53$

$2x - 4y = -248$

Solve the second equation for x, $x = 2y - 124$. Substitute $2y - 124$ for x in the first equation.

$2(2y - 124) + 3y = 53$

$7y - 248 = 53$

$7y - 248 + 248 = 53 + 248$

$7y = 301$

$\dfrac{7y}{7} = \dfrac{301}{7}$

$y = 43$

Substitute 43 for y in the first equation.

$2x + 3(43) = 53$

$2x + 129 - 129 = 53 - 129$

$2x = -76$

$\dfrac{2x}{2} = -\dfrac{76}{2}$

$x = -38$

The solution is $(-38, 43)$.

30. $y = 4x - 7$

$2y + 12 = 4x + y + 5$

Substitute $4x - 7$ for y in the second equation.

$2(4x - 7) + 12 = 4x + 4x - 7 + 5$

$8x - 14 + 12 = 8x - 2$

$8x - 2 - 8x = 8x - 2 - 8x$

$-2 = -2$

This is an identity. The solutions are all ordered pairs (x, y) that satisfy $y = 4x - 7$.

31. $y = -5x$

$x - y = 1$

Substitute $-5x$ for y in the second equation.

$x - (-5x) = 1$

$6x = 1$

$\dfrac{6x}{6} = \dfrac{1}{6}$

$x = \dfrac{1}{6}$

Substitute $\dfrac{1}{6}$ for x in the first equation.

$y = -5\left(\dfrac{1}{6}\right)$

$y = -\dfrac{5}{6}$

The solution is $\left(\dfrac{1}{6}, -\dfrac{5}{6}\right)$.

32. $5x - 10y = 21$
$\qquad 5x = -3$

Solve the second equation for x, $x = -\dfrac{3}{5}$.

Substitute $-\dfrac{3}{5}$ for x in the first equation.

$5\left(-\dfrac{3}{5}\right) - 10y = 21$

$-3 - 10y + 3 = 21 + 3$

$-10y = 24$

$\dfrac{-10y}{-10} = \dfrac{24}{-10}$

$y = -\dfrac{12}{5}$

The solution is $\left(-\dfrac{3}{5}, -\dfrac{12}{5}\right)$.

33. $x + 7y = 3$
$4x + y = 9$
Solve the first equation for x, $x = 3 - 7y$.
Substitute $3 - 7y$ for x in the second equation.
$4(3 - 7y) + y = 9$
$12 - 28y + y = 9$
$12 - 27y - 12 = 9 - 12$
$-27y = -3$

$\dfrac{-27y}{-27} = \dfrac{-3}{-27}$

$y = \dfrac{1}{9}$

Substitute $\dfrac{1}{9}$ for y in the first equation.

$x + 7\left(\dfrac{1}{9}\right) = 3$

$x + \dfrac{7}{9} - \dfrac{7}{9} = 3 - \dfrac{7}{9}$

$x = \dfrac{20}{9}$

The solution is $\left(\dfrac{20}{9}, \dfrac{1}{9}\right)$.

34. $5x + 6y = 669$
$\quad x - 2y = 1705$
Solve the second equation for x,
$x = 2y + 1705$. Substitute $2y + 1705$ for x in
the first equation.

$5(2y + 1705) + 6y = 669$
$10y + 8525 + 6y = 669$
$16y + 8525 - 8525 = 669 - 8525$
$16y = -7856$

$\dfrac{16y}{16} = \dfrac{-7856}{16}$

$y = -491$
Substitute -491 for y in the second equation.
$x - 2(-491) = 1705$
$x + 982 - 982 = 1705 - 982$
$x = 723$
The solution is $(723, -491)$.

35. $\quad x - y = 5$
$6x + 2y = 9$
Solve the first equation for x, $x = y + 5$.
Substitute $y + 5$ for x in the second equation.
$6(y + 5) + 2y = 9$
$6y + 30 + 2y = 9$
$8y + 30 - 30 = 9 - 30$
$8y = -21$

$\dfrac{8y}{8} = -\dfrac{21}{8}$

$y = -\dfrac{21}{8}$

Substitute $-\dfrac{21}{8}$ for y in the first equation.

$x - \left(-\dfrac{21}{8}\right) = 5$

$x + \dfrac{21}{8} - \dfrac{21}{8} = \dfrac{40}{8} - \dfrac{21}{8}$

$x = \dfrac{19}{8}$

The solution is $\left(\dfrac{19}{8}, -\dfrac{21}{8}\right)$.

36. $x = 324y$
$3x - 810y = 90$
Substitute $324y$ for x in the second equation.
$3(324y) - 810y = 90$
$972y - 810y = 90$
$162y = 90$

$\dfrac{162y}{162} = \dfrac{90}{162}$

$y = \dfrac{5}{9}$

Substitute $\dfrac{5}{9}$ for y in the first equation.

$x = 324\left(\dfrac{5}{9}\right)$

$x = 180$

The solution is $\left(180, \dfrac{5}{9}\right)$.

37. $y = -2x + 7$
$2x + y = 0$
Substitute $-2x + 7$ for y in the second equation.
$2x + (-2x + 7) = 0$
$7 = 0$
This is a contradiction. There is no solution.

38. $2x - y = 60$
$5x - 3y = 60$
Solve the first equation for y, $y = 2x - 60$.
Substitute $2x - 60$ for y in the second equation.
$5x - 3(2x - 60) = 60$
$5x - 6x + 180 = 60$
$-x + 180 - 180 = 60 - 180$
$-x = -120$
$\dfrac{-x}{-1} = \dfrac{-120}{-1}$
$x = 120$
Substitute 120 for x in the first equation.
$2(120) - y = 60$
$240 - y - 240 = 60 - 240$
$-y = -180$
$\dfrac{-y}{-1} = \dfrac{-180}{-1}$
$y = 180$
The solution is (120, 180).

39.

$$\begin{array}{c|c} 4x + 2y = 3 \\ \hline 4\left(\frac{5}{6}\right) + 2\left(-\frac{1}{6}\right) & 3 \\ \frac{10}{3} - \frac{1}{3} & \\ 3 & \end{array}$$

$$\begin{array}{c|c} x - y = -1 \\ \hline \frac{5}{6} - \left(-\frac{1}{6}\right) & -1 \\ \frac{5}{6} + \frac{1}{6} & \\ 1 & \end{array}$$

The second equation is false. Therefore, $\left(\dfrac{5}{6}, -\dfrac{1}{6}\right)$ is not a solution.

40.

$$\begin{array}{c|c} 3x + y = -1 \\ \hline 3(-1.6) + 3.8 & -1 \\ -4.8 + 3.8 & \\ -1 & \end{array}$$

$$\begin{array}{c|c} 5x + 5y = 11 \\ \hline 5(-1.6) + 5(3.8) & 11 \\ -8 + 19 & \\ 11 & \end{array}$$

Both equations are true. Therefore, (–1.6, 3.8) is a solution.

41.

$$\begin{array}{c|c} 5x + 7y = -6 \\ \hline 5(-4) + 7(2) & -6 \\ -20 + 14 & \\ -6 & \end{array}$$

$$\begin{array}{c|c} 2x - 7y = -22 \\ \hline 2(-4) - 7(2) & -22 \\ -8 - 14 & \\ -22 & \end{array}$$

Both equations are true. Therefore, (–4, 2) is a solution.

42.

$$\begin{array}{c|c} x + y = 1 \\ \hline \frac{13}{9} + \left(-\frac{4}{9}\right) & 1 \\ \frac{9}{9} & \\ 1 & \end{array}$$

$$\frac{8x - y = 12}{8\left(\frac{13}{9}\right) - \left(-\frac{4}{9}\right) \;\Big|\; 12}$$

$$\frac{104}{9} + \frac{4}{9} \;\Big|$$

$$\frac{108}{9} \;\Big|$$

$$12 \;\Big|$$

Both equations are true. Therefore, $\left(\dfrac{13}{9}, -\dfrac{4}{9}\right)$ is a solution.

43.
$$\frac{2x - 1 = 5}{2(3) - 1 \;\Big|\; 5}$$
$$6 - 1 \;\Big|$$
$$5 \;\Big|$$

$$\frac{3y + 5 = 2}{3(-2) + 5 \;\Big|\; 2}$$
$$-6 + 5 \;\Big|$$
$$-1 \;\Big|$$

The second equation is false. Therefore, $(3, -2)$ is not a solution.

44.
$$\frac{x + 10 = 5}{-5 + 10 \;\Big|\; 5}$$
$$5 \;\Big|$$

$$\frac{3y + 11 = 2}{3(-3) + 11 \;\Big|\; 2}$$
$$-9 + 11 \;\Big|$$
$$2 \;\Big|$$

Both equations are true. Therefore, $(-5, -3)$ is a solution.

45.
$$\frac{x + 2y = 2}{-0.44 + 2(1.22) \;\Big|\; 2}$$
$$-0.44 + 2.44 \;\Big|$$
$$2 \;\Big|$$

$$\frac{2x + 4y = 5}{2(-0.44) + 4(1.22) \;\Big|\; 5}$$
$$-0.88 + 4.88 \;\Big|$$
$$4 \;\Big|$$

The second equation is false. Therefore, $(-0.44, 1.22)$ is not a solution.

46.
$$\frac{2x + 5y = -19}{2(-2) + 5(-3) \;\Big|\; -19}$$
$$-4 - 15 \;\Big|$$
$$-19 \;\Big|$$

$$\frac{3x - 7y = 10}{3(-2) - 7(-3) \;\Big|\; 10}$$
$$-6 + 21 \;\Big|$$
$$15 \;\Big|$$

The second equation is false. Therefore, $(-2, -3)$ is not a solution.

47. $2x - 3 = 1$
$2x = 4$
$x = 2$ m is undefined

$4x - 3y = 6$
$3y = 4x - 6$
$y = \dfrac{4}{3}x - 2$ $m = \dfrac{4}{3}, \; b = -2$

The lines are only intersecting because one is vertical and the other has a slope of $\dfrac{4}{3}$.

48. $5x - 4y = 8$
$4y = 5x - 8$
$y = \dfrac{5}{4}x - 2$ $m = \dfrac{5}{4}, \; b = -2$

$4x - 5y = -15$
$5y = 4x + 15$
$y = \dfrac{4}{5}x + 3$ $m = \dfrac{4}{5}, \; b = 3$

The lines are only intersecting because they have different slopes and their product is not -1.

49. $y = 5(x - 1)$
$y = 5x - 5$ $m = 5, b = -5$

$2(x - 2) = 7x - y + 3$
$2x - 2 = 7x - y + 3$
$y = 5x + 5$ $m = 5, b = 5$
The lines are parallel because they have the same slopes, but different y-intercepts.

50. $5(x + 1) = 15$
$5x + 5 = 15$
$5x = 10$
$x = 2$

$3y + 1 = 10$
$3y = 9$
$y = 3$
The lines are intersecting and perpendicular because one is vertical and the other is horizontal.

51. $y = -2x + 3$ $m = -2, b = 3$
$2(x + y) = y + 3$
$2x + 2y = y + 3$
$y = -2x + 3$ $m = -2, b = 3$
The lines are coinciding because their slopes are the same and their y-intercepts are the same.

52. $y = x + 3$ $m = 1, b = 3$
$y = -x - 4$ $m = -1, b = -4$
The lines are intersecting and perpendicular because the product of their slopes is -1.

53. $5y = 20$
$y = 4$ $m = 0, b = 4$
$2(y + 3) = -2$
$2y + 6 = -2$
$2y = -8$
$y = -4$ $m = 0, b = -4$
The lines are parallel because they are both horizontal with different y-intercepts.

54. Let x = number of pounds of 15%
     y = number of pounds of 35%
$x + y = 200$
$0.15x + 0.35y - 0.27(200)$
Multiply the first equation by -0.15 and add to the second equation.

$-0.15x - 0.15y = -30$
$0.15x + 0.35y = 54$
$\overline{\hphantom{0.15x + 0.35y = }0.2y = 24}$
$\dfrac{0.2y}{0.2} = \dfrac{24}{0.2}$
$y = 120$
Substitute 120 for y in the first equation.
$x + 120 = 200$
$x + 120 - 120 = 200 - 120$
$x = 80$
The amounts should be as follows:
80 pounds of 15% brass
120 pounds of 35% brass
$120 - 80 = 40$
40 pounds more of the 35% alloy will be used.

55. Let x = amount for 1-bedroom
     y = amount for 2-bedroom
$8x + 12y = 7300$
$y = x + 150$
Substitute $x + 150$ for y in the first equation.
$8x + 12(x + 150) = 7300$
$8x + 12x + 1800 = 7300$
$20x + 1800 - 1800 = 7300 - 1800$
$20x = 5500$
$\dfrac{20x}{20} = \dfrac{5500}{20}$
$x = 275$
Substitute 275 for x in the second equation.
$y = 275 + 150 = 425$
$8x = 8(275) = 2200$
$12y = 12(425) = 5100$
She will receive $2200 for all of the 1-bedroom apartments and $5100 for all of the 2-bedroom apartments.

56. Let x = measure of smaller angle
     y = measure of larger angle
$x + y = 180$
$y = x + 10$
Substitute $x + 10$ for y in the first equation.
$x + x + 10 = 180$
$2x + 10 - 10 = 180 - 10$
$2x = 170$
$\dfrac{2x}{2} = \dfrac{170}{2}$
$x = 85$
The smaller angle measures 85°.

57. Let x = side length of smaller square
    y = side length of larger square
$y = x + 5$
$4x + 4y = 100$
Substitute $x + 5$ for y in the second equation.
$4x + 4(x + 5) = 100$
$4x + 4x + 20 = 100$
$8x + 20 - 20 = 100 - 20$
$8x = 80$
$\dfrac{8x}{8} = \dfrac{80}{8}$
$x = 10$
$y = 10 + 5 = 15$
$(10)^2 = 100$
$(15)^2 = 225$
The areas of the squares will be 100 in^2 and 225 in^2.

58. Let x = number of women surveyed
    y = number of men surveyed
$x + y = 700$
$0.7x + 0.4y = 400$
Multiply the first equation by -0.4 and add to the second equation.
$-0.4x - 0.4y = -280$
$\underline{0.7x + 0.4y = 400}$
$0.3x = 120$
$\dfrac{0.3x}{0.3} = \dfrac{120}{0.3}$
$x = 400$
Substitute 400 for x in the first equation.
$400 + y = 700$
$400 + y - 400 = 700 - 400$
$y = 300$
$400 - 300 = 100$
There were 100 more women surveyed.

59. Let x = measure of smaller angle
    y = measure of larger angle
$x + y = 180$
$y = 3x + 12$
Substitute $3x + 12$ for y in the first equation.
$x + 3x + 12 = 180$
$4x + 12 - 12 = 180 - 12$
$4x = 168$
$\dfrac{4x}{4} = \dfrac{168}{4}$
$x = 42$
Substitute 42 for x in second equation.

$y = 3(42) + 12$
$y = 138$
The angles measure $42°$ and $138°$.

60. Let x = number of cc of 15%
    y = number of cc of 25%
$x + y = 30$
$0.15x + 0.25y = 0.21(30)$
Multiply the first equation by -0.15 and add to the second equation.
$-0.15x - 0.15y = -4.5$
$\underline{0.15x + 0.25y = 6.3}$
$0.1y = 1.8$
$\dfrac{0.1y}{0.1} = \dfrac{1.8}{0.1}$
$y = 18$
Substitute 18 for y in the first equation.
$x + 18 = 30$
$x + 18 - 18 = 30 - 18$
$x = 12$
She should use 12 cc of 15% and 18 cc of 25%.

61. Let x = walking rate
    y = jogging rate
$3.5 = \dfrac{1}{2}x + \dfrac{1}{3}y$
$4 = \dfrac{2}{3}x + \dfrac{1}{3}y$
Multiply the first equation by -6 and the second equation by 6. Add the new equations.
$-21 = -3x - 2y$
$\underline{24 = 4x + 2y}$
$3 = x$
Substitute 3 for x in the second equation.
$4 = \dfrac{2}{3}(3) + \dfrac{1}{3}y$
$4 - 2 = 2 + \dfrac{1}{3}y - 2$
$2 = \dfrac{1}{3}y$
$3(2) = 3\left(\dfrac{1}{3}y\right)$
$6 = y$
He walked at 3 mph and jogged at 6 mph.

62. Let x = number of pounds of gourmet coffee
    y = number of pounds of chocolate
$x + y = 50$
$8.5x + 12.5y = 9.5(50)$
Multiply the first equation by -8.5 and add to the second equation.
$$-8.5x - 8.5y = -425$$
$$\underline{8.5x + 12.5y = 475}$$
$$4y = 50$$
$$\frac{4y}{4} = \frac{50}{4}$$
$$y = 12.5$$
Substitute 12.5 for y in the first equation.
$x + 12.5 = 50$
$x + 12.5 - 12.5 = 50 - 12.5$
$x = 37.5$
The shop should mix 37.5 pounds of gourmet coffee with 12.5 pounds of dutch chocolate.

63. Let x = number of hours on line
    y = number of hours as clerk
$10.75x + 6.5y = 181$
$x + y = 20$
Multiply the second equation by -6.5 and add to the first equation.
$$10.75x + 6.5y = 181$$
$$\underline{-6.5x - 6.5y = -130}$$
$$4.25x = 51$$
$$\frac{4.25x}{4.25} = \frac{51}{4.25}$$
$$x = 12$$
Substitute 12 for x in the second equation.
$12 + y = 20$
$12 + y - 12 = 20 - 12$
$y = 8$
She should work 12 hours at \$10.75 per hour and 8 hours at \$6.50 per hour.

64. Let x = number of hours of shorter trip
    y = distance
$y = x(400 + 40)$
$y = (x + 1)(400 - 40)$
Substitute $x(400 + 40)$ for y in the second equation.

$x(400 + 40) = (x + 1)(400 - 40)$
$440x = 360x + 360$
$440x - 360x = 360x + 360 - 360x$
$80x = 360$
$$\frac{80x}{80} = \frac{360}{80}$$
$x = 4.5$
Substitute 4.5 for x in the first equation.
$y = 4.5(400 + 40)$
$y = 1980$
The distance traveled was 1980 miles.

65. Let x = number of miles driven
$U(x) = 39.95 + 0.15x$
$B(x) = 19.95 + 0.22x$

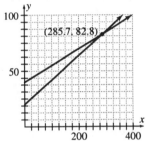

The graphs intersect at approximately (285.7, 82.8). You must drive at least 286 miles in order for U Rent It to be less costly.

66. Let x = number of items
$C(x) = 75 + 2.5x$
$R(x) = 6.5x$

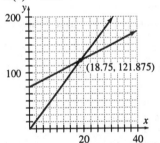

The graphs intersect at (18.75, 121.875). The shopkeeper must produce and sell at least 19 items in order not to lose money.

67. a. $c(x) = 75 + 150 + 0.75x$
    $c(x) = 0.75x + 225$

   b. $r(x) = 2x$

c.

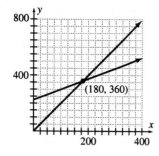

The break-even point is (180, 360).
Revenue equals cost at 180 bows.

d. No, 150 is less than 180.
Yes, 200 is greater than 180.
She must sell at least 180 bows.

Chapter 6 Test

1.

$$\underline{ y = -3.5x + 15.5 }$$

$$-64.3 \,\Big|\, -3.5(22.8) + 15.5$$

$$\Big|\, -79.8 + 15.5$$

$$\Big|\, -64.3$$

$$\underline{ 15x + 10y = -301 }$$

$$15(22.8) + 10(-64.3) \,\Big|\, -301$$

$$342 - 643 \,\Big|$$

$$-301 \,\Big|$$

Both equations are true.
Therefore, (22.8, –64.3) is a solution.

2. $2y - 8 = 0$
      $y = 4$

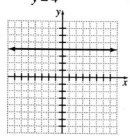

The graphs coincide. The solutions are all
ordered pairs (x, y) that satisfy $y = 4$.

3. $y = \dfrac{1}{2}x + 3$

      $y = \dfrac{1}{2}x - 5$

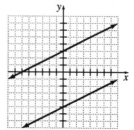

The graphs are parallel. There is no solution.

4. $y = 5x - 9$
      $4x + 8y = 16$

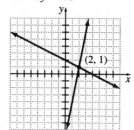

The solution is (2, 1).

5. $x + 2y = 6$
      $5x - 11y = -54$
      Solve the first equation for x, $x = 6 - 2y$.
      Substitute $6 - 2y$ for x in the second
      equation.
      $5(6 - 2y) - 11y = -54$
      $30 - 10y - 11y = -54$
      $30 - 21y = -54$
      $30 - 21y - 30 = -54 - 30$
      $-21y = -84$
      $\dfrac{-21y}{-21} = \dfrac{-84}{-21}$
      $y = 4$
      Substitute 4 for y in the first equation.
      $x + 2(4) = 6$
      $x + 8 - 8 = 6 - 8$
      $x = -2$
      The solution is (–2, 4).

6. $3x + 6y = 12$
$x + 2y = -5$
Solve the second equation for x, $x = -5 - 2y$.
Substitute $-5 - 2y$ for x in the first equation.
$3(-5 - 2y) + 6y = 12$
$-15 - 6y + 6y = 12$
$-15 = 12$
This is a contradiction. There is no solution.

7. $x + 6y = -2$
$x - 2y = 1$
Solve the second equation for x, $x = 2y + 1$.
Substitute $2y + 1$ for x in the first equation.
$2y + 1 + 6y = -2$
$8y + 1 - 1 = -2 - 1$
$8y = -3$
$\dfrac{8y}{8} = \dfrac{-3}{8}$
$y = -\dfrac{3}{8}$

Substitute $-\dfrac{3}{8}$ for y in the second equation.

$x - 2\left(-\dfrac{3}{8}\right) = 1$

$x + \dfrac{3}{4} - \dfrac{3}{4} = 1 - \dfrac{3}{4}$

$x = \dfrac{1}{4}$

The solution is $\left(\dfrac{1}{4}, -\dfrac{3}{8}\right)$.

8. $3x + 7y = 11$
$6x + 14y = 2$
Multiply the first equation by -2 and add to the second equation.
$-6x - 14y = -22$
$\underline{6x + 14y = 2}$
$\qquad\quad 0 = -20$
This is a contradiction. There is no solution.

9. $2x + 9y = 16$
$5y = 8 - x$
Multiply the second equation by -2 and write in standard form. Add to the first equation.

$2x + 9y = 16$
$\underline{-2x - 10y = -16}$
$\quad\ -y = 0$
$\dfrac{-y}{-1} = \dfrac{0}{-1}$
$y = 0$
Substitute 0 for y in the second equation.
$5(0) = 8 - x$
$0 = 8 - x$
$0 - 8 = 8 - x - 8$
$-8 = -x$
$\dfrac{-8}{-1} = \dfrac{-x}{-1}$
$8 = x$
The solution is $(8, 0)$.

10. $5x + 3y = 11$
$2x + 5y = 7$
Multiply the first equation by -2 and the second equation by 5. Add the new equations.
$-10x - 6y = -22$
$\underline{10x + 25y = 35}$
$\qquad\ 19y = 13$
$\dfrac{19y}{19} = \dfrac{13}{19}$
$y = \dfrac{13}{19}$
Substitute $\dfrac{13}{19}$ for y in the first equation.
$5x + 3\left(\dfrac{13}{19}\right) = 11$
$5x + \dfrac{39}{19} - \dfrac{39}{19} = 11 - \dfrac{39}{19}$
$5x = \dfrac{170}{19}$
$\dfrac{1}{5}(5x) = \dfrac{1}{5}\left(\dfrac{170}{19}\right)$
$x = \dfrac{34}{19}$
The solution is $\left(\dfrac{34}{19}, \dfrac{13}{19}\right)$.

11. $y = -8x + 1$ $m = -8, b = 1$
$y = x - 8$ $m = 1, b = -8$
The lines are only intersecting because the slopes are not equal and their product is not equal to -1.

12. $9(y - 2) + 2x = 7x$
$9y - 18 + 2x = 7x$
$9y - 18 = 5x$
$9y = 5x + 18$
$y = \frac{5}{9}x + 2$ $m = \frac{5}{9}, b = 2$
$4x = 9(y + 4) - x$
$4x = 9y + 36 - x$
$9y = 5x - 36$
$y = \frac{5}{9}x - 4$ $m = \frac{5}{9}, b = -4$
The lines are parallel because their slopes are equal and their y-intercepts are not equal.

13. $3x - y = 2$
$y = 3x - 2$ $m = 3, b = -2$
$y - x = 2(x - 1)$
$y - x = 2x - 2$
$y = 3x - 2$ $m = 3, b = -2$
The lines are coinciding because their slopes are the same and their y-intercepts are the same.

14. $5y - 2 = 3y + 2$
$2y = 4$
$y = 2$
$3x = 2(x + 2)$
$3x = 2x + 4$
$x = 4$
The lines are intersecting and perpendicular because one is horizontal and the other is vertical.

15. $C(x) = 5500 + 65x$
$R(x) = 125x$

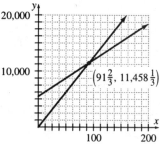

The graphs intersect at $\left(91\frac{2}{3}, 11{,}458\frac{1}{3}\right)$.

The break-even point is at $91\frac{2}{3}$ items.

After producing and selling 92 items, the factory's revenue will be greater than its cost.

16. Let x = Kenny's rate
$\quad\quad y$ = distance to Dolly
$y = 3x$
$y = 2(x + 4)$
Substitute $3x$ for y in the second equation.
$3x = 2(x + 4)$
$3x = 2x + 8$
$3x - 2x = 2x + 8 - 2x$
$x = 8$
$y = 3(8) = 24$
Kenny's rate was 8 mph and the distance to Dolly was 24 miles.

17. Let x = number of nanoseconds to perform one addition
$\quad\quad y$ = number of nanoseconds to perform one multiplication
$6x + 8y = 58$
$10x + 5y = 55$
Divide the first equation by 2 and the second equation by 5. Then multiply the new second equation by -4 and add it to the new first equation.
$3x + 4y = 29$
$\underline{-8x - 4y = -44}$
$\quad\quad -5x = -15$
$\quad\quad \dfrac{-5x}{-5} = \dfrac{-15}{-5}$
$\quad\quad\quad x = 3$

Substitute 3 for x in the first equation.
$6(3) + 8y = 58$
$18 + 8y = 58$
$18 + 8y - 18 = 58 - 18$
$8y = 40$
$\dfrac{8y}{8} = \dfrac{40}{8}$
$y = 5$
The computer takes 3 nanoseconds per addition and 5 nanoseconds per multiplication.

18. Let x = number of pounds of raisins
    y = number of pounds of peanuts
    $x + y = 30$
    $1.25x + 2y = 1.5(30)$
    Multiply the first equation by -1.25 and add to the second equation.
    $-1.25x - 1.25y = -37.5$
    $\underline{1.25x + 2y = 45}$
    $0.75y = 7.5$
    $\dfrac{0.75y}{0.75} = \dfrac{7.5}{0.75}$
    $y = 10$
    Substitute 10 for y in the first equation.
    $x + 10 = 30$
    $x + 10 - 10 = 30 - 10$
    $x = 20$
    He should mix 20 pounds of raisins with 10 pounds of peanuts.

19. Let x = measure of small angle
    y = measure of large angle
    $x + y = 90$
    $y = 2x$
    Substitute $2x$ for y in the first equation.
    $x + 2x = 90$
    $3x = 90$
    $\dfrac{3x}{3} = \dfrac{90}{3}$
    $x = 30$
    $y = 2(30) = 60$
    The angles measure $30°$ and $60°$.

20. Let x = number of cc of 50% solution
    y = number of cc of 10% solution
    $x + y = 100$
    $0.5x + 0.1y = 0.3(100)$
    Multiply the first equation by -0.1 and add to the second equation.

$-0.1x - 0.1y = -10$
$\underline{0.5x + 0.1y = 30}$
$0.4x = 20$
$\dfrac{0.4x}{0.4} = \dfrac{20}{0.4}$
$x = 50$
Substitute 50 for x in the first equation.
$50 + y = 100$
$50 + y - 50 = 100 - 50$
$y = 50$
She should mix 50 cc of 50% solution and 50 cc of 10% solution.

21. One solution exists; no solution exists; an infinite number of solutions exist.
    Answers will vary.

Chapter 7

7.1 Experiencing Algebra the Exercise Way

1. $3(x + 1) > -(5x - 4)$ is linear because it simplifies to $3x + 3 > -5x + 4$.

3. $4x^2 + 1 < 3x + 2$ is nonlinear because the variable x has an exponent of 2.

5. $\frac{5}{7}(a + 2) \geq \frac{3}{7}a + \frac{2}{3}$ is linear because it simplifies to $\frac{5}{7}a + \frac{10}{7} \geq \frac{3}{7}a + \frac{2}{3}$.

7. $0.6x + 2.7 \leq 5.2 - 1.9x$ is linear.

9. $x + 4(x - 8) > 2(3x + 1)$ is linear because it simplifies to $5x - 32 > 6x + 2$.

11. $\sqrt{2x - 7} \leq x + 3$ is nonlinear because the radical expression contains a variable.

13. $\frac{5}{x} + 2x > 0$ is nonlinear because the variable x appears in the denominator.

15. $2p + 6 < 0$ is linear.

17. $x \geq 6$

The solution set is $[6, \infty)$.

19. $z < 12$

The solution set is $(-\infty, 12)$.

21. $b > -\frac{13}{5}$

The solution set is $\left(-\frac{13}{5}, \infty\right)$.

23. $x \leq 4\frac{2}{3}$

The solution set is $\left(-\infty, 4\frac{2}{3}\right]$.

25. $P \leq 12.59$

The solution set is $(-\infty, 12.59]$.

27. $q > -6.7$

The solution set is $(-6.7, \infty)$.

29. $5 < x < 13$

The solution set is $(5, 13)$.

31. $8 > x > 2$
$2 < x < 8$

The solution set is $(2, 8)$.

33. $-4 < d \leq 0$

The solution set is $(-4, 0]$.

35. $-1 \leq m < 6$

The solution set is $[-1, 6)$.

37. $2 \leq t \leq 7$

The solution set is $[2, 7]$.

39. $\dfrac{2}{5} < s \le 3\dfrac{1}{3}$

The solution set is $\left(\dfrac{2}{5},\ 3\dfrac{1}{3}\right]$.

41. $-2.5 \le q < 3.5$

The solution set is $[-2.5,\ 3.5)$.

43. $\dfrac{4}{5} \le x \le 4.5$

The solution set is $\left[\dfrac{4}{5},\ 4.5\right]$.

45.

#	Inequality	Number Line	Interval Notation
a.	$x < 4.5$		$(-\infty,\ 4.5)$
b.	$x \le 3$		$(-\infty,\ 3]$
c.	$x < -2$		$(-\infty,\ -2)$
d.	$z \ge 5.7$		$[5.7,\ \infty)$
e.	$z > -7$		$(-7,\ \infty)$
f.	$z > 2$		$(2,\ \infty)$
g.	$2 < y < 8$		$(2,\ 8)$
h.	$-3 \le y < 2$		$[-3,\ 2)$
i.	$0 \le y \le 9$		$[0,\ 9]$

47. Let x = number of miles driven
$39.95 + 0.2x \le 150$

49. Let x = amount of weekly sales
$800 < 450 + 0.05x$

51. Let x = number of people
$150 \le 25 + 12.5x \le 200$

7.1 Experiencing Algebra the Calculator Way

1. $x < -2$

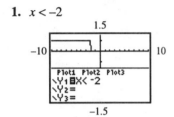

2. $x > -2$

3. $x \le -2$

4. $x \ge -2$

5. $-2.5 \le x \le 2.5$

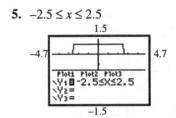

6. $-2.5 \le x < 2.5$

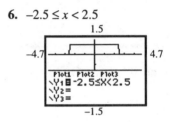

7. $-2.5 < x < 2.5$

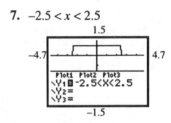

7.2 Experiencing Algebra the Exercise Way

1.

x	$6(x - 12)$	$3x$	
23	$6(23 - 12) = 66$	$3(23) = 69$	$66 < 69$
24	$6(24 - 12) = 72$	$3(24) = 72$	$72 \not< 72$
25	$6(25 - 12) = 78$	$3(25) = 75$	$78 \not< 75$

The integers less than 24 are solutions of the inequality.

3.

x	$3(x + 5) + 3$	$3x + 18$	
-1	$3(-1 + 5) + 3 = 15$	$3(-1) + 18 = 15$	$15 \not> 15$
0	$3(0 + 5) + 3 = 18$	$3(0) + 18 = 18$	$18 \not> 18$
1	$3(1 + 5) + 3 = 21$	$3(1) + 18 = 21$	$21 \not> 21$

The expressions are equal. The inequality is never true. There is no solution.

5.

x	$3x + 3$	$x + 2$	
-1	$3(-1) + 3 = 0$	$-1 + 2 = 1$	$0 \not> 1$
0	$3(0) + 3 = 3$	$0 + 2 = 2$	$3 > 2$
1	$3(1) + 3 = 6$	$1 + 2 = 3$	$6 > 3$

The integers equal to and greater than 0 are solutions of the inequality.

7.

x	$2(4x + 2) + 2x - 7$	$2(5x + 4)$	
-1	$2[4(-1) + 2] + 2(-1) - 7 = -13$	$2[5(-1) + 4] = -2$	$-13 < -2$
0	$2[4(0) + 2] + 2(0) - 7 = -3$	$2[5(0) + 4] = 8$	$-3 < 8$
1	$2[4(1) + 2] + 2(1) - 7 = 7$	$2[5(1) + 4] = 18$	$7 < 18$

The difference is always 11. The inequality is always true. The solutions are all integers.

9. $\dfrac{2}{3}x - \dfrac{2}{3} \geq -\dfrac{3}{4}x - \dfrac{7}{2}$

$Y1 = \dfrac{2}{3}x - \dfrac{2}{3}$

$Y2 = -\dfrac{3}{4}x - \dfrac{7}{2}$

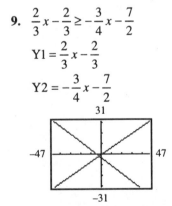

The intersection is $(-2, -2)$.
The solution set is all x that satisfies $x \geq -2$.

11. $2(x + 3) - (x - 1) > 8 - (x + 9)$
$Y1 = 2(x + 3) - (x - 1)$
$Y2 = 8 - (x + 9)$

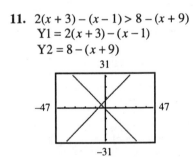

The intersection is $(-4, 3)$. The solution set is all x that satisfies $x > -4$.

13. $0.4x - 3.2 \leq -0.6x - 0.2$
$Y1 = 0.4x - 3.2$
$Y2 = -0.6x - 0.2$

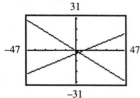

The intersection is $(3, -2)$.
The solution set is all x that satisfies $x \leq 3$.

15. $2(x + 1) > 3(x - 1) - x$
$Y1 = 2(x + 1)$
$Y2 = 3(x - 1) 1 - x$

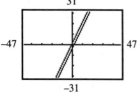

The lines are parallel and the graph of Y1 is always above that of Y2.
The solution set is all real numbers.

17. $y - (3x + 1) > -x - (x + 1)$
$Y1 = x - (3x + 1)$
$Y2 = -x - (x + 1)$

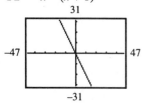

The lines are coinciding. Since the inequality does not include the equal symbol, there is no solution.

19. $4x + 12 > 0$
$4x + 12 - 12 > 0 - 12$
$4x > -12$
$\dfrac{4x}{4} = \dfrac{-12}{4}$
$x > -3$
The solution set is $(-3, \infty)$.

21. $6x - 8 < -32$
$6x - 8 + 8 < -32 + 8$
$6x < -24$
$\dfrac{6x}{6} < \dfrac{-24}{6}$
$x < -4$
The solution set is $(-\infty, -4)$.

23. $-3x + 12 \geq 12$
$-3x + 12 - 12 \geq 12 - 12$
$-3x \geq 0$
$\dfrac{-3x}{-3} \leq \dfrac{0}{-3}$
$x \leq 0$
The solution set is $(-\infty, 0]$.

25. $-7x - 12 < -26$
$-7x - 12 + 12 < -26 + 12$
$-7x < -14$
$\dfrac{-7x}{-7} > \dfrac{-14}{-7}$
$x > 2$
The solution set is $(2, \infty)$.

27. $15.17 < 5.9x - 4.3$
$15.17 + 4.3 < 5.9x - 4.3 + 4.3$
$19.47 < 5.9x$
$\dfrac{19.47}{5.9} < \dfrac{5.9x}{5.9}$
$3.3 < x$
$x > 3.3$
The solution set is $(3.3, \infty)$.

29. $6.1 > -0.55a + 6.1$
$6.1 - 6.1 > -0.55a + 6.1 - 6.1$
$0 > -0.55a$
$\dfrac{0}{-0.55} < \dfrac{-0.55a}{-0.55}$
$0 < a$
$a > 0$
The solution set is $(0, \infty)$.

31. $2.07z + 4.12 \geq 16.54$
$2.07z + 4.12 - 4.12 \geq 16.54 - 4.12$
$2.07z \geq 12.42$
$\dfrac{2.07z}{2.07} \geq \dfrac{12.42}{2.07}$
$z \geq 6$
The solution set is $[6, \infty)$.

33. $-9.2b - 4.3 \le 70.6$
$-9.2b - 4.3 + 4.3 \le 70.6 + 4.3$
$-9.2b \le 74.9$
$\dfrac{-9.2b}{-9.2} \ge \dfrac{74.9}{-9.2}$
$b \ge -\dfrac{749}{92}$

The solution set is $\left[-\dfrac{749}{92}, \infty \right)$.

35. $-\dfrac{5}{9}b + \dfrac{11}{12} < \dfrac{23}{36}$
$36\left(-\dfrac{5}{9}b + \dfrac{11}{12} \right) < 36\left(\dfrac{23}{36} \right)$
$-20b + 33 < 23$
$-20b + 33 - 33 < 23 - 33$
$-20b < -10$
$\dfrac{-20b}{-20} > \dfrac{-10}{-20}$
$b > \dfrac{1}{2}$

The solution set is $\left(\dfrac{1}{2}, \infty \right)$.

37. $156z - 210 > 47z + 662$
$156z - 210 + 210 > 47z + 662 + 210$
$156z > 47z + 872$
$156z - 47z > 47z + 872 - 47z$
$109z > 872$
$\dfrac{109z}{109} > \dfrac{872}{109}$
$z > 8$
The solution set is $(8, \infty)$.

39. $-1.05x - 15.41 < 2.55x - 47.09$
$-1.05x - 15.41 + 15.41$
$\qquad < 2.55x - 47.09 + 15.41$
$-1.05x < 2.55x - 31.68$
$-1.05x - 2.55x < 2.55x 1 - 31.68 - 2.55x$
$-3.6x < -31.68$
$\dfrac{-3.6x}{-3.6} > \dfrac{-31.68}{-3.6}$
$x > 8.8$
The solution set is $(8.8, \infty)$.

41. $11x + \dfrac{1}{4} \le 2x + \dfrac{7}{36}$
$36\left(11x + \dfrac{1}{4} \right) \le 36\left(2x + \dfrac{7}{36} \right)$
$396x + 9 \le 72x + 7$
$396x + 9 - 9 \le 72x + 7 - 9$
$396x \le 72x - 2$
$396x - 72x \le 72x - 2 - 72x$
$324x \le -2$
$\dfrac{324x}{324} \le \dfrac{-2}{324}$
$x \le -\dfrac{1}{162}$

The solution set is $\left(-\infty, -\dfrac{1}{162} \right]$.

43. $\dfrac{2}{5}b - 12 < \dfrac{2}{3}b + 20$
$15\left(\dfrac{2}{5}b - 12 \right) < 15\left(\dfrac{2}{3}b + 20 \right)$
$6b - 180 < 10b + 300$
$6b - 180 + 180 < 10b 1 + 300 + 180$
$6b < 10b + 480$
$6b - 10b < 10b + 480 - 10b$
$-4b < 480$
$\dfrac{-4b}{-4} > \dfrac{480}{-4}$
$b > -120$
The solution set is $(-120, \infty)$.

45. $4x - (3x + 5) < x - 5$
$4x - 3x - 5 < x - 5$
$x - 5 < x - 5$
$x - 5 - x < x - 5 - x$
$-5 < -5$
Since this is a contradiction, there is no solution.

47. $4x - (3x + 5) \le x - 5$
$4x - 3x - 5 \le x - 5$
$x - 5 \le x - 5$
$x - 5 - x \le x - 5 - x$
$-5 \le -5$
Since this is an identity, the solution set is all real numbers.

49. $0.05x + 10.5 < 0.15x - 0.25$
$0.05x\, 1+ 10.5 - 10.5 < 0,.15x - 0.25 - 10.5$
$0.05x < 0.15x - 10.75$
$0.05x - 0.15x < 0.15x - 10.75 - 0.15x$
$-0.1x < -10.75$
$\dfrac{-0.1x}{-0.1} > \dfrac{-10.75}{-0.1}$
$x > 107.5$
The solution set is $(107.5, \infty)$.

51. Let x = sixth test score
$\dfrac{73 + 97 + 82 + 89 + 95 + x}{6} \geq 87$

$\dfrac{436 + x}{6} \geq 87$

$6\left(\dfrac{436 + x}{6}\right) \geq 6(87)$

$436 + x \geq 522$
$436 + x - 436 \geq 522 - 436$
$x \geq 86$
His score must be 86 or greater.

53. Let x = number of hours worked
$9.75x + 5200 \leq 7500$
$9.75x + 5200 - 5200 \leq 7500 - 5200$
$9.75x \leq 2300$
$\dfrac{9.75x}{9.75} \leq \dfrac{2300}{9.75}$
$x \leq 235.9$
He can work 235 hours or less.

55. Let x = number of people
$165x \leq 2000$
$\dfrac{165x}{165} \leq \dfrac{2000}{165}$
$x \leq 12.1$
The number of people can be 12 or less.

57. Let x = number of calories for breakfast
$x + 150$ = number of calories for lunch
$2x + 50$ = number of calories for dinner
$x + x + 150 + 2x + 50 \leq 1200$
$4x + 200 \leq 1200$
$4x + 200 - 200 \leq 1200 - 200$
$4x \leq 1000$
$\dfrac{4x}{4} \leq \dfrac{1000}{4}$
$x \leq 250$
$x + 150 \leq 400$
$2x + 50 \leq 550$
She can have no more than 250 calories for breakfast, 400 calories for lunch, and 550 calories for dinner.

59. Let x = number of meat plates
$8.5x + 6.5(120 - x) \leq 950$
$8.5x + 780 - 6.5x \leq 950$
$2x + 780 - 780 \leq 950 - 780$
$2x \leq 170$
$\dfrac{2x}{2} \leq \dfrac{170}{2}$
$x \leq 85$
She can order 85 or fewer meat plates.

61. Let x = width
$\dfrac{3}{4}x + 30$ = length
$2x + \dfrac{3}{4}x + 30 - 4 \leq 185$
$2.75x + 26 \leq 185$
$2.75x + 26 - 26 \leq 185 - 26$
$2.75x \leq 159$
$\dfrac{2.75x}{2.75} \leq \dfrac{159}{2.75}$
$x \leq 57.8$
The width must be 57 or less feet.

63. Let x = amount of weekly sales
$300 + 0.05(x - 1000) > 500 + 0.05(x - 2000)$
$300 + 0.05x - 50 > 500 + 0.05x - 100$
$10.05x + 250 > 0.05x\, 1+ 400$
$0.05x + 250 - 0.05x > 0.05x + 400 - 0.05x$
$250 > 400$
This is a contradiction. There is no solution. There are no values of sales where plan A will pay more than plan B.

65. $C(n) > 1 \times 10^{12}$

$(4.73 \times 10^{10})n + (7.66 \times 10^{10}) > 1 \times 10^{12}$

$(4.73 \times 10^{10})n + (7.66 \times 10^{10}) - (7.66 \times 10^{10})$
$> 1 \times 10^{12} - (7.66 \times 10^{10})$

$(4.73 \times 10^{10})n > 9.234 \times 10^{11}$

$\dfrac{(4.73 \times 10^{10})n}{4.73 \times 10^{10}} > \dfrac{9.234 \times 10^{11}}{4.73 \times 10^{10}}$

$n > 1.9522 \times 10^{1}$
$n > 19.522$
$1970 + 20 = 1990$
The years are 1990 to the present.

7.2 Experiencing Algebra the Calculator Way

Answers will vary.

7.3 Experiencing Algebra the Exercise Way

1. $2x + 1.7y > x - 4.6$ is linear.
In standard form,
$2x + 1.7y - x > x - 4.6 - x$
$x + 1.7y > -4.6$
$a = 1, b = 1.7, c = -4.6$

3. $y < x^2 + 2x - 3$ is nonlinear because the x
term is squared.

5. $y > \sqrt{x} + 9$ is nonlinear because the radical
expression contains a variable.

7. $\dfrac{x}{2} - \dfrac{y}{6} > \dfrac{1}{12}$ is linear. In standard form,

$12\left(\dfrac{x}{2} - \dfrac{y}{6}\right) > 12\left(\dfrac{1}{12}\right)$

$6x - 2y > 1$
$a = 6, b = -2, c = 1$

9. $4x + 16 \le y$ is linear. In standard form,
$4x + 16 - 16 \le y - 16$
$4x \le y - 16$
$4x - y \le y - 16 - y$
$4x - y \le -16$
$a = 4, b = -1, c = -16$

11. $y \le -\dfrac{2}{5}x + \dfrac{7}{15}$ is linear.

$15(y) \le 15\left(-\dfrac{2}{5}x + \dfrac{7}{15}\right)$

$15y \le -6x + 7$
$15y + 6x \le -6x + 7 + 6x$
$6x + 15y \le 7$
$a = 6, b = 15, c = 7$

13. $2x + y < 3$
$2x + y - 2x < 3 - 2x$
$y < 3 - 2x$
$y < -2x + 3$

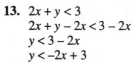

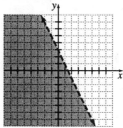

15. $x - 3y \ge 6$
$5x - 3y - 5x \ge 6 - 5x$
$-3y \ge 6 - 5x$
$\dfrac{-3y}{-3} \le \dfrac{-5x + 6}{-3}$

$y \le \dfrac{5}{3}x - 2$

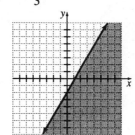

17. $y < -\dfrac{3}{4}x + 4$

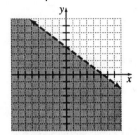

19. $y \geq 2.8x - 1.6$

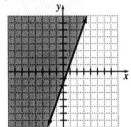

21. $2y > 3x - 5$

$y > \dfrac{3x - 5}{2}$

$y > \dfrac{3}{2}x - \dfrac{5}{2}$

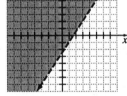

23. $3x + y - 4 \leq x + 2y - 3$
$3x + y - 4 + 4 \leq x + 2y - 3 + 4$
$3x + y \leq x + 2y + 1$
$3x + y - 3x \leq x + 2y + 1 - 3x$
$y \leq -2x + 2y + 1$
$y - 2y \leq -2x + 2y + 1 - 2y$
$-y \leq -2x + 1$
$\dfrac{-y}{-1} \geq \dfrac{-2x + 1}{-1}$
$y \geq 2x - 1$

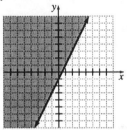

25. $5y > -20$

$\dfrac{5y}{5} > \dfrac{-20}{5}$

$y > -4$

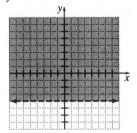

27. $3x + 6 \leq 9$
$3x + 6 - 6 \leq 9 - 6$
$3x \leq 3$
$\dfrac{3x}{3} \leq \dfrac{3}{3}$
$x \leq 1$

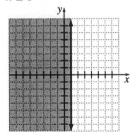

29. $3x + 5y \geq 12$

$3x + 5y - 3x \geq 12 - 3x$

$5y \geq -3x + 12$

$\dfrac{5y}{5} \geq \dfrac{-3x + 12}{5}$

$y \geq -\dfrac{3}{5}x + \dfrac{12}{5}$

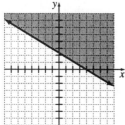

31. $-x - y < 7$

$-x - y + x < 7 + x$

$-y < x + 7$

$\dfrac{-y}{-1} > \dfrac{x + 7}{-1}$

$y > -x - 7$

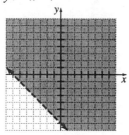

33. $-3x - 9y > 27$

$-3x - 9y + 3x > 27 + 3x$

$-9y > 3x + 27$

$\dfrac{-9y}{-9} < \dfrac{3x + 27}{-9}$

$y < -\dfrac{1}{3}x - 3$

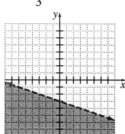

35. $\dfrac{x}{8} + \dfrac{y}{3} \leq 1$

$\dfrac{x}{8} + \dfrac{y}{3} - \dfrac{x}{8} \leq 1 - \dfrac{x}{8}$

$\dfrac{y}{3} \leq -\dfrac{x}{8} + 1$

$3\left(\dfrac{y}{3}\right) \leq 3\left(-\dfrac{x}{8} + 1\right)$

$y \leq -\dfrac{3}{8}x + 3$

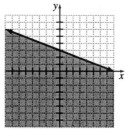

37. $-\dfrac{4}{7}x + \dfrac{2}{3}y \geq \dfrac{10}{21}$

$-\dfrac{4}{7}x + \dfrac{2}{3}y + \dfrac{4}{7}x \geq \dfrac{10}{21} + \dfrac{4}{7}x$

$\dfrac{2}{3}y \geq \dfrac{4}{7}x + \dfrac{10}{21}$

$\dfrac{3}{2}\left(\dfrac{2}{3}y\right) \geq \dfrac{3}{2}\left(\dfrac{4}{7}x + \dfrac{10}{21}\right)$

$y \geq \dfrac{6}{7}x + \dfrac{5}{7}$

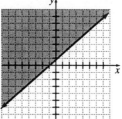

39. $1.8x - 3.2y > 0$
$1.8x - 3.2y - 1.8x > 0 - 1.8x$
$-3.2y > -1.8x$
$\dfrac{-3.2y}{-3.2} < \dfrac{-1.8x}{-3.2}$
$y < 0.5625x$

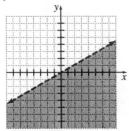

41. $4.6y < 3.5y + 5.94$
$4.6y - 3.5y < 3.5y + 5.94 - 3.5y$
$1.1y < 5.94$
$\dfrac{1.1y}{1.1} < \dfrac{5.94}{1.1}$
$y < 5.4$

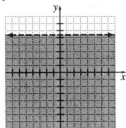

43. $x - y > 9$
$x - y - x > 9 - x$
$-y > -x + 9$
$\dfrac{-y}{-1} < \dfrac{-x + 9}{-1}$
$y < x - 9$

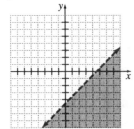

45. $-x - y > 9$
$-x - y + x > 9 + x$
$-y > x + 9$
$\dfrac{-y}{-1} < \dfrac{x + 9}{-1}$
$y < -x - 9$

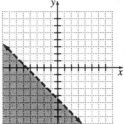

47. $y \le x$

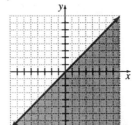

49. Let x = number of village pieces
$\quad\quad y$ = number of angel ornaments
$25x + 12y \le 225$
Solve for y.
$25x + 12y - 25x \le 225 - 25x$
$12y \le -25x + 225$
$\dfrac{12y}{12} \le \dfrac{-25x + 225}{12}$
$y \le -\dfrac{25}{12}x + \dfrac{75}{4}$

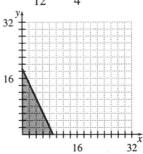

The ordered pairs in the shaded region represent the possible combinations.
Yes, the point $(4, 7)$ is in the shaded region.

Therefore she can buy 4 village pieces and 7 angel ornaments.

No, the point (7, 9) is not in the shaded region. Therefore, she cannot buy 7 village pieces and 9 angel ornaments.

51. Let x = width
     y = length
$2x + 2y \leq 220$
Solve for y.
$2x + 2y - 2x \leq 220 - 2x$
$2y \leq -2x + 220$
$\dfrac{2y}{2} \leq \dfrac{-2x + 220}{2}$
$y \leq -x + 110$

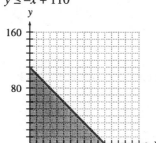

The ordered pairs in the shaded region represent the possible combinations.
No, the point (60, 70) is not in the shaded region. Therefore, she will not have enough for a 70 by 60 foot room.
Yes, the point (40, 60) is in the shaded region. Therefore, she will have enough for a 40 by 60 foot room.

53. Let x = number of days
     y = number of items sold
$15x + 12y \geq 400$
Solve for y.
$15x + 12y - 15x \geq 400 - 15x$
$12y \geq -15x + 400$
$\dfrac{12y}{12} \geq \dfrac{-15x + 400}{12}$
$y \geq -\dfrac{5}{4}x + \dfrac{100}{3}$

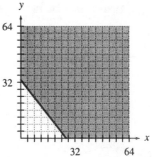

The ordered pairs in the shaded region represent the possible combinations.
No, the point (3, 20) is not in the shaded region. Therefore, he will not earn at least $400 working 3 days and selling 20 items.
Yes, the point (5, 30) is in the shaded region. Therefore, he will earn at least $400 working 5 days and selling 30 items.

55. Let x = number of adults
     y = number of children
$4.5x + 2y \geq 250$
Solve for y.
$4.5x + 2y - 4.5x \geq 250 - 4.5x$
$2y \geq -4.5x + 250$
$\dfrac{2y}{2} \geq \dfrac{-4.5x + 250}{2}$
$y \geq -2.25x + 125$

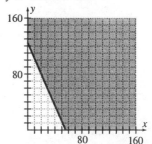

The ordered pairs in the shaded region represent the possible combinations.
No, the point (40, 25) is not in the shaded region. Therefore, they will not make their goal with 40 adults and 25 children.
Yes, the point (42, 45) is in the shaded region. Therefore, they will make their goal with 42 adults and 45 children.

7.3 Experiencing Algebra the Calculator Way

1. $y < 3x - \pi$

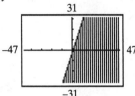

yes; no; yes; no

2. $y > x + e$

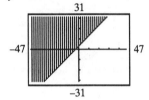

no; no; yes; yes

3. $y \geq -\dfrac{\pi x}{e} + \dfrac{15}{e}$

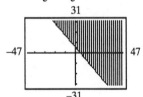

no; yes; yes; yes

4. $y \leq -\dfrac{\sqrt{6}}{3}x + \dfrac{8\sqrt{3}}{3}$

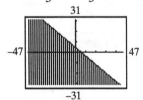

yes; no; yes; no

5. $y < 2\pi x$

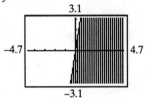

yes; no; yes; yes

6. $y > e - x$

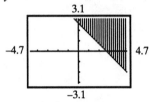

no; no; yes; no

7.4 Experiencing Algebra the Exercise Way

1. Solve the first inequality for y.
$$5x - 3y > 15$$
$$5x - 3y - 5x > 15 - 5x$$
$$-3y > -5x + 15$$
$$\frac{-3y}{-3} < \frac{-5x + 15}{-3}$$
$$y < \frac{5}{3}x - 5$$

Solve the second inequality for y.
$$4x + y > 12$$
$$4x + y - 4x > 12 - 4x$$
$$y > -4x + 12$$

Find the intersection.
$$\begin{aligned} 5x - 3y &= 15 \\ 12x - 3y &= 36 \\ \hline 17x \quad\;\; &= 51 \end{aligned}$$
$$\frac{17x}{17} = \frac{51}{17}$$
$$x = 3$$
$$y = -4x + 12$$
$$y = -4(3) + 12$$
$$y = 0$$

The intersection is $(3, 0)$.

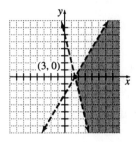

3. Solve the first inequality for y.
$$2x + 4y \geq 8$$
$$2x + 4y - 2x \geq 8 - 2x$$
$$4y \geq -2x + 8$$
$$\frac{4y}{4} \geq \frac{-2x + 8}{4}$$
$$y \geq -\frac{1}{2}x + 2$$
Solve the second inequality for y.
$$3x + y \leq -8$$
$$3x + y - 3x \leq -8 - 3x$$
$$y \leq -3x - 8$$
Find the intersection.
$$2x + 4y = 8$$
$$-12x - 4y = 32$$
$$\overline{-10x \qquad = 40}$$
$$\frac{-10x}{-10} = \frac{40}{-10}$$
$$x = -4$$
$$y = -3x - 8$$
$$y = -3(-4) - 8$$
$$y = 12 - 8$$
$$y = 4$$
The intersection is $(-4, 4)$.

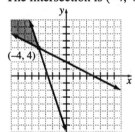

5. $y > 3x - 1$
$\quad y > -2x + 4$
Find the intersection.
$$3x - 1 = -2x + 4$$
$$3x - 1 + 1 = -2x + 4 + 1$$
$$3x = -2x + 5$$
$$3x + 2x = -2x + 5 + 2x$$
$$5x = 5$$
$$\frac{5x}{5} = \frac{5}{5}$$
$$x = 1$$
$$y = 3x - 1$$
$$y = 3(1) - 1$$
$$y = 2$$
The intersection is $(1, 2)$.

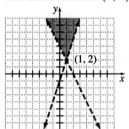

7. $y < \dfrac{3}{4}x + 1$

$\quad y < -\dfrac{2}{3}x + 1$

Find the intersection.
$$\frac{3}{4}x + 1 = -\frac{2}{3}x + 1$$
$$\frac{3}{4}x + 1 - 1 = -\frac{2}{3}x + 1 - 1$$
$$\frac{3}{4}x = -\frac{2}{3}x$$
$$\frac{3}{4}x + \frac{2}{3}x = -\frac{2}{3}x + \frac{2}{3}x$$
$$\frac{17}{12}x = 0$$
$$\frac{12}{17}\left(\frac{17}{12}x\right) = \frac{12}{17}(0)$$
$$x = 0$$

$$y = \frac{3}{4}x + 1$$

$$y = \frac{3}{4}(0) + 1$$

$$y = 1$$

The intersection is (0, 1).

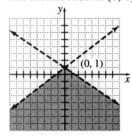

9. Solve the first inequality for y.

$$3x + 2y < 12$$

$$3x + 2y - 3x < 12 - 3x$$

$$2y < -3x + 12$$

$$\frac{2y}{2} < \frac{-3x + 12}{2}$$

$$y < -\frac{3}{2}x + 6$$

Solve the second inequality for y.

$$y - 2 < 1$$

$$y - 2 + 2 < 1 + 2$$

$$y < 3$$

Find the intersection.

$$y = 3$$

$$3x + 2y = 12$$

$$3x + 2(3) = 12$$

$$3x + 6 - 6 = 12 - 6$$

$$3x = 6$$

$$\frac{3x}{3} = \frac{6}{3}$$

$$x = 2$$

The intersection is (2, 3).

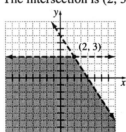

11. Solve the first inequality for y.

$$3y - 2 > 2y + 2$$

$$3y - 2 + 2 > 2y + 2 + 2$$

$$3y > 2y + 4$$

$$3y - 2y > 2y + 4 - 2y$$

$$y > 4$$

Solve the second inequality for x.

$$x - 1 < 0$$

$$x - 1 + 1 < 0 + 1$$

$$x < 1$$

Find the intersection.

$$x = 1$$

$$y = 4$$

The intersection is (1, 4).

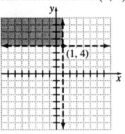

13. $y \le 3x - 5$

Solve the second inequality for y.

$$x > 4 - 2y$$

$$x - 4 > 4 - 2y - 4$$

$$x - 4 > -2y$$

$$\frac{x - 4}{-2} < \frac{-2y}{-2}$$

$$-\frac{1}{2}x + 2 < y$$

or

$$y > -\frac{1}{2}x + 2$$

Find the intersection.

$$3x - 5 = -\frac{1}{2}x + 2$$

$$3x - 5 + 5 = -\frac{1}{2}x + 2 + 5$$

$$3x = -\frac{1}{2}x + 7$$

$$3x + \frac{1}{2}x = -\frac{1}{2}x + 7 + \frac{1}{2}x$$

$$3.5x = 7$$

$$\frac{3.5x}{3.5} = \frac{7}{3.5}$$

$$x = 2$$

$$y = 3x - 5$$

$$y = 3(2) - 5$$

$$y = 1$$

The intersection is (2, 1).

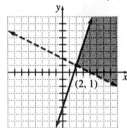

15. $y \le \frac{1}{2}x + 3$

Solve the second inequality for *y*.

$$x + 2y < 6$$

$$x + 2y - x < 6 - x$$

$$2y < -x + 6$$

$$\frac{2y}{2} < \frac{-x + 6}{2}$$

$$y < -\frac{1}{2}x + 3$$

Find the intersection.

$$\frac{1}{2}x + 3 = -\frac{1}{2}x + 3$$

$$\frac{1}{2}x + 3 - 3 = -\frac{1}{2}x + 3 - 3$$

$$\frac{1}{2}x = -\frac{1}{2}x$$

$$\frac{1}{2}x + \frac{1}{2}x = -\frac{1}{2}x + \frac{1}{2}x$$

$$x = 0$$

$$y = \frac{1}{2}x + 3$$

$$y = \frac{1}{2}(0) + 3$$

$$y = 3$$

The intersection is (0, 3).

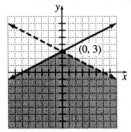

17. $y \le 10 - x$

Solve the second inequality for *y*.

$$y > \frac{1}{2}x + 3$$

$$y > 0$$

$$x < 3$$

Find the intersections and label the graph.

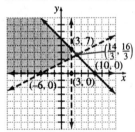

19. Solve the first inequality for *y*.

$$x - 2y > 7$$

$$x - 2y - x > 7 - x$$

$$-2y > -x + 7$$

$$\frac{-2y}{-2} < \frac{-x + 7}{-2}$$

$$y < \frac{1}{2}x - \frac{7}{2}$$

Solve the second inequality for *y*.

$5x + 10y < -3$

$5x + 10y - 5x < -3 - 5x$

$10y < -5x - 3$

$\dfrac{10y}{10} < \dfrac{-5x - 3}{10}$

$y < -\dfrac{1}{2}x - \dfrac{3}{10}$

$Y1 = \dfrac{1}{2}x - \dfrac{7}{2}$

$Y2 = -\dfrac{1}{2} - \dfrac{3}{10}$

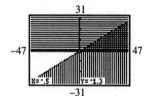

21. Solve the first inequality for y.

$3x - 5y \geq 5$

$3x - 5y - 3x \geq 5 - 3x$

$-5y \geq -3x + 5$

$\dfrac{-5y}{-5} \leq \dfrac{-3x + 5}{-5}$

$y = \dfrac{3}{5}x - 1$

Solve the second inequality for y.

$10y + 15 > 2$

$10y + 15 - 15 > 2 - 15$

$10y > -13$

$\dfrac{10y}{10} > \dfrac{-13}{10}$

$y > -1.3$

$Y1 = \dfrac{3}{5}x - 1$

$Y2 = -1.3$

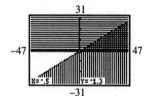

23. $y \geq x - 3$

Solve the second inequality for y.

$10x - 5y \geq 14$

$10x - 5y - 10x \geq 14 - 10x$

$-5y \geq -10x + 14$

$\dfrac{-5y}{-5} \leq \dfrac{-10x + 14}{-5}$

$y \leq 2x - \dfrac{14}{5}$

$Y1 = x - 3$

$Y2 = 2x - \dfrac{14}{5}$

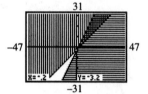

25. Solve the first inequality for y.

$2x + 9y < 102$

$2x + 9y - 2x < 102 - 2x$

$9y < -2x + 102$

$\dfrac{9y}{9} < \dfrac{-2x + 102}{9}$

$y < -\dfrac{2}{9}x + \dfrac{34}{3}$

Solve the second inequality for y.

$5x - 11y < -147$

$5x - 11y - 5x < -147 - 5x$

$-11y < -5x - 147$

$\dfrac{-11y}{-11} > \dfrac{-5x - 147}{-11}$

$y > \dfrac{5}{11}x + \dfrac{147}{11}$

$Y1 = -\dfrac{2}{9}x + \dfrac{34}{3}$

$Y2 = \dfrac{5}{11}x + \dfrac{147}{11}$

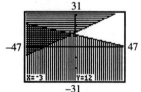

27. Solve the first inequality for y.

$$3.2x + 4.2y > 368$$

$$3.2x + 4.2y - 3.2x > 368 - 3.2x$$

$$4.2y > -3.2x + 368$$

$$\frac{4.2y}{4.2} > \frac{-3.2x + 368}{4.2}$$

$$y = -\frac{16}{21}x + \frac{1840}{21}$$

Solve the second inequality for y.

$$4.4x - 2.1y > 128$$

$$4.4x - 2.1y - 4.4x > 128 - 4.4x$$

$$-2.1y > -4.4x + 128$$

$$\frac{-2.1y}{-2.1} < \frac{-4.4x + 128}{-2.1}$$

$$y < \frac{44}{21}x - \frac{1280}{21}$$

$$Y1 = -\frac{16}{21}x + \frac{1840}{21}$$

$$Y2 = \frac{44}{21}x - \frac{1280}{21}$$

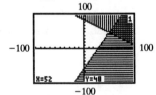

29. Solve the first inequality for y.

$$0.05x + 0.10y < 0.75$$

$$0.05x + 0.10y - 0.05x < 0.75 - 0.05x$$

$$0.10y < -0.05x + 0.75$$

$$\frac{0.10y}{0.10} < \frac{-0.05x + 0.75}{0.10}$$

$$y < -0.5x + 7.5$$

Solve the second inequality for y.

$$x + y > 11$$

$$x + y - x > 11 - x$$

$$y > -x + 11$$

$$Y1 = -0.5x + 7.5$$

$$Y2 = -x + 11$$

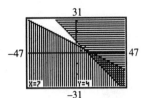

31. $2x - 3y > -6$

$-2x + 3y > -6$

or

$$y < \frac{2}{3}x + 2$$

$$y > \frac{2}{3}x - 2$$

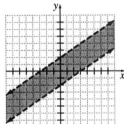

The solution is all ordered pairs contained in the shaded region.

33. $y > -3x + 2$

$y \leq -3x$

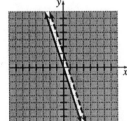

There are no ordered pairs that satisfy this system of linear inequalities.

35. $x + 4 > 0$

$3x + 2 < -4$

  or

$x > -4$

$x < -2$

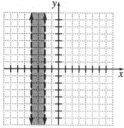

The solution is all ordered pairs contained in the shaded region.

37. $y < 4 - 2x$
$2x + y \leq -1$
or
$y < -2x + 4$
$y \leq -2x - 1$

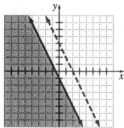

The solution is all ordered pairs below and including the boundary line $2x + y = -1$.

39. $3y \leq x + 6$
$3y - x \geq 6$
　　or
$y \leq \dfrac{1}{3}x + 2$

$y \geq \dfrac{1}{3}x + 2$

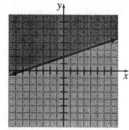

The solution is all ordered pairs on the line

$3y = x + 6$ or $y = \dfrac{1}{3}x + 2$.

41. $y < 7$
$2y - 3 \leq y + 4$
　　or
$y < 7$
$y \leq 7$

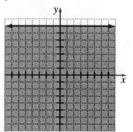

The solution is all ordered pairs below the line $y = 7$.

43. Let $x = $ width
　　　$y = $ length
$2x + 2y \leq 100$
$y \geq x + 10$
$x \geq 0$
$y \geq 0$
Solve the first inequality for y.
$y \leq -x + 50$

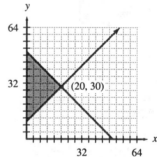

Possible answer:
The dimensions could be 10 feet by 30 feet.

45. Let $x = $ amount invested at 6%
　　　$y = $ amount invested at 8%
$x + y \leq 3000$
$0.06x + 0.08y \geq 200$
$x \geq 0$
$y \geq 0$
Solve the first two inequalities for y.
$y \leq -x + 3000$
$y \geq -0.75x + 2500$

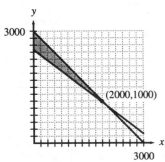

Possible answer:
The investments could be $1000 at 6% and $2000 at 8%.

47. Let x = number of hours at first job
    y = number of hours at second job
$x + y \leq 20$
$6.5x + 8.25y \geq 150$
$x \geq 0$
$y \geq 0$
Solve the first two inequalities for y.
$y \leq -x + 20$

$$y \geq -\frac{6.5}{8.25}x + \frac{150}{8.25}$$

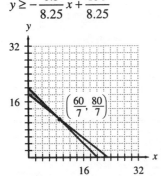

Possible answer:
He could work 6 hours on the first job and 14 hours on the second job.
No, the point (7, 10) is not in the shaded region. Therefore he could not work 7 hours on the first job and 10 hours on the second job.
Yes, the point (5, 15) is in the shaded region. Therefore, he could work 5 hours on the first job and 15 hours on the second job.

49. Let x = number of servings of lasagna
    y = number of servings of veal
$1.75x + 2.25y \leq 200$
$x \geq 50$
$y \geq 25$
Solve the first inequality for y.

$$y \leq -\frac{7}{9}x + \frac{800}{9}$$

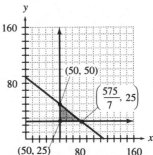

Possible answer:
A combination could be 75 servings of lasagna and 30 servings of veal.
Yes, the point (60, 35) is in the shaded region. Therefore, she could prepare 60 servings of lasagna and 35 servings of veal parmigiana.
No, the point (60, 50) is not in the shaded region. Therefore, she should not prepare 60 servings of lasagna and 50 servings of veal parmigiana.

7.4 Experiencing Algebra the Calculator Way

1.
```
X▶Frac
          -1/3
Y▶Frac
          13/3
```

2.
```
X▶Frac
          38/15
Y▶Frac
          31/15
```

3.
```
X▸Frac
           -100/3
Y▸Frac
            -58/3
```

4.
```
X▸Frac
          168/265
Y▸Frac
           72/265
```

Chapter 7 Review

Reflections

1–7. Answers will vary.

Exercises

1. $3x^2 - 2x + 1 < 0$ is nonlinear because the variable x is squared.

2. $5x - 4 > x + 7$ is linear.

3. $\dfrac{2}{3}x + \dfrac{4}{5} \leq \dfrac{7}{15}$ is linear.

4. $\sqrt{x} - 3 \geq 6$ is nonlinear because the radical expression contains a variable.

5. $x - 3 \geq 6$ is linear.

6. $\dfrac{1}{x} - 3x \geq 6$ is nonlinear because the variable x appears in the denominator.

7. $1.5z - 12.6 < 14.7z$ is linear.

8. $3(a + 2) < 15a - (2a + 1)$ is linear because it simplifies to $3a + 6 < 13a - 1$.

9. $x < 3$

The solution set is $(-\infty, 3)$.

10. $x > -2$

The solution set is $(-2, \infty)$.

11. $x \leq -5$

The solution set is $(-\infty, -5]$.

12. $x \geq -3.5$

The solution set is $[-3.5, \infty)$.

13. $-2 < a < 4$

The solution set is $(-2, 4)$.

14. $-1 < b \leq 0$

The solution set is $(-1, 0]$.

15. $3 \leq c \leq 5.5$

The solution set is $[3, 5.5]$.

16. $2\dfrac{1}{2} < d < 8$

The solution set is $\left(2\dfrac{1}{2}, 8\right)$.

17. $-2.3 \leq f \leq -1\dfrac{1}{3}$

The solution set is $\left[-2.3, -1\dfrac{1}{3}\right]$.

18. Let x = number of minutes
$(30 + 25)x \geq 300$

19. Let x = cost of base material

$$x + 2x + \frac{1}{4}(2x) < 200,000$$

20. Let x = number of pieces
$$60 + 0.18x < 0.32x$$

21.

x	$4x + 7$	$2x - 5$	
-7	$4(-7) + 7 = -21$	$2(-7) - 5 = -19$	$-21 < -19$
-6	$4(-6) + 7 = -17$	$2(-6) - 5 = -17$	$-17 \not< -17$
-5	$4(-5) + 7 = -13$	$2(-5) - 5 = -15$	$-13 \not< -15$

The integers less than -6 are solutions of the inequality.

22.

x	$2.4x - 9.6$	4.8	
5	$2.4(5) - 9.6 = 2.4$	4.8	$2.4 \not> 4.8$
6	$2.4(6) - 9.6 = 4.8$	4.8	$4.8 \not> 4.8$
7	$2.4(7) - 9.6 = 7.2$	4.8	$7.2 > 4.8$

The integers greater than 6 are solutions of the inequality.

23.

x	$\frac{3}{5}x - \frac{7}{10}$	$\frac{1}{5}x + \frac{1}{2}$	
2	$\frac{3}{5}(2) - \frac{7}{10} = \frac{1}{2}$	$\frac{1}{5}(2) + \frac{1}{2} = \frac{9}{10}$	$\frac{1}{2} \le \frac{9}{10}$
3	$\frac{3}{5}(3) - \frac{7}{10} = \frac{11}{10}$	$\frac{1}{5}(3) + \frac{1}{2} = \frac{11}{10}$	$\frac{11}{10} \le \frac{11}{10}$
4	$\frac{3}{5}(4) - \frac{7}{10} = \frac{17}{10}$	$\frac{1}{5}(4) + \frac{1}{2} = \frac{13}{10}$	$\frac{17}{10} \not\le \frac{13}{10}$

The integers less than or equal to 3 are solutions of the inequality.

24.

x	$\frac{1}{2}x - 2$	$-\frac{1}{3}x - \frac{11}{3}$	
-3	$\frac{1}{2}(-3) - 2 = \frac{7}{2}$	$-\frac{1}{3}(-3) - \frac{11}{3} = -\frac{8}{3}$	$-3 \not\ge -\frac{8}{3}$
-2	$\frac{1}{2}(-2) - 2 = -3$	$-\frac{1}{3}(-2) - \frac{11}{3} = -3$	$-3 \ge -3$
-1	$\frac{1}{2}(-1) - 2 = -\frac{5}{2}$	$-\frac{1}{3}(-1) - \frac{11}{3} = -\frac{10}{3}$	$-\frac{5}{2} \ge -\frac{10}{3}$

The integers greater than or equal to -2 are solutions of the inequality.

25. $2x - 2 > -x + 4$
$Y1 = 2x - 2$
$Y2 = -x + 4$

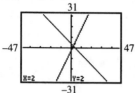

The intersection is (2, 2).
$x > 2, (2, \infty)$

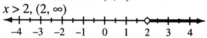

26. $1.2x + 0.72 \le -2.1x + 8.64$
$Y1 = 1.2x + 0.72$
$Y2 = -2.1x + 8.64$

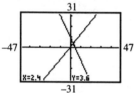

The intersection is (2.4, 3.6)
$x \le 2.4$
$(-\infty, 2.4]$

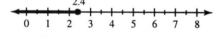

27. $(x + 6) - 3(x + 1) < (2x + 5) - 2(2x + 3)$
$Y1 = (x + 6) - 3(x + 1)$
$Y2 = (2x + 5) - 2(2x + 3)$

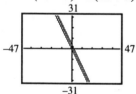

The lines are parallel and the graph of Y1 is always above the graph of Y2. Therefore, there is no solution.

28. $(x + 3) + (x + 1) \ge 3(x + 1) - (x - 1)$
$Y1 = (x + 3) + (x + 1)$
$Y2 = 3(x + 1) - (x - 1)$

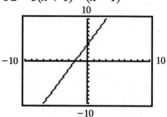

The lines are coinciding. Since the inequality includes the equal symbol the solution is all real numbers.
$(-\infty, \infty)$

29. $412 + x > 671$
$412 + x - 412 > 671 - 412$
$x > 259$
The solution set is $(259, \infty)$.

30. $y - \dfrac{7}{13} < \dfrac{11}{39}$

$y - \dfrac{7}{13} + \dfrac{7}{13} < \dfrac{11}{39} + \dfrac{7}{13}$

$y < \dfrac{11}{39} + \dfrac{21}{39}$

$y < \dfrac{32}{39}$

The solution set is $\left(-\infty, \dfrac{32}{39} \right)$.

31. $3x + 7 < 4x + 21$
$3x + 7 - 3x < 4x + 21 - 3x$
$7 < x + 21$
$7 - 21 < x + 21 - 21$
$-14 < x$
$x > -14$
The solution set is $(-14, \infty)$.

32. $14 + 2x < 2x$
$14 + 2x - 2x < 2x - 2x$
$14 < 0$
This is a contradiction.
There is no solution.

33. $8.7x + 4.33 \leq -2.4x - 33.41$

$8.7x + 4.33 - 4.33 \leq -2.4x - 3.41 - 4.33$

$8.7x \leq -2.4x - 37.74$

$8.7x + 2.4x \leq -2.4x - 37.74 + 2.4x$

$11.1x \leq -37.74$

$\dfrac{11.1x}{11.1} \leq \dfrac{-37.74}{11.1}$

$x \leq -3.4$

The solution set is $(-\infty, -3.4]$.

34. $6.8z - 9.52 \geq 0$

$6.8z - 9.52 + 9.52 \geq 0 + 9.52$

$6.8z \geq 9.52$

$\dfrac{6.8z}{6.8} \geq \dfrac{9.52}{6.8}$

$z \geq 1.4$

The solution set is $[1.4, \infty)$.

35. $2(x + 5) - (x + 6) < 2(x + 2)$

$2x + 10 - x - 6 < 2x + 4$

$x + 4 < 2x + 4$

$x + 4 - 4 < 2x + 4 - 4$

$x < 2x$

$x - x < 2x - x$

$x < x$

$0 < x$

$x > 0$

The solution set is $(0, \infty)$.

36. $3(x - 4) + 2(x + 1) > 5x + 10$

$3x - 12 + 2x + 2 > 5x + 10$

$5x - 10 > 5x + 10$

$5x - 10 - 5x > 5x + 10 - 5x$

$-10 > 10$

This is a contradiction.

There is no solution.

37. Let x = number of miles driven

$49.95 + 0.18x \leq 150$

$49.95 + 0.18x - 49.95 \leq 150 - 49.95$

$0.18x \leq 100.05$

$\dfrac{0.18x}{0.18} \leq \dfrac{100.05}{0.18}$

$x \leq 555.8\overline{3}$

The number of miles driven should be less than or equal to 555 miles.

38. Let x = amount of sixth month's sales

$\dfrac{2100 + 1300 + 1650 + 1250 + 1725 + x}{6}$
> 1500

$\dfrac{8025 + x}{6} > 1500$

$6\left(\dfrac{8025 + x}{6}\right) > 6(1500)$

$8025 + x > 9000$

$8025 + x - 8025 > 9000 - 8025$

$x > 975$

His sales should be greater than \$975.

39. Let x = width

$2x + 2(x + 4) \leq 40$

$2x + 2x + 8 \leq 40$

$4x + 8 \leq 40$

$4x + 8 - 8 \leq 40 - 8$

$4x \leq 32$

$\dfrac{4x}{4} \leq \dfrac{32}{4}$

$x \leq 8$

The width can be no more than 8 feet.

40. $x + 2y < 12$ is linear.

$x + 2y < 12$ is in standard form.

$a = 1, b = 2, c = 12$

41. $y < \dfrac{2}{3}x + \dfrac{5}{9}$ is linear.

$9(y) < 9\left(\dfrac{2}{3}x + \dfrac{5}{9}\right)$

$9y < 6x + 5$

$-5 < 6x - 9y$

$6x - 9y > -5$ is in standard form.

$a = 6, b = -9, c = -5$.

42. $x^2 + y^2 \geq 1$ is nonlinear because the x-variable and y-variable are squared.

43. $0.3x + 2.9 > 1.4y$ is linear.

$0.3x - 1.4y > -2.9$

$10(0.3x - 1.4y) > 10(-2.9)$

$3x - 14y > -29$ is in standard form.

$a = 3, b = -14, c = -29$

44. $y \geq \sqrt{x} - 1.44$ is nonlinear because the radical expression contains a variable.

45. $y < x^2 + 9$ is nonlinear because the x-variable is squared.

46. $12x + 6y < 48$
$12x + 16y - 12x < 48 - 12x$
$6y < -12x + 48$
$\dfrac{6}{6} < \dfrac{-12x + 48}{6}$
$y < -2x + 8$

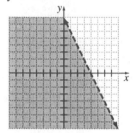

47. $y > \dfrac{3}{5}x - 6$

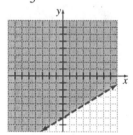

48. $4y \le x + 12$
$\dfrac{4y}{4} \le \dfrac{x + 12}{4}$
$y \le \dfrac{1}{4}x + 3$

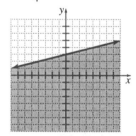

49. $y + 9 \ge 12$
$y + 9 - 9 \ge 12 - 9$
$y \ge 3$

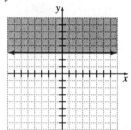

50. $5y - 2 < y - 10$
$5y - 2 + 2 < y - 10 + 2$
$5y < y - 8$
$5y - y < y - 8 - y$
$4y < -8$
$\dfrac{4y}{4} < \dfrac{-8}{4}$
$y < -2$

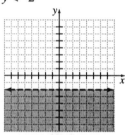

51. $2x + 16 > x + 18$
$2x + 16 - 16 > x + 18 - 16$
$2x > x + 2$
$2x - x > x + 2 - x$
$x > 2$

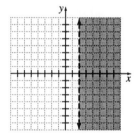

52. $8x - 12y > 24$

$8x - 12y - 8x > 24 - 8x$

$-12y > -8x + 24$

$\dfrac{-12y}{-12} < \dfrac{-8x + 24}{-12}$

$y < \dfrac{2}{3}x - 2$

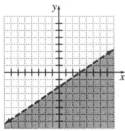

53. $4.4x + 1.1y \geq 12.1$

$4.4x + 1.1y - 4.4x \geq 12.1 - 4.4x$

$1.1y \geq -4.4x + 12.1$

$\dfrac{1.1y}{1.1} \geq \dfrac{-4.4x + 12.1}{1.1}$

$y \geq -4x + 11$

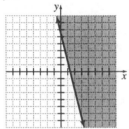

54. $7.4x - 14.8y \leq 29.6$

$7.4x - 14.8y - 7.4x \leq 29.6 - 7.4x$

$-14.8y \leq -7.4x + 29.6$

$\dfrac{-14.8y}{-14.8} \geq \dfrac{-7.4x + 29.6}{-14.8}$

$y \geq 0.5x - 2$

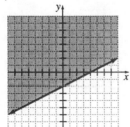

55. $\dfrac{x}{12} + \dfrac{y}{8} > -\dfrac{5}{4}$

$\dfrac{x}{12} + \dfrac{y}{8} - \dfrac{x}{12} > -\dfrac{5}{4} - \dfrac{x}{12}$

$\dfrac{y}{8} > -\dfrac{x}{12} - \dfrac{5}{4}$

$8\left(\dfrac{y}{8}\right) > 8\left(-\dfrac{x}{12} - \dfrac{5}{4}\right)$

$y > -\dfrac{2}{3}x - 10$

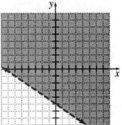

56. $-x - y < 2$

$-x - y + x < 2 + x$

$-y < x + 2$

$\dfrac{-y}{-1} > \dfrac{x + 2}{-1}$

$y > -x - 2$

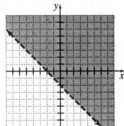

57. $y > -9x + 6$

58. Let x = number of rhododendrons
  y = number of azaleas
$4x + 6y \leq 85$
Solve the inequality for y.
$4x + 6y - 4x \leq 85 - 4x$
$6y \leq -4x + 85$
$\dfrac{6y}{6} \leq \dfrac{-4x + 85}{6}$
$y \leq -\dfrac{2}{3}x + \dfrac{85}{6}$

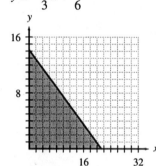

The possible combinations are in the shaded region.
Possible answer: She could buy 15 rhododendrons and 2 azaleas or 10 rhododendrons and 5 azaleas.

59. Let x = number correct
  y = number incorrect
$5x - 3y \geq 80$
Solve the inequality for y.
$5x - 3y - 5x \geq 80 - 5x$
$-3y \geq -5x + 80$
$\dfrac{-3y}{-3} \leq \dfrac{-5x + 80}{-3}$
$y \leq \dfrac{5}{3}x - \dfrac{80}{3}$

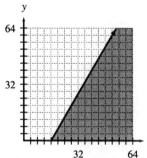

The possible combinations are in the shaded region.

Possible answer: She could get 40 correct and 3 incorrect or 30 correct and 0 wrong.

60. Solve the first inequality for y.
$2x + y > 10$
$2x + y - 2x > 10 - 2x$
$y > -2x + 10$
The second inequality is solved for y.
$y < 3x - 5$

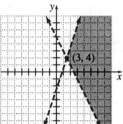

61. Solve the first inequality for x.
$3(x + 2) + 1 > -5$
$3x + 6 + 1 > -5$
$3x + 7 > -5$
$3x + 7 - 7 > -5 - 7$
$3x > -12$
$\dfrac{3x}{3} > \dfrac{-12}{3}$
$x > -4$
Solve the second inequality for y.
$2x - y < -8$
$2x - y - 2x < -8 - 2x$
$-y < -2x - 8$
$\dfrac{-y}{-1} > \dfrac{-2x - 8}{-1}$
$y > 2x + 8$

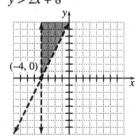

62. Solve the first inequality for x.
$x + 6 < 4$
$x + 6 - 6 < 4 - 6$
$x < -2$
Solve the second inequality for y.

$2(y + 2) > y + 3$

$2y + 4 > y + 3$

$2y + 4 - 4 > y + 3 - 4$

$2y > y - 1 - y$

$y > -1$

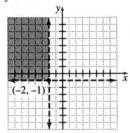

$(-2, -1)$

63. Solve the first inequality for y.

$x - 2y \geq 12$

$x - 2y - x \geq 12 - x$

$-2y \geq -x + 12$

$\dfrac{-2y}{-2} \leq \dfrac{-x + 12}{-2}$

$y \leq \dfrac{1}{2}x - 6$

Solve the second inequality for y.

$2x + 3y < -6$

$2x + 3y - 2x < -6 - 2x$

$3y < -2x - 6$

$\dfrac{3y}{3} < \dfrac{-2x - 6}{3}$

$y < -\dfrac{2}{3}x - 2$

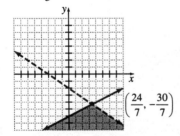

$\left(\dfrac{24}{7}, -\dfrac{30}{7}\right)$

64. $y \geq 2(x + 2)$

Solve the second inequality for y.

$2x - y \geq 5$

$2x - y - 2x \geq 5 - 2x$

$-y \geq -2x + 5$

$\dfrac{-y}{-1} \leq \dfrac{-2x + 5}{-1}$

$y \leq 2x - 5$

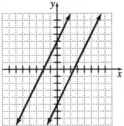

There is no solution.

65. $y > \dfrac{3}{2}x - 6$

Solve the second inequality for y.

$3x - 2y > 12$

$3x - 2y - 3x > 12 - 3x$

$-2y > -3x + 12$

$\dfrac{-2y}{-2} < \dfrac{-3x + 12}{-2}$

$y < \dfrac{3}{2}x - 6$

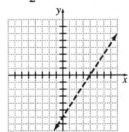

There is no solution.

66. $y < 3x - 6$

$y < -3$

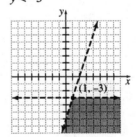

$(1, -3)$

67. Solve the first inequality for y.

$2y - 3 > 1$

$2y - 3 + 3 > 1 + 3$

$2y > 4$

$\dfrac{2y}{2} > \dfrac{4}{2}$

$y > 2$

Solve the second inequality for y.

$5(y - 4) > 10$

$5y - 20 > 10$

$5y - 20 + 20 > 10 + 20$

$5y > 30$

$\dfrac{5y}{5} > \dfrac{30}{5}$

$y > 6$

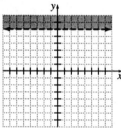

68. Solve the first inequality for x.

$2x - 11 < 3$

$2x - 11 + 11 < 3 + 11$

$2x < 14$

$\dfrac{2x}{2} < \dfrac{14}{2}$

$x < 7$

Solve the second inequality for x.

$14 - 2x > x - 7$

$14 - 2x - 14 > x - 7 - 14$

$-2x > x - 21$

$-2x - x > x - 21 - x$

$-3x > -21$

$\dfrac{-3x}{-3} < \dfrac{-21}{-3}$

$x < 7$

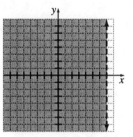

69. Solve the first inequality for y.

$y - 1 < 4$

$y - 1 + 1 < 4 + 1$

$y < 5$

Solve the second inequality for y.

$3y - 2 < 13$

$3y - 2 + 2 < 13 + 2$

$3y < 15$

$\dfrac{3y}{3} < \dfrac{15}{3}$

$y < 5$

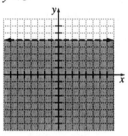

70. Solve the first inequality for y.

$2x + 3y \le 6$

$2x + 3y - 2x \le 6 - 2x$

$3y \le -2x + 6$

$\dfrac{3y}{3} \le \dfrac{-2x + 6}{3}$

$y \le -\dfrac{2}{3}x + 2$

Solve the second inequality for y.

$x + y \le 1$

$x + y - x \le 1 - x$

$y \le -x + 1$

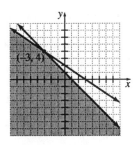

71. $y \leq 2x - 15$
$y > -3x + 10$

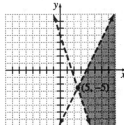

72. $y > 2x + 1$
Solve the second inequality for y.
$x > 2x + 7$
$x - 7 > 2y + 7 - 7$
$x - 7 > 2y$
$$\frac{x-7}{2} > \frac{2y}{2}$$
$$\frac{1}{2}x - \frac{7}{2} > y$$
$$y < \frac{1}{2}x - \frac{7}{2}$$

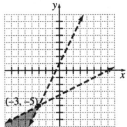

73. Solve the first inequality for y.
$2y + 6 \leq 3y + 5$
$2y + 6 - 2y \leq 3y + 5 - 2y$
$6 \leq y + 5$
$6 - 5 \leq y + 5 - 5$
$1 \leq y$
$y \geq 1$

Solve the second inequality for x.
$2(x - 3) + 1 > 5$
$2x - 6 + 1 > 5$
$2x - 5 > 5$
$2x - 5 + 5 > 5 + 5$
$2x > 10$
$$\frac{2x}{2} > \frac{10}{2}$$
$x > 5$

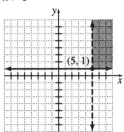

74. $y < 6 - x$
$y < 2x + 1$
$x \geq 0$
$y \geq 0$

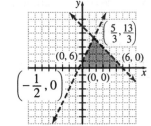

75. $y < 25 - \frac{1}{4}x$
$y < \frac{2}{3}x + 5$
$x \geq 0$
$y \geq 0$

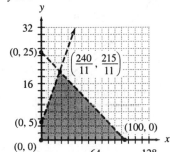

76. Let x = number of rhododendrons
$\quad\quad y$ = number of azaleas
$4x + 6y \leq 85$
$y + 4 \leq x$
$x \geq 0$
$y \geq 0$
Solve the first two inequalities for y.

$y \leq -\dfrac{2}{3}x + \dfrac{85}{6}$

$y \leq x - 4$

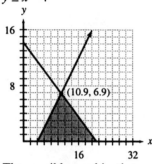

The possible combinations are in the shaded region.
Possible answer: She could purchase 15 rhododendrons and 2 azaleas.

77. Let x = amount invested at 5%
$\quad\quad y$ = amount invested at 6%
$0.05x + 0.06y \geq 225$
$x + y \leq 4000$
$x \geq 0$
$y \geq 0$
Solve the first two inequalities for y.

$y \geq -\dfrac{5}{6}x + 3750$

$y \leq -x + 4000$

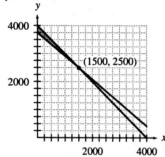

The possible combinations are in the shaded region.
Possible answer: You could invest $200 at 5% and $3800 at 6%.

Chapter 7 Mixed Review

1. $x < -2$

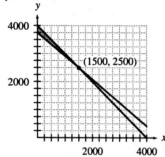

$(-\infty, -2)$

2. $x > 7$

$(7, \infty)$

3. $-1 < x < 3$

$(-1, 3)$

4. $x \geq 2.6$

$[2.6, \infty)$

5. $x \leq 3\dfrac{2}{3}$

$\left(-\infty, 3\dfrac{2}{3}\right]$

6. $-2.4 < x \leq 4\dfrac{1}{3}$

$\left(-2.4, 4\dfrac{1}{3}\right]$

7.

x	$5x + 3$	$2x - 9$	
–5	$5(–5) + 3 = –22$	$2(–5) – 9 = –19$	$–22 < –19$
–4	$4(–4) + 3 = –17$	$2(–4) – 9 = –17$	$–17 \not< –17$
–3	$5(–3) + 3 = –12$	$2(–3) – 9 = –15$	$–12 \not< –15$

The integers less than –4 are solutions to the inequality.

8.

x	$5.6x - 15.3$	$1.3x + 19.1$	
7	$5.6(7) – 15.3 = 23.9$	$1.3(7) + 19.1 = 28.2$	$23.9 \not> 28.2$
8	$5.6(8) – 15.3 = 29.5$	$1.3(8) + 19.1 = 29.5$	$29.5 \not> 29.5$
9	$5.6(9) – 15.3 = 35.1$	$1.3(9) + 19.1 = 30.8$	$35.1 > 30.8$

The integers greater than 8 are solutions to the inequality.

9.

x	$\frac{1}{6}x + \frac{23}{3}$	$\frac{13}{6} - \frac{5}{3}x$	
–4	$\frac{1}{6}(–3) + \frac{23}{3} = 7$	$\frac{13}{6} - \frac{5}{3}(–4) = 8.8\overline{3}$	$7 \le 8.8\overline{3}$
–3	$\frac{1}{6}(–3) + \frac{23}{3} = 7.1\overline{6}$	$\frac{13}{6} - \frac{5}{3}(–3) = 7.1\overline{6}$	$7.1\overline{6} \le 7.1\overline{6}$
–2	$\frac{1}{6}(–2) + \frac{23}{3} = 7.\overline{3}$	$\frac{13}{6} - \frac{5}{3}(–2) = 5.5$	$7.\overline{3} \not\le 5.5$

The integers less than or equal to –3 are solutions to the inequality.

10.

x	$\frac{3}{7}x + \frac{9}{5}$	$\frac{4}{5}x - \frac{17}{5}$	
13	$\frac{3}{7}(13) + \frac{9}{5} = \frac{258}{35}$	$\frac{4}{5}(13) - \frac{17}{5} = 7$	$\frac{258}{35} \ge 7$
14	$\frac{3}{7}(14) + \frac{9}{5} = \frac{39}{5}$	$\frac{4}{5}(14) - \frac{17}{5} = \frac{39}{5}$	$\frac{39}{5} \ge \frac{39}{5}$
15	$\frac{3}{7}(15) + \frac{9}{5} = \frac{288}{35}$	$\frac{4}{5}(15) - \frac{17}{5} = \frac{43}{5}$	$\frac{288}{35} \not\ge \frac{43}{5}$

The integers less than or equal to 14 are solutions to the inequality.

11. $5x - 13 > 3(1 - x)$
$Y1 = 5x - 13$
$Y2 = 3(1 - x)$

The intersection is (2, –3).
$x > 2$

$(2, \infty)$

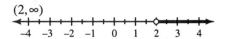

12. $2.1x + 31.71 \le 8.19 - 3.5x$
$Y1 = 2.1x + 31.71$
$Y2 = 8.19 - 3.5x$

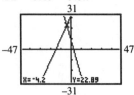

The intersection is $(-4.2, 22.89)$.
$x \le -4.2$
$(-\infty, -4.2]$

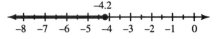

13. $3(x - 1) + (x - 1) < 2(x - 1) + 2x - 1$
$Y1 = 3(x - 1) + (x - 1)$
$Y2 = 2(x - 1) + 2x - 1$

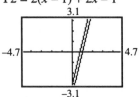

The lines are parallel and the graph of Y1 is always below the graph of Y2. Therefore, the solution is all real numbers.
$(-\infty, \infty)$

14. $4(x + 2) - 3(x - 5) \le 5x + 8 - 4(x + 3)$
$Y1 = 4(x + 2) - 3(x - 5)$
$Y2 = 5x + 8 - 4(x + 3)$

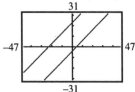

The lines are parallel and the graph of Y1 is always above the graph of Y2. Therefore, there is no solution.

15. $173 - x < 359$
$173 - x - 173 < 359 - 173$
$-x < 186$
$\dfrac{-x}{-1} > \dfrac{186}{-1}$
$x > -186$
The solution set is $(-186, \infty)$.

16. $z - \dfrac{4}{17} > \dfrac{15}{34}$
$z - \dfrac{4}{17} + \dfrac{4}{17} > \dfrac{15}{34} + \dfrac{4}{17}$
$z > \dfrac{15}{34} + \dfrac{8}{34}$
$z > \dfrac{23}{34}$
The solution set is $\left(\dfrac{23}{34}, \infty \right)$.

17. $5x + 4 > 3x + 18$
$5x + 4 - 4 > 3x + 18 - 4$
$5x > 3x + 14$
$5x - 3x > 3x + 18 - 3x$
$2x > 14$
$\dfrac{2x}{2} > \dfrac{14}{2}$
$x > 7$
The solution set is $(7, \infty)$.

18. $8x < 8x - 16$
$8x - 8x < 8x - 16 - 8x$
$0 < -16$
This is a contradiction. There is no solution.

19. $3.9x + 19.88 \ge -1.9x + 4.76$
$3.5x + 19.88 - 19.88 \ge -1.9x + 4.76 - 19.88$
$3.5x \ge -1.9x - 15.12$
$3.5x + 1.9x \ge -1.9x - 15.12 + 1.9x$
$5.4x \ge -15.12$
$\dfrac{5.4x}{5.4} \ge \dfrac{-15.12}{5.4}$
$x \ge -2.8$
The solution set is $[-2.8, \infty)$.

20. $2.6y + 9.62 \le 0$
$2.6y + 9.62 - 9.62 \le 0 - 9.62$
$2.6y \le -9.62$
$\dfrac{2.6y}{2.6} \le \dfrac{-9.62}{2.6}$
$y \le -3.7$
The solution set is $(-\infty, -3.7]$.

21. $3(x + 3) < 4(x + 1) - 2(x - 2)$
$3x + 9 < 4x + 4 - 2x + 4$
$3x + 9 < 2x + 8$
$3x + 9 - 9 < 2x + 8 - 9$
$3x < 2x - 1$
$3x - 2x < 2x - 1 - 2x$
$x < -1$
The solution set is $(-\infty, -1)$.

22. $3(x - 3) + 2(x + 2) < 5x + 7$
$5x - 9 + 2x + 4 < 5x + 7$
$5x - 5 < 5x + 7$
$5x - 5 - 5x < 5x + 7 - 5x$
$-5 < 7$
This is always true.
The solution set is $(-\infty, \infty)$.

23. Solve the first inequality for y.
$2y + 2 > y$
$2y + 2 - y > y - y$
$y + 2 > 0$
$y + 2 - 2 > 0 - 2$
$y > -2$
The second inequality is:
$y < -x + 1$

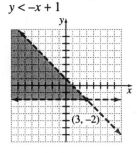

24. Solve the first inequality for x.
$2x + 3 > x$
$2x + 3 - 3 > x - 3$
$2x > x - 3$
$2x - x > x - 3 - x$
$x > -3$
Solve the second inequality for x.

$2x < 6$
$\dfrac{2x}{2} < \dfrac{6}{2}$
$x < 3$

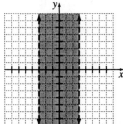

25. $y \ge 2x - 8$
Solve the second inequality for y.
$3x + y < 8$
$3x + y - 3x < 8 - 3x$
$y < -3x + 8$

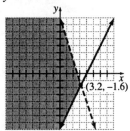

$(3.2, -1.6)$

26. Solve the first inequality for x.
$2(x - 1) - 4 > 0$
$2x - 2 - 4 > 0$
$2x - 6 > 0$
$2x - 6 + 6 > 0 + 6$
$2x > 6$
$\dfrac{2x}{2} > \dfrac{6}{2}$
$x > 3$
Solve the second equation for y.
$2x + y \le 3$
$2x + y - 2x \le 3 - 2x$
$y \le -2x + 3$

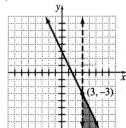

$(3, -3)$

27. Solve the first inequality for y.
$x + y < 9 - x$
$x + y - x < 9 - x - x$
$y < -2x + 9$
Solve the second inequality for y.
$2x + y > 5$
$2x + y - 2x > 5 - 2x$
$y > -2x + 5$

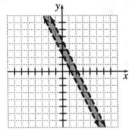

28. $y < -4x - 3$
Solve the second inequality for y.
$4x + 2y < y - 3$
$4x + 2y - 4x < y - 3 - 4x$
$2y < y - 4x - 3$
$2y - y < y - 4x - 3 - y$
$y < -4x - 3$

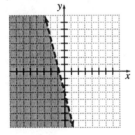

29. Solve the first inequality for x.
$3x + 4 \geq 2x$
$3x + 4 - 4 \geq 2x - 4$
$3x \geq 2x - 4$
$3x - 2x \geq 2x - 4 - 2x$
$x \geq -4$
Solve the second inequality for x.
$x + 7 \leq 3$
$x + 7 - 7 \leq 3 - 7$
$x \leq -4$

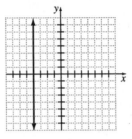

30. Solve the first equation for y.
$3y + 7 > 2y + 5$
$3y + 7 - 7 > 2y + 5 - 7$
$3y > 2y - 2$
$3y - 2y > 2y - 2 - 2y$
$y > -2$
Solve the second inequality for y.
$y + 9 < 7$
$y + 9 - 9 < 7 - 9$
$y < -2$

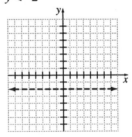

31. Solve the first inequality for x.
$3(x - 3) - 1 > 8$
$3x - 9 - 1 > 8$
$3x - 10 > 8$
$3x - 10 + 10 > 8 + 10$
$3x > 18$
$\dfrac{3x}{3} > \dfrac{18}{3}$
$x > 6$
Solve the second inequality for y.
$3y - 10 < 15 - 2y$
$3y - 10 + 10 < 15 - 2y + 10$
$3y < 25 - 2y$
$3y + 2y < 25 - 2y + 2y$
$5y < 25$
$\dfrac{5y}{5} < \dfrac{25}{5}$
$y < 5$

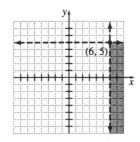

32. $y < 3x + 7$
Solve the second inequality for y.
$x - y > -1$
$x - y - x > -1 - x$
$-y > -x - 1$
$\dfrac{-y}{-1} < \dfrac{-x - 1}{-1}$
$y < x + 1$

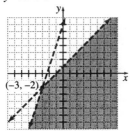

33. Solve the first inequality for y.
$x - 5y \le 15$
$x - 5y - x \le 15 - x$
$-5y \le -x + 15$
$\dfrac{-5y}{-5} \ge \dfrac{-x + 15}{-5}$
$y \ge \dfrac{1}{5}x - 3$
Solve the second inequality for y.
$x + 5y \le -5$
$x + 5y - x \le -5 - x$
$5y \le -x - 5$
$\dfrac{5y}{5} \le \dfrac{-x - 5}{5}$
$y \le -\dfrac{1}{5}x - 1$

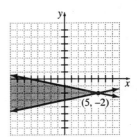

34. $y < -4x + 6$
$y > 3x - 8$

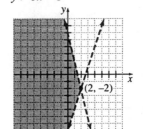

35. $y > x$
Solve the second inequality for y.
$x > 2 - y$
$x - 2 > 2 - y - 2$
$x - 2 > -y$
$\dfrac{x - 2}{-1} < \dfrac{-y}{-1}$
$-x + 2 < y$
$y > -x + 2$

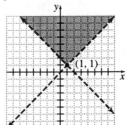

36. Solve the first inequality for x.
$x + 9 \le 3$
$x + 9 - 9 \le 3 - 9$
$x \le -6$
Solve the second inequality for y.
$3y > 2y - 1$
$3y - 2y > 2y - 1 - 2y$
$y > -1$

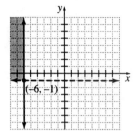

37. $y < -2x + 7$
$y < 2x$
$x \geq 0$
$y \geq 0$

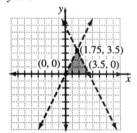

38. $y > \dfrac{3}{4}x - 4$

$y > -\dfrac{2}{3}x + 3$

$x \geq 0$
$y \geq 0$

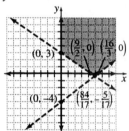

39. $9x - 5y < 45$
$9x - 5y - 9x < 45 - 9x$
$-5y < -9x + 45$
$\dfrac{-5y}{-5} > \dfrac{-9x + 45}{-5}$

$y > \dfrac{9}{5}x - 9$

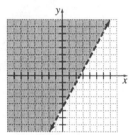

40. $y > \dfrac{4}{3}x - 5$

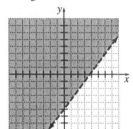

41. $3y \leq 2x + 9$
$\dfrac{3y}{3} \leq \dfrac{2x + 9}{3}$

$y \leq \dfrac{2}{3}x + 3$

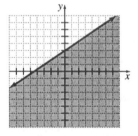

42. $y + 13 \geq 8$
$y + 13 - 13 \geq 8 - 13$
$y \geq -5$

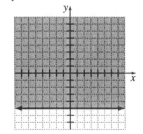

43. $-2y - 6 < y - 11$

$-2y - 6 + 6 < y - 11 + 6$

$-2y < y - 5$

$-2y - y < y - 5 - y$

$-3y < -5$

$\dfrac{-3y}{-3} > \dfrac{-5}{-3}$

$y > \dfrac{5}{3}$

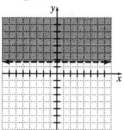

44. $x - 13 > 3x + 19$

$x - 13 + 13 > 3x + 19 + 13$

$x > 3x + 32$

$x - 3x > 3x + 32 - 3x$

$-2x > 32$

$\dfrac{-2x}{-2} < \dfrac{32}{-2}$

$x < -16$

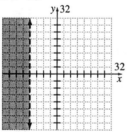

45. $x - 7y > 21$

$x - 7y - x > 21 - x$

$-7y > -x + 21$

$\dfrac{-7y}{-7} < \dfrac{-x + 21}{-7}$

$y < \dfrac{1}{7}x - 3$

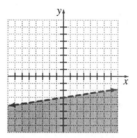

46. $2.7x + 5.4y \ge 16.2$

$2.7x + 5.4y - 2.7x \ge 16.2 - 2.7x$

$5.4y \ge -2.7x + 16.2$

$\dfrac{5.4y}{5.4} \ge \dfrac{-2.7x + 16.2}{5.4}$

$y \ge -0.5x + 3$

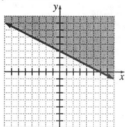

47. $4.8x - 1.8y \le 14.4$

$4.8x - 1.8y - 4.8x \le 14.4 - 4.8x$

$-1.8y \le -4.8x + 14.4$

$\dfrac{-1.8y}{-1.8} \ge \dfrac{-4.8x + 14.4}{-1.8}$

$y \ge \dfrac{8}{3}x - 8$

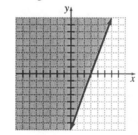

48. $\dfrac{x}{12} + \dfrac{y}{9} > -\dfrac{1}{3}$

$\dfrac{x}{12} + \dfrac{y}{9} - \dfrac{x}{12} > -\dfrac{1}{3} - \dfrac{x}{12}$

$\dfrac{y}{9} > -\dfrac{x}{12} - \dfrac{1}{3}$

$9\left(\dfrac{y}{9}\right) > 9\left(-\dfrac{x}{12} - \dfrac{1}{3}\right)$

$y > -\dfrac{3}{4}x - 3$

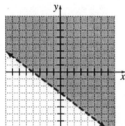

49. $x - y < 7$

$x - y - x < 7 - x$

$-y < -x + 7$

$\dfrac{-y}{-1} > \dfrac{-x+7}{-1}$

$y > x - 7$

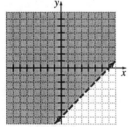

50. $y > -5x + 7$

51. Let x = number of acres of oats

y = number of acres of wheat

$y \geq 2x$

$x + y \leq 540$

$x \geq 0$

$y \geq 0$

Solve the second inequality for y.

$x + y \leq 540$

$x + y - x \leq 540 - x$

$y \leq -x + 540$

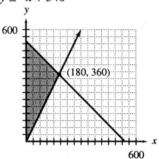

The possible combinations are represented by the shaded region.

Possible answer:

He could plant 100 acres of oats and 400 acres of wheat.

52. Let x = rent for efficiency apartment

y = rent for regular apartment

$3x + 5y \geq 6000$

$y \geq x + 75$

$x \geq 0$

$y \geq 0$

Solve the first inequality for y.

$3x + 5y \geq 6000$

$3x + 5y - 3x \geq 6000 - 3x$

$5y \geq -3x + 6000$

$\dfrac{5y}{5} \geq \dfrac{-3x + 6000}{5}$

$y \geq -\dfrac{3}{5}x + 1200$

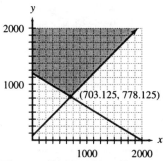

The possible combinations are represented by the shaded region.
Possible answer:
The rents could be $1000 dollars for an efficiency and $1500 for a regular apartment.

53. Let x = number of packs
$255 + 2.5x < 12.5x$
Solve the inequality for x.
$255 + 2.5x - 2.5x < 12.5x - 2.5x$
$255 < 10x$
$\dfrac{255}{10} < \dfrac{10x}{10}$
$25.5 < x$
$x > 25.5$
To make a profit at least 26 packs must be sold.
Possible answer:
There were 27 packs sold.

54. Let x = amount of bill for the sixth month
$\dfrac{45 + 36 + 52 + 48 + 31 + x}{6} < 42$
Solve the inequality for x.
$\dfrac{212 + x}{6} < 42$
$6\left(\dfrac{212 + x}{6}\right) < 6(42)$
$212 + x < 252$
$212 + x - 212 < 252 - 212$
$x < 40$
Her sixth phone bill must be less than $40.
Possible answer:
Her phone bill should be $35.

55. Let x = length
$18x > 600$
Solve the inequality for x.
$\dfrac{18x}{18} > \dfrac{600}{18}$
$x > 33\dfrac{1}{3}$

The length must be at least $33\dfrac{1}{3}$ inches.

Possible answer:
The length is 40 inches.

56. Let x = number of wins
    y = number of ties
$3x + y \geq 25$
Solve the inequality for y.
$3x + y - 3x \geq 25 - 3x$
$y \geq -3x + 25$

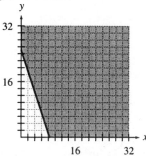

The possible combinations are represented in the shaded region.
Possible answer:
They could have 10 wins and 5 ties.

57. Let x = number of liters of the 25% solution
    y = number of liters of the 40% solution
$0.25x + 0.40y \leq 0.30(x + y)$
Solve the inequality for y.
$0.25x + 0.40y \leq 0.30x + 0.30y$
$0.25x + 0.40y - 0.25x$
$\leq 0.30x + 0.30y - 0.25x$
$0.40y \leq 0.05x + 0.30y$
$0.40y - 0.30y \leq 0.05x + 0.30y - 0.30y$
$0.1y \leq 0.05x$
$\dfrac{0.1y}{0.1} \leq \dfrac{0.05x}{0.1}$
$y \leq 0.5x$

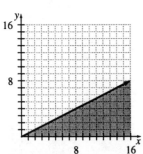

The possible combinations are represented by the shaded region.
Possible answer:
There should be 10 liters of 25% solution and 2 liters of 40% solution.

Chapter 7 Test

1. $2x - 3y > x + 8$ is linear because it simplifies to $y < \dfrac{1}{3}x - \dfrac{8}{3}$.

2. $4x^2 + 2x < x - 9$ is nonlinear because the x-variable is squared.

3. $5(x - 3) \geq 4 - (x + 1)$ is linear because it simplifies to $x \geq 3$.

4. $\dfrac{1}{2}x - 4 \geq y + \dfrac{3}{8}$ is linear.

5. $7(x + 2) - 3(x + 1) > 4(x + 8)$
   $7x + 14 - 3x - 3 > 4x + 32$
   $4x + 11 > 4x + 32$
   $4x + 11 - 4x > 4x + 32 - 4x$
   $11 > 32$
   This is a contradiction. There is no solution.

6. $5x + 9 < 2x - 3$
   $5x + 9 - 9 < 2x - 3 - 9$
   $5x < 2x - 12$
   $5x - 2x < 2x - 12 - 2x$
   $3x < -12$
   $\dfrac{3x}{3} < \dfrac{-12}{3}$
   $x < -4$

   The solution set is $(-\infty, -4)$.

7. $\dfrac{4}{5}(x - 10) < \dfrac{1}{5}(4x + 5) + 1$
   $\dfrac{4}{5}x - 8 < \dfrac{4}{5}x + 2$
   $\dfrac{4}{5}x - 8 - \dfrac{4}{5}x < \dfrac{4}{5}x + 2 - \dfrac{4}{5}x$
   $-8 < 2$
   This is always true. The solution is all real numbers.

   The solution set is $(-\infty, \infty)$.

8. $5a - 7 \geq 8a + 1$
   $5a - 7 + 7 \geq 8a + 1 + 7$
   $5a \geq 8a + 8$
   $5a - 8a \geq 8a + 8 - 8a$
   $-3a \geq 8$
   $\dfrac{-3a}{-3} \leq \dfrac{8}{-3}$
   $a \leq -\dfrac{8}{3}$

   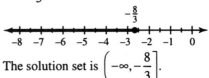

   The solution set is $\left(-\infty, -\dfrac{8}{3}\right]$.

9. $y < -2x + 5$

   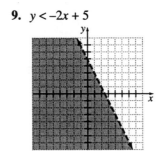

10. Solve the inequality for x.
    $x + 3 > 2x - 1$
    $x + 3 - x > 2x - 1 - x$
    $3 > x - 1$
    $3 + 1 > x - 1 + 1$
    $4 > x$
    $x < 4$

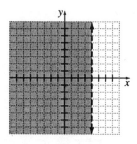

11. Solve the first inequality for y.

$x - y > 4$

$x - y - x > 4 - x$

$-y > -x + 4$

$\dfrac{-y}{-1} < \dfrac{-x + 4}{-1}$

$y < x - 4$

Solve the second inequality for y.

$x + 2y > 4$

$x + 2y - x > 4 - x$

$2y > -x + 4$

$\dfrac{2y}{2} > \dfrac{-x + 4}{2}$

$y > -\dfrac{1}{2}x + 2$

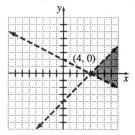

12. $y \geq 4x - 5$

$\quad y > -3x + 4$

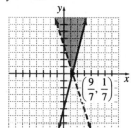

13. $y < 2x + 6$

$\quad y < 2x - 1$

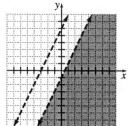

14. Solve the first inequality for y.

$3y - 1 > 2$

$3y - 1 + 1 > 2 + 1$

$3y > 3$

$\dfrac{3y}{3} > \dfrac{3}{3}$

$y > 1$

Solve the second inequality for x.

$x - 3 \leq -5$

$x - 3 + 3 \leq -5 + 3$

$x \leq -2$

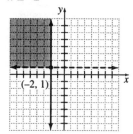

15. Let x = third score

$\dfrac{83 + 72 + x}{3} \geq 80$

Solve the inequality for x.

$\dfrac{155 + x}{3} \geq 80$

$3\left(\dfrac{155 + x}{3}\right) \geq 3(80)$

$155 + x \geq 240$

$155 + x - 155 \geq 240 - 155$

$x \geq 85$

The score must be at least 85.

Possible answer:

The score is 90 points.

16. Let x = number of good deeds
$\quad\quad y$ = number of activity sheets
$5x + 2y \geq 20$
Solve the inequality for y.
$5x + 2y - 5x \geq 20 - 5x$
$2y \geq -5x + 20$
$\dfrac{2y}{2} \geq \dfrac{-5x + 20}{2}$
$y \geq -\dfrac{5}{2}x + 10$

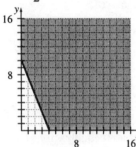

The solutions are represented by the shaded region.
Possible answers:
The cub scout had 5 good deeds and 5 activity sheets.

17. Let x = amount invested at 4%
$\quad\quad y$ = amount invested at 8%
$0.04x + 0.08y \geq 400$
$x + y \leq 6000$
$x \geq 0$
$y \geq 0$
Solve the first two inequalities for y
$y \geq -0.5x + 5000$
$y \leq -x + 6000$

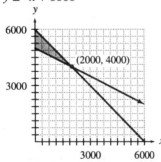

The possible combinations are represented by the shaded region.
Possible answer:
You could invest $500 at 4%
and $5500 at 8%

18. Answers will vary.

Chapters 1–7 Cumulative Review

1. $-(-9) = 9$

2. $-|-9| = -(9) = -9$

3. $\sqrt[3]{-\dfrac{27}{64}} = \dfrac{\sqrt[3]{-27}}{\sqrt[3]{64}} = \dfrac{-3}{4} = -\dfrac{3}{4}$

4. $\left(\dfrac{-3}{8}\right)^{-2} = \left(\dfrac{8}{-3}\right)^2 = \dfrac{(8)^2}{(-3)^2} = \dfrac{64}{9}$

5. $\left(-\dfrac{9}{14}\right)\left(\dfrac{7}{3}\right) = -\dfrac{3 \cdot 3 \cdot 7}{2 \cdot 7 \cdot 3} = -\dfrac{3}{2}$

6. $-\dfrac{3}{8} \div \left(-1\dfrac{2}{3}\right) = -\dfrac{3}{8} \div \left(-\dfrac{5}{3}\right)$
$\quad = \left(-\dfrac{3}{8}\right)\left(-\dfrac{3}{5}\right) = \dfrac{9}{40}$

7. $12(-3) \div (-6)(2) \div (-2)$
$\quad = (-36) \div (-6)(2) \div (-2)$
$\quad = (6)(2) \div (-2)$
$\quad = 12 \div (-2)$
$\quad = -6$

8. $40 + 16 \div 8 - \sqrt{3^2 + 7 \cdot 5 + 5}$
$\quad = 40 + 16 \div 8 - \sqrt{9 + 7 \cdot 5 + 5}$
$\quad = 40 + 2 - \sqrt{9 + 35 + 5}$
$\quad = 42 - \sqrt{49}$
$\quad = 42 - 7$
$\quad = 35$

9. $\dfrac{2(5^2 - 10) + 4^2 - 1}{8^2 - 2(32)}$

$= \dfrac{2(25 - 10) + 4^2 - 1}{8^2 - 2(32)}$

$= \dfrac{2(15) + 16 - 1}{64 - 2(32)}$

$= \dfrac{30 + 16 - 1}{64 - 64}$

$= \dfrac{45}{0} = $ undefined

10. $6x - 2(4x - 1) = 6x - 8x + 2 = -2x + 2$

11. $4[2(x - 3) + 1] - [5(2x - 4) - 6]$
$= 4(2x - 5) - (10x - 26)$
$= 8x - 20 - 10x + 26$
$= -2x + 6$

12.

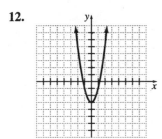

13. The domain of the relation is all real numbers. The range of the relation is all real numbers ≥ -3.

14. The relation is a function since all possible vertical lines cross the graph a maximum of one time.

15. The relative minimum value is -3 at $x = 0$. There is no relative maximum.

16. The relation is increasing for all $x > 0$, and decreasing for all $x < 0$.

17. For the x-intercept, solve $0 = 2x^2 - 3$.
$0 = 2x^2 - 3$
$3 = 2x^2$
$\dfrac{3}{2} = x^2$
$\pm\sqrt{\dfrac{3}{2}} = x, \; x \approx \pm 1.225$

The x-intercepts are $(-1.225, 0)$ and $(1.225, 0)$.

For the y-intercept, solve $y = 2(0)^2 - 3$.
$y = 2(0)^2 - 3$
$y = 2(0) - 3$
$y = 0 - 3$
$y = -3$

The y-intercept is $(0, -3)$.

18. $7x - 3 = 5$
$7x = 8$
$x = \dfrac{8}{7}$

19. $(x + 3) - 2(3x + 4) = 5$
$x + 3 - 6x - 8 = 5$
$-5x - 5 = 5$
$-5x = 10$
$x = \dfrac{10}{-5}$
$x = -2$

20. $1.2(x + 3) - 4(0.3x + 0.15) = 3$
$1.2x + 3.6 - 1.2x - 0.6 = 3$
$3 = 3$
All real numbers are solutions.

21. $\quad \dfrac{2}{3}x + \dfrac{1}{6} = 2\left(\dfrac{1}{3}x - \dfrac{1}{2}\right)$

$\dfrac{2}{3}x + \dfrac{1}{6} = \dfrac{2}{3}x - 1$

$\dfrac{2}{3}x - \dfrac{2}{3}x + \dfrac{1}{6} = \dfrac{2}{3}x - \dfrac{2}{3}x - 1$

$\dfrac{1}{6} = -1$

The equation has no solution.

22. $6x - (4x + 2) = 3(2x + 6)$
$6x - 4x - 2 = 6x + 18$
$2x - 2 = 6x + 18$
$-2 = 4x + 18$
$-20 = 4x$
$-\dfrac{20}{4} = x$
$-5 = x$

23. $5x + 4 > x - 8$
$5x > x - 12$
$4x > -12$
$x > -3$
$(-3, \infty)$

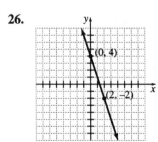

24. $-12x + 4 \geq -2x + 8$
$4 \geq 10x + 8$
$-4 \geq 10x$
$-\dfrac{4}{10} \geq x$
$-\dfrac{2}{5} \geq x$
$\left(-\infty, -\dfrac{2}{5}\right]$

25. $4(x + 3) - 3(x - 2) \leq x - 5$
$4x + 12 - 3x + 6 \leq x - 5$
$x + 18 \leq x - 5$
$18 \leq -5$
The statement is false; there is no solution.

26.

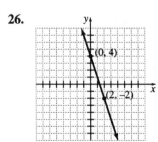

27.

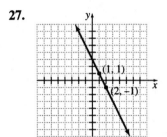

28.

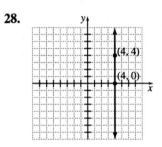

29.

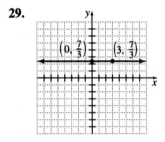

30. Graph $y = 4x - 3$ with a dashed line.
Use $(0, 0)$ as a test point.
$0 < 4(0) - 3?$
$0 < -3$ False
Shade the half-plane not containing $(0, 0)$.

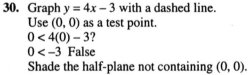

31. $2x + 3y = 21$
$3y = -2x + 21$
$y = -\dfrac{2}{3}x + 7$

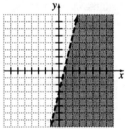

$3x - 2y = 2$

$-2y = -3x + 2$

$y = \dfrac{3}{2}x - 1$

Since $\left(-\dfrac{2}{3}\right)\left(\dfrac{3}{2}\right) = -1$, the graph of the pair of equations is intersecting and perpendicular lines.

32. Subtract the equations.

$y = 3x + 4$

$\underline{-y = -2x + 5}$

$0 = x + 9$

$-9 = x$

$y = 3(-9) + 4 = -27 = 4 = -23$

The solution is $(-9, -23)$.

33. Substitute the expression $-2x + 4$ for y in the first equation.

$4x + 2(-2x + 4) = 8$

$4x - 4x + 8 = 8$

$8 = 8$

This is an identity. The solution is all ordered pairs which satisfy $y = -2x + 4$.

34. Graph $y = -x$ with a solid line and $y = 2x + 4$ with a dashed line. Shade above the line $y = -x$ and below the line $y = 2x + 4$.

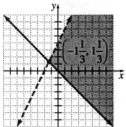

One solution is $(1, 0)$.

35. $m = \dfrac{y_2 - y_1}{x_2 - x_1} = \dfrac{-2 - 3}{-1 - (-2)} = \dfrac{-5}{1} = -5$

36. $m = \dfrac{y_2 - y_1}{x_2 - x_1} = \dfrac{3 - (-5)}{6 - 6} = \dfrac{8}{0}$

The slope is undefined.

37. $y = mx + b$

$y = 2x + 3$

The slope is 2.

38. $y = mx + b$

$y = 0x + 1.4$

The slope is 0.

39. $m = \dfrac{y_2 - y_1}{x_2 - x_1} = \dfrac{-2 - (-3)}{-1 - 2} = \dfrac{1}{-3} = -\dfrac{1}{3}$

$y - y_1 = m(x - x_1)$

$y - (-3) = -\dfrac{1}{3}(x - 2)$

$y + 3 = -\dfrac{1}{3}x + \dfrac{2}{3}$

$y = -\dfrac{1}{3}x - \dfrac{7}{3}$

40. $2x + 3y = 1$

$3y = -2x + 1$

$y = -\dfrac{2}{3}x + \dfrac{1}{3}$

The slope of the graph is $-\dfrac{2}{3}$.

$y - y_1 = m(x - x_1)$

$y - 4 = -\dfrac{2}{3}(x - 5)$

$y - 4 = -\dfrac{2}{3}x + \dfrac{10}{3}$

$y = -\dfrac{2}{3}x + \dfrac{22}{3}$

41. $5,340,000 = 5.34 \times 10^6$

42. $1.2 \times 10^{-4} = 0.00012$

43. $-4.783 \text{ E}{-5} = -0.00004783$

44. $P = 2L + 2W$

$P - 2W = 2L$

$\dfrac{P - 2W}{2} = L$

$L = \dfrac{P}{2} - W$

45. $f(x) = x^2 + 2x - 1$

$f(-3) = (-3)^2 + 2(-3) - 1$

$f(-3) = 9 + (-6) - 1$

$f(-3) = 9 - 6 - 1$

$f(-3) = 2$

46. Let x be the amount that Lance invests.

$0.11x = 2775$

$x = \dfrac{2775}{0.11}$

$x \approx 25{,}227.273$

Lance should invest \$25,227.28.

47. The sum of the measures of the angles is $90°$. Let x be the measure of the smaller angle. Then the measure of the larger angle is $2x + 25$.

$x + (2x + 25) = 90$

$3x + 25 = 90$

$3x = 65$

$x = \dfrac{65}{3} = 21\dfrac{2}{3}$

$2x + 25 = 2\left(\dfrac{65}{3}\right) + 25 = \dfrac{205}{3} = 68\dfrac{1}{3}$

The smaller angle measures $21\dfrac{2}{3}°$ and the larger angle measures $68\dfrac{1}{3}°$.

48. Let x be the amount of Hazelnut Coffee that Mike uses. Since he is making 10 pounds total, the amount of Cinnamon Coffee that Mike uses is $10 - x$ pounds.

$7.5x + 6.75(10 - x) = 7(10)$

$100[7.5x + 6.75(10 - x)] = 100(70)$

$750x + 675(10 - x) = 7000$

$750x + 6750 - 675x = 7000$

$75x = 250$

$x = \dfrac{10}{3} = 3\dfrac{1}{3}$

$10 - \dfrac{10}{3} = \dfrac{20}{3} = 6\dfrac{2}{3}$

Mike should mix $3\dfrac{1}{3}$ pounds of Hazelnut Coffee and $6\dfrac{2}{3}$ pounds of Cinnamon Coffee.

49. Let x be April's score on her last test.

$\dfrac{82 + 88 + 80 + 95 + x}{5} \geq 85$

$345 + x \geq 425$

$x \geq 80$

April must score at least 80 on the last test to get a B in her Algebra class.

50. $c(x) = 35 + 1.50x$

To find the cost for 12 hours, we find $c(12)$.

$c(12) = 35 + 1.50(12)$

$c(12) = 35 + 18$

$c(12) = 53$

The cost of renting a chainsaw for 12 hours is \$53.

Chapter 8

8.1 Experiencing Algebra the Exercise Way

1. Yes; $2a + 5$ is a polynomial.

3. Yes; $x^2 - 3x + 2$ is a polynomial.

5. No; $x^{1/2} - 6x - 7$ is not a polynomial because $x^{1/2}$ is not a monomial.

7. Yes; $5x^2 + 12xy + 2y^2$ is a polynomial.

9. Yes; $\sqrt{5}x - \sqrt{3}y$ is a polynomial.

11. No; $5\sqrt{x} - 3\sqrt{y}$ is not a polynomial because the first and second addends are not monomials.

13. Yes; $\frac{3}{5}x^3 - \frac{2}{3}x^2 + 4x - \frac{7}{10}$ is a polynomial.

15. No; $\frac{4}{x^2} + \frac{1}{x} - \frac{5}{7}$ is not a polynomial because the first and second addends are not monomials.

17. No; $a^{-2} + 17a^{-1} + 13$ is not a polynomial because the first and second addends are not monomials.

19. Yes; $0.07b^2 - 2.6b + 13.908$ is a polynomial.

21. $3a + 4b - 5c$ has three terms. Therefore, the expression is a trinomial.

23. $2z^2$ has one term. Therefore, it is a monomial.

25. $x - y$ has two terms. Therefore, the expression is a binomial.

27. $4p^4 - 2p^3 + 11p - 57$ has four terms. Therefore, the expression is a polynomial.

29. $6x^2 - 12 + 8x - 5x^2 + x - 17 = x^2 + 9x - 29$
There are three terms. Therefore, the expression is a trinomial.

31. $3b - 4 + 7b = 10b - 4$
There are two terms. Therefore, the expression is a binomial.

33. $x + 2x + 3x + 4x + 5x = 15x$
There is one term. Therefore, it is a monomial.

35. $\frac{1}{2}x + \frac{2}{3}y - \frac{3}{4}z$ has three terms. Therefore, the expression is a trinomial.

37. $2 - 15c$
2 or $2x^0$ has a degree of 0.
$-15c$ or $-15c^1$ has a degree of 1.
The degree of the polynomial is 1.

39. 123 or $123x^0$ has a degree of 0.
The degree of the polynomial is 0.

41. $5 + 5x - 4x - x = 5$
5 or $5x^0$ has a degree of 0.
The degree of the polynomial is 0.

43. $7x^5 + 2x^2y^2 - 12$
$7x^5$ has a degree of 5.
$2x^2y^2$ has a degree of 2 + 2 or 4.
-12 or $-12x^0$ has a degree of 0.
The degree of the polynomial is 5.

45. $\pi r^2 + 2\pi rh$
πr^2 has a degree of 2.
$2\pi rh$ or $2\pi r^1h^1$ has a degree of 1 + 1 or 2.
The degree of the polynomial is 2.

47. $4x^2yz^{12} - 8xy^2z^9 + 3x^3y^3z$
$4x^2yz^{12}$ has a degree of 2 + 1 + 12 or 15.
$-8xy^2z^9$ has a degree of 1 + 2 + 9 or 12.
$3x^3y^3z$ has a degree of 3 + 3 + 1 or 7.
The degree of the polynomial is 15.

49. $5 - 2a + 3a^2 + a^3 = a^3 + 3a^2 - 2a + 5$

51. $\dfrac{3}{5}x + \dfrac{4}{5}x^3 - \dfrac{7}{15}x - \dfrac{8}{15}x^4$

$= -\dfrac{8}{15}x^4 + \dfrac{4}{5}x^3 + \dfrac{9}{15}x - \dfrac{7}{15}x$

$= -\dfrac{8}{15}x^4 + \dfrac{4}{5}x^3 + \dfrac{2}{15}x$

53. $0.1x^3 - 1.72 + 4.6x^2 + 3.06x^4$

$= 3.06x^4 + 0.1x^3 + 4.6x^2 - 1.72$

55. $7b^2 - 6b^3 + 11 - 2b = 11 - 2b + 7b^2 - 6b^3$

57. $\dfrac{2}{7}x + \dfrac{1}{7}x^5 - \dfrac{3}{14}x - \dfrac{5}{14}x^3$

$= \dfrac{4}{14}x - \dfrac{3}{14}x - \dfrac{5}{14}x^3 + \dfrac{1}{7}x^5$

$= \dfrac{1}{14}x - \dfrac{5}{14}x^3 + \dfrac{1}{7}x^5$

59. $0.5x^7 - 2.77 + 3.2x^3 + 9.76x^5$

$= -2.77 + 3.2x^3 + 9.76x^5 + 0.5x^7$

61. a. $3x^2 - 4x + 1 = 3(4)^2 - 4(4) + 1$

$= 48 - 16 + 1$

$= 33$

The result is 33.

b. $3x^2 - 4x + 1 = 3(-2)^2 - 4(-2) + 1$

$= 12 + 8 + 1$

$= 21$

The result is 21.

c. $3x^2 - 4x + 1 = 3(0)^2 - 4(0) + 1$

$= 0 - 0 + 1$

$= 1$

The result is 1.

63. a. $x^2 + 4xy + y^2 = (2)^2 + 4(2)(3) + (3)^2$

$= 4 + 24 + 9$

$= 37$

The result is 37.

b. $x^2 + 4xy + y^2$

$= (-2)^2 + 4(-2)(-3) + (-3)^2$

$= 4 + 24 + 9$

$= 37$

The result is 37.

c. $x^2 + 4xy + y^2 = (0)^2 + 4(0)(0) + (0)^2$

$= 0 + 0 + 0$

$= 0$

The result is 0.

65. a. $1.3m^3 - 2.5m^2 + 3.7m - 4.9$

$= 1.3(1)^3 - 2.5(1)^2 + 3.7(1) - 4.9$

$= 1.3 - 2.5 + 3.7 - 4.9$

$= -2.4$

The result is -2.4.

b. $1.3m^3 - 2.5m^2 + 3.7m - 4.9$

$= 1.3(2.5)^3 - 2.5(2.5)^2 + 3.7(2.5) - 4.9$

$= 20.3125 - 15.625 + 9.25 - 4.9$

$= 9.0375$

The result is 9.0375.

c. $1.3m^3 - 2.5m^2 + 3.7m - 4.9$

$= 1.3(-1.5)^3 - 2.5(-1.5)^2$

$\qquad + 3.7(-1.5) - 4.9$

$= -4.3875 - 5.625 - 5.55 - 4.9$

$= -20.4625$

The result is -20.4625.

67. a. $\dfrac{2}{3}x^2 - x - 3 = \dfrac{2}{3}\left(-\dfrac{3}{2}\right)^2 - \left(-\dfrac{3}{2}\right) - 3$

$= \dfrac{3}{2} + \dfrac{3}{2} - 3$

$= 0$

The result is 0.

b. $\dfrac{2}{3}x^2 - x - 3 = \dfrac{2}{3}\left(\dfrac{1}{4}\right)^2 - \dfrac{1}{4} - 3$

$= \dfrac{1}{24} - \dfrac{1}{4} - 3$

$= -\dfrac{77}{24}$

The result is $-\dfrac{77}{24}$.

c. $\dfrac{2}{3}x^2 - x - 3 = \dfrac{2}{3}(3)^2 - 3 - 3$

      $= 6 - 3 - 3$

      $= 0$

      The result is 0.

69.   Let x = measure of first side
      $2x$ = measure of second side
     $x + 1$ = measure of third side
     $x + 2x + x + 1 = 4x + 1$
     The perimeter measures $4x + 1$ units.
     $4x + 1 = 4(8) + 1 = 33$
     The perimeter measures 33 inches.

71.   Let x = width
      x = length
      3 = height
     $2x(x) + 2(x)(3) + 2(x)(3) = 2x^2 + 12x$
     The surface area measures
     $2x^2 + 12x$ square units.
     $2(1.5)^2 + 12(1.5) = 22.5$
     The surface area measures 22.5 m^2.

73.   Area of $A = x(x)$
     Area of $B = x(12)$
     Area of $C = 3(7)$
     $x(x) + x(12) + 3(7) = x^2 + 12x + 21$
     The total area is $x^2 + 12x + 21$ m^2.

75.   Let l = length of yard
      w = width of yard
     $lw - 15(20) = lw - 300$
     The area not covered by the patio is
     $lw - 300$ ft^2.
     $120(75) - 300 = 8700$
     The area not covered by the patio is
     8700 ft^2.

77.   Let x = length of one leg
      y = length of other leg
     $x^2 + y^2$
     The square of the length of the hypotenuse is
     $x^2 + y^2$.
     $x^2 + y^2 = 4^2 + 7^2 = 65$
     The square of the length of the hypotenuse is
     65 ft^2. The hypotenuse is equal to $\sqrt{65}$ feet.

79.   Let x = length of one side
     $12x^2$
     The replacement cost is $12x^2$ dollars.
     $12x^2 = 12(12)^2 = 1728$
     The replacement cost is \$1728.

81.   Let x = length of one side

     a.   Revenue $= 200 + 12x^2$

     b.   Cost $= 75 + 2.75x^2$

     c. $200 + 12x^2 = 200 + 12(15)^2 = 2900$
        The revenue is \$2900.
        $75 + 2.75x^2 = 75 + 2.75(15)^2 = 693.75$
        The cost is \$693.75.
        $2900 - 693.75 = 2206.25$
        The net return is \$2206.25.

8.1 Experiencing Algebra the Calculator Way

1. $3x^3 - 8x^2 + 9x - 12$
     when $x = \{-3, -1, 1, 3\}$
     The calculator returns the list
     $\{-192, -32, -8, 24\}$.

2. $4a^3 + 2a^2b + 2ab^2 + b^3$ for
     $\{(-5, 2), (-3, 0), (-1, -2)\}$
     The calculator returns the list
     $\{-432, -108, -24\}$.

3. $\dfrac{1}{2}x^2 + \dfrac{3}{5}xy - \dfrac{1}{4}y^3$ for the (x, y) pairs
     $\left(\dfrac{1}{2}, \dfrac{2}{5}\right), \left(\dfrac{1}{4}, \dfrac{3}{5}\right), \left(-\dfrac{1}{2}, 5\right), \left(0, \dfrac{2}{5}\right)$
     The calculator returns the list
     $\{0.229, 0.06725, -32.625, -0.016\}$, or, as
     fractions, $\left\{\dfrac{229}{1000}, \dfrac{269}{4000}, -\dfrac{261}{8}, -\dfrac{2}{125}\right\}$.

8.2 Experiencing Algebra the Exercise Way

1. Possible answer:

x	$y = x^3 + 2x^2 - 5x - 6$	y
-2	$y = (-2)^3 + 2(-2)^2 - 5(-2) - 6 = 4$	4
-1	$y = (-1)^3 + 2(-1)^2 - 5(-1) - 6 = 0$	0
0	$y = (0)^3 + 2(0)^2 - 5(0) - 6 = -6$	-6
1	$y = (1)^3 + 2(1)^2 - 5(1) - 6 = -8$	-8
2	$y = (2)^3 + 2(2)^2 - 5(2) - 6 = 0$	0

3. Possible answer:

x	$y = x^2 + 4x + 1$	y
-2	$y = (-2)^2 + 4(-2) + 1 = -3$	-3
-1	$y = (-1)^2 + 4(-1) + 1 = -2$	-2
0	$y = (0)^2 + 4(0) + 1 = 1$	1
1	$y = (1)^2 + 4(1) + 1 = 6$	6
2	$y = (2)^2 + 4(2) + 1 = 13$	13

5. $Y1 = 2x - 5$

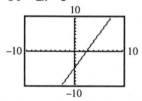

The range is all real numbers.

7. $Y1 = -x^2 + 6x - 4$

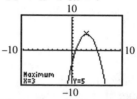

The range is all real numbers less than or equal to 5.

9. $Y1 = 2x^2 - 8x + 3$

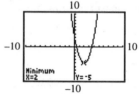

The range is all real numbers greater than or equal to -5.

11. $Y1 = x^3 + 3x^2 - 10x - 24$

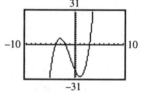

The y-coordinates have no absolute maximum or absolute minimum value. The range is all real numbers.

13. $Y1 = -x^4 - x^3 + 11x^2 + 9x - 18$

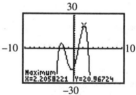

The y-coordinates have an absolute maximum at approximately 21. There is no absolute minimum value. The range is all real numbers less than 21.

15. $f(x) = -2x^3 + x^2 - 5x + 8$

$f(-2) = -2(-2)^3 + (-2)^2 - 5(-2) + 8$

$f(-2) = 16 + 4 + 10 + 8$

$f(-2) = 38$

17. $f(x) = -2x^3 + x^2 - 5x + 8$

$f(2) = -2(2)^3 + (2)^2 - 5(2) + 8$

$f(2) = -16 + 4 - 10 + 8$

$f(2) = -14$

19. $g(x) = 2.7x^3 - 1.5x^2 + 3.5x - 6.7$

$g(-1.7)$

$= 2.7(-1.7)^3 - 1.5(-1.7)^2 + 3.5(-1.7) - 6.7$

$g(-1.7) = -13.2651 - 4.335 - 5.95 - 6.7$

$g(-1.7) = -30.2501$

21. $g(x) = 2.7x^3 - 1.5x^2 + 3.5x - 6.7$

$g(2564.5) = 2.7(2564.5)^3 - 1.5(2564.5)^2$
$\qquad\qquad\quad + 3.5(2564.5) - 6.7$

$g(2564.5) \approx 4.5528 \times 10^{10}$

23. $g(x) = 2.7x^3 - 1.5x^2 + 3.5x - 6.7$

$g(1.5)$

$= 2.7(1.5)^3 - 1.5(1.5)^2 + 3.5(1.5) - 6.7$

$g(1.5) = 9.1125 - 3.375 + 5.25 - 6.7$

$g(1.5) = 4.2875$

25. $g(x) = 2.7x^3 - 1.5x^2 + 3.5x - 6.7$

$g(1.1995) = 2.7(1.1995)^3 - 1.5(1.1995)^2$
$\qquad\qquad\quad + 3.5(1.1995) - 6.7$

$g(1.1995) \approx -0.0001799$

27. $h(x) = \dfrac{1}{2}x^3 - \dfrac{3}{4}x^2 + \dfrac{3}{8}x - \dfrac{5}{8}$

$h\left(-\dfrac{3}{2}\right)$

$= \dfrac{1}{2}\left(-\dfrac{3}{2}\right)^3 - \dfrac{3}{4}\left(-\dfrac{3}{2}\right)^2 + \dfrac{3}{8}\left(-\dfrac{3}{2}\right) - \dfrac{5}{8}$

$h\left(-\dfrac{3}{2}\right) = -\dfrac{27}{16} - \dfrac{27}{16} - \dfrac{9}{16} - \dfrac{10}{16}$

$h\left(-\dfrac{3}{2}\right) = -\dfrac{73}{16}$

29. $h(x) = \dfrac{1}{2}x^3 - \dfrac{3}{4}x^2 + \dfrac{3}{8}x - \dfrac{5}{8}$

$h\left(\dfrac{5}{2}\right) = \dfrac{1}{2}\left(\dfrac{5}{2}\right)^3 - \dfrac{3}{4}\left(\dfrac{5}{2}\right)^2 + \dfrac{3}{8}\left(\dfrac{5}{2}\right) - \dfrac{5}{8}$

$h\left(\dfrac{5}{2}\right) = \dfrac{125}{16} - \dfrac{75}{16} + \dfrac{15}{16} - \dfrac{10}{16}$

$h\left(\dfrac{5}{2}\right) = \dfrac{55}{16}$

31. a. $R(x) = 150x - 5x^2$

$R(5) = 150(5) - 5(5)^2 = 625$

$R(10) = 150(10) - 5(10)^2 = 1000$

$R(15) = 150(15) - 5(15)^2 = 1125$

$R(20) = 150(20) - 5(20)^2 = 1000$

$R(25) = 150(25) - 5(25)^2 = 625$

$R(30) = 150(30) - 5(30)^2 = 0$

When 5, 10, 15, 20, 25, and 30 watches are ordered, the revenue is \$625, \$1000, \$1125, \$1000, \$625, and \$0 respectively.

b. $Y1 = 150x - 5x^2$

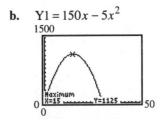

The range is all real numbers less than or equal to 1125.

c. This range shows us that the maximum revenue would be \$1125.

33. $S(x) = -0.011x^2 + 1.91x - 72.54$

$S(63) = -0.011(63)^2 + 1.91(63) - 72.54 = 4.131$

$S(65) = -0.011(65)^2 + 1.91(65) - 72.54 = 5.135$

$S(70) = -0.011(70)^2 + 1.91(70) - 72.54 = 7.26$

$S(75) = -0.011(75)^2 + 1.91(75) - 72.54 = 8.835$

$S(80) = -0.011(80)^2 + 1.91(80) - 72.54 = 9.86$

$S(85) = -0.011(85)^2 + 1.91(85) - 72.54 = 10.335$

Age (x)	63	65	70	75	80	85
Days ($S(x)$)	4.131	5.135	7.26	8.835	9.86	10.335

a. The function predicted a stay of 10 days for an 85-year-old woman. This is not a very good prediction of the actual data of an 8-day stay for an 85-year-old woman.

b. The function predicted a stay of 7 days for a 70-year-old woman. This is a fairly good prediction of the actual data of a 6-day stay, but not as good for a 9-day stay.

c. Answers will vary.

35. $s(t) = -16t^2 + 270t$

$Y1 = -16x^2 + 270x$

1500

The domain is all real numbers between 0 and 8. This means the rocket is actually in the air for 8 seconds before it explodes. The range is all real numbers less than or equal to approximately 1139. That is, the maximum height of the rocket is 1139 ft.

8.2 Experiencing Algebra the Calculator Way

1. The domain is all real numbers greater than or equal to zero. The number of years, x, the salesperson spends in the same territory must be greater than or equal to zero.

2. The range is all real numbers between 0 and approximately 456,518.

$Y1 = -144,100 + 354,050x - 52,176x^2$

500,000

3. $S(x) = -144,100 + 354,050x - 52,176x^2$

$S(4.4)$

$= -144,100 + 354,050(4.4) - 52,176(4.4)^2$

$S(4.4) = 403,592.64$

Her estimated sales are $403,592.64. This is very close to the actual sales of $402,000.

4.

x	Actual Sales	Predicted Sales, $S(x)$	Difference ($)
0.8	$102,000	$105,747.36	–3747.36
1.1	$195,000	$182,222.04	12,777.96
1.4	$265,000	$249,305.04	15,694.96
1.5	$237,000	$269,579	–32,597
2.2	$345,000	$382,278.16	–37,278.16
3.0	$533,000	$448,466	84,534
3.6	$473,000	$454,279.04	18,720.96
4.4	$402,000	$403,592.64	–1592.64
4.5	$298,000	$392,561	–94,561
5.5	$263,000	$224,851	38,149

Answers will vary.

8.3 Experiencing Algebra the Exercise Way

1. $y = 1 - x - x^2 - x^3$ is nonquadratic because $1 - x - x^2 - x^3$ is a third-degree polynomial (x is cubed).

3. $f(x) = 8x + 11$ is nonquadratic because $a = 0$.

5. $g(x) = 0.5x^2 + 2.6x - 8.4$ is quadratic. It is in standard form.

7. $y = -2x^2 - 5x - 7$ is quadratic. It is written in standard form.

9. $a = \pi r^2$ is quadratic. It can be written in standard form $a = \pi r^2 + 0r + 0$.

11. $S = 6e^2$ is quadratic. It can be written in standard form $S = 6e^2 + 0e + 0$.

13. $y = \dfrac{7}{x^2} + 3x - 12$ is nonquadratic because $\dfrac{7}{x^2} + 3x - 12$ is not a polynomial (the squared variable term is in the denominator of a fraction).

15. $y = 2x^2 - 2x + 5$ is quadratic. It is written in standard form.

	function	a	b	c	graph wide/narrow	graph concave upward/ downward	graph vertex	axis of symmetry
17.	$y = -5x^2 + 10x + 1$	-5	10	1	narrow	downward	$(1, 6)$	$x = 1$
19.	$y = 0.6x^2 + 6x - 2$	0.6	6	-2	wide	upward	$(-5, -17)$	$x = -5$
21.	$y = 2x^2 + 3x + 5$	2	3	5	narrow	upward	$(-0.75, 3.875)$	$x = -0.75$
23.	$y = -\frac{1}{4}x^2 + x - 3$	$-\frac{1}{4}$	1	-3	wide	downward	$(2, -2)$	$x = 2$
25.	$f(x) = x^2 + 8x + 1$	1	8	1	neither	upward	$(-4, -15)$	$x = -4$

27. $f(x) = 2x^2 + 5x - 7$

Determine the vertex.

$$x = -\frac{b}{2a} = -\frac{5}{2(2)} = -\frac{5}{4}$$

$$f\left(-\frac{5}{4}\right) = 2\left(-\frac{5}{4}\right)^2 + 5\left(-\frac{5}{4}\right) - 7 = -\frac{81}{8}$$

The vertex is $\left(-\frac{5}{4}, -\frac{81}{8}\right)$.

The axis of symmetry is the line $x = -\frac{5}{4}$.

x	$f(x) = 2x^2 + 5x - 7$	$f(x)$
-3	$f(-3) = 2(-3)^2 + 5(-3) - 7$ $f(-3) = -4$	-4
-2	$f(-2) = 2(-2)^2 + 5(-2) - 7$ $f(-2) = -9$	-9
0	$f(0) = 2(0)^2 + 5(0) - 7$ $f(0) = -7$	-7
1	$f(1) = 2(1)^2 + 5(1) - 7$ $f(1) = 0$	0

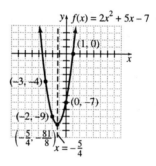

29. $y = \frac{1}{6}x^2 + 3x + 12$

Determine the vertex.

$$x = -\frac{b}{2a} = -\frac{3}{2\left(\frac{1}{6}\right)} = -9$$

$$y = \frac{1}{6}(-9)^2 + 3(-9) + 12 = -\frac{3}{2}$$

The vertex is $\left(-9, -\frac{3}{2}\right)$.

The axis of symmetry is the line $x = -9$.

x	$y = \frac{1}{6}x^2 + 3x + 12$	y
-14	$y = \frac{1}{6}(-14)^2 + 3(-14) + 12$ $y = 2\frac{2}{3}$	$2\frac{2}{3}$
-12	$y = \frac{1}{6}(-12)^2 + 3(-12) + 12$ $y = 0$	0
-6	$y = \frac{1}{6}(-6)^2 + 3(-6) + 12$ $y = 0$	0
-3	$y = \frac{1}{6}(-3)^2 + 3(-3) + 12$ $y = 4\frac{1}{2}$	$4\frac{1}{2}$

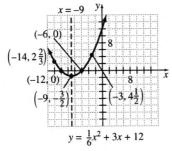

$y = \frac{1}{6}x^2 + 3x + 12$

31. $h(x) = 14 + 5x - x^2$

Determine the vertex.

$$x = -\frac{b}{2a} = -\frac{5}{2(-1)} = \frac{5}{2}$$

$$h\left(\frac{5}{2}\right) = 14 + 5\left(\frac{5}{2}\right) - \left(\frac{5}{2}\right)^2 = \frac{81}{4}$$

The vertex is $\left(\frac{5}{2}, \frac{81}{4}\right)$.

The axis of symmetry is the line $x = \frac{5}{2}$.

x	$h(x) = 14 + 5x - x^2$	$h(x)$
0	$h(0) = 14 + 5(0) - (0)^2$ $h(0) = 14$	14
1	$h(1) = 14 + 5(1) - (1)^2$ $h(1) = 18$	18
4	$h(4) = 14 + 5(4) - (4)^2$ $h(4) = 18$	18
5	$h(5) = 14 + 5(5) - (5)^2$ $h(5) = 14$	14

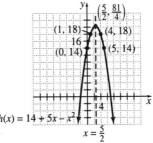

$h(x) = 14 + 5x - x^2$

33. $y = -2x^2 + 8x - 3$

Determine the vertex.

$$x = -\frac{b}{2a} = -\frac{8}{2(-2)} = 2$$

$$y = -2(2)^2 + 8(2) - 3 = 5$$

The vertex is $(2, 5)$.

The axis of symmetry is the line $x = 2$.

x	$y = -2x^2 + 8x - 3$	y
0	$y = -2(0)^2 + 8(0) - 3$ $y = -3$	-3
1	$y = -2(1)^2 + 8(1) - 3$ $y = 3$	3
3	$y = -2(3)^2 + 8(3) - 3$ $y = 3$	3
4	$y = -2(4)^2 + 8(4) - 3$ $y = -3$	-3

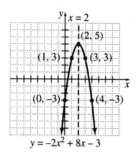

$$y = -2x^2 + 8x - 3$$

35. $g(x) = 0.8x^2 - 1.2x$

Determine the vertex.

$$x = -\frac{b}{2a} = -\frac{(-1.2)}{2(0.8)} = 0.75$$

$g(0.75) = 0.8(0.75)^2 - 1.2(0.75) = -0.45$

The vertex is (0.75, –0.45).

The axis of symmetry is the line $x = 0.75$.

x	$g(x) = 0.8x^2 - 1.2x$	$g(x)$
–1	$g(-1) = 0.8(-1)^2 - 1.2(-1)$ $g(-1) = 2$	2
0	$g(0) = 0.8(0)^2 - 1.2(0)$ $g(0) = 0$	0
2	$g(2) = 0.8(2)^2 - 1.2(2)$ $g(2) = 0.8$	0.8
4	$g(4) = 0.8(4)^2 - 1.2(4)$ $g(4) = 8$	8

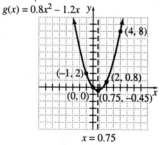

37. $y = 0.4x^2$

Determine the vertex.

$$x = -\frac{b}{2a} = -\frac{0}{2(0.4)} = 0$$

$y = 0.4(0)^2 = 0$

The vertex is (0, 0).

The axis of symmetry is the line $x = 0$.

x	$y = 0.4x^2$	y
–5	$y = 0.4(-5)^2$ $y = 10$	10
–2	$y = 0.4(-2)^2$ $y = 1.6$	1.6
2	$y = 0.4(2)^2$ $y = 1.6$	1.6
5	$y = 0.4(5)^2$ $y = 10$	10

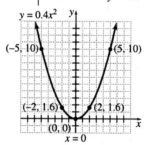

39. $y = 3x^2 - 3$

Determine the vertex.

$$x = -\frac{b}{2a} = -\frac{0}{2(3)} = 0$$

$y = 3(0)^2 - 3 = -3$

The vertex is (0, –3).

The axis of symmetry is the line $x = 0$.

x	$y = 3x^2 - 3$	y
-2	$y = 3(-2)^2 - 3$ $y = 9$	9
-1	$y = 3(-1)^2 - 3$ $y = 0$	0
1	$y = 3(1)^2 - 3$ $y = 0$	0
2	$y = 3(2)^2 - 3$ $y = 9$	9

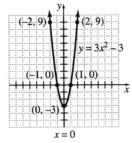

41. $f(x) = -0.5x^2 + 3x$

Determine the vertex.

$$x = -\frac{b}{2a} = -\frac{3}{2(-0.5)} = 3$$

$f(3) = -0.5(3)^2 + 3(3) = 4.5$

The vertex is $(3, 4.5)$.

The axis of symmetry is the line $x = 3$.

x	$f(x) = -0.5x^2 + 3x$	$f(x)$
-2	$f(-2) = -0.5(-2)^2 + 3(-2)$ $f(-2) = -8$	-8
0	$f(0) = -0.5(0)^2 + 3(0)$ $f(0) = 0$	0
6	$f(6) = -0.5(6)^2 + 3(6)$ $f(6) = 0$	0
8	$f(8) = -0.5(8)^2 + 3(8)$ $f(8) = -8$	-8

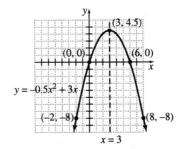

43. $s(t) = -16t^2 + 12t + 24$

$Y1 = -16x^2 + 12x + 24$

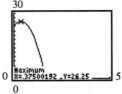

The vertex is approximately $(0.375, 26.25)$.
Therefore, the maximum height is
26.25 feet.
$s(t) = 0$ at $t \approx 1.7$
Therefore, the apple will hit the ground after
approximately 1.7 seconds.

45. $A(x) = 140x - x^2$

$Y1 = 140x - x^2$

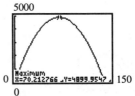

The vertex is $(70, 4900)$.
The room has a maximum area of 4900 ft^2
when the width is 70 feet.

47. $s(x) = 5x^2 - 80x + 400$

$Y1 = 5x^2 - 80x + 400$

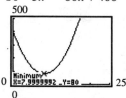

The vertex is (8, 80).
At a height of 8 inches the square of the hypotenuse will be minimized at 80 in^2.

49. $R(x) = 46x - x^2$

$Y1 = 46x - x^2$

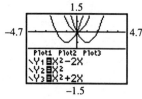

The vertex is (23, 529).
At $x = \$23$ the revenue will be at a maximum of $529.

8.3 Experiencing Algebra the Calculator Way

1. $Y1 = x^2 - 2x$

$Y2 = x^2$

$Y3 = x^2 + 2x$

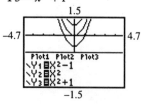

2. $Y1 = x^2 - 1$

$Y2 = x^2$

$Y3 = x^2 + 1$

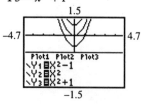

3. $Y1 = 0.3x^2$

$Y2 = x^2$

$Y3 = 3x^2$

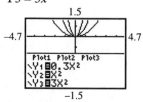

4. $Y1 = -3x^2$

$Y2 = 3x^2$

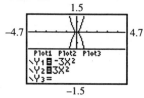

5. $Y1 = 3x^2$

$Y2 = 3x^2 + 1$

$Y3 = 3x^2 - 2$

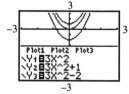

6. $Y1 = x^2 + 1$

$Y2 = x^2 + 2x + 1$

$Y3 = x^2 - 2x + 1$

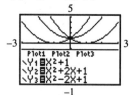

Answers may vary.

Chapter 8 Review

Reflections

1.–7. Answers will vary.

Exercises

1. Yes; x is a monomial.

2. Yes; $5x - 3$ is a binomial.

3. No; $\sqrt{x} + 2$ is not a polynomial because $\sqrt{x}$ is not a monomial.

4. Yes; $3x^3 - 4x^2 + x - 1$ is a polynomial.

5. No; $\dfrac{3}{a} - 2a + 1$ is not a polynomial because $\dfrac{3}{a}$ is not a monomial.

6. Yes; $3a^4 - 2a^2 + 5$ is a trinomial.

7. $5x + 3x^3 - 2$
   $5x$ or $5x^1$ has a degree of 1.
   $3x^3$ has a degree of 3.
   -2 or $-2x^0$ has a degree of 0.
   The degree of the polynomial is 3.

8. $2x^2y + 3xy - 5$
   $2x^2y$ has a degree of $2 + 1$ or 3.
   $3xy$ has a degree of $1 + 1$ or 2.
   -5 or $-5x^0$ has a degree of 0.
   The degree of the polynomial is 3.

9. $x + 9$
   x has a degree of 1.
   9 or $9x^0$ has a degree of 0.
   The degree of the polynomial is 1.

10. $0.5a - 3.1a^2 + 9.6a + 3.1a^2 = 10.1a$
    $10.1a$ has a degree of 1.
    The degree of the polynomial is 1.

11. $12x^2 + 30x + 3$
    $12x^2$ has a degree of 2.
    $30x$ has a degree of 1.
    3 has a degree of 0.
    The degree of the polynomial is 2.

12. $5y^2 + 11y^4 + 12 - 6y + 9y^3$
    $= 11y^4 + 9y^3 + 5y^2 - 6y + 12$

13. $5 - p = -p + 5$

14. $\dfrac{1}{4}z^4 + \dfrac{1}{2}z^2 + z + \dfrac{1}{3}z^3 + 1$
    $= \dfrac{1}{4}z^4 + \dfrac{1}{3}z^3 + \dfrac{1}{2}z^2 + z + 1$

15. $0.6b - 2.3b^5 + 1.8 - 9.1b^3$
    $= -2.3b^5 - 9.1b^3 + 0.6b + 1.8$

16. $2x^3 + 11x^2 - 21x - 90$
    $= 2(3)^3 + 11(3)^2 - 21(3) - 90$
    $= 54 + 99 - 63 - 90$
    $= 0$
    The result is 0.

17. $2x^3 + 11x^2 - 21x - 90$
    $= 2(0)^3 + 11(0)^2 - 21(0) - 90$
    $= 0 + 0 - 0 - 90$
    $= -90$
    The result is -90.

18. $2x^3 + 11x^2 - 21x - 90$
    $= 2(1)^3 + 11(1)^2 - 21(1) - 90$
    $= 2 + 11 - 21 - 90$
    $= -98$
    The result is -98.

19. $2x^3 + 11x^2 - 21x - 90$
    $= 2(-6)^3 + 11(-6)^2 - 21(-6) - 90$
    $= -432 + 396 + 126 - 90$
    $= 0$
    The result is 0.

20. $2x^3 + 11x^2 - 21x - 90$
    $= 2\left(-\dfrac{5}{2}\right)^3 + 11\left(-\dfrac{5}{2}\right)^2 - 21\left(-\dfrac{5}{2}\right) - 90$
    $= -\dfrac{125}{4} + \dfrac{275}{4} + \dfrac{210}{4} - \dfrac{360}{4}$
    $= 0$
    The result is 0.

21. $a^3 + 2a^2b - 3ab^2 - b^3$
$= (0)^3 + 2(0)^2(1) - 3(0)(1)^2 - (1)^3$
$= 0 + 0 - 0 - 1$
$= -1$
The result is -1.

22. $a^3 + 2a^2b - 3ab^2 - b^3$
$= (-1)^3 + 2(-1)^2(0) - 3(-1)(0)^2 - 0^3$
$= -1 + 0 - 0 - 0$
$= -1$
The result is -1.

23. $a^3 + 2a^2b - 3ab^2 - b^3$
$= 0^3 + 2(0)^2(0) - 3(0)(0)^2 - 0^3$
$= 0 + 0 - 0 - 0$
$= 0$
The result is 0.

24. $a^3 + 2a^2b - 3ab^2 - b^3$
$= (1)^3 + 2(1)^2(1) - 3(1)(1)^2 - (1)^3$
$= 1 + 2 - 3 - 1$
$= -1$
The result is -1.

25. $a^3 + 2a^2b - 3ab^2 - b^3$
$= (-1)^3 + 2(-1)^2(1) - 3(-1)(1)^2 - (1)^3$
$= -1 + 2 + 3 - 1$
$= 3$
The result is 3.

26. $a^3 + 2a^2b - 3ab^2 - b^3$
$= (-1)^3 + 2(-1)^2(-1) - 3(-1)(-1)^2 - (-1)^3$
$= -1 - 2 + 3 + 1$
$= 1$
The result is 1.

27. Let w = width
w^2 = length
The perimeter is:
$2w + 2w^2$
$2(7) + 2(7)^2 = 112$
The perimeter is 112 yards.

28. Let x = length of first side
$3x$ = length of second side
$x^2 + 1$ = length of third side
The perimeter is:
$x + 3x + x^2 + 1 = x^2 + 4x + 1$
$(4)^2 + 4(4) + 1 = 33$
The perimeter is 33 inches.

29. Area of square $= a^2$
Area of rectangle $= 17a$
Area of triangle $= \dfrac{1}{2}(6)a = 3a$
The total area is
$a^2 + 17a + 3a = a^2 + 20a$ in^2.

30. Shaded area
= Rectangle area − Triangle area
$= 10x - \dfrac{1}{2}x(x)$
The shaded area is $-\dfrac{1}{2}x^2 + 10x$ in^2.

31. Area of lawn not covered by the garden
= Area of square − area of triangle
$= z^2 - \dfrac{1}{2}yx$
$z^2 - \dfrac{1}{2}xy = (80)^2 - \dfrac{1}{2}(10)(6) = 6370$
The area is 6370 ft^2.

32. Let w = width
l = length

a. $500 + 40lw$

b. $275 + 12lw$

c. Revenue = $500 + 40(25)(15) = \$15,500$
Cost = $275 + 12(25)(15) = \$4775$
Return = $15,500 - 4775 = \$10,725$
The revenue is \$15,500, the cost is \$4775 and the net return is \$10,725.

33.

x	$y = x^3 - x^2 - 6x$	y
-3	$y = (-3)^3 - (-3)^2 - 6(-3) = -18$	-18
-2	$y = (-2)^3 - (-2)^2 - 6(-2) = 0$	0
-1	$y = (-1)^3 - (-1)^2 - 6(-1) = 4$	4
0	$y = 0^3 - 0^2 - 6(0) = 0$	0
1	$y = 1^3 - 1^2 - 6(1) = -6$	-6
2	$y = 2^3 - 2^2 - 6(2) = -8$	-8
3	$y = 3^3 - 3^2 - 6(3) = 0$	0

34. $Y1 = -3x + 5$

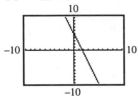

The range is all real numbers.

35. $Y1 = 2x^2 - 2x - 12$

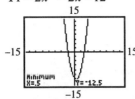

The range is all real numbers greater than or equal to -12.5.

36. $Y1 = x^3 + 2x^2 - 5x - 6$

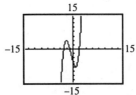

The y-coordinates have no absolute maximum or absolute minimum. Therefore, the range is all real numbers.

37. $Y1 = x^4 + 2x^3 - 5x^2 - 6x$

There is an absolute minimum at approximately $y = -9$. The range is all real numbers greater than or equal to -9.

38. $f(x) = 3x^3 - x^2 + 2x - 4$

$f(-2) = 3(-2)^3 - (-2)^2 + 2(-2) - 4$

$f(-2) = -24 - 4 - 4 - 4$

$f(-2) = -36$

39. $f(x) = 3x^3 - x^2 + 2x - 4$

$f(0) = 3(0)^3 - 0^2 + 2(0) - 4$

$f(0) = 0 - 0 + 0 - 4$

$f(0) = -4$

40. $f(x) = 3x^3 - x^2 + 2x - 4$

$f(2) = 3(2)^3 - 2^2 + 2(2) - 4$

$f(2) = 24 - 4 + 4 - 4$

$f(2) = 20$

41. $f(x) = 3x^3 - x^2 + 2x - 4$

$f\left(-\dfrac{1}{2}\right) = 3\left(-\dfrac{1}{2}\right)^3 - \left(-\dfrac{1}{2}\right)^2 + 2\left(-\dfrac{1}{2}\right) - 4$

$f\left(-\dfrac{1}{2}\right) = -\dfrac{3}{8} - \dfrac{1}{4} - 1 - 4$

$f\left(-\dfrac{1}{2}\right) = -\dfrac{3}{8} - \dfrac{2}{8} - \dfrac{8}{8} - \dfrac{32}{8}$

$f\left(-\dfrac{1}{2}\right) = -\dfrac{45}{8}$

42. $f(x) = 3x^3 - x^2 + 2x - 4$

$f(1.7) = 3(1.7)^3 - (1.7)^2 + 2(1.7) - 4$

$f(1.7) = 14.739 - 2.89 + 3.4 - 4$

$f(1.7) = 11.249$

43. a. $R(x) = 12x - 0.5x^2$

$R(5) = 12(5) - 0.5(5)^2 = 47.5$

$R(10) = 12(10) - 0.5(10)^2 = 70$

$R(15) = 12(15) - 0.5(15)^2 = 67.5$

$R(20) = 12(20) - 0.5(20)^2 = 40$

$R(25) = 12(25) - 0.5(25)^2 = -12.5$

The revenue for 5, 10, 15, 20, and 25 shirts is $47.50, $70, $67.50, $40, and $-$12.50, respectively.
Answers will vary.

b. $C(x) = 4x$
$C(5) = 4(5) = 20$
$C(10) = 4(10) = 40$
$C(15) = 4(15) = 60$
$C(20) = 4(20) = 80$
The cost for 5, 10, 15, and 20 shirts is $20, $40, $60, and $80, respectively.

c. Profit from
5 shirts: $47.50 - 20 = 27.50$
10 shirts: $70 - 40 = 30$
15 shirts: $67.50 - 60 = 7.50$
20 shirts: $40 - 80 = -40$
The profit from 5, 10, 15, and 20 shirts is $27.50, $30, $7.50, and $-$40, respectively.

d. Answers will vary.

44. $y = 15,800 + 2.2x - 0.001x^2$

Gallons (x)	900	1000	1100	1200	1300	1400
Cost (y)	16,970	17,000	17,010	17,000	16,970	16,920

Answers will vary.

45. $y = x^2 + x + 1$ is quadratic.

46. $y = x^3 - x - 1$ is nonquadratic because $x^3 - x - 1$ is a third-degree polynomial (x is cubed).

47. $y = \dfrac{5}{x^2} + x + 1$ is nonquadratic because $\dfrac{5}{x^2} + x + 1$ is not a polynomial (the squared variable term is in the denominator of a fraction).

48. $y = x^2 + 4x + 4$.

	function	a	b	c	graph wide/narrow	graph concave upward/downward	graph vertex	axis of symmetry
49.	$y = -\frac{1}{4}x^2 + \frac{1}{2}x + 1$	$-\frac{1}{4}$	$\frac{1}{2}$	1	wide	downward	$\left(1, \frac{5}{4}\right)$	$x = 1$
50.	$f(x) = -2x^2 + 4x$	-2	4	0	narrow	downward	$(1, 2)$	$x = 1$
51.	$g(x) = \frac{1}{3}x^2 + x$	$\frac{1}{3}$	1	0	wide	upward	$\left(-\frac{3}{2}, -\frac{3}{4}\right)$	$x = -\frac{3}{2}$
52.	$y = 3x^2 - 3x + 1$	3	-3	1	narrow	upward	$\left(\frac{1}{2}, \frac{1}{4}\right)$	$x = \frac{1}{2}$

53. $f(x) = x^2 + 2x - 8$
Determine the vertex.
$$x = -\frac{b}{2a} = -\frac{2}{2(1)} = -1$$
$$f(-1) = (-1)^2 + 2(-1) - 8 = -9$$
The vertex is $(-1, -9)$.
The axis of symmetry is the line $x = -1$.

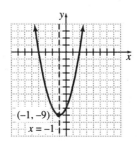

54. $y = -\dfrac{1}{2}x^2 + x - 2$
Determine the vertex.
$$x = -\frac{b}{2a} = -\frac{1}{2\left(-\frac{1}{2}\right)} = 1$$
$$y = -\frac{1}{2}(1)^2 + 1 - 2 = -1\frac{1}{2}$$
The vertex is $\left(1, -1\frac{1}{2}\right)$.
The axis of symmetry is the line $x = 1$.

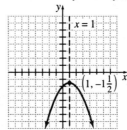

55. $h(x) = 2x^2 - 8$
Determine the vertex.
$$x = -\frac{b}{2a} = -\frac{0}{2(2)} = 0$$
$$h(0) = 2(0)^2 - 8 = -8$$
The vertex is $(0, -8)$.
The axis of symmetry is the line $x = 0$.

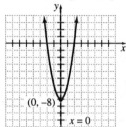

56. $A(w) = w^2 + 8w$

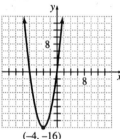

The vertex is $(-4, -16)$.
No; the vertex has no physical meaning.

57. $R(x) = 30x - 0.50x^2$

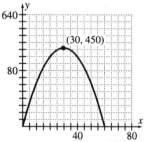

The vertex is $(30, 450)$.
The revenue is at a maximum of $450 when 30 photos are ordered.

58. $s(t) = -16t^2 + 60t + 120$
Determine the vertex.
$$x = -\frac{b}{2a} = -\frac{60}{2(-16)} = 1.875$$
$$s(1.875) = -16(1.875)^2 + 60(1.875) + 120$$
$$s(1.875) = 176.25$$
The vertex is $(1.875, 176.25)$.
The egg reaches a maximum height of 176.25 feet at 1.875 seconds.
$$Y1 = -16x^2 + 60x + 120$$
$$Y2 = 0$$

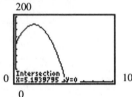

The egg will reach the ground in approximately 5.19 seconds.

Chapter 8 Mixed Review

1. $2x^3 - 3x^2 - 29x - 30$
$= 2(5)^3 - 3(5)^2 - 29(5) - 30$
$= 250 - 75 - 145 - 30$
$= 0$
The result is 0.

2. $2x^3 - 3x^2 - 29x - 30$
$= 2(0)^3 - 3(0)^2 - 29(0) - 30$
$= 0 - 0 - 0 - 30$
$= -30$
The result is -30.

3. $2x^3 - 3x^2 - 29x - 30$
$= 2(1)^3 - 3(1)^2 - 29(1) - 30$
$= 2 - 3 - 29 - 30$
$= -60$
The result is -60.

4. $2x^3 - 3x^2 - 29x - 30$
$= 2(-2)^3 - 3(-2)^2 - 29(-2) - 30$
$= -16 - 12 + 58 - 30$
$= 0$
The result is 0.

5. $2x^3 - 3x^2 - 29x - 30$

$$= 2\left(-\frac{3}{2}\right)^3 - 3\left(-\frac{3}{2}\right)^2 - 29\left(-\frac{3}{2}\right) - 30$$

$$= -\frac{27}{4} - \frac{27}{4} + \frac{174}{4} - \frac{120}{4}$$

$$= 0$$

The result is 0.

6. $2a^3 + 4a^2b - 2ab^2 + b^3$

$$= 2(-1)^3 + 4(-1)^2(0) - 2(-1)(0)^2 + 0^3$$

$$= -2 + 0 - 0 + 0$$

$$= -2$$

The result is -2.

7. $2a^3 + 4a^2b - 2ab^2 + b^3$

$$= 2(1)^3 + 4(1)^2(1) - 2(1)(1)^2 + 1^3$$

$$= 2 + 4 - 2 + 1$$

$$= 5$$

The result is 5.

8. $2a^3 + 4a^2b - 2ab^2 + b^3$

$$= 2(-1)^3 + 4(-1)^2(1) - 2(-1)(1)^2 + 1^3$$

$$= -2 + 4 + 2 + 1$$

$$= 5$$

The result is 5.

9. $2a^3 + 4a^2b - 2ab^2 + b^3$

$$= 2(-1)^3 + 4(-1)^2(-1) - 2(-1)(-1)^2 + (-1)^3$$

$$= -2 - 4 + 2 - 1$$

$$= -5$$

The result is -5.

10. a. binomial; degree of each term is 0, 2; degree of polynomial is 2; $3x^2 + 5$

   b. polynomial; degree of each term is 2, 3, 0, 1; degree of polynomial is 3; $-5a^3 + 15a^2 + a + 4$

   c. polynomial; degree of each term is 4, 1, 0, 5, 2; degree of polynomial is 5; $x^5 + 5x^4 - 3x^2 + x - 2$

11. a. $3b^2 + 13b - 4$; trinomial; degree of each term is 2, 1, 0; degree of polynomial is 2.

   b. $3x^2y - 4xy + 3xy^2 + 5 - 4x^2y^2$; polynomial; degree of each term is 3, 2, 3, 0, 4; degree of polynomial is 4.

   c. $6xyz$; monomial; degree of term is 3; degree of polynomial is 3.

12. $f(x) = 2x^3 - 3x^2 - 23x + 12$

$$f(-3) = 2(-3)^3 - 3(-3)^2 - 23(-3) + 12$$

$$f(-3) = -54 - 27 + 69 + 12$$

$$f(-3) = 0$$

The result is 0.

13. $f(x) = 2x^3 - 3x^2 - 23x + 12$

$$f(0) = 2(0)^3 - 3(0)^2 - 23(0) + 12$$

$$f(0) = 0 - 0 - 0 + 12$$

$$f(0) = 12$$

The result is 12.

14. $f(x) = 2x^3 - 3x^2 - 23x + 12$

$$f(4) = 2(4)^3 - 3(4)^2 - 23(4) + 12$$

$$f(4) = 128 - 48 - 92 + 12$$

$$f(4) = 0$$

The result is 0.

15. $f(x) = 2x^3 - 3x^2 - 23x + 12$

$$f\left(\frac{1}{2}\right) = 2\left(\frac{1}{2}\right)^3 - 3\left(\frac{1}{2}\right)^2 - 23\left(\frac{1}{2}\right) + 12$$

$$f\left(\frac{1}{2}\right) = \frac{1}{4} - \frac{3}{4} - \frac{46}{4} + \frac{48}{4}$$

$$f\left(\frac{1}{2}\right) = 0$$

The result is 0.

16. $f(x) = 2x^3 - 3x^2 - 23x + 12$

$$f(2.2) = 2(2.2)^3 - 3(2.2)^2 - 23(2.2) + 12$$

$$f(2.2) = 21.296 - 14.52 - 50.6 + 12$$

$$f(2.2) = -31.824$$

The result is -31.824.

17.

x	$y = 2x^3 + 2x^2 - 12x$	y
–3	$y = 2(-3)^3 + 2(-3)^2 - 12(-3) = 0$	0
–2	$y = 2(-2)^3 + 2(-2)^2 - 12(-2) = 16$	16
–1	$y = 2(-1)^3 + 2(-1)^2 - 12(-1) = 12$	12
0	$y = 2(0)^3 + 2(0)^2 - 12(0) = 0$	0
1	$y = 2(1)^3 + 2(1)^2 - 12(1) = -8$	–8
2	$y = 2(2)^3 + 2(2)^2 - 12(2) = 0$	0
3	$y = 2(3)^3 + 2(3)^2 - 12(3) = 36$	36

18. $Y1 = 4x - 2$

The range is all real numbers.

19. $Y1 = 3x^2 + 3x - 6$

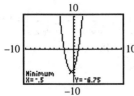

The range is all real numbers greater than or equal to –6.75.

20. $Y1 = x^3 - 3x^2 - 13x + 15$

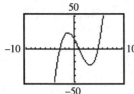

The y-coordinates have no absolute

maximum and no absolute minimum. The range is all real numbers.

21. $Y1 = x^4 - 3x^3 - 13x^2 + 15x$

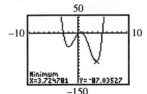

The y-coordinates have an absolute minimum at approximately –87.04. The range is all real numbers greater than or equal to –87.04.

22. $f(x) = 2x^2 + 7x - 4$

Determine the vertex.

$$x = -\frac{b}{2a} = -\frac{7}{2(2)} = -\frac{7}{4}$$

$$f\left(-\frac{7}{4}\right) = 2\left(-\frac{7}{4}\right)^2 + 7\left(-\frac{7}{4}\right) - 4 = -\frac{81}{8}$$

The vertex is $\left(-\frac{7}{4}, -\frac{81}{8}\right)$.

The axis of symmetry is the line $x = -\frac{7}{4}$.

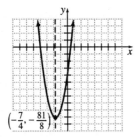

$\left(-\frac{7}{4}, -\frac{81}{8}\right)$

23. $y = \frac{1}{4}x^2 - x + 3$

Determine the vertex.

$x = -\dfrac{b}{2a} = -\dfrac{(-1)}{2\left(\frac{1}{4}\right)} = 2$

$y = \dfrac{1}{4}(2)^2 - 2 + 3 = 2$

The vertex is (2, 2).
The axis of symmetry is the line $x = 2$.

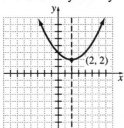

(2, 2)

24. $h(x) = -x^2 + 9$

Determine the vertex.

$x = -\dfrac{b}{2a} = -\dfrac{0}{2(-1)} = 0$

$h(0) = -0^2 + 9 = 9$

The vertex is (0, 9).
The axis of symmetry is the line $x = 0$.

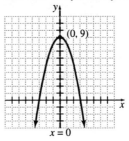

(0, 9)

$x = 0$

	function	a	b	c	graph wide/narrow	graph concave upward/downward	graph vertex	axis of symmetry
25.	$y = \frac{1}{3}x^2 + \frac{2}{3}x + 1$	$\frac{1}{3}$	$\frac{2}{3}$	1	wide	upward	$\left(-1, \frac{2}{3}\right)$	$x = -1$
26.	$f(x) = -3x^2 + 6x$	-3	6	0	narrow	downward	(1, 3)	$x = 1$
27.	$y = -\frac{1}{4}x^2 + x + 3$	$-\frac{1}{4}$	1	3	wide	downward	(2, 4)	$x = 2$
28.	$g(x) = 2x^2 + 4x - 6$	2	4	-6	narrow	upward	$(-1, -8)$	$x = -1$

29. Let w = width

$w^3 + 5$ = length

The perimeter is:

$2w + 2(w^3 + 5) = 2w^3 + 2w + 10$

The polynomial for the perimeter is
$2w^3 + 2w + 10$.

$2(3)^3 + 2(3) + 10 = 70$

The perimeter is 70 feet.

30. Let x = length of first side

$2x - 3$ = length of second side

$x^2 - 7$ = length of third side

The perimeter is:

$x + 2x - 3 + x^2 - 7 = x^2 + 3x - 10$

The polynomial for the perimeter is
$x^2 + 3x - 10$.

$(10)^2 + 3(10) - 10 = 120$

The perimeter is 120 cm.

31. Area of first $= \dfrac{1}{2}(5)x$

Area of second $= x(x + 2)$

Area of third $= \dfrac{1}{2}x(5 + 8)$

Total area is:

$\dfrac{5}{2}x + x(x + 2) + \dfrac{13}{2}x = x^2 + 11x$

The polynomial for the total area is
$x^2 + 11x$ square units.

32. Shaded area

= Area of rectangle + area of semicircle

$= 3x(x) + \dfrac{1}{2}\pi\left(\dfrac{x}{2}\right)^2$

$= 3x^2 + \dfrac{\pi}{8}x^2$

$= \dfrac{24 + \pi}{8}x^2$

The polynomial for the shaded area is
$\dfrac{24 + \pi}{8}x^2$ square units.

33. Area of yard not covered by pool

= Area of rectangle – area of circle

$= xy - \pi z^2$

The polynomial is $xy - \pi z^2$.

$xy - \pi z^2 = 80(50) - \pi(8)^2$

$= 4000 - 64\pi$

≈ 3798.9

The area measures approximately 3798.9 ft^2.

34. a. Let l = length

w = width

The cost is $1.5lw$ dollars.

b. $800 - 1.5lw$ dollars

c. $800 - 1.5(15)(10) = 575$

The net return will be $575.

35. a.

x	$P(x) = 10x + 2x^2$	$P(x)$
0	$P(0) = 10(0) + 2(0)^2 = 0$	0
1	$P(1) = 10(1) + 2(1)^2 = 12$	12
2	$P(2) = 10(2) + 2(2)^2 = 28$	28
3	$P(3) = 10(3) + 2(3)^2 = 48$	48
4	$P(4) = 10(4) + 2(4)^2 = 72$	72
5	$P(5) = 10(5) + 2(5)^2 = 100$	100
6	$P(6) = 10(6) + 2(6)^2 = 132$	132
7	$P(7) = 10(7) + 2(7)^2 = 168$	168

b. Answers will vary.

36. a. $y = -0.25 + 0.5x + 0.0082z - 0.0000081z^2$

$y = -0.25 + 0.5(3.71) + 0.0082(750) - 0.0000081(750)^2$
$y = 3.19875$
The predicted GPA is 3.19875.

b.

GMAT score, z	500	550	600	650	700	750
GPA in MBA, y	3.575	3.55975	3.504	3.40775	3.271	3.09375

c. As the entrance exam score goes up, the student's predicted performance goes down.

37. $s(t) = -0.8t^2 + 1500$

$Y1 = -0.8x^2 + 1500$
$Y2 = 0$

The object will reach the ground in approximately 43.3 seconds.

Chapter 8 Test

1. $123x^2y^3z$ is a monomial because it has one term.

2. $3a^3 + 5a^2b + 7ab^2 + 9b^3$ is a polynomial because it has 4 terms.

3. $2 - c$ is a binomial because it has 2 terms.

4. The term with the largest degree is $-0.5x^5$. Therefore, the degree of the polynomial is 5.

5. The term with the largest degree is $5x^2y^3$. Therefore the degree of the polynomial is $2 + 3$ or 5.

6. $15 + 3x^4 - 7x + x^5 + 9x^2 + 21x$
$= x^5 + 3x^4 + 9x^2 + 14x + 15$

7. $a - \dfrac{2}{3}a^3 - \dfrac{5}{6}a^2 - \dfrac{4}{9} = -\dfrac{4}{9} + a - \dfrac{5}{6}a^2 - \dfrac{2}{3}a^3$

8. $a^2 + 3ab^3 - 7b^2 - b - 6$
$= (0)^2 + 3(0)(3)^3 - 7(3)^2 - 3 - 6$
$= 0 + 0 - 63 - 3 - 6$
$= -72$
The result is -72.

9. $a^2 + 3ab^3 - 7b^2 - b - 6$
$= (-2)^2 + 3(-2)(0)^3 - 7(0)^2 - 0 - 6$
$= 4 - 0 - 0 - 0 - 6$
$= -2$
The result is -2.

10. $a^2 + 3ab^3 - 7b^2 - b - 6$
$= (2)^2 + 3(2)(-3)^3 - 7(-3)^2 - (-3) - 6$
$= 4 - 162 - 63 + 3 - 6$
$= -224$
The result is -224.

11. Let l = length
$\qquad w$ = width
The total cost is $16.50 multiplied by the area and added to $75.
The total cost will be $16.5lw + 75$ dollars.
$$16.5(5)\left(7\dfrac{1}{3}\right) + 75 = \$680$$
The total cost will be $680.

12. $g(x) = 3x^2 + 7x - 6$
$g(-3) = 3(-3)^2 + 7(-3) - 6$
$g(-3) = 27 - 21 - 6$
$g(-3) = 0$
The result is 0.

13. $g(x) = 3x^2 + 7x - 6$
$g(0) = 3(0)^2 + 7(0) - 6$
$g(0) = 0 + 0 - 6$
$g(0) = -6$
The result is -6.

14. $g(x) = 3x^2 + 7x - 6$
$g(1) = 3(1)^2 + 7(1) - 6$
$g(1) = 3 + 7 - 6$
$g(1) = 4$
The result is 4.

15. $y = \dfrac{1}{2}x^2 - 2x - 6$

16. $x = -\dfrac{b}{2a}$
$x = -\dfrac{(-2)}{2\left(\frac{1}{2}\right)}$
$x = 2$
$y = \dfrac{1}{2}(2)^2 - 2(2) - 6$
$y = 2 - 4 - 6$
$y = -8$
The vertex is $(2, -8)$.

17. The y-coordinates have an absolute minimum at -8. The range is all real numbers greater than or equal to -8.

18. Yes; the graph of the relation passes the vertical line test.

	function	a	b	c	graph wide/narrow	graph concave upward/downward	graph vertex	axis of symmetry
19.	$y = \frac{1}{2}x^2 + 2x + 3$	$\frac{1}{2}$	2	3	wide	upward	$(-2, 1)$	$x = -2$
20.	$y = 3x^2 - 3x + \frac{1}{4}$	3	–3	$\frac{1}{4}$	narrow	upward	$\left(\frac{1}{2}, -\frac{1}{2}\right)$	$x = \frac{1}{2}$

21. Answers will vary.

Chapter 9

9.1 Experiencing Algebra the Exercise Way

1. $4^3 a^2 = 4 \cdot 4 \cdot 4 \cdot a \cdot a$

3. $(-3x)^4 = (-3x)(-3x)(-3x)(-3x)$

5. $a^3 b^0 c^5 = a^3 \cdot 1 \cdot c^5 = a \cdot a \cdot a \cdot c \cdot c \cdot c \cdot c \cdot c$

7. $\left(\dfrac{3x}{4}\right)^3 = \left(\dfrac{3x}{4}\right)\left(\dfrac{3x}{4}\right)\left(\dfrac{3x}{4}\right)$

9. $5(x+y)^2 = 5(x+y)(x+y)$

11. $p^{-3} = \dfrac{1}{p^3}$

13. $\dfrac{1}{q^{-5}} = q^5$

15. $\dfrac{p^{-3}}{q^{-5}} = \dfrac{q^5}{p^3}$

17. $p^{-3} q^5 = \dfrac{q^5}{p^3}$

19. $\dfrac{4^3 m^{-2}}{3^4 n^{-3}} = \dfrac{4^3 n^3}{3^4 m^2} = \dfrac{64 n^3}{81 m^2}$

21. $\dfrac{4^{-3} m^{-2}}{3^{-4} n^{-3}} = \dfrac{3^4 n^3}{4^3 m^2} = \dfrac{81 n^3}{64 m^2}$

23. $\dfrac{-4(m-n)^{-1}}{5(m+n)^{-2}} = \dfrac{-4(m+n)^2}{5(m-n)}$

25. $\dfrac{4^{-1}(m-n)}{5^{-1}(m+n)^2} = \dfrac{5(m-n)}{4(m+n)^2}$

27. $a^{-3} + 2a^{-2} - 3a^{-1} + 4a^0 = \dfrac{1}{a^3} + \dfrac{2}{a^2} - \dfrac{3}{a} + 4$

29. $x^5 \cdot x^8 = x^{5+8} = x^{13}$

31. $y^{-5} \cdot y^{13} = y^{-5+13} = y^8$

33. $z^{-9} \cdot z^4 = z^{-9+4} = z^{-5} = \dfrac{1}{z^5}$

35. $p^{-2} \cdot p^{-7} = p^{-2+(-7)} = p^{-9} = \dfrac{1}{p^9}$

37. $(x+y)^4 (x+y)^{-4} = (x+y)^{4+(-4)}$
$= (x+y)^0 = 1$

39. $(x+3)^2 (x+3) = (x+3)^{2+1} = (x+3)^3$

41. $\dfrac{p^{11}}{p^6} = p^{11-6} = p^5$

43. $\dfrac{y^{-4}}{y^{-5}} = y^{-4-(-5)} = y^1 = y$

45. $\dfrac{b^6}{b^8} = b^{6-8} = b^{-2} = \dfrac{1}{b^2}$

47. $\dfrac{(2x-3)^8}{(2x-3)^3} = (2x-3)^{8-3} = (2x-3)^5$

49. $\dfrac{(p+q)^2}{(p+q)^{-1}} = (p+q)^{2-(-1)} = (p+q)^3$

51. $\dfrac{(4-x)^{-3}}{(4-x)^2}$
$= (4-x)^{-3-2}$
$= (4-x)^{-5}$
$= \dfrac{1}{(4-x)^5}$

53. $\dfrac{(z-5)^{-4}}{(z-5)^{-1}}$

$= (z-5)^{-4-(-1)}$

$= (z-5)^{-3}$

$= \dfrac{1}{(z-5)^3}$

55. $(a^5)^6 = a^{5\cdot6} = a^{30}$

57. $(c^{-4})^{-2} = c^{(-4)(-2)} = c^8$

59. $[(x+y)^3]^2 = (x+y)^{3\cdot2} = (x+y)^6$

61. $[(a-b)^4]^{-1}$

$= (a-b)^{4(-1)}$

$= (a-b)^{-4}$

$= \dfrac{1}{(a-b)^4}$

63. $(x^2)^0 = x^{2\cdot0} = x^0 = 1$

65. $(5m)^4 = 5^4 m^4 = 625m^4$

67. $\left(\dfrac{b}{d}\right)^4 = \dfrac{b^4}{d^4}$

69. $(pq)^{21} = p^{21}q^{21}$

71. $\left(\dfrac{3b}{c}\right)^4 = \dfrac{3^4 b^4}{c^4} = \dfrac{81b^4}{c^4}$

73. $(-2c)^6 = (-2)^6 c^6 = 64c^6$

75. $\left(\dfrac{4x}{y}\right)^{-3} = \left(\dfrac{y}{4x}\right)^3 = \dfrac{y^3}{4^3 x^3} = \dfrac{y^3}{64x^3}$

77. $\left(\dfrac{-3p}{q}\right)^{-4} = \left(\dfrac{q}{-3p}\right)^4 = \dfrac{q^4}{(-3)^4 p^4} = \dfrac{q^4}{81p^4}$

79. $(p^5 q^7)^3 = p^{5\cdot3}q^{7\cdot3} = p^{15}q^{21}$

81. $[(2a)^2]^5$

$= (2a)^{2\cdot5}$

$= (2a)^{10}$

$= 2^{10}a^{10}$

$= 1024a^{10}$

83. $\left[\left(\dfrac{x}{2y}\right)^2\right]^3$

$= \left(\dfrac{x}{2y}\right)^{2\cdot3}$

$= \left(\dfrac{x}{2y}\right)^6$

$= \dfrac{x^6}{2^6 y^6}$

$= \dfrac{x^6}{64y^6}$

85. Let x = length of original side
    $5x$ = length of enlarged side
The original area is x^2 square units.
The enlarged area is $(5x)^2 = 5^2 x^2 = 25x^2$ square units.
The enlarged area is 25 times bigger.
$x^2 = 6^2 = 36$
$25x^2 = 25(6)^2 = 25(36) = 900$
If the original side is 6 feet, the original area and enlarged area are 36 ft^2 and 900 ft^2, respectively.

87. Let x = length of original side
    $4x$ = length of enlarged side
The volume of the original bin is x^3 cubic units.
The volume of the enlarged bin is
$(4x)^3 = 4^3 x^3 = 64x^3$ cubic units.
The volume of the enlarged bin is 64 times greater.
$x^3 = (1.5)^3 = 3.375$
$64x^3 = 64(3.375) = 216$
If the original side measures 1.5 feet, then the volumes of the original bin and enlarged bin are 3.375 ft^3 and 216 ft^3, respectively.

9.1 Experiencing Algebra the Calculator Way

1. $x^3 \cdot x^2 = x^6$
   $Y1 = x^3 \cdot x^2$
   $Y2 = x^6$

   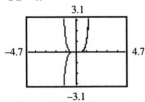

   They are not equivalent.

2. $x^3 \cdot x^2 = x^5$
   $Y1 = x^3 \cdot x^2$
   $Y2 = x^5$

   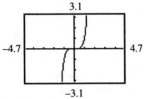

   They are equivalent.

3. $\dfrac{x^3}{x^{-2}} = x^5$

   $Y1 = \dfrac{x^3}{x^{-2}}$

   $Y2 = x^5$

   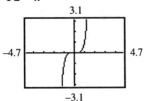

   They are equivalent.

4. $\dfrac{x^3}{x^{-2}} = x$

   $Y1 = \dfrac{x^3}{x^{-2}}$

   $Y2 = x$

   They are not equivalent.

5. $\dfrac{x^3}{x^2} = x^5$

   $Y1 = \dfrac{x^3}{x^2}$

   $Y2 = x^5$

   They are not equivalent.

6. $\dfrac{x^3}{x^2} = x$

   $Y1 = \dfrac{x^3}{x^2}$

   $Y2 = x$

   They are equivalent.

9.2 Experiencing Algebra the Exercise Way

1.
$$9x^2 - 17x + 31$$
$$2x^4 \qquad + 3x^2 \qquad\quad + 12$$
$$5x^3 \qquad\quad - 17x + 11$$
$$\underline{2x^4 + 4x^3 - 7x^2 - \;8x - 26}$$
$$4x^4 + 9x^3 + 5x^2 - 42x + 28$$

3. $(5x^4 + 6x + 3x^3 - 2x^2 - 12) + (4x^4 + 21 - 8x^2 - 9x)$
$$= 5x^4 + 6x + 3x^3 - 2x^2 - 12 + 4x^4 + 21 - 8x^2 - 9x$$
$$= 9x^4 + 3x^3 - 10x^2 - 3x + 9$$

5. $(5x^2 y - 3xy^2 + 6y^3) + (15x^3 - 8x^2 y + 3xy^2)$
$$= 5x^2 y - 3xy^2 + 6y^3 + 15x^3 - 8x^2 y + 3xy^2$$
$$= -3x^2 y + 6y^3 + 15x^3$$

7. $(6 - 7a^3 + 3a^2 - 5a) + (6a + 8a^3 + 2) + (5a^2 - 8a - 9)$
$$= 6 - 7a^3 + 3a^2 - 5a + 6a + 8a^3 + 2 + 5a^2 - 8a - 9$$
$$= a^3 + 8a^2 - 7a - 1$$

9. $\left(\dfrac{2}{3}y^4 + \dfrac{1}{6}y^3 + 3y^2 - \dfrac{1}{3}y + \dfrac{5}{9}\right) + \left(\dfrac{7}{3}y^3 - \dfrac{8}{9}y^2 + \dfrac{5}{6}y - 3\right)$
$$= \dfrac{2}{3}y^4 + \dfrac{1}{6}y^3 + 3y^2 - \dfrac{1}{3}y + \dfrac{5}{9} + \dfrac{7}{3}y^3 - \dfrac{8}{9}y^2 + \dfrac{5}{6}y - 3$$
$$= \dfrac{2}{3}y^4 + \left(\dfrac{1}{6} + \dfrac{7}{3}\right)y^3 + \left(3 - \dfrac{8}{9}\right)y^2 + \left(-\dfrac{1}{3} + \dfrac{5}{6}\right)y + \left(\dfrac{5}{9} - 3\right)$$
$$= \dfrac{2}{3}y^4 + \dfrac{5}{2}y^3 + \dfrac{19}{9}y^2 + \dfrac{1}{2}y - \dfrac{22}{9}$$

11. $(12.07x^3 + 8.6x^2 - 3.19x + 14) + (6.7x^3 - 9.83x^2 + 7x - 4.265)$
$$= 12.07x^3 + 8.6x^2 - 3.19x + 14 + 6.7x^3 - 9.83x^2 + 7x - 4.265$$
$$= 18.77x^3 - 1.23x^2 + 3.81x + 9.735$$

13.
$$4756a^3 - \;3219a^2 - 1816a + 2083$$
$$\underline{361a^3 + 54217a^2 + \quad 0a + \quad 12}$$
$$5117a^3 + 50998a^2 - 1816a + 2095$$

15. $(3a + 4b) + (5c + 6d)$
$$= 3a + 4b + 5c + 6d$$

17. $(5z^3 + 27z^2 - 35z + 42) - (3z^3 + 16z - 72)$
$$= 5z^3 + 27z^2 - 35z + 42 - 3z^3 - 16z + 72$$
$$= 2z^3 + 27z^2 - 51z + 114$$

19. $(a^5 - 9) - (a^5 - a^4 + a^3 - a^2 + a - 9)$
$$= a^5 - 9 - a^5 + a^4 - a^3 + a^2 - a + 9$$
$$= a^4 - a^3 + a^2 - a$$

21. $(16x^2 - 32 + 9x) - (12x + 7 + 9x^2)$
$= 16x^2 - 32 + 9x - 12x - 7 - 9x^2$
$= 7x^2 - 3x - 39$

23. $(13a^3 - 6a^2 + 11) - (12a - 3 + 18a^3)$
$= 13a^3 - 6a^2 + 11 - 12a + 3 - 18a^3$
$= -5a^3 - 6a^2 - 12a + 14$

25.
$$42x^3 + 17x^2y + 3xy^2 + 23y^3$$
$$\underline{0x^3 - 47x^2y + 0xy^2 - 12y^3}$$
$$42x^3 - 30x^2y + 3xy^2 + 11y^3$$

27. $(4a + 7c) - (2b + 6d)$
$= 4a + 7c - 2b - 6d$
$= 4a - 2b + 7c - 6d$

29. $\left(\dfrac{5}{7}x^2 + \dfrac{8}{21}x - \dfrac{11}{14}\right) - \left(\dfrac{1}{2}x^2 + \dfrac{5}{6}x + \dfrac{19}{42}\right)$
$= \dfrac{5}{7}x^2 + \dfrac{8}{21}x - \dfrac{11}{14} - \dfrac{1}{2}x^2 - \dfrac{5}{6}x - \dfrac{19}{42}$
$= \left(\dfrac{5}{7} - \dfrac{1}{2}\right)x^2 + \left(\dfrac{8}{21} - \dfrac{5}{6}\right)x + \left(-\dfrac{11}{14} - \dfrac{19}{42}\right)$
$= \dfrac{3}{14}x^2 - \dfrac{19}{42}x - \dfrac{26}{21}$

31.
$$21.2x^3 + 0.9x^2y - 13.22xy^2 + 81.07y^3$$
$$\underline{-12.2x^3 - 0.1x^2y - 0.78xy^2 - 13.07y^3}$$
$$9x^3 + 0.8x^2y - 14xy^2 + 68y^3$$

33.
$$5062z^2 - 106z + 8295$$
$$\underline{- \ \ 379z^2 - 4297z - 1108}$$
$$4683z^2 - 4403z + 7187$$

35. $-6p^3q \cdot 7p^{-3}q^{-4}$
$= -42p^{3+(-3)}q^{1+(-4)}$
$= -42p^0q^{-3}$
$= -\dfrac{42}{q^3}$

37. $(1.4x^2) \cdot (4.3x^3y^{-2})$
$= 6.02x^{2+3}y^{-2}$
$= \dfrac{6.02x^5}{y^2}$

39. $\left(\dfrac{3}{7}x^2y^{-1}\right) \cdot \left(\dfrac{14}{15}x^{-1}y^4\right)$
$= \dfrac{2}{5}x^{2-1}y^{-1+4}$
$= \dfrac{2}{5}xy^3$

41. $4a^2(3a^2 - 5ab + b^2)$
$= 4a^2(3a^2) - 4a^2(5ab) + 4a^2(b^2)$
$= 12a^{2+2} - 20a^{2+1}b + 4a^2b^2$
$= 12a^4 - 20a^3b + 4a^2b^2$

43. $-3x^{-1}(-2x^3 + 3x^2 - x + 1)$
$= -3x^{-1}(-2x^3) - 3x^{-1}(3x^2) - 3x^{-1}(-x) - 3x^{-1}(1)$
$= 6x^{-1+3} - 9x^{-1+2} + 3x^{-1+1} - 3x^{-1}$
$= 6x^2 - 9x^1 + 3x^0 - 3\left(\dfrac{1}{x}\right)$
$= 6x^2 - 9x + 3 - \dfrac{3}{x}$

45. $(3a^2 + 2ab + b^2) \cdot (-ab)$

$= 3a^2(-ab) + 2ab(-ab) + b^2(-ab)$

$= -3a^{2+1}b - 2a^{1+1}b^{1+1} - ab^{2+1}$

$= -3a^3b - 2a^2b^2 - ab^3$

47. $\dfrac{3}{5}a^2(5a - 15a^2 + 25a^3)$

$= \dfrac{3}{5}a^2(5a) - \dfrac{3}{5}a^2(15a^2) + \dfrac{3}{5}a^2(25a^3)$

$= 3a^{2+1} - 9a^{2+2} + 15a^{2+3}$

$= 3a^3 - 9a^4 + 15a^5$

49. $\dfrac{24x^5}{3x^3}$

$= \dfrac{24}{3} \cdot \dfrac{x^5}{x^3}$

$= \dfrac{24}{3} \cdot x^{5-3}$

$= 8x^2$

51. $\dfrac{13x^4y^2}{2x^4y^7}$

$= \dfrac{13}{2} \cdot \dfrac{x^4}{x^4} \cdot \dfrac{y^2}{y^7}$

$= \dfrac{13}{2} \cdot x^{4-4} \cdot y^{2-7}$

$= \dfrac{13}{2}x^0y^{-5}$

$= \dfrac{13}{2}(1)\left(\dfrac{1}{y^5}\right)$

$= \dfrac{13}{2y^5}$

53. $\dfrac{122a^{-3}b^3}{11a^2b^{-2}}$

$= \dfrac{122}{11} \cdot \dfrac{a^{-3}}{a^2} \cdot \dfrac{b^3}{b^{-2}}$

$= \dfrac{122}{11} \cdot a^{-3-2} \cdot b^{3-(-2)}$

$= \dfrac{122}{11}a^{-5}b^5$

$= \dfrac{122}{11} \cdot \dfrac{1}{a^5} \cdot b^5$

$= \dfrac{122b^5}{11a^5}$

55. $\dfrac{(2x^2)^3}{6x}$

$= \dfrac{2^3x^{2\cdot3}}{6x}$

$= \dfrac{8x^6}{6x}$

$= \dfrac{8}{6} \cdot \dfrac{x^6}{x}$

$= \dfrac{4x^5}{3}$

57. $\left(\dfrac{2x^2}{6x}\right)^3$

$= \left(\dfrac{2}{6} \cdot \dfrac{x^2}{x}\right)^3$

$= \left(\dfrac{x}{3}\right)^3$

$= \dfrac{x^3}{3^3}$

$= \dfrac{x^3}{27}$

59. $\dfrac{(6x)^6}{(6x)^4}$

$= (6x)^{6-4}$

$= (6x)^2$

$= 6^2 x^2$

$= 36x^2$

61. $\dfrac{(3a^{-2}b)^3}{6ab^2}$

$= \dfrac{3^3 a^{-2\cdot3} b^3}{6ab^2}$

$= \dfrac{27 a^{-6} b^3}{6ab^2}$

$= \dfrac{27}{6} \cdot \dfrac{a^{-6}}{a} \cdot \dfrac{b^3}{b^2}$

$= \dfrac{9b}{2a^7}$

63. $\left(\dfrac{3a^2b}{6ab^2}\right)^3$

$= \left(\dfrac{3}{6} \cdot \dfrac{a^2}{a} \cdot \dfrac{b}{b^2}\right)^3$

$= \left(\dfrac{a}{2b}\right)^3$

$= \dfrac{a^3}{2^3 b^3}$

$= \dfrac{a^3}{8b^3}$

65. $\left(\dfrac{3a^2b}{6ab^2}\right)^{-3}$

$= \left(\dfrac{6ab^2}{3a^2b}\right)^3$

$= \left(\dfrac{6}{3} \cdot \dfrac{a}{a^2} \cdot \dfrac{b^2}{b}\right)^3$

$= \left(\dfrac{2b}{a}\right)^3$

$= \dfrac{2^3 b^3}{a^3}$

$= \dfrac{8b^3}{a^3}$

67. $\dfrac{(x^2 y^{-3} z)^4}{(2xyz^2)^2}$

$= \dfrac{x^{2\cdot4} y^{-3\cdot4} z^4}{2^2 x^2 y^2 z^{2\cdot2}}$

$= \dfrac{x^8 y^{-12} z^4}{4x^2 y^2 z^4}$

$= \dfrac{1}{4} \cdot \dfrac{x^8}{x^2} \cdot \dfrac{y^{-12}}{y^2} \cdot \dfrac{z^4}{z^4}$

$= \dfrac{x^6}{4y^{14}}$

69. $\dfrac{7,309,800 a^9}{0.000031 a^3}$

$= \dfrac{7,309,800}{0.000031} \cdot \dfrac{a^9}{a^3}$

$= (2.358 \times 10^{11}) a^6$

71. $\dfrac{2.35 \cdot 10^{-4} x^5}{4.7 \cdot 10^8 x^2}$

$= \dfrac{2.35 \cdot 10^{-4}}{4.7 \cdot 10^8} \cdot \dfrac{x^5}{x^2}$

$= (5 \times 10^{-13}) x^3$

73. $\dfrac{6x^3 + 12x^2 - 18x}{3x}$

$= \dfrac{6x^3}{3x} + \dfrac{12x^2}{3x} - \dfrac{18x}{3x}$

$= 2x^{3-1} + 4x^{2-1} - 6x^{1-1}$

$= 2x^2 + 4x - 6$

75. $\dfrac{4x^2 - 2x + 12}{8x}$

$= \dfrac{4x^2}{8x} - \dfrac{2x}{8x} + \dfrac{12}{8x}$

$= \dfrac{1}{2}x^{2-1} - \dfrac{1}{4}x^{1-1} + \dfrac{3}{2x}$

$= \dfrac{1}{2}x - \dfrac{1}{4} + \dfrac{3}{2x}$

77. $\dfrac{3.72x^4 - 6.96x^2 + 1.08}{1.2x^2}$

$= \dfrac{3.72x^4}{1.2x^2} - \dfrac{6.96x^2}{1.2x^2} + \dfrac{1.08}{1.2x^2}$

$= 3.1x^{4-2} - 5.8x^{2-2} + \dfrac{0.9}{x^2}$

$= 3.1x^2 - 5.8 + \dfrac{0.9}{x^2}$

79. $\dfrac{9x^4 + 6x^3y - 18x^2y^2 - 24xy^3 + 72y^4}{-3xy}$

$= \dfrac{9x^4}{-3xy} + \dfrac{6x^3y}{-3xy} - \dfrac{18x^2y^2}{-3xy} - \dfrac{24xy^3}{-3xy} + \dfrac{72y^4}{-3xy}$

$= -3x^{4-1}y^{-1} - 2x^{3-1}y^{1-1} + 6x^{2-1}y^{2-1} + 8x^{1-1}y^{3-1} - 24x^{-1}y^{4-1}$

$= -\dfrac{3x^3}{y} - 2x^2 + 6xy + 8y^2 - \dfrac{24y^3}{x}$

81. Let x = number of pots

   a. $C(x) = 200 + (2.50 + 2.00)x - 0.05x$
   $C(x) = 200 + 4.45x$

   b. $R(x) = 13.5x$

   c. $P(x) = R(x) - C(x)$
   $P(x) = 13.5x - (200 + 4.45x)$
   $P(x) = 9.05x - 200$

d. $A(x) = \dfrac{9.05x - 200}{x}$

$A(x) = 9.05 - \dfrac{200}{x}$

e. $A(20) = 9.05 - \dfrac{200}{20} = -0.95$

$A(30) = 9.05 - \dfrac{200}{30} \approx 2.38$

The average profit per pot if 20 and 30 are produced is –$0.95 and $2.38, respectively.

83. Let width = 15 feet
    length = $(15 + x)$ feet
    The polynomial for the perimeter is:
    $2(15) + 2(15 + x)$
    $= 30 + 30 + 2x$
    $= 60 + 2x$
    $60 + 2x = 60 + 2(8) = 76$
    The perimeter is 76 feet.

85. Surface area of cube: $6x^2$

    Surface area of sphere: $4\pi\left(\dfrac{x}{2}\right)^2 = \pi x^2$

    Difference: $6x^2 - \pi x^2$
    $= 6x^2 - \pi x^2 = 6(8)^2 - \pi(8)^2$
    ≈ 182.9
    The difference is approximately 182.9 in^2.

87. Area of rectangle: $8(2x) = 16x$
    Area of square: x^2 square inches
    Area of shade: $16x - x^2$ square inches

89. Let r = radius
    $4r + 5$ = length
    Area of two semi-circles: πr^2
    Area of rectangle: $2r(4r + 5)$
    Total area:
    $\pi r^2 + 2r(4r + 5)$
    $= \pi r^2 + 8r^2 + 10r$
    $= (\pi + 8)r^2 + 10r$ square feet

r (feet)	$(\pi+8)r^2+10r$	Area (feet)^2
8	$(\pi+8)8^2+10(8)\approx 793.06$	793.06
9	$(\pi+8)9^2+10(9)\approx 992.47$	992.47
10	$(\pi+8)10^2+10(10)\approx 1214.16$	1214.16
11	$(\pi+8)11^2+10(11)\approx 1458.13$	1458.13
12	$(\pi+8)12^2+10(12)\approx 1724.39$	1724.39

91. a. $x=$ height
$4x=$ base
Area is:
$\frac{1}{2}(4x)x=2x^2$ square feet

b. $2x=$ height
$4x+10=$ base
Area is:
$\frac{1}{2}(4x+10)(2x)=4x^2+10x$ square
feet

c. Ratio $=\dfrac{\text{Planned area}}{\text{Current area}}$

$=\dfrac{4x^2+10x}{2x^2}$

$=\dfrac{4x^2}{2x^2}+\dfrac{10x}{2x^2}$

$=2+\dfrac{5}{x}$

d. Current Area $=2x^2$
$=2(10)^2$
$=200$
The current area is 200 ft^2.
Planned area $=4x^2+10x$
$=4(10)^2+10(10)$
$=400+100$
$=500$
The planned area is 500 ft^2.

Ratio $=2+\dfrac{5}{x}$

$=2+\dfrac{5}{10}$

$=2+\dfrac{1}{2}$

$=\dfrac{5}{2}$

The ratio is $\dfrac{5}{2}$.

93. Let r = radius of heel

Area of two heels = $2\pi r^2$

Pressure from each heel = $\dfrac{110}{2\pi r^2}$

$= \dfrac{55}{\pi r^2}$ pounds per square inch

r (inch)	$\dfrac{55}{\pi r^2}$	Pounds per square inch
0.25	$\dfrac{55}{\pi(0.25)^2} \approx 280.1$	280.1
0.50	$\dfrac{55}{\pi(0.5)^2} \approx 70.0$	70.0
0.75	$\dfrac{55}{\pi(0.75)^2} \approx 31.1$	31.1
1.00	$\dfrac{55}{\pi(1)^2} \approx 17.5$	17.5

For a radius of 1 inch, the woman's pressure on the floor approximates the elephant's.

95. Let w = width

$2w$ = length

Area = $2w^2$

Cost per square foot = $\dfrac{80,000}{2w^2} = \dfrac{40,000}{w^2}$

w (feet)	$\dfrac{40,000}{w^2}$	Cost per square foot
20	$\dfrac{40,000}{(20)^2} = 100$	100
25	$\dfrac{40,000}{(25)^2} = 64$	64
30	$\dfrac{40,000}{(30)^2} \approx 44.44$	44.44
35	$\dfrac{40,000}{(35)^2} \approx 32.65$	32.65
40	$\dfrac{40,000}{(40)^2} = 25$	25
45	$\dfrac{40,000}{(45)^2} \approx 19.75$	19.75
50	$\dfrac{40,000}{(50)^2} = 16$	16

A home with a size of 35 ft by 70 ft would cost approximately \$33 per square foot.

9.2 Experiencing Algebra the Calculator Way

1. $Y1 = (3x^4 - x^3 + 2x^2 - 5)$
$\qquad - (2x^4 + x^3 + x^2 - 4)$
$Y2 = x^4 + x^2 - 9$

No; they are not equivalent.

2. $Y1 = (3x^4 - x^3 + 2x^2 - 5)$
$\qquad - (2x^4 + x^3 + x^2 - 4)$
$Y2 = x^4 - 2x^3 + x^2 - 1$

Yes; they are equivalent.

3. $Y1 = 2x^2(x^2 - 3x - 2)$
$Y2 = 2x^4 - 6x^3 - 4x^2$

Yes, they are equivalent.

4. $Y1 = \left(\dfrac{1}{2}x\right)\left(2x + \dfrac{1}{3}\right)$

$Y2 = x^2 - \dfrac{1}{6}x$

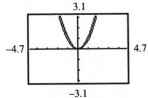

No; they are not equivalent.

5. $Y1 = -0.4x(7 - 5x + x^2)$

$Y2 = 2.8x + 2x^2 - 0.4x^3$

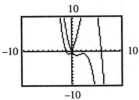

No; they are not equivalent.

6. $Y1 = \dfrac{2x^3 - 6x^2 + 2x}{2x}$

$Y2 = x^2 - 3x + 1$

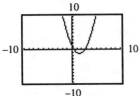

Yes; they are equivalent.

7. $Y1 = \dfrac{-6x^2 + 9x}{-3x}$

$Y2 = 2x + 3$

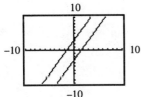

No; they are not equivalent.

8. $Y1 = \dfrac{-8x^4 + 4x^3 - 8x^2 + 4x}{4x}$

$Y2 = -2x^3 + x^2 - 2x + 4$

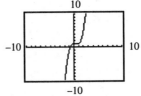

No; they are not equivalent.

9. $Y1 = \dfrac{3x^4 - 3x^3 + 3x}{3x}$

$Y2 = x^3 - x^2 + 1$

Yes; they are equivalent.

9.3 Experiencing Algebra the Exercise Way

1. $(x + 3)(x + 8)$
$= x(x + 8) + 3(x + 8)$
$= x^2 + 8x + 3x + 24$
$= x^2 + 11x + 24$

3. $(2x + 1)(3x - 4)$
$= 2x(3x - 4) + 1(3x - 4)$
$= 6x^2 - 8x + 3x - 4$
$= 6x^2 - 5x - 4$

5. $(4 - x)(x + 2)$
$= 4(x + 2) - x(x + 2)$
$= 4x + 8 - x^2 - 2x$
$= -x^2 + 2x + 8$

7. $(5 + x)(3 + 2x)$
$= 5(3 + 2x) + x(3 + 2x)$
$= 15 + 10x + 3x + 2x^2$
$= 2x^2 + 13x + 15$

9. $(2x + 5)(3y - 2)$
$= 2x(3y - 2) + 5(3y - 2)$
$= 6xy - 4x + 15y - 10$

11. $(3x + 4y)(x - 2y)$
$= 3x(x - 2y) + 4y(x - 2y)$
$= 3x^2 - 6xy + 4xy - 8y^2$
$= 3x^2 - 2xy - 8y^2$

13. $(a - 2.4)(5a + 3.8)$
$= a(5a + 3.8) - 2.4(5a + 3.8)$
$= 5a^2 + 3.8a - 12a - 9.12$
$= 5a^2 - 8.2a - 9.12$

15. $(2x + 1.1)(3y + 3.2)$
$= 2x(3y + 3.2) + 1.1(3y + 3.2)$
$= 6xy + 6.4x + 3.3y + 3.52$

17. $\left(a + \dfrac{2}{3}\right)\left(a + \dfrac{1}{3}\right)$

$= a\left(a + \dfrac{1}{3}\right) + \dfrac{2}{3}\left(a + \dfrac{1}{3}\right)$

$= a^2 + \dfrac{1}{3}a + \dfrac{2}{3}a + \dfrac{2}{9}$

$= a^2 + a + \dfrac{2}{9}$

19. $\left(2x + \dfrac{1}{4}\right)\left(8x - \dfrac{1}{6}\right)$

$= 2x\left(8x - \dfrac{1}{6}\right) + \dfrac{1}{4}\left(8x - \dfrac{1}{6}\right)$

$= 16x^2 - \dfrac{1}{3}x + 2x - \dfrac{1}{24}$

$= 16x^2 + \dfrac{5}{3}x - \dfrac{1}{24}$

21. $(2x^2 - 3)(x^2 + 4)$
$= 2x^2(x^2 + 4) - 3(x^2 + 4)$
$= 2x^4 + 8x^2 - 3x^2 - 12$
$= 2x^4 + 5x^2 - 12$

23. $(4x^2 + 3)(2x + 1)$
$= 4x^2(2x + 1) + 3(2x + 1)$
$= 8x^3 + 4x^2 + 6x + 3$

25. $(2x + 3y^2)(3x^2 - 5y)$
$= 2x(3x^2 - 5y) + 3y^2(3x^2 - 5y)$
$= 6x^3 - 10xy + 9x^2y^2 - 15y^3$

27. $(3a^2 + 5b^3)(a^2 + b^3)$
$= 3a^2(a^2 + b^3) + 5b^3(a^2 + b^3)$
$= 3a^4 + 3a^2b^3 + 5a^2b^3 + 5b^6$
$= 3a^4 + 8a^2b^3 + 5b^6$

29. $(x + 1)(x^2 - x + 1)$
$= x(x^2 - x + 1) + 1(x^2 - x + 1)$
$= x^3 - x^2 + x + x^2 - x + 1$
$= x^3 + 1$

31. $(3x - 2)(2x^2 - 5x - 3)$
$= 3x(2x^2 - 5x - 3) - 2(2x^2 - 5x - 3)$
$= 6x^3 - 15x^2 - 9x - 4x^2 + 10x + 6$
$= 6x^3 - 19x^2 + x + 6$

33. $(x^2 + x + 1)(x^2 + 2x + 3)$
$= x^2(x^2 + 2x + 3) + x(x^2 + 2x + 3) + 1(x^2 + 2x + 3)$
$= x^4 + 2x^3 + 3x^2 + x^3 + 2x^2 + 3x + x^2 + 2x + 3$
$= x^4 + 3x^3 + 6x^2 + 5x + 3$

35. $(a+b+c)^2$

$= (a+b+c)(a+b+c)$

$= a(a+b+c)+b(a+b+c)+c(a+b+c)$

$= a^2 + ab + ac + ab + b^2 + bc + ac + bc + c^2$

$= a^2 + b^2 + c^2 + 2ab + 2bc + 2ac$

37. $(z+3)^3$

$= (z+3)(z+3)(z+3)$

$= [z(z+3)+3(z+3)](z+3)$

$= (z^2 + 3z + 3z + 9)(z+3)$

$= (z^2 + 6z + 9)(z+3)$

$= (z^2 + 6z + 9)(z) + (z^2 + 6z + 9)(3)$

$= z^3 + 6z^2 + 9z + 3z^2 + 18z + 27$

$= z^3 + 9z^2 + 27z + 27$

39. $(3a-2b)^3$

$= (3a-2b)(3a-2b)(3a-2b)$

$= [3a(3a-2b)-2b(3a-2b)](3a-2b)$

$= (9a^2 - 6ab - 6ab + 4b^2)(3a-2b)$

$= (9a^2 - 12ab + 4b^2)(3a-2b)$

$= (9a^2 - 12ab + 4b^2)(3a) + (9a^2 - 12ab + 4b^2)(-2b)$

$= 27a^3 - 36a^2 b + 12ab^2 - 18a^2 b + 24ab^2 - 8b^3$

$= 27a^3 - 54a^2 b + 36ab^2 - 8b^3$

41. $(x-5)(x+5)$

$= x^2 - 5^2$

$= x^2 - 25$

43. $(3m+7)(3m-7)$

$= (3m)^2 - 7^2$

$= 9m^2 - 49$

45. $(2a+3b)(2a-3b)$

$= (2a)^2 - (3b)^2$

$= 4a^2 - 9b^2$

47. $(4x-1.5)(4x+1.5)$

$= (4x)^2 - (1.5)^2$

$= 16x^2 - 2.25$

49. $\left(\dfrac{2}{5}x - 1\right)\left(\dfrac{2}{5}x + 1\right)$

$= \left(\dfrac{2}{5}x\right)^2 - 1^2$

$= \dfrac{4}{25}x^2 - 1$

51. $\left(\dfrac{1}{3}x + \dfrac{4}{5}\right)\left(\dfrac{1}{3}x - \dfrac{4}{5}\right)$

$= \left(\dfrac{1}{3}x\right)^2 - \left(\dfrac{4}{5}\right)^2$

$= \dfrac{1}{9}x^2 - \dfrac{16}{25}$

53. $(x^2 + 7)(x^2 - 7)$

$= (x^2)^2 - 7^2$

$= x^4 - 49$

55. $(2x^3 + 5y)(2x^3 - 5y)$

$= (2x^3)^2 - (5y)^2$

$= 4x^6 - 25y^2$

57. $(m + 7)^2$

$= m^2 + 2(m)(7) + 7^2$

$= m^2 + 14m + 49$

59. $(x - y)^2$

$= x^2 - 2xy + y^2$

61. $(2p + 9q)^2$

$= (2p)^2 + 2(2p)(9q) + (9q)^2$

$= 4p^2 + 36pq + 81q^2$

63. $(6c - 5)^2$

$= (6c)^2 - 2(6c)(5) + (5)^2$

$= 36c^2 - 60c + 25$

65. $(3x^3 + 2)^2$

$= (3x^3)^2 + 2(3x^3)(2) + 2^2$

$= 9x^6 + 12x^3 + 4$

67. $(2x^2 - 3y^3)^2$

$= (2x^2)^2 - 2(2x^2)(3y^3) + (3y^3)^2$

$= 4x^4 - 12x^2y^3 + 9y^6$

69. a.   length: $(18 - 2x)$ in.
width: $(12 - 2x)$ in.
height: (x) in.

b.   Volume: $x(18 - 2x)(12 - 2x)$

$= (18x - 2x^2)(12 - 2x)$

$= 18x(12 - 2x) - 2x^2(12 - 2x)$

$= 216x - 36x^2 - 24x^2 + 4x^3$

$= 4x^3 - 60x^2 + 216x$

The volume is $4x^3 - 60x^2 + 216x$ in^3.

c.   Surface Area:

$2(x)(12 - 2x) + 2(x)(18 - 2x) + (12 - 2x)(18 - 2x)$

$= 24x - 4x^2 + 36x - 4x^2 + 216 - 24x - 36x + 4x^2$

$= -4x^2 + 216$

The surface area is $-4x^2 + 216$ in^2.

d.   Volume:

$4(2)^3 - 60(2)^2 + 216(2) = 224$

Surface area:

$-4(2)^2 + 216 = 200$

The volume is 224 in^3 and the surface area is 200 in^2.

71. a. Area of outer circle:

(πx^2) ft^2

b. Area of inner circle:

$[\pi(x-5)^2]$ ft^2

c. Difference of areas:

$\pi x^2 - \pi(x-5)^2$

$= \pi x^2 - \pi(x^2 - 10x + 25)$

$= \pi x^2 - \pi x^2 + 10\pi x - 25\pi$

$= 10\pi x - 25\pi$

The area of the deck is

$(10\pi x - 25\pi)$ ft^2.

d. Area:

$10\pi(22) - 25\pi$

$= 220\pi - 25\pi$

$= 195\pi$

The area is 195π ft^2.

73. a. The old volume is x^3 ft^3.

b. New volume:

$x(x+5)(x+8)$

$= (x^2 + 5x)(x+8)$

$= x^2(x+8) + 5x(x+8)$

$= x^3 + 8x^2 + 5x^2 + 40x$

$= x^3 + 13x^2 + 40x$

The new volume is

$(x^3 + 13x^2 + 40x)$ ft^3.

c. Ratio of new volume to old volume:

$\dfrac{x^3 + 13x^2 + 40x}{x^3}$

$= \dfrac{x(x^2 + 13x + 40)}{x(x^2)}$

$= \dfrac{x^2 + 13x + 40}{x^2}$

The ratio is $\dfrac{x^2 + 13x + 40}{x^2}$

d. Ratio:

$\dfrac{(12)^2 + 13(12) + 40}{(12)^2}$

$= \dfrac{340}{144}$

$= \dfrac{85}{36}$

The ratio is $\dfrac{85}{36}$.

9.3 Experiencing Algebra the Calculator Way

1. Y1 $= (2x-1)(2x+1)$

Y2 $= 2x^2 - 1$

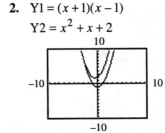

They are not equivalent.

$(2x-1)(2x+1)$

$= (2x)^2 - 1^2$

$= 4x^2 - 1$

The correct expression

is $4x^2 - 1$.

2. Y1 $= (x+1)(x-1)$

Y2 $= x^2 + x + 2$

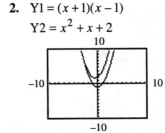

They are not equivalent.

$(x+1)(x-1)$

$= x^2 - 1^2$

$= x^2 - 1$

The correct expression

is $x^2 - 1$.

3. $Y1 = (x-2)(x-1)$
$Y2 = x^2 - 3x + 2$

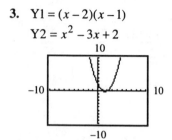

They are equivalent.

4. $Y1 = (0.5x + 1)(4x - 0.8)$
$Y2 = 2x^2 + 3.6x - 0.8$

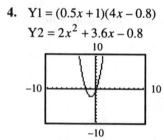

They are equivalent.

5. $Y1 = \left(\frac{1}{2}x - 3\right)\left(2x + \frac{1}{3}\right)$

$Y2 = x^2 - \frac{35}{6}x + 1$

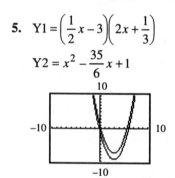

They are not equivalent.

$\left(\frac{1}{2}x - 3\right)\left(2x + \frac{1}{3}\right)$

$= \frac{1}{2}x\left(2x + \frac{1}{3}\right) - 3\left(2x + \frac{1}{3}\right)$

$= x^2 + \frac{1}{6}x - 6x - 1$

$= x^2 - \frac{35}{6}x - 1$

The correct expression

is $x^2 - \frac{35}{6}x - 1$.

6. $Y1 = (x^2 + 4)(x^2 - 1)$
$Y2 = x^4 + 3x^2 - 4$

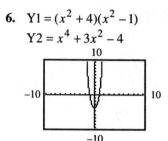

They are equivalent.

7. $Y1 = (x^2 + 1)(2x - 1)$
$Y2 = 2x^3 - x^2 + 2x - 1$

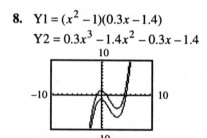

They are equivalent.

8. $Y1 = (x^2 - 1)(0.3x - 1.4)$
$Y2 = 0.3x^3 - 1.4x^2 - 0.3x - 1.4$

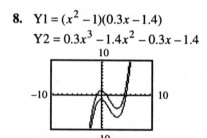

They are not equivalent.
$(x^2 - 1)(0.3x - 1.4)$
$= x^2(0.3x - 1.4) - 1(0.3x - 1.4)$
$= 0.3x^3 - 1.4x^2 - 0.3x + 1.4$
The correct expression is
$0.3x^3 - 1.4x^2 - 0.3x + 1.4$.

9.4 Experiencing Algebra the Exercise Way

1. $60a^2b^4c^3$ and $50a^3bc^2$
$60 = 2^2 \cdot 3 \cdot 5$
$50 = 2 \cdot 5^2$
The GCF for the coefficients is $2 \cdot 5 = 10$.
The GCF for the variable factors is a^2bc^2.
The GCF for $60a^2b^4c^3$ and $50a^3bc^2$ is
$10a^2bc^2$.

3. $252x^3y^4$ and $180x^2z$

$252 = 2^2 \cdot 3^2 \cdot 7$

$180 = 2^2 \cdot 3^2 \cdot 5$

The GCF for the coefficients is $2^2 \cdot 3^2 = 36$.
The GCF for the variable factors is x^2.
The GCF for $252x^3y^4$ and $180x^2z$ is
$36x^2$.

5. $45pq^4$ and $135r^4s$

$45 = 1 \cdot 3^2 \cdot 5$

$135 = 3^3 \cdot 5$

The GCF for the coefficients is $3^2 \cdot 5 = 45$.
There are no common variable factors. The
GCF for $45pq^4$ and $135r^4s$ is 45.

7. $63xyz$ and $98xyz$

$63 = 3^2 \cdot 7$

$98 = 2 \cdot 7^2$

The GCF for the coefficients is 7. The GCF
for the variable factors is xyz. The GCF for
$63xyz$ and $98xyz$ is $7xyz$.

9. $60a^2b^3c^3$, $90ab^2c$, and $150a^2b^4c^2$

$60 = 2^2 \cdot 3 \cdot 5$

$90 = 2 \cdot 3^2 \cdot 5$

$150 = 2 \cdot 3 \cdot 5^2$

The GCF for the coefficients is $2 \cdot 3 \cdot 5 = 30$.
The GCF for the variable factors is ab^2c.
The GCF for $60a^2b^3c^3$, $90ab^2c$, and
$150a^2b^4c^2$ is $30ab^2c$.

11. $4x + 12y$

$= 4 \cdot x + 4 \cdot 3y$

$= 4(x + 3y)$

13. $8x^3 - 4x^2 + 12x - 24$

$= 4 \cdot 2x^3 - 4 \cdot x^2 + 4 \cdot 3x - 4 \cdot 6$

$= 4(2x^3 - x^2 + 3x - 6)$

15. $3a^4 - 5a^3 + 7a^2$

$= a^2 \cdot 3a^2 - a^2 \cdot 5a + a^2 \cdot 7$

$= a^2(3a^2 - 5a + 7)$

17. $-3x^5 - 9x^4 - 12x^3$

$= -3x^3 \cdot x^2 + (-3x^3) \cdot 3x + (-3x^3) \cdot 4$

$= -3x^3(x^2 + 3x + 4)$

19. $7x^4y^2 - 3x^2y^2 + 9x^2y^4$

$= x^2y^2 \cdot 7x^2 - x^2y^2 \cdot 3 + x^2y^2 \cdot 9y^2$

$= x^2y^2(7x^2 - 3 + 9y^2)$

21. $8a^5b^3c + 4a^4b^2 + 16a^3c$

$= 4a^3 \cdot 2a^2b^3c + 4a^3 \cdot ab^2 + 4a^3 \cdot 4c$

$= 4a^3(2a^2b^3c + ab^2 + 4c)$

23. $66u^3v^4 - 88u^4v^3$

$= 22u^3v^3 \cdot 3v + 22u^3v^3 \cdot 4u$

$= 22u^3v^3(3v + 4u)$

25. $3x^3 + 5y^4$

There is no greatest common factor,
therefore the binomial does not factor.

27. $5x(x + 3) - 4(x + 3)$

$= (x + 3)(5x - 4)$

29. $x(2x + y) + 2y(2x + y)$

$= (2x + y)(x + 2y)$

31. $6x^2 + 10x + 21x + 35$

$= (6x^2 + 10x) + (21x + 35)$

$= 2x(3x + 5) + 7(3x + 5)$

$= (3x + 5)(2x + 7)$

33. $x^2 + 8x + x + 8$

$= (x^2 + 8x) + (x + 8)$

$= x(x + 8) + 1(x + 8)$

$= (x + 8)(x + 1)$

35. $2a^2 + 3a - 2a - 3$

$= (2a^2 + 3a) + (-2a - 3)$

$= a(2a + 3) - 1(2a + 3)$

$= (2a + 3)(a - 1)$

37. $2x^2 + xy + 4xy + 2y^2$

$= (2x^2 + xy) + (4xy + 2y^2)$

$= x(2x + y) + 2y(2x + y)$

$= (2x + y)(x + 2y)$

39. $x^2 + xy - xy - y^2$

$= (x^2 + xy) + (-xy - y^2)$

$= x(x + y) - y(x + y)$

$= (x + y)(x - y)$

41. $10xy - 55y + 24x - 132$

$= (10xy - 55y) + (24x - 132)$

$= 5y(2x - 11) + 12(2x - 11)$

$= (2x - 11)(5y + 12)$

43. $12ac + 3bc + 4ad + bd$

$= (12ac + 3bc) + (4ad + bd)$

$= 3c(4a + b) + d(4a + b)$

$= (4a + b)(3c + d)$

45. $2x^2y^2 + 3xy - 8xy - 12$

$= (2x^2y^2 + 3xy) + (-8xy - 12)$

$= xy(2xy + 3) - 4(2xy + 3)$

$= (2xy + 3)(xy - 4)$

47. $-x^2 - 3x - xy - 3y$

$= -1(x^2 + 3x + xy + 3y)$

$= -1[(x^2 + 3x) + (xy + 3y)]$

$= -1[x(x + 3) + y(x + 3)]$

$= -1(x + 3)(x + y)$

49. $x^4 + x^2y^2 + 2x^2y^2 + 2y^4$

$= (x^4 + x^2y^2) + (2x^2y^2 + 2y^4)$

$= x^2(x^2 + y^2) + 2y^2(x^2 + y^2)$

$= (x^2 + y^2)(x^2 + 2y^2)$

51. $ac + bc + ad + bd$

$= (ac + bc) + (ad + bd)$

$= c(a + b) + d(a + b)$

$= (a + b)(c + d)$

53. $8x^2 + 4x + 24x + 12$

$= 4(2x^2 + x + 6x + 3)$

$= 4[(2x^2 + x) + (6x + 3)]$

$= 4[x(2x + 1) + 3(2x + 1)]$

$= 4(2x + 1)(x + 3)$

55. $u^4 + u^3v - 2u^3v - 2u^2v^2$

$= u^2(u^2 + uv - 2uv - 2v^2)$

$= u^2[(u^2 + uv) + (-2uv - 2v^2)]$

$= u^2[u(u + v) - 2v(u + v)]$

$= u^2(u + v)(u - 2v)$

57. $6a^4 + 6a^3b^2 + 6a^3b + 6a^2b^3$

$= 6a^2(a^2 + ab^2 + ab + b^3)$

$= 6a^2[(a^2 + ab^2) + (ab + b^3)]$

$= 6a^2[a(a + b^2) + b(a + b^2)]$

$= 6a^2(a + b^2)(a + b)$

59. $-4x^3 - 12x^2 - 2x^2 - 6x$

$= -2x(2x^2 + 6x + x + 3)$

$= -2x[(2x^2 + 6x) + (x + 3)]$

$= -2x[2x(x + 3) + 1(x + 3)]$

$= -2x(x + 3)(2x + 1)$

61. $4x^2 + 6xz + 6xz + 9z^2$

$= (4x^2 + 6xz) + (6xz + 9z^2)$

$= 2x(2x + 3z) + 3z(2x + 3z)$

$= (2x + 3z)(2x + 3z)$

$= (2x + 3z)^2$

63. $36a^2 - 30ab - 30ab + 25b^2$

$= (36a^2 - 30ab) + (-30ab + 25b^2)$

$= 6a(6a - 5b) - 5b(6a - 5b)$

$= (6a - 5b)(6a - 5b)$

$= (6a - 5b)^2$

65. $5m^3 + 5m^2n + 5m^2n + 5mn^2$
$= 5m(m^2 + mn + mn + n^2)$
$= 5m[(m^2 + mn) + (mn + n^2)]$
$= 5m[m(m + n) + n(m + n)]$
$= 5m(m + n)(m + n)$
$= 5m(m + n)^2$

67. $2x^3 + 3x + 8x^2 + 12$
$= (2x^3 + 3x) + (8x^2 + 12)$
$= x(2x^2 + 3) + 4(2x^2 + 3)$
$= (2x^2 + 3)(x + 4)$

69. $5a^2x + 2b^2x + 15a^2y + 6b^2y$
$= (5a^2x + 2b^2x) + (15a^2y + 6b^2y)$
$= x(5a^2 + 2b^2) + 3y(5a^2 + 2b^2)$
$= (5a^2 + 2b^2)(x + 3y)$

71. a.　Let x = amount received on the first goal
$x + (x + 1) + (x + 2) + (x + 3)$
$\quad + (x + 4) + (x + 5) + (x + 6) + (x + 7)$
$\quad + (x + 8)$
$= 9x + 36$
She will receive $(9x + 36)$ dollars.

b.　$9x + 36 = 9(x + 4)$

c.　The binomial $(x + 4)$ is the average amount in dollars that she will receive for each of her 9 goals.

d.　$9(x + 4) = 9(10 + 4) = 126$
She will receive $126.
Check:
$9 \text{ days} \cdot \dfrac{\$14}{\text{day}} = \$126$

73. a.　$n^2 - n = n(n - 1)$

b.　$n^2 - n = (21)^2 - 21 = 420$
$n(n - 1) = 21(21 - 1) = 420$
Each equals 420 times.

c.　The factored expression was easier to evaluate. Answers will vary.

75.　$x^2 + 7x - 3x - 21$
$= (x^2 + 7x) + (-3x - 21)$
$= x(x + 7) - 3(x + 7)$
$= (x + 7)(x - 3)$
The rectangle's length is $(x + 7)$ units and width is $(x - 3)$ units.

9.4 Experiencing Algebra the Calculator Way

1.　$30 = 2 \cdot 3 \cdot 5$

2.　$108 = 2^2 \cdot 3^3$

3.　$525 = 3 \cdot 5^2 \cdot 7$

4.　$1287 = 3^2 \cdot 11 \cdot 13$

5.　$1547 = 7 \cdot 13 \cdot 17$

6.　$4500 = 2^2 \cdot 3^2 \cdot 5^3$

9.5 Experiencing Algebra the Exercise Way

1.　$x^2 + 14x + 45$
$a = 1, b = 14, c = 45$

factor	factor	sum of factors
1	45	46
3	15	18
5	9	14　←b

$x^2 + 14x + 45 = (x + 5)(x + 9)$

3.　$y^2 - 15y + 56$
$a = 1, b = -15, c = 56$

factor	factor	sum of factors
−1	−56	−57
−2	−28	−30
−4	−14	−18
−7	−8	−15　←b

$y^2 - 15y + 56 = (y - 7)(y - 8)$

5. $p^2 - 9p - 36$

$a = 1, b = -9, c = -36$

factor	factor	sum of factors
−1	36	35
1	−36	−35
−2	18	16
2	−18	−16
−3	12	9
3	−12	−9 ← b
−4	9	5
4	−9	−5
−6	6	0

$p^2 - 9p - 36 = (p + 3)(p - 12)$

7. $z^2 + 6z + 12$

$a = 1, b = 6, c = 12$

factor	factor	sum of factors
1	12	13
2	6	8
3	4	7

The possible factors of 12 do not add to 6.
Therefore $z^2 + 6z + 12$ does not factor.

9. $x^4 + 25x^2 + 144$

$a = 1, b = 25, c = 144$

factor	factor	sum of factors
1	144	145
2	72	74
3	48	51
4	36	40
6	24	30
8	18	26
9	16	25 ← b
12	12	24

$x^4 + 25x^2 + 144 = (x^2 + 9)(x^2 + 16)$

11. $x^4 - 2x^2 - 3$

$a = 1, b = -2, c = -3$

factor	factor	sum of factors
−1	3	2
1	−3	−2 ← b

$x^4 - 2x^2 - 3 = (x^2 + 1)(x^2 - 3)$

13. $3a^2 + 48a + 165 = 3(a^2 + 16a + 55)$

$a = 1, b = 16, c = 55$

factor	factor	sum of factors
1	55	56
5	11	16 ← b

$3a^2 + 48a + 165 = 3(a + 5)(a + 11)$

15. $4c^2 + 44c - 104$

$= 4(c^2 + 11c - 26)$

$a = 1, b = 11, c = -26$

factor	factor	sum of factors
−1	26	25
1	−26	−25
−2	13	11 ← b
2	−13	−11

$4c^2 + 44c - 104 = 4(c - 2)(c + 13)$

17. $x^2 - 11xy + 24y^2$

$a = 1, b = -11, c = 24$

factor	factor	sum of factors
−1	−24	−25
−2	−12	−14
−3	−8	−11 ← b
−4	−6	−10

$x^2 - 11xy + 24 = (x - 3y)(x - 8y)$

19. $x^2 + 11xy - 12y^2$
$a = 1, b = 11, c = -12$

factor	factor	sum of factors	
−1	12	11	← b
1	−12	−11	
−2	6	4	
2	−6	−4	
−3	4	1	
3	−4	−1	

$x^2 + 11xy - 12y^2 = (x - y)(x + 12y)$

21. $-3a^2 - 15ab - 18b^2 = -3(a^2 + 5ab + 6b^2)$
$a = 1, b = 5, c = 6$

factor	factor	sum of factors	
1	6	7	
2	3	5	← b

$-3a^2 - 15ab - 18b^2 = -3(a + 2b)(a + 3b)$

23. $-2x^2 + 14xy + 36y^2$
$= -2(x^2 - 7xy - 18y^2)$
$a = 1, b = -7, c = -18$

factor	factor	sum of factors	
−1	18	17	
1	−18	−17	
−2	9	7	
2	−9	−7	← b
−3	6	3	
3	−6	−3	

$-2x^2 + 14xy + 36y^2 = -2(x + 2y)(x - 9y)$

25. $a = 3, b = 10, c = 3, ac = 9$
$3x^2 + 10x + 3$
$= 3x^2 + x + 9x + 3$
$= x(3x + 1) + 3(3x + 1)$
$= (3x + 1)(x + 3)$

27. $a = 2, b = -15, c = 7, ac = 14$
$2x^2 - 15x + 7$
$= 2x^2 - 14x - x + 7$
$= 2x(x - 7) - 1(x - 7)$
$= (x - 7)(2x - 1)$

29. $a = 3, b = -1, c = -2, ac = -6$
$3x^2 - x - 2$
$= 3x^2 - 3x + 2x - 2$
$= 3x(x - 1) + 2(x - 1)$
$= (x - 1)(3x + 2)$

31. $a = 5, b = 9, c = -2, ac = -10$
$5m^2 + 9m - 2$
$= 5m^2 + 10m - m - 2$
$= 5m(m + 2) - 1(m + 2)$
$= (m + 2)(5m - 1)$

33. $a = 2, b = 7, c = -3, ac = -6$
$2m^2 + 7m - 3$ does not factor because there are no factors of −6 that add to 7.

35. $a = 4, b = 25, c = 6, ac = 24$
$4a^2 + 25a + 6$
$= 4a^2 + 24a + a + 6$
$= 4a(a + 6) + 1(a + 6)$
$= (a + 6)(4a + 1)$

37. $a = 9, b = -13, c = 4, ac = 36$
$9d^2 - 13d + 4$
$= 9d^2 - 9d - 4d + 4$
$= 9d(d - 1) - 4(d - 1)$
$= (d - 1)(9d - 4)$

39. $a = 6, b = -23, c = -4, ac = -24$
$6x^2 - 23x - 4$
$= 6x^2 - 24x + x - 4$
$= 6x(x - 4) + 1(x - 4)$
$= (x - 4)(6x + 1)$

41. $a = 8, b = 7, c = -18, ac = -144$
$8y^2 + 7y - 18$
$= 8y^2 + 16y - 9y \quad 18$
$= 8y(y + 2) - 9(y + 2)$
$= (y + 2)(8y - 9)$

43. $a = 6$, $b = 17$, $c = 12$, $ac = 72$

$6b^2 + 17b + 12$

$= 6b^2 + 9b + 8b + 12$

$= 3b(2b + 3) + 4(2b + 3)$

$= (2b + 3)(3b + 4)$

45. $a = 20$, $b = -31$, $c = 12$, $ac = 240$

$20x^2 - 31x + 12$

$= 20x^2 - 16x - 15x + 12$

$= 4x(5x - 4) - 3(5x - 4)$

$= (5x - 4)(4x - 3)$

47. $a = 18$, $b = -9$, $c = -20$, $ac = -360$

$18x^2 - 9x - 20$

$= 18x^2 - 24x + 15x - 20$

$= 6x(3x - 4) + 5(3x - 4)$

$= (3x - 4)(6x + 5)$

49. $a = 6$, $b = -19$, $c = -7$, $ac = -42$

$18p^2 - 57p - 21$

$= 3(6p^2 - 19p - 7)$

$= 3(6p^2 - 21p + 2p - 7)$

$= 3[3p(2p - 7) + 1(2p - 7)]$

$= 3(2p - 7)(3p + 1)$

51. $a = 2$, $b = 11$, $c = 9$, $ac = 18$

$2x^4 + 11x^2 + 9$

$= 2x^4 + 9x^2 + 2x^2 + 9$

$= x^2(2x^2 + 9) + 1(2x^2 + 9)$

$= (2x^2 + 9)(x^2 + 1)$

53. $a = 4$, $b = 13$, $c = -12$, $ac = -48$

$4m^4 + 13m^2 - 12$

$= 4m^4 + 16m^2 - 3m^2 - 12$

$= 4m^2(m^2 + 4) - 3(m^2 + 4)$

$= (m^2 + 4)(4m^2 - 3)$

55. $a = 6$, $b = -5$, $c = -56$, $ac = -336$

$-6x^2 + 5x + 56$

$= -1(6x^2 - 5x - 56)$

$= -1(6x^2 - 21x + 16x - 56)$

$= -1[3x(2x - 7) + 8(2x - 7)]$

$= -1(2x - 7)(3x + 8)$

57. $a = 6$, $b = 5$, $c = -6$, $ac = -36$

$6x^2 + 5xy - 6y^2$

$= 6x^2 + 9xy - 4xy - 6y^2$

$= 3x(2x + 3y) - 2y(2x + 3y)$

$= (2x + 3y)(3x - 2y)$

59. $a = 4$, $b = -39$, $c = 56$, $ac = 224$

$4u^2 - 39uv + 56v^2$

$= 4u^2 - 32uv - 7uv + 56v^2$

$= 4u(u - 8v) - 7v(u - 8v)$

$= (u - 8v)(4u - 7v)$

61. $a = 9$, $b = 13$, $c = 4$, $ac = 36$

$9x^4 + 13x^2y^2 + 4y^4$

$= 9x^4 + 9x^2y^2 + 4x^2y^2 + 4y^4$

$= 9x^2(x^2 + y^2) + 4y^2(x^2 + y^2)$

$= (x^2 + y^2)(9x^2 + 4y^2)$

63. $a = 10$, $b = 1$, $c = -21$, $ac = -210$

$10x^2y^2 + xy - 21$

$= 10x^2y^2 + 15xy - 14xy - 21$

$= 5xy(2xy + 3) - 7(2xy + 3)$

$= (2xy + 3)(5xy - 7)$

65. Let w = original width

$w + 8$ = original length

a. $w^2 + 14w + 24$

$= w^2 + 12w + 2w + 24$

$= w(w + 12) + 2(w + 12)$

$= (w + 12)(w + 2)$

The increased width is $(w + 2)$ in. and the increased length is $(w + 12)$ in.

b. $(w + 2) - w = 2$

The width was increased by 2 in.

c. $(w + 12) - (w + 8) = 4$

The length was increased by 4 in.

67. Let x = length of original triangle's small side

a. $x^2 + \dfrac{21}{2}x + 20$

$= \dfrac{1}{2}(2x^2 + 21x + 40)$

$= \dfrac{1}{2}(2x^2 + 16x + 5x + 40)$

$= \dfrac{1}{2}[2x(x+8) + 5(x+8)]$

$= \dfrac{1}{2}(x+8)(2x+5)$

The base is $(x + 8)$ in. and the height is $(2x + 5)$ in.
The lengths of the legs are $(x + 8)$ in. and $(2x + 5)$ in.

b. They were each increased by 8 in.

c. $(2x + 5) - 8 = 2x - 3$
The expression is $(2x - 3)$ in.

9.5 Experiencing Algebra the Calculator Way

1. $a = 96$, $b = -16$, $c = -2$,

$$96 \rightarrow A: \ -16 \rightarrow B: \ -2 \rightarrow C$$
$$-2$$

$Y1 = \dfrac{AC}{x}$

$Y2 = x + Y1 = B$

X	Y₁	Y₂
7	-27.43	0
8	-24	1
9	-21.33	0
10	-19.2	0
11	-17.45	0
12	-16	0
13	-14.77	0

X=7

$96x^2 - 16x - 2$

$= 96x^2 - 24x + 8x - 2$

$- 24x(4x - 1) + 2(4x - 1)$

$= (4x - 1)(24x + 2)$

$= 2(4x - 1)(12x + 1)$

2. $a = 24$, $b = 7$, $c = -55$

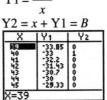
$$24 \rightarrow A: 7 \rightarrow B: \ -55 \rightarrow C$$
$$-55$$

$Y1 = \dfrac{AC}{x}$

$Y2 = x + Y1 = B$

X	Y₁	Y₂
39	-33.85	0
40	-33	1
41	-32.2	0
42	-31.43	0
43	-30.7	0
44	-30	0
45	-29.33	0

X=39

$24x^2 + 7x - 55$

$= 24x^2 + 40x - 33x - 55$

$= 8x(3x + 5) - 11(3x + 5)$

$= (3x + 5)(8x - 11)$

3. $a = 32$, $b = 102$, $c = 81$

$$32 \rightarrow A: 102 \rightarrow B: 81 \rightarrow C$$
$$81$$

$Y1 = \dfrac{AC}{x}$

$Y2 = x + Y1 = B$

X	Y₁	Y₂
47	55.149	0
48	54	1
49	52.898	0
50	51.84	0
51	50.824	0
52	49.846	0
53	48.906	0

X=47

$32x^2 + 102x + 81$

$= 32x^2 + 48x + 54x + 81$

$= 16x(2x + 3) + 27(2x + 3)$

$= (2x + 3)(16x + 27)$

4. $a = 72$, $b = -99$, $c = 34$

$$72 \rightarrow A: \ -99 \rightarrow B: 34 \rightarrow C$$
$$34$$

$Y1 = \dfrac{AC}{x}$

$Y2 = x + Y1 = B$

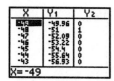

$$72x^2 - 99x + 34$$
$$= 72x^2 - 48x - 51x + 34$$
$$= 24x(3x - 2) - 17(3x - 2)$$
$$= (3x - 2)(24x - 17)$$

5. $a = 4$, $b = 109$, $c = 225$

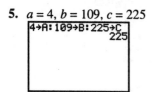

$$Y1 = \frac{AC}{x}$$
$$Y2 = x + Y1 = B$$

$$4p^4 + 109p^2 + 225$$
$$= 4p^4 + 100p^2 + 9p^2 + 225$$
$$= 4p^2(p^2 + 25) + 9(p^2 + 25)$$
$$= (p^2 + 25)(4p^2 + 9)$$

9.6 Experiencing Algebra the Exercise Way

1. $x^2 - 100$
$$= x^2 - (10)^2$$
$$= (x + 10)(x - 10)$$

3. $121 - c^2$
$$= (11)^2 - c^2$$
$$= (11 + c)(11 - c)$$

5. $49a^2 - 4$
$$= (7a)^2 - 2^2$$
$$= (7a + 2)(7a - 2)$$

7. $25 - 4y^2$
$$= 5^2 - (2y)^2$$
$$= (5 + 2y)(5 - 2y)$$

9. $16u^2 - 9v^2$
$$= (4u)^2 - (3v)^2$$
$$= (4u + 3v)(4u - 3v)$$

11. $7z^2 - 28$
$$= 7(z^2 - 4)$$
$$= 7(z^2 - 2^2)$$
$$= 7(z + 2)(z - 2)$$

13. $25 + 4p^2$ will not factor. This is not a difference of squares, but a sum of squares.

15. $x^4 - 625$
$$= (x^2)^2 - (25)^2$$
$$= (x^2 + 25)(x^2 - 25)$$
$$= (x^2 + 25)(x^2 - 5^2)$$
$$= (x^2 + 25)(x + 5)(x - 5)$$

17. $256 - z^4$
$$= (16)^2 - (z^2)^2$$
$$= (16 + z^2)(16 - z^2)$$
$$= (16 + z^2)(4^2 - z^2)$$
$$= (16 + z^2)(4 + z)(4 - z)$$

19. $x^8 - 1$
$$= (x^4)^2 - 1^2$$
$$= (x^4 + 1)(x^4 - 1)$$
$$= (x^4 + 1)[(x^2)^2 - 1^2]$$
$$= (x^4 + 1)(x^2 + 1)(x^2 - 1)$$
$$= (x^4 + 1)(x^2 + 1)(x^2 - 1^2)$$
$$= (x^4 + 1)(x^2 + 1)(x + 1)(x - 1)$$

21. $x^2 + 4x + 4$
$$= x^2 + 2(x)(2) + 2^2$$
$$= (x + 2)^2$$

23. $16z^2 + 40z + 25$
$$= (4z)^2 + 2(4z)(5) + 5^2$$
$$= (4z + 5)^2$$

25. $x^2 + 13x + 169$ does not factor. The first and last terms are perfect squares, x^2 and 13^2. However, the middle term is not $2(x)(13)$.

27. $x^2 - 10x + 25$
$= x^2 - 2(x)(5) + 5^2$
$= (x - 5)^2$

29. $36z^2 - 60z + 25$
$= (6z)^2 - 2(6z)(5) + 5^2$
$= (6z - 5)^2$

31. $c^2 - 16d + 16d^2$ does not factor. The first and last terms are perfect squares, c^2 and $(4d)^2$. However, the middle term is not $2(c)(4d)$.

33. $a^4 - 32a^2 + 256$
$= (a^2)^2 - 2(a^2)(16) + (16)^2$
$= (a^2 - 16)^2$
$= (a^2 - 16)(a^2 - 16)$
$= (a^2 - 4^2)(a^2 - 4^2)$
$= (a + 4)(a - 4)(a + 4)(a - 4)$
$= (a + 4)^2(a - 4)^2$

35. $16x^4 - 72x^2 + 81$
$= (4x^2)^2 - 2(4x^2)(9) + 9^2$
$= (4x^2 - 9)^2$
$= (4x^2 - 9)(4x^2 - 9)$
$= [(2x)^2 - 3^2][(2x)^2 - 3^2]$
$= (2x + 3)(2x - 3)(2x + 3)(2x - 3)$
$= (2x + 3)^2(2x - 3)^2$

37. $3x^2 + 24x + 48$
$= 3(x^2 + 8x + 16)$
$= 3[x^2 + 2(x)(4) + 4^2]$
$= 3(x + 4)^2$

39. $2a^2 - 12a + 18$
$= 2(a^2 - 6a + 9)$
$= 2[a^2 - 2(1)(3) + 3^2]$
$= 2(a - 3)^2$

41. $p^3 + 2p^2q + pq^2$
$= p(p^2 + 2pq + q^2)$
$= p[p^2 + 2(p)(q) + q^2]$
$= p(p + q)^2$

43. $2p^5 - 4p^3q^2 + 2pq^4$
$= 2p(p^4 - 2p^2q^2 + q^4)$
$= 2p[(p^2)^2 - 2(p^2)(q^2) + (q^2)^2]$
$= 2p(p^2 - q^2)^2$
$= 2p(p^2 - q^2)(p^2 - q^2)$
$= 2p(p + q)(p - q)(p + q)(p - q)$
$= 2p(p + q)^2(p - q)^2$

45. $m^5 + 2m^3n^2 + mn^4$
$= m(m^4 + 2m^2n^2 + n^4)$
$= m[(m^2)^2 + 2(m^2)(n^2) + (n^2)^2]$
$= m(m^2 + n^2)^2$

47. $x^3 - 27$
$= x^3 - 3^3$
$= (x - 3)(x^2 + 3x + 3^2)$
$= (x - 3)(x^2 + 3x + 9)$

49. $a^3 + 64$
$= a^3 + 4^3$
$= (a + 4)(a^2 - 4a + 4^2)$
$= (a + 4)(a^2 - 4a + 16)$

51. $27x^3 + 64y^3$
$= (3x)^3 + (4y)^3$
$= (3x + 4y)[(3x)^2 - (3x)(4y) + (4y)^2]$
$= (3x + 4y)(9x^2 - 12xy + 16y^2)$

53. $8p^3 - 125q^3$

$= (2p)^3 - (5q)^3$

$= (2p - 5q)[(2p)^2 + (2p)(5q) + (5q)^2]$

$= (2p - 5q)(4p^2 + 10pq + 25q^2)$

55. $p^4 + 64pq^3$

$= p(p^3 + 64q^3)$

$= p[p^3 + (4q)^3]$

$= p(p + 4q)[p^2 - p(4q) + (4q)^2]$

$= p(p + 4q)(p^2 - 4pq + 16q^2)$

57. $81u^4 - 3uv^3$

$= 3u(27u^3 - v^3)$

$= 3u[(3u)^3 - v^3]$

$= 3u(3u - v)[(3u)^2 + (3u)(v) + v^2]$

$= 3u(3u - v)(9u^2 + 3uv + v^2)$

59. $3y^3 + 3y^2 + 3y$

$= 3y(y^2 + y + 1)$

61. $10abc^2 + 15abc - 20ab$

$= 5ab(2c^2 + 3c - 4)$

63. $-40a^2 - 24ab + 48ac$

$= -8a(5a + 3b - 6c)$

65. $108x^5 - 75x^3$

$= 3x^3(36x^2 - 25)$

$= 3x^3[(6x)^2 - 5^2]$

$= 3x^3(6x + 5)(6x - 5)$

67. $-200x^3y + 32xy^3$

$= -8xy(25x^2 - 4y^2)$

$= -8xy[(5x)^2 - (2y)^2]$

$= -8xy(5x + 2y)(5x - 2y)$

69. $x^3 - 16x^2 + 64x$

$= x(x^2 - 16x + 64)$

$= x[x^2 - 2(x)(8) + 8^2]$

$= x(x - 8)^2$

71. $12u^2v + 36uv^2 + 27v^3$

$= 3v(4u^2 + 12uv + 9v^2)$

$= 3v[(2u)^2 + 2(2u)(3v) + (3v)^2]$

$= 3v(2u + 3v)^2$

73. $3x^4 - 48x^2 + 7x^2 - 112$

$= (3x^4 - 48x^2) + (7x^2 - 112)$

$= 3x^2(x^2 - 16) + 7(x^2 - 16)$

$= (x^2 - 16)(3x^2 + 7)$

$= (x^2 - 4^2)(3x^2 + 7)$

$= (x + 4)(x - 4)(3x^2 + 7)$

75. $4p^4 - 37p^2 + 9$

$ac = 36$

factors	sum
−36, −1	−37

$= 4p^4 - 36p^2 - p^2 + 9$

$= 4p^2(p^2 - 9) - 1(p^2 - 9)$

$= (p^2 - 9)(4p^2 - 1)$

$= (p^2 - 3^2)[(2p)^2 - 1^2]$

$= (p + 3)(p - 3)(2p + 1)(2p - 1)$

77. $2 + 8y - 42y^2$

$= -42y^2 + 8y + 2$

$= -2(21y^2 - 4y - 1)$

$ac = -21$

factors	sum
−7, 3	−4

$= -2(21y^2 - 7y + 3y - 1)$

$= -2[7y(3y - 1) + 1(3y - 1)]$

$= -2(3y - 1)(7y + 1)$

79. $4u^2v^2 + 36uv + 56$

$= 4(u^2v^2 + 9uv + 14)$

$ac = 14$

factors	sum
2, 7	9

$= 4(u^2v^2 + 2uv + 7uv + 14)$

$= 4[uv(uv + 2) + 7(uv + 2)]$

$= 4(uv + 2)(uv + 7)$

81. $32x^3 - 64x^2 - 28x^2 + 56x$

$= 4x(8x^2 - 16x - 7x + 14)$

$= 4x[8x(x - 2) - 7(x - 2)]$

$= 4x(x - 2)(8x - 7)$

83. $12x^4 + 26x^3 - 30x^2$

$= 2x^2(6x^2 + 13x - 15)$

$ac = -90$

factors	sum
18, –5	13

$= 2x(6x^2 + 18x - 5x - 15)$

$= 2x[6x(x + 3) - 5(x + 3)]$

$= 2x(x + 3)(6x - 5)$

85. $x^6 - 5x^3y^3 + 3x^3y^3 - 15y^6$

$= x^3(x^3 - 5y^3) + 3y^3(x^3 - 5y^3)$

$= (x^3 - 5y^3)(x^3 + 3y^3)$

87. $x^4 - 5x^3 + 8x^2 - 40x$

$= x(x^3 - 5x^2 + 8x - 40)$

$= x[x^2(x - 5) + 8(x - 5)]$

$= x(x - 5)(x^2 + 8)$

89. $1 - k^8$

$= 1^2 - (k^4)^2$

$= (1 + k^4)(1 - k^4)$

$= (1 + k^4)[1 - (k^2)^2]$

$= (1 + k^4)(1 + k^2)(1 - k^2)$

$= (1 + k^4)(1 + k^2)(1 + k)(1 - k)$

91. a. Large area – small area

$= x^2 - (15)^2$

$= x^2 - 225$

The garden area is $(x^2 - 225)$ ft^2.

b. $x^2 - 225$

$= x^2 - (15)^2$

$= (x + 15)(x - 15)$

The rectangular plot has dimensions of $(x + 15)$ ft by $(x - 15)$ ft.

c. $x^2 - 225 = (100)^2 - 225 = 9775$

$85(100) = 8500$

Yes, the plot that is 100 ft on each side has a larger garden area than the 85 by 100 foot plot.

9.6 Experiencing Algebra the Calculator Way

Students should check exercises using the computer program.
Answers will vary.

Chapter 9 Review

Reflections

   1–5. Answers will vary.

Exercises

   1. $-5c^2 = -5 \cdot c \cdot c$

   2. $(-5c)^2 = (-5c)(-5c)$

   3. $4(x + y)^2z^0 = 4(x + y)^2(1)$

      $= 4(x + y)(x + y)$

   4. $\left(\dfrac{2}{3x}\right)^4 = \left(\dfrac{2}{3x}\right)\left(\dfrac{2}{3x}\right)\left(\dfrac{2}{3x}\right)\left(\dfrac{2}{3x}\right)$

   5. $\dfrac{x(y - z)^{-1}}{z^{-3}} = \dfrac{xz^3}{y - z}$

6. $3y^{-3} + 2y^{-2} + y^{-1} + 10y^0$

$= \dfrac{3}{y^3} + \dfrac{2}{y^2} + \dfrac{1}{y} + 10$

7. $a^9 \cdot a^{-4} = a^{9+(-4)} = a^5$

8. $(p+q)^{-3}(p+q) = (p+q)^{-3+1}$

$= (p+q)^{-2} = \dfrac{1}{(p+q)^2}$

9. $\dfrac{t^{12}}{t^9} = t^{12-9} = t^3$

10. $\dfrac{m^{-2}}{m^{-4}} = m^{-2-(-4)} = m^2$

11. $\dfrac{(x+3y)^{-5}}{(x+3y)^{-4}} = (x+3y)^{-5-(-4)}$

$= (x+3y)^{-1} = \dfrac{1}{x+3y}$

12. $\dfrac{72x^4y^5z}{24x^2y^4z^3} = \dfrac{72}{24} \cdot \dfrac{x^4}{x^2} \cdot \dfrac{y^5}{y^4} \cdot \dfrac{z}{z^3}$

$= 3x^{4-2}y^{5-4}z^{1-3} = \dfrac{3x^2y}{z^2}$

13. $(z^4)^3 = z^{4 \cdot 3} = z^{12}$

14. $(w^{-3})^2 = w^{(-3)(2)} = w^{-6} = \dfrac{1}{w^6}$

15. $(p^{-2})^{-4} = p^{(-2)(-4)} = p^8$

16. $(a^3)^0 = a^{3 \cdot 0} = a^0 = 1$

17. $(-2a)^4 = (-2)^4 a^4 = 16a^4$

18. $(-2a)^5 = (-2)^5 a^5 = -32a^5$

19. $\left(\dfrac{-2x}{3z}\right)^3 = \dfrac{(-2)^3 x^3}{3^3 z^3} = -\dfrac{8x^3}{27z^3}$

20. $\left(\dfrac{4a}{5b}\right)^{-2} = \left(\dfrac{5b}{4a}\right)^2 = \dfrac{5^2 b^2}{4^2 a^2} = \dfrac{25b^2}{16a^2}$

21. $(a^2b^3c)^3 = a^{2 \cdot 3}b^{3 \cdot 3}c^3 = a^6b^9c^3$

22. $\left[(3x)^3\right]^2 = (3^3 x^3)^2 = (27x^3)^2 = (27)^2 x^{3 \cdot 2}$

$= 729x^6$

23. $\left[\left(\dfrac{m}{4n}\right)^2\right]^2 = \left(\dfrac{m^2}{4^2 n^2}\right)^2 = \left(\dfrac{m^2}{16n^2}\right)^2$

$= \dfrac{m^{2 \cdot 2}}{(16)^2 n^{2 \cdot 2}} = \dfrac{m^4}{256n^4}$

24. Let $x = $ length of original side

$\dfrac{1}{4}x = $ length of reduced side

Area of original square: x^2 square units
Area of reduced square:

$\left(\dfrac{1}{4}x\right)^2 = \dfrac{x^2}{16}$ square units

The smaller area is $\dfrac{1}{16}$ of the original area.

25. Let $r = $ original radius
$2r = $ increased radius

Area of original garden: πr^2 square units
Area of enlarged garden:
$\pi(2r)^2 = 4\pi r^2$ square units
The enlarged garden has an area that is 4
times that of the original garden.
$4\pi r^2 = 4\pi(6)^2 = 144\pi$

The garden has an area of 144π ft^2.

26.

$$
\begin{array}{l}
5x^4 + 3x^3 \qquad\quad + 6x - 3 \\
\qquad 4x^3 + 5x^2 \qquad\quad + 7 \\
\;\; x^4 \qquad\quad - 3x^2 + \;\; x - 9 \\
\underline{\qquad x^3 + 2x^2 - 7x + 1} \\
6x^4 + 8x^3 + 4x^2 + 0x - 4 \\
\qquad\qquad \text{or} \\
6x^4 + 8x^3 + 4x^2 - 4
\end{array}
$$

27. $(4y^2 + 3y - 7) + (2y + 8) + (5y^2 - 4y + 1)$

$= 4y^2 + 3y - 7 + 2y + 8 + 5y^2 - 4y + 1$

$= 9y^2 + y + 2$

28. $(3.57z^3 - 2.08z^2 + 8.77z - 1.99) + (4.73 - 2.98z + 5.64z^2)$

$= 3.57z^3 - 2.08z^2 + 8.77z - 1.99 + 4.73 - 2.98z + 5.64z^2$

$= 3.57z^3 + 3.56z^2 + 5.79z + 2.74$

29. $(4x^3 + 8x^2 - 6x + 2) - (2x^3 + 12x^2 - x + 9)$

$= 4x^3 + 8x^2 - 6x + 2 - 2x^3 - 12x^2 + x - 9$

$= 2x^3 - 4x^2 - 5x - 7$

30. $(5a^4 + a^3 + a^2 + a + 1) - (a^4 + 2a^3 + 3a^2 + 4a + 5)$

$= 5a^4 + a^3 + a^2 + a + 1 - a^4 - 2a^3 - 3a^2 - 4a - 5$

$= 4a^4 - a^3 - 2a^2 - 3a - 4$

31. $(65z^4 + 27z^2 + 36) - (16z^3 + 8z + 12)$

$= 65z^4 + 27z^2 + 36 - 16z^3 - 8z - 12$

$= 65z^4 - 16z^3 + 27z^2 - 8z + 24$

32. $\left(\dfrac{5}{8}b^4 + \dfrac{7}{8}b^3 - \dfrac{3}{4}b^2 + \dfrac{1}{2}b - \dfrac{1}{4} \right) - \left(\dfrac{1}{2}b^4 + \dfrac{3}{8}b^2 + \dfrac{1}{8}b \right)$

$= \dfrac{5}{8}b^4 + \dfrac{7}{8}b^3 - \dfrac{3}{4}b^2 + \dfrac{1}{2}b - \dfrac{1}{4} - \dfrac{1}{2}b^4 - \dfrac{3}{8}b^2 - \dfrac{1}{8}b$

$= \left(\dfrac{5}{8} - \dfrac{1}{2} \right)b^4 + \dfrac{7}{8}b^3 + \left(-\dfrac{3}{4} - \dfrac{3}{8} \right)b^2 + \left(\dfrac{1}{2} - \dfrac{1}{8} \right)b - \dfrac{1}{4}$

$= \dfrac{1}{8}b^4 + \dfrac{7}{8}b^3 - \dfrac{9}{8}b^2 + \dfrac{3}{8}b - \dfrac{1}{4}$

33. $-3a^3b(5a^{-2}b^3)$

$= -3(5)a^{3-2}b^{1+3}$

$= -15ab^4$

34. $6x^3(3x^2 + 2x - 7)$
$= 6x^3(3x^2) + 6x^3(2x) + 6x^3(-7)$
$= 18x^{3+2} + 12x^{3+1} - 42x^3$
$= 18x^5 + 12x^4 - 42x^3$

35. $7a^{-2}(3a^6 + a^4 - 2a^2)$
$= 7a^{-2}(3a^6) + 7a^{-2}(a^4) + 7a^{-2}(-2a^2)$
$= 21a^{-2+6} + 7a^{-2+4} - 14a^{-2+2}$
$= 21a^4 + 7a^2 - 14$

36. $(-6.9x^3z^{-4})(3.4xz^3)$
$= (-6.9)(3.4)x^{3+1}z^{-4+3}$
$= -23.46x^4z^{-1}$
$= -\dfrac{23.46x^4}{z}$

37. $\dfrac{144x^6y^3z^4}{2x^5y^2z^6}$
$= \dfrac{144}{2}x^{6-5}y^{3-2}z^{4-6}$
$= 72x^1y^1z^{-2}$
$= \dfrac{72xy}{z^2}$

38. $\dfrac{(2a^2)^3}{24a^5}$
$= \dfrac{2^3a^{2\cdot3}}{24a^5}$
$= \dfrac{8a^6}{24a^5}$
$= \dfrac{8}{24}a^{6-5}$
$= \dfrac{a}{3}$ or $\dfrac{1}{3}a$

39. $\left(\dfrac{27a^5b^3}{9a^2b^7}\right)^2$
$= \left(\dfrac{27}{9}a^{5-2}b^{3-7}\right)^4$
$= \left(\dfrac{3a^3}{b^4}\right)^2$
$= \dfrac{3^2a^{3\cdot2}}{b^{4\cdot2}}$
$= \dfrac{9a^6}{b^8}$

40. $\dfrac{824,500,000y^{-5}}{0.0097y^7}$
$= 85,000,000,000y^{-5-7}$
$= \dfrac{85,000,000,000}{y^{12}}$ or $\dfrac{8.5\times10^{10}}{y^{12}}$

41. $\dfrac{15b^4 - 10b^3 - 25b^2 + 5b}{-5b}$

$= \dfrac{15b^4}{-5b} - \dfrac{10b^3}{-5b} - \dfrac{25b^2}{-5b} + \dfrac{5b}{-5b}$

$= -3b^{4-1} + 2b^{3-1} + 5b^{2-1} - 1b^{1-1}$

$= -3b^3 + 2b^2 + 5b - 1$

42. $\dfrac{6ab^2 + 18a^2b}{3a^2b^2}$

$= \dfrac{6ab^2}{3a^2b^2} + \dfrac{18a^2b}{3a^2b^2}$

$= 2a^{1-2}b^{2-2} + 6a^{2-2}b^{1-2}$

$= 2a^{-1}b^0 + 6a^0b^{-1}$

$= \dfrac{2}{a} + \dfrac{6}{b}$

43. a. $C(x) = 10 + 3.5x$

   b. $R(x) = (10 - 0.25x)x$

   c. $P(x) = R(x) - C(x)$
$P(x) = (10 - 0.25x)x - (10 + 3.5x)$
$P(x) = 10x - 0.25x^2 - 10 - 3.5x$
$P(x) = -0.25x^2 + 6.5x - 10$

   d. $A(x) = \dfrac{-0.25x^2 + 6.5x - 10}{x}$

$A(x) = -\dfrac{0.25x^2}{x} + \dfrac{6.5x}{x} - \dfrac{10}{x}$

$A(x) = -0.25x + 6.5 - \dfrac{10}{x}$

   e. $x = 13$

44. Let length = 20 meters
width = $(20 - x)$ meters
Perimeter: $2(20) + 2(20 - x)$
$= 40 + 40 - 2x$
$= 80 - 2x$
The perimeter is $(80 - 2x)$ meters.

45. Area = Patio area − Box area
Area = $x(x + 8) - 3^2$
Area = $x^2 + 8x - 9$
$9^2 + 8(9) - 9 = 144$
The patio needs 144 ft^2 of carpeting.

46. $(y + 9)^2$
$= y^2 + 2y(9) + 9^2$
$= y^2 + 18y + 81$

47. $(x^3 - 5)^2$
$= (x^3)^2 - 2(x^3)(5) + 5^2$
$= x^6 - 10x^3 + 25$

48. $(5x - 2)(x + 11)$
$= 5x(x + 11) - 2(x + 11)$
$= 5x^2 + 55x - 2x - 22$
$= 5x^2 + 53x - 22$

49. $(7 - z)(7 + z)$
$= 7^2 - z^2$
$= 49 - z^2$

50. $(y - 1.8)(y + 3.4)$
$= y(y + 3.4) - 1.8(y + 3.4)$
$= y^2 + 3.4y - 1.8y - 1.8(3.4)$
$= y^2 + 1.6y - 6.12$

51. $(z^2 - 10)(z^2 + 10)$
$= (z^2)^2 - (10)^2$
$= z^4 - 100$

52. $\left(\dfrac{4}{5}x - \dfrac{1}{2}\right)\left(\dfrac{4}{5}x + \dfrac{1}{2}\right)$

$= \left(\dfrac{4}{5}x\right)^2 - \left(\dfrac{1}{2}\right)^2$

$= \dfrac{16}{25}x^2 - \dfrac{1}{4}$

53. $(b-4)^3$

$= (b-4)(b-4)(b-4)$

$= [b(b-4)-4(b-4)](b-4)$

$= (b^2 - 4b - 4b + 16)(b-4)$

$= (b^2 - 8b + 16)(b-4)$

$= (b^2 - 8b + 16)b + (b^2 - 8b + 16)(-4)$

$= b^3 - 8b^2 + 16b - 4b^2 + 32b - 64$

$= b^3 - 12b^2 + 48b - 64$

54. $(2x+1)(3x^2 + 5x - 4)$

$= 2x(3x^2 + 5x - 4) + 1(3x^2 + 5x - 4)$

$= 6x^3 + 10x^2 - 8x + 3x^2 + 5x - 4$

$= 6x^3 + 13x^2 - 3x - 4$

55. $(a+b+c)^2$

$= (a+b+c)(a+b+c)$

$= a(a+b+c) + b(a+b+c) + c(a+b+c)$

$= a^2 + ab + ac + ab + b^2 + bc + ac + bc + c^2$

$= a^2 + b^2 + c^2 + 2ab + 2bc + 2ac$

56. $(a-3)(a^2 + 3a + 9)$

$= a(a^2 + 3a + 9) - 3(a^2 + 3a + 9)$

$= a^3 + 3a^2 + 9a - 3a^2 - 9a - 27$

$= a^3 - 27$

57. a. length: $(x+3)$ in.

width: $(x-3)$ in.

height: x in.

b. Volume:

$x(x+3)(x-3)$

$= x(x^2 - 3^2)$

$= x(x^2 - 9)$

$= x^3 - 9x$

The volume is $(x^3 - 9x)$ in^3.

c. Surface area:

$2(x+3)(x-3) + 2x(x+3) + 2x(x-3)$

$= 2(x^2 - 3^2) + 2x(x) + 2x(3) + 2x(x) - 2x(3)$

$= 2x^2 - 18 + 2x^2 + 6x + 2x^2 - 6x$

$= 6x^2 - 18$

The surface area is $(6x^2 - 18)$ in^2.

58. a.   current length $= 3x$
current width $= x$
Area $= x(3x) = 3x^2$
The current area is $3x^2$ ft^2.

b.   new length $= 3x + 9$
new width $= 2x$
Area:
$2x(3x + 9)$
$= 2x(3x) + 2x(9)$
$= 6x^2 + 18x$
The new area is $(6x^2 + 18x)$ ft^2.

c.   Ratio of planned area to current area:
$$\frac{6x^2 + 18x}{3x^2}$$
$$= \frac{3x(2x + 6)}{3x(x)}$$
$$= \frac{2x + 6}{x}$$
The ratio is $\dfrac{2x + 6}{x}$.

59. $20a^6 - 28a^4 + 44a^2$
$= 4a^2(5a^4 - 7a^2 + 11)$

60. $22u^3v^2 + 22u^2v^3$
$= 22u^2v^2(u + v)$

61. $3x^3 + 3x + x^2 + 1$
$= (3x^3 + 3x) + (x^2 + 1)$
$= 3x(x^2 + 1) + 1(x^2 + 1)$
$= (x^2 + 1)(3x + 1)$

62. $7a^4 + 7a^2b^2 + 7a^2b^2 + 7b^4$
$= 7(a^4 + a^2b^2 + a^2b^2 + b^4)$
$= 7[(a^4 + a^2b^2) + (a^2b^2 + b^4)]$
$= 7[a^2(a^2 + b^2) + b^2(a^2 + b^2)]$
$- 7(a^2 + b^2)(a^2 + b^2)$
$= 7(a^2 + b^2)^2$

63. $15ac + 18ad + 20bc + 24bd$
$= (15ac + 18ad) + (20bc + 24bd)$
$= 3a(5c + 6d) + 4b(5c + 6d)$
$= (5c + 6d)(3a + 4b)$

64. a. $\dfrac{1}{2}n^2 + \dfrac{1}{2}n$
$= \dfrac{1}{2}n(n + 1)$

b. $\dfrac{1}{2}n^2 + \dfrac{1}{2}n = \dfrac{1}{2}(12)^2 + \dfrac{1}{2}(12) = 78$
$\dfrac{1}{2}n(n + 1) = \dfrac{1}{2}(12)(12 + 1) = 78$
The sum is 78.

c.   The factored expression is easier to evaluate. Answers will vary.

65. $2x^2 - 6x + 5x - 15$
$= (2x^2 - 6x) + (5x - 15)$
$= 2x(x - 3) + 5(x - 3)$
$= (x - 3)(2x + 5)$
The width is $(x - 3)$ units and the length is $(2x + 5)$ units.

66. Total pay:
$x + (x + 10) + (x + 20) + (x + 30) + (x + 40) + (x + 50) + (x + 60)$
$= 7x + 210$
$= 7(x + 30)$
The employee earns an average of
$(x + 30)$ dollars for each of 7 months.
$7(960 + 30) = 6930$
The salary is $6930.

67. $z^2 + 2z - 99$
$a = 1, b = 2, c = -99$

factor	factor	sum of factors	
–1	99	98	
1	–99	–98	
–3	33	30	
3	–33	–30	
–9	11	2	← b
9	–11	–2	

$z^2 + 2z - 99 = (z - 9)(z + 11)$

68. $p^2 + 5pq - 66q^2$
$a = 1, b = 5, c = -66$

factor	factor	sum of factors	
–1	66	65	
1	–66	–65	
–2	33	31	
2	–33	–31	
–3	22	19	
3	–22	–19	
–6	11	5	← b
6	–11	–5	

$p^2 + 5pq - 66q^2 = (p - 6q)(p + 11q)$

69. $6a^2 + 96a + 234$
$= 6(a^2 + 16a + 39)$
$a = 1, b = 16, c = 39$

factor	factor	sum of factors	
1	39	40	
3	13	16	← b

$6a^2 + 96a + 234 = 6(a + 3)(a + 13)$

70. $x^4 + 8x^2 + 15$
$a = 1, b = 8, c = 15$

factor	factor	sum of factors	
1	15	16	
3	5	8	← b

$x^4 + 8x^2 + 15 = (x^2 + 3)(x^2 + 5)$

71. $4q^3 - 28q^2 - 240q$

$= 4q(q^2 - 7q - 60)$

$a = 1, b = -7, c = -60$

factor	factor	sum of factors	
−1	60	59	
1	−60	−59	
−2	30	28	
2	−30	−28	
−3	20	17	
3	−20	−17	
−4	15	11	
4	−15	−11	
−5	12	7	
5	−12	−7	← *b*
−6	10	4	
6	−10	−4	

$4q^3 - 28q^2 - 240q = 4q(q + 5)(q - 12)$

72. $x^2y^2 - 4xy - 117$

$a = 1, b = -4, c = -117$

factor	factor	sum of factors	
−1	117	116	
1	−117	−116	
−3	39	36	
3	−39	−36	
−9	13	4	
9	−13	−4	← *b*

$x^2y^2 - 4xy - 117 = (xy + 9)(xy - 13)$

73. $-7x^2 + 98x - 168$

$= -7(x^2 - 14x + 24)$

$a = 1, b = -14, c = 24$

factor	factor	sum of factors	
−1	−24	−25	
−2	−12	−14	← *b*
−3	−8	−11	
−4	−6	−10	

$-7x^2 + 98x - 168 = -7(x - 2)(x - 12)$

74. $a = 2, b = -11, c = 5, ac = 10$

$2x^2 - 11x + 5$

$= 2x^2 - 10x - x + 5$

$= 2x(x - 5) - 1(x - 5)$

$= (x - 5)(2x - 1)$

75. $a = 6, b = 17, c = 5, ac = 30$

$6x^2 + 17x + 5$

$= 6x^2 + 15x + 2x + 5$

$= 3x(2x + 5) + 1(2x + 5)$

$= (2x + 5)(3x + 1)$

76. $a = 4, b = 13, c = 3, ac = 12$

$28a^2b^2 + 91ab + 21$

$= 7(4a^2b^2 + 13ab + 3)$

$= 7(4a^2b^2 + 12ab + ab + 3)$

$= 7[4ab(ab + 3) + 1(ab + 3)]$

$= 7(ab + 3)(4ab + 1)$

77. $a = 15, b = 34, c = -16, ac = -240$

$-45x^3 - 102x^2 + 48x$

$= -3x(15x^2 + 34x - 16)$

$= -3x(15x^2 + 40x - 6x - 16)$

$= -3x[5x(3x + 8) - 2(3x + 8)]$

$= -3x(3x + 8)(5x - 2)$

78. a. $2x^2 + 10x + \dfrac{21}{2}$

$$= \frac{1}{2}(4x^2 + 20x + 21)$$

$$= \frac{1}{2}(4x^2 + 6x + 14x + 21)$$

$$= \frac{1}{2}[2x(2x+3) + 7(2x+3)]$$

$$= \frac{1}{2}(2x+3)(2x+7)$$

The base is $(2x + 7)$ units and the height is $(2x + 3)$ units.

b. $2x + 7 - 2x = 7$
The base was increased by 7 units.

c. $2x + 3 - x = x + 3$
The height was increased by $(x + 3)$ units.

79. a. $x^2 + 17x + 30$
$= x^2 + 2x + 15x + 30$
$= x(x + 2) + 15(x + 2)$
$= (x + 2)(x + 15)$
The new length is $(x + 15)$ units and the new width is $(x + 2)$ units.

b. $x + 15 - x = 15$
It was increased by 15 units.

c. $x + 2 - (x - 6) = 8$
It was increased by 8 units.

80. $x^2 - 169$
$= x^2 - (13)^2$
$= (x + 13)(x - 13)$

81. $625 - a^2$
$= (25)^2 - a^2$
$= (25 + a)(25 - a)$

82. $12x^2 - 75$
$= 3(4x^2 - 25)$
$= 3[(2x)^2 - 5^2]$
$= 3(2x + 5)(2x - 5)$

83. $p^2 - q^2$
$= (p + q)(p - q)$

84. $p^2 + q^2$ will not factor.
This is not the difference of squares, but the sum of squares.

85. $9x^2 - 25y^2$
$= (3x)^2 - (5y)^2$
$= (3x + 5y)(3x - 5y)$

86. $16x^4 - 81$
$= (4x^2)^2 - 9^2$
$= (4x^2 + 9)(4x^2 - 9)$
$= (4x^2 + 9)[(2x)^2 - 3^2]$
$= (4x^2 + 9)(2x + 3)(2x - 3)$

87. $p^2 + 12p + 36$
$= p^2 + 2p(6) + 6^2$
$= (p + 6)^2$

88. $q^2 - 16q + 64$
$= q^2 - 2q(8) + 8^2$
$= (q - 8)^2$

89. $c^3 + 27$
$= c^3 + 3^3$
$= (c + 3)(c^2 - 3c + 3^2)$
$= (c + 3)(c^2 - 3c + 9)$

90. $8z^3 - 125$
$= (2z)^3 - 5^3$
$= (2z - 5)[(2z)^2 + (2z)(5) + 5^2]$
$= (2z - 5)(4z^2 + 10z + 25)$

91. $5h^3 + 40k^3$
$= 5(h^3 + 8k^3)$
$= 5[h^3 + (2k)^3]$
$= 5(h + 2k)[h^2 + h(2k) + (2k)^2]$
$= 5(h + 2k)(h^2 + 2hk + 4k^2)$

92. $-12x^3 + 60x^2y - 75xy^2$
$$= -3x(4x^2 - 20xy + 25y^2)$$
$$= -3x[(2x)^2 - 2(2x)(5y) + (5y)^2]$$
$$= -3x(2x - 5y)^2$$

93. $7x^4 + 7x^3 + 7x^2$
$$= 7x^2(x^2 + x + 1)$$

94. $12x^3 - 243x$
$$= 3x(4x^2 - 81)$$
$$= 3x[(2x)^2 - 9^2]$$
$$= 3x(2x + 9)(2x - 9)$$

95. $32x^3 + 32x^2 + 8x$
$$= 8x(4x^2 + 4x + 1)$$
$$= 8x[(2x)^2 + 2(2x)(1) + 1^2]$$
$$= 8x(2x + 1)^2$$

96. $24x^3 - 14x^2 - 90x$
$$= 2x(12x^2 - 7x - 45)$$
$$= 2x(12x^2 - 27x + 20x - 45)$$
$$= 2x[3x(4x - 9) + 5(4x - 9)]$$
$$= 2x(4x - 9)(3x + 5)$$

97. $256x^4 - 288x^2 + 81$
$$= (16x^2)^2 - 2(16x^2)(9) + 9^2$$
$$= (16x^2 - 9)^2$$
$$= (16x^2 - 9)(16x^2 - 9)$$
$$= [(4x)^2 - 3^2][(4x)^2 - 3^2]$$
$$= (4x + 3)(4x - 3)(4x + 3)(4x - 3)$$
$$= (4x + 3)^2(4x - 3)^2$$

98. $36x^4 - 25x^2 + 4$
$$= 36x^4 - 16x^2 - 9x^2 + 4$$
$$= 4x^2(9x^2 - 4) - 1(9x^2 - 4)$$
$$= (9x^2 - 4)(4x^2 - 1)$$
$$= [(3x)^2 - (2)^2][(2x)^2 - (1)^2]$$
$$= (3x + 2)(3x - 2)(2x + 1)(2x - 1)$$

99. $2x^4 + 14x^3 - 8x^2 - 56x$
$$= 2x(x^3 + 7x^2 - 4x - 28)$$
$$= 2x[x^2(x + 7) - 4(x + 7)]$$
$$= 2x(x + 7)(x^2 - 4)$$
$$= 2x(x + 7)(x - 2)(x + 2)$$

100. a. Area of land – Area of base
$$= x(4x) - 5^2$$
$$= 4x^2 - 25$$
The land area not covered is
$(4x^2 - 25)$ ft^2.

b. $4x^2 - 25$
$$= (2x)^2 - 5^2$$
$$= (2x + 5)(2x - 5)$$
The dimensions would be $(2x + 5)$ ft by $(2x - 5)$ ft.

c. $(2 \cdot 80 + 5)$ ft by $(2 \cdot 80 - 5)$ ft
165 ft by 155 ft
The dimensions would be 165 ft by 155 ft.

Chapter 9 Mixed Review

1. $z^2 + 9z - 90$
$$= z^2 + 15z - 6z - 90$$
$$= z(z + 15) - 6(z + 15)$$
$$= (z + 15)(z - 6)$$

2. $a^2 - 18a + 72$
$$= a^2 - 6a - 12a + 72$$
$$= a(a - 6) - 12(a - 6)$$
$$= (a - 6)(a - 12)$$

3. $x^2 + 14xy + 45y^2$
$$= x^2 + 9xy + 5xy + 45y^2$$
$$= x(x + 9y) + 5y(x + 9y)$$
$$= (x + 9y)(x + 5y)$$

4. $5a^2 + 70a + 245$
$$= 5(a^2 + 14a + 49)$$
$$= 5[a^2 + 2a(7) + 7^2]$$
$$= 5(a + 7)^2$$

5. $2a^2 + 8ab + 12b^2$
$= 2(a^2 + 4ab + 6b^2)$

6. $x^4 + 10x^2 + 21$
$= x^4 + 3x^2 + 7x^2 + 21$
$= x^2(x^2 + 3) + 7(x^2 + 3)$
$= (x^2 + 3)(x^2 + 7)$

7. $3q^3 - 33q^2 - 126q$
$= 3q(q^2 - 11q - 42)$
$= 3q(q^2 - 3q + 14q - 42)$
$= 3q[q(q - 3) + 14(q - 3)]$
$= 3q(q - 3)(q + 14)$

8. $-6x^2 + 42x + 360$
$= -6(x^2 - 7x - 60)$
$= -6(x^2 + 5x - 12x - 60)$
$= -6[x(x + 5) - 12(x + 5)]$
$= -6(x + 5)(x - 12)$

9. $10 + 7x + x^2$
$= x^2 + 7x + 10$
$= x^2 + 2x + 5x + 10$
$= x(x + 2) + 5(x + 2)$
$= (x + 2)(x + 5)$

10. $x^2 - 289$
$= x^2 - (17)^2$
$= (x + 17)(x - 17)$

11. $x^3 - 1$
$= x^3 - 1^3$
$= (x - 1)[x^2 + x(1) + 1^2]$
$= (x - 1)(x^2 + x + 1)$

12. $4x^2 - 64$
$= 4(x^2 - 16)$
$= 4(x^2 - 4^2)$
$= 4(x + 4)(x - 4)$

13. $u^2 + v^2$ does not factor. It is the sum of two squares, not the difference of two squares.

14. $36x^2 - 49y^2$
$= (6x)^2 - (7y)^2$
$= (6x + 7y)(6x - 7y)$

15. $81x^4 - 1$
$= (9x^2)^2 - 1^2$
$= (9x^2 + 1)(9x^2 - 1)$
$= (9x^2 + 1)[(3x)^2 - 1^2]$
$= (9x^2 + 1)(3x + 1)(3x - 1)$

16. $p^2 + 22p + 121$
$= p^2 + 2p(11) + (11)^2$
$= (p + 11)^2$

17. $q^2 - 30q + 225$
$= q^2 - 2q(15) + (15)^2$
$= (q - 15)^2$

18. $27a^3 + 64b^3$
$= (3a)^3 + (4b)^3$
$= (3a + 4b)[(3a)^2 + 3a(4b) + (4b)^2]$
$= (3a + 4b)(9a^2 + 12ab + 16b^2)$

19. $27a^2b^2 - 72ab + 48$
$= 3(9a^2b^2 - 24ab + 16)$
$= 3[(3ab)^2 - 2(3ab)(4) + 4^2]$
$= 3(3ab - 4)^2$

20. $-50x^3 + 120x^2y - 72xy^2$
$= -2x(25x^2 - 60xy + 36y^2)$
$= -2x[(5x)^2 - 2(5x)(6y) + (6y)^2]$
$= -2x(5x - 6y)^2$

21. $8x^4 - 2x^3 + 6x^2 - 12x$
$= 2x(4x^3 - x^2 + 3x - 6)$

22. $35u^3v^2 + 25u^2v^3$
$= 5u^2v^2(7u + 5v)$

23. $2x^3 + 10x + x^2 + 5$
$= 2x(x^2 + 5) + 1(x^2 + 5)$
$= (x^2 + 5)(2x + 1)$

24. $m^2 - 2mn - 8mn + 16n^2$
$= m(m - 2n) - 8n(m - 2n)$
$= (m - 2n)(m - 8n)$

25. $4a^4 + 8a^2b^2 + 8a^2b^2 + 16b^4$
$= 4(a^4 + 2a^2b^2 + 2a^2b^2 + 4b^4)$
$= 4[a^2(a^2 + 2b^2) + 2b^2(a^2 + 2b^2)]$
$= 4(a^2 + 2b^2)(a^2 + 2b^2)$
$= 4(a^2 + 2b^2)^2$

26. $2x^2 - 13x + 11$
$= 2x^2 - 2x - 11x + 11$
$= 2x(x - 1) - 11(x - 1)$
$= (x - 1)(2x - 11)$

27. $24ac + 20ad + 18bc + 15bd$
$= 4a(6c + 5d) + 3b(6c + 5d)$
$= (6c + 5d)(4a + 3b)$

28. $7x^2 - 19x - 6$
$= 7x^2 - 21x + 2x - 6$
$= 7x(x - 3) + 2(x - 3)$
$= (x - 3)(7x + 2)$

29. $10x^2 - 11x - 6$
$= 10x^2 - 15x + 4x - 6$
$= 5x(2x - 3) + 2(2x - 3)$
$= (2x - 3)(5x + 2)$

30. $36a^2 + 66a + 24$
$= 6(6a^2 + 11a + 4)$
$= 6(6a^2 + 3a + 8a + 4)$
$= 6[3a(2a + 1) + 4(2a + 1)]$
$= 6(2a + 1)(3a + 4)$

31. $-30x^2 - 28x + 32$
$= -2(15x^2 + 14x - 16)$
$= -2(15x^2 + 10x - 24x - 16)$
$= -2[5x(3x + 2) - 8(3x + 2)]$
$= -2(3x + 2)(5x - 8)$

32. $12x^4 + 13x^2 + 3$
$= 12x^4 + 4x^2 + 9x^2 + 3$
$= 4x^2(3x^2 + 1) + 3(3x^2 + 1)$
$= (3x^2 + 1)(4x^2 + 3)$

33. $54x^3 + 36x^2 + 6x$
$= 6x(9x^2 + 6x + 1)$
$= 6x\left[(3x)^2 + 2(3x)(1) + 1^2\right]$
$= 6x(3x + 1)^2$

34. $81x^4 - 72x^2 + 16$
$= (9x^2)^2 - 2(9x^2)(4) + 4^2$
$= (9x^2 - 4)^2$
$= \left[(3x)^2 - 2^2\right]^2$
$= [(3x + 2)(3x - 2)]^2$
$= (3x + 2)^2(3x - 2)^2$

35. $4x^4 - 61x^2 + 225$
$= (4x^4 - 36x^2) - (25x^2 + 225)$
$= 4x^2(x^2 - 9) - 25(x^2 - 9)$
$= (4x^2 - 25)(x^2 - 9)$
$= [(2x)^2 - (5)^2][x^2 - (3)^2]$
$= (2x + 5)(2x - 5)(x + 3)(x - 3)$

36. $12x^3 + 18x^2 - 30x^2 - 45x$
$= 3x(4x^2 + 6x - 10x - 15)$
$= 3x[2x(2x + 3) - 5(2x + 3)]$
$= 3x(2x + 3)(2x - 5)$

37. $3x^4 + 15x^3 - 27x^2 - 135x$
$= 3x(x^3 + 5x^2 - 9x - 45)$
$= 3x[x^2(x + 5) - 9(x + 5)]$
$= 3x(x + 5)(x^2 - 9)$
$= 3x(x + 5)(x + 3)(x - 3)$

38. $m^{-7}n^5 = \dfrac{n^5}{m^7}$

39. $\dfrac{5^{-3}t^3}{4^{-3}s^4} = \dfrac{4^3t^3}{5^3s^4} = \dfrac{64t^3}{125s^4}$

40. $\dfrac{(d+2e)^3}{(d-2e)^{-2}} = (d+2e)^2 (d-2e)^2$

41. $2y^{-4} + 4y^{-3} + 6y^{-2} + 8y^{-1} + 10y^0$

$= \dfrac{2}{y^4} + \dfrac{4}{y^3} + \dfrac{6}{y^2} + \dfrac{8}{y} + 10$

42. $(2y^2 + 4y - 3) + (9y + 7) + (8y^2 - 12y + 15)$

$= 2y^2 + 4y - 3 + 9y + 7 + 8y^2 - 12y + 15$

$= 10y^2 + y + 19$

43. $\left(\dfrac{3}{8}a^3 + \dfrac{3}{4}a^2 - \dfrac{5}{8}a + \dfrac{1}{4}\right) + \left(\dfrac{3}{4}a^3 + \dfrac{7}{16}a - \dfrac{5}{8}a^2 + \dfrac{11}{16}\right)$

$= \dfrac{3}{8}a^3 + \dfrac{3}{4}a^2 - \dfrac{5}{8}a + \dfrac{1}{4} + \dfrac{3}{4}a^3 + \dfrac{7}{16}a - \dfrac{5}{8}a^2 + \dfrac{11}{16}$

$= \left(\dfrac{3}{8} + \dfrac{3}{4}\right)a^3 + \left(\dfrac{3}{4} - \dfrac{5}{8}\right)a^2 + \left(-\dfrac{5}{8} + \dfrac{7}{16}\right)a + \left(\dfrac{1}{4} + \dfrac{11}{16}\right)$

$= \dfrac{9}{8}a^3 + \dfrac{1}{8}a^2 - \dfrac{3}{16}a + \dfrac{15}{16}$

44. $(4.9z^3 - 6.82z^2 + 12z - 11.07) + (4.6 - 1.83z + 4.9z^2)$

$= 4.9z^3 - 6.82z^2 + 12z - 11.07 + 4.6 - 1.83z + 4.9z^2$

$= 4.9z^3 - 1.92z^2 + 10.17z - 6.47$

45. $(2x^3 + 6x^2 - 9x + 13) - (4x^3 + 17x^2 - x + 6)$

$= 2x^3 + 6x^2 - 9x + 13 - 4x^3 - 17x^2 + x - 6$

$= -2x^3 - 11x^2 - 8x + 7$

46. $(6a^4 + 3a^2 + 4a + 5) - (5a^4 + 4a^3 + 3a^2 + 2a + 1)$

$= 6a^4 + 3a^2 + 4a + 5 - 5a^4 - 4a^3 - 3a^2 - 2a - 1$

$= a^4 - 4a^3 + 2a + 4$

47. $(117z^4 + 43z^2 + 88) - (18z^3 + 50z + 32)$

$= 117z^4 + 43z^2 + 88 - 18z^3 - 50z - 32$

$= 117z^4 - 18z^3 + 43z^2 - 50z + 56$

48. $11x^5 y^6 (13x^{-2} y^{-6})$

$= (11)(13)x^{5-2} y^{6-6}$

$= 143x^3 y^0$

$= 143x^3$

49. $-6x(2x^4 - 4x^3 + 6x^2 + 8x - 10)$
$= -12x^{1+4} + 24x^{1+3} - 36x^{1+2} - 48x^{1+1} + 60x$
$= -12x^5 + 24x^4 - 36x^3 - 48x^2 + 60x$

50. $9a^{-1}(4a^5 + 2a^3 - 3a)$
$= 36a^{-1+5} + 18a^{-1+3} - 27a^{-1+1}$
$= 36a^4 + 18a^2 - 27a^0$
$= 36a^4 + 18a^2 - 27$

51. $(m - 11)(m + 11)$
$= m(m + 11) - 11(m + 11)$
$= m^2 + 11m - 11m - 121$
$= m^2 - 121$

52. $(z - 8)^2$
$= z^2 - 2(z)(8) + 8^2$
$= z^2 - 16z + 64$

53. $(4a - 7)(a + 13)$
$= 4a(a + 13) - 7(a + 13)$
$= 4a^2 + 52a - 7a - 91$
$= 4a^2 + 45a - 91$

54. $(13 - x)(13 + x)$
$= (13)^2 - x^2$
$= 169 - x^2$

55. $(b^3 + 4)^2$
$= (b^3)^2 + 2(b^3)(4) + 4^2$
$= b^6 + 8b^3 + 16$

56. $(t + 2)^3$
$= (t + 2)(t + 2)(t + 2)$
$= [t(t + 2) + 2(t + 2)](t + 2)$
$= (t^2 + 2t + 2t + 4)(t + 2)$
$= (t^2 + 4t + 4)(t + 2)$
$= (t^2 + 4t + 4)t + (t^2 + 4t + 4)(2)$
$= t^3 + 4t^2 + 4t + 2t^2 + 8t + 8$
$= t^3 + 6t^2 + 12t + 8$

57. $(7x - 2)(x^2 + 4x - 3)$
$= 7x(x^2 + 4x - 3) - 2(x^2 + 4x - 3)$
$= 7x^3 + 28x^2 - 21x - 2x^2 - 8x + 6$
$= 7x^3 + 26x^2 - 29x + 6$

58. $(p + q + r)^2$
$= (p + q + r)(p + q + r)$
$= p(p + q + r) + q(p + q + r) + r(p + q + r)$
$= p^2 + pq + pr + pq + q^2 + qr + pr + qr + r^2$
$= p^2 + q^2 + r^2 + 2pq + 2qr + 2pr$

59. $(b - 4)(b^2 + 4b + 16)$
$= b(b^2 + 4b + 16) - 4(b^2 + 4b + 16)$
$= b^3 + 4b^2 + 16b - 4b^2 - 16b - 64$
$= b^3 - 64$

60. $\dfrac{c^9}{c^6} = c^{9-6} = c^3$

61. $\dfrac{(a + 4b)^{-6}}{(a + 4b)^{-2}} = (a + 4b)^{-6-(-2)} = (a + 4b)^{-4}$
$= \dfrac{1}{(a + 4b)^4}$

62. $\dfrac{81x^5 y^7 z}{27x^2 y^6 z^2} = \dfrac{81}{27} x^{5-2} y^{7-6} z^{1-2} = \dfrac{3x^3 y}{z}$

63. $\dfrac{(3a^3)^3}{54a^7} = \dfrac{3^3 a^{3 \cdot 3}}{54a^7} = \dfrac{27a^9}{54a^7} = \dfrac{27}{54} a^{9-7} = \dfrac{a^2}{2}$

64. $\left(\dfrac{21a^6 b^2}{7a^3 b^6}\right)^2 = \left(\dfrac{3a^3}{b^4}\right)^2 = \dfrac{3^2 a^{3 \cdot 2}}{b^{4 \cdot 2}} = \dfrac{9a^6}{b^8}$

65. $\dfrac{14b^4 - 21b^3 - 35b^2 + 28b}{-7b}$
$= \dfrac{14b^4}{-7b} - \dfrac{21b^3}{-7b} - \dfrac{35b^2}{-7b} + \dfrac{28b}{-7b}$
$= -2b^{4-1} + 3b^{3-1} + 5b^{2-1} - 4b^{1-1}$
$= -2b^3 + 3b^2 + 5b - 4$

66. $\dfrac{16cd^2 + 8c^2d}{4c^2d^2}$

$= \dfrac{16cd^2}{4c^2d^2} + \dfrac{8c^2d}{4c^2d^2}$

$= 4c^{1-2}d^{2-2} + 2c^{2-2}d^{1-2}$

$= \dfrac{4}{c} + \dfrac{2}{d}$

67. $s^2 \cdot s^8 = s^{2+8} = s^{10}$

68. $y^5 \cdot y^{-8} = y^{5-8} = y^{-3} = \dfrac{1}{y^3}$

69. $(m+n)^5(m+n) = (m+n)^{5+1} = (m+n)^6$

70. $(b^5)^7 = b^{5\cdot7} = b^{35}$

71. $(d^6)^{-3} = d^{6\cdot(-3)} = d^{-18} = \dfrac{1}{d^{18}}$

72. $(x^0)^8 = x^{0\cdot8} = x^0 = 1$

73. $(-3d)^6 = (-3)^6 d^6 = 729d^6$

74. $(-3d)^3 = (-3)^3 d^3 = -27d^3$

75. $\left(\dfrac{m}{n}\right)^{-6} = \left(\dfrac{n}{m}\right)^6 = \dfrac{n^6}{m^6}$

76. $(mn)^{-6} = \left(\dfrac{1}{mn}\right)^6 = \dfrac{1}{m^6n^6}$

77. $(x^4y^5z)^4 = x^{4\cdot4}y^{5\cdot4}z^4 = x^{16}y^{20}z^4$

78. $[(4x)^2]^3 = (4^2x^2)^3 = (16x^2)^3 = (16)^3 x^{2\cdot3}$
    $= 4096x^6$

79. $\left[\left(\dfrac{c}{2d}\right)^3\right]^2 = \left(\dfrac{c^3}{2^3d^3}\right)^2$

$= \left(\dfrac{c^3}{8d^3}\right)^2 = \dfrac{c^{3\cdot2}}{8^2d^{3\cdot2}}$

$= \dfrac{c^6}{64d^6}$

80. $8x^2 - 2x - 3$
    $= 8x^2 - 6x + 4x - 3$
    $= 2x(4x - 3) + 1(4x - 3)$
    $= (4x - 3)(2x + 1)$
    The width is $(2x + 1)$ units and the length is $(4x - 3)$ units.

81. a. $6x^2 - 10x - 4$
    $= 6x^2 + 2x - 12x - 4$
    $= 2x(3x + 1) - 4(3x + 1)$
    $= (3x + 1)(2x - 4)$
    The new length is $(3x + 1)$ inches and the new width is $(2x - 4)$ inches.

  b. $3x + 1 - x = 2x + 1$
    The length was increased by $(2x + 1)$ inches.

  c. $2x - 4 - (x - 2) = x - 2$
    The width was increased by $(x - 2)$ inches.

82. a. $x^2 + 12x + 32$
    $= \dfrac{1}{2}(2x^2 + 24x + 64)$
    $= \dfrac{1}{2}(2x^2 + 8x + 16x + 64)$
    $= \dfrac{1}{2}[2x(x + 4) + 16(x + 4)]$
    $= \dfrac{1}{2}(x + 4)(2x + 16)$
    The base is $(2x + 16)$ units and the height is $(x + 4)$ units.

  b. $2x + 16 - 2x = 16$
    The base was increased by 16 units.

c. $x + 4 - x = 4$
The height was increased by 4 units.

83. a. length: $(2x + 3)$ in.
width: x in.
height: x in.

b. Volume:
$x(x)(2x + 3)$
$= x^2(2x + 3)$
$= 2x^3 + 3x^2$
The volume is $(2x^3 + 3x^2)$ in^3.

c. Surface area:
$2x(2x + 3) + 2x(2x + 3) + 2x^2$
$= 4x^2 + 6x + 4x^2 + 6x + 2x^2$
$= 10x^2 + 12x$
The surface area is $(10x^2 + 12x)$ in^2.

d. Volume:
$2x^3 + 3x^2$
$= 2(8)^3 + 3(8)^2$
$= 1024 + 192$
$= 1216$
The volume is 1216 in^3.

e. Surface area:
$10x^2 + 12x$
$= 10(8)^2 + 12(8)$
$= 640 + 96$
$= 736$
The surface area is 736 in^2.

84. a. height: x cm
base: $(2x + 5)$ cm
Area:
$\frac{1}{2}(2x + 5)(x)$
$= \frac{1}{2}x(2x) + \frac{1}{2}x(5)$
$= x^2 + \frac{5}{2}x$
The area is $\left(x^2 + \frac{5}{2}x\right)$ cm^2.

b. $x^2 + \frac{5}{2}x$
$= (12)^2 + \frac{5}{2}(12)$
$= 144 + 30$
$= 174$
The area is 174 cm^2.

85. a. old width: x ft
old length: x ft
Area $= x^2$
The current area is x^2 ft^2.

b. new width: $2x$
new length: $2x + 5$
Area $= 2x(2x + 5)$
Area $= 4x^2 + 10x$
The new area will be $(4x^2 + 10x)$ ft^2.

c. Ratio of enlarged area to original area:
$\frac{4x^2 + 10x}{x^2}$
$= \frac{x(4x + 10)}{x(x)}$
$= \frac{4x + 10}{x}$
The ratio is $\frac{4x + 10}{x}$.

Chapter 9 Test

1. $81a^3 + 54a^2 + 9a$
$= 9a(9a^2 + 6a + 1)$
$= 9a[(3a)^2 + 2(3a)(1) + 1^2]$
$= 9a(3a + 1)^2$

2. $p^3 + 125$
$= p^3 + 5^3$
$= (p + 5)(p^2 - 5p + 5^2)$
$= (p + 5)(p^2 - 5p + 25)$

3. $-8a^4b^2 - 36a^3b^3 - 16a^2b^4$
$= -4a^2b^2(2a^2 + 9ab + 4b^2)$
$= -4a^2b^2(2a^2 + 8ab + 1ab + 4b^2)$
$= -4a^2b^2[2a(a + 4b) + b(a + 4b)]$
$= -4a^2b^2(a + 4b)(2a + b)$

4. $a(a^2 + b^2) - 5b(a^2 + b^2)$
$= (a^2 + b^2)(a - 5b)$

5. $15x^2 - 21xy + 10xy - 14y^2$
$= 3x(5x - 7y) + 2y(5x - 7y)$
$= (5x - 7y)(3x + 2y)$

6. $64a^2 - 49b^2$
$= (8a)^2 - (7b)^2$
$= (8a + 7b)(8a - 7b)$

7. $25x^2 - 70x + 49$
$= (5x)^2 - 2(5x)(7) + 7^2$
$= (5x - 7)^2$

8. $3x^3 - 27x^2 + 24x$
$= 3x(x^2 - 9x + 8)$
$= 3x(x^2 - x - 8x + 8)$
$= 3x[x(x - 1) - 8(x - 1)]$
$= 3x(x - 1)(x - 8)$

9. $x^2 - 4xy - 21y^2$
$= x^2 + 3xy - 7xy - 21y^2$
$= x(x + 3y) - 7y(x + 3y)$
$= (x + 3y)(x - 7y)$

10. $14x^2 + 25x + 9$
$= 14x^2 + 7x + 18x + 9$
$= 7x(2x + 1) + 9(2x + 1)$
$= (2x + 1)(7x + 9)$

11. $4x^4 + 27x^2 - 7$
$= 4x^4 + 28x^2 - x^2 - 7$
$= 4x^2(x^2 + 7) - 1(x^2 + 7)$
$= (x^2 + 7)(4x^2 - 1)$
$= (x^2 + 7)[(2x)^2 - 1^2]$
$= (x^2 + 7)(2x + 1)(2x - 1)$

12. $x^2 + 8x + 14$ will not factor.

13. $(2x - 1)(2x - 1)^8 = (2x - 1)^{1+8} = (2x - 1)^9$

14. $3x^2y^0z^{-3} = \dfrac{3x^2(1)}{z^3} = \dfrac{3x^2}{z^3}$

15. $(a + b)^3(a - b)^{-5} = \dfrac{(a + b)^3}{(a - b)^5}$

16. $(-2x^2y^{-1})^3$
$= (-2)^3 x^{2 \cdot 3} y^{(-1)(3)}$
$= -8x^6 y^{-3}$
$= \dfrac{-8x^6}{y^3}$

17. $\dfrac{(3c - 7)^3}{(3c - 7)^5}$
$= (3c - 7)^{3-5}$
$= (3c - 7)^{-2}$
$= \dfrac{1}{(3c - 7)^2}$

18. $\left[\dfrac{(2p)^3}{3q}\right]^{-2} = \left(\dfrac{2^3 p^3}{3q}\right)^{-2} = \left(\dfrac{8p^3}{3q}\right)^{-2}$
$= \left(\dfrac{3q}{8p^3}\right)^2 = \dfrac{3^2 q^2}{8^2 p^{3 \cdot 2}} = \dfrac{9q^2}{64p^6}$

19. $(3y^2 + 16 - 7y + 5y^3) + (7y + 6y^2 + 4y^4 - 11)$

$\quad = 3y^2 + 16 - 7y + 5y^3 + 7y + 6y^2 + 4y^4 - 11$

$\quad = 4y^4 + 5y^3 + 9y^2 + 5$

20. $(7x^5 + 23x^3 + 17x^2 - 39) - (2x^4 - 4x^2 - 2x - 9)$

$\quad = 7x^5 + 23x^3 + 17x^2 - 39 - 2x^4 + 4x^2 + 2x + 9$

$\quad = 7x^5 - 2x^4 + 23x^3 + 21x^2 + 2x - 30$

21. $(-2p^3 q^{-5} r^2)(5.7p^6 q^7 r)$

$\quad = (-2)(5.7)p^{3+6} q^{-5+7} r^{2+1}$

$\quad = -11.4p^9 q^2 r^3$

22. $-4t(2t^3 - 3t^2 - 8t + 6)$

$\quad = -8t^{1+3} + 12t^{1+2} + 32t^{1+1} - 24t$

$\quad = -8t^4 + 12t^3 + 32t^2 - 24t$

23. $(9 - 5d)(9 + 5d)$

$\quad = (9)^2 - (5d)^2$

$\quad = 81 - 25d^2$

24. $(3x + 4)(5x - 7)$

$\quad = 3x(5x - 7) + 4(5x - 7)$

$\quad = 15x^2 - 21x + 20x - 28$

$\quad = 15x^2 - x - 28$

25. $(4z - 3)(2z^2 - z + 5)$

$\quad = 4z(2z^2 - z + 5) - 3(2z^2 - z + 5)$

$\quad = 8z^3 - 4z^2 + 20z - 6z^2 + 3z - 15$

$\quad = 8z^3 - 10z^2 + 23z - 15$

26. $(x + 3)^2$

$\quad = x^2 + 2(x)(3) + 3^2$

$\quad = x^2 + 6x + 9$

27. $\dfrac{24a^2 b^{-3} c}{3ab^2 c^{-2}}$

$\quad = \dfrac{24}{3} a^{2-1} b^{-3-2} c^{1-(-2)}$

$\quad = 8ab^{-5} c^3$

$\quad = \dfrac{8ac^3}{b^5}$

28. $\dfrac{(a^{-2} b^4)^3}{a^2 b^5 c}$

$\quad = \dfrac{a^{(-2)\cdot 3} b^{4\cdot 3}}{a^2 b^5 c}$

$\quad = \dfrac{a^{-6} b^{12}}{a^2 b^5 c}$

$\quad = a^{-6-2} b^{12-5} c^{-1}$

$\quad = a^{-8} b^7 c^{-1}$

$\quad = \dfrac{b^7}{a^8 c}$

29. $\dfrac{15x^6 + 25x^5 - 5x^3}{5x^2}$

$\quad = \dfrac{15x^6}{5x^2} + \dfrac{25x^5}{5x^2} - \dfrac{5x^3}{5x^2}$

$\quad = 3x^{6-2} + 5x^{5-2} - 1x^{3-2}$

$\quad = 3x^4 + 5x^3 - x$

30. a. width: $(x + 4)$ in.
height: x in.
length: $(2x)$ in.
volume:
$x(2x)(x + 4)$
$= 2x^2(x + 4)$
$= 2x^3 + 8x^2$
The volume is $(2x^3 + 8x^2)$ in^3.

b. Surface area:
$2x(x + 4) + 2x(2x) + 2(2x)(x + 4)$
$= 2x^2 + 8x + 4x^2 + 4x^2 + 16x$
$= 10x^2 + 24x$
The surface area is $(10x^2 + 24x)$ in^2.

c. Volume:
$2x^3 + 8x^2$
$= 2(5)^3 + 8(5)^2$
$= 250 + 200$
$= 450$
The volume is 450 in^3.
Surface area:
$10x^2 + 24x$
$= 10(5)^2 + 24(5)$
$= 250 + 120$
$= 370$
The surface area is 370 in^2.

31. a. Factor by grouping

b. He did not factor completely.
$(4x^2 - 9)$ can be factored further.

c. $12x^3 + 28x^2 - 27x - 63$
$= (12x^3 + 28x^2) + (-27x - 63)$
$= 4x^2(3x + 7) - 9(3x + 7)$
$= (3x + 7)(4x^2 - 9)$
$= (3x + 7)[(2x)^2 - 3^2]$
$= (3x + 7)(2x + 3)(2x - 3)$

Chapter 10

10.1 Experiencing Algebra the Exercise Way

1. $3x^3 - 2x^2 + x = 5$ or $3x^3 - 2x^2 + x - 5 = 0$ is a cubic polynomial equation.

3. $3\sqrt{y} + y - 4 = 0$ is not a polynomial because y is the radicand of $\sqrt{y}$.

5. $\frac{1}{4}x^4 + 3x^2 - \frac{3}{4} = 0$ is a polynomial equation.

7. $4(x-2)(x+7) = 16$ or $4x^2 + 20x - 72 = 0$ is a quadratic polynomial equation.

9. $3x^{-2} - 5x = 4x^2$ is not a polynomial because x has an exponent of -2.

11. $1.7x^2 + 3.2x = 5.7$ or $1.7x^2 + 3.2x - 5.7 = 0$ is a quadratic polynomial equation.

13. $x^2 + 8 = 6x$
$Y1 = x^2 + 8$
$Y2 = 6x$

X	Y₁	Y₂
0	8	0
1	9	6
2	12	12
3	17	18
4	24	24
5	33	30
6	44	36

X=0

The solutions are 2 and 4.

15. $x^3 = x$
$Y1 = x^3$
$Y2 = x$

X	Y₁	Y₂
-4	-64	-4
-3	-27	-3
-2	-8	-2
-1	-1	-1
0	0	0
1	1	1
2	8	2

X=-4

The solutions are -1, 1, and 0.

17. $x^2 - 7 = x^2 + 3$
$Y1 = x^2 - 7$
$Y2 = x^2 + 3$

X	Y₁	Y₂
0	-7	3
1	-6	4
2	-3	7
3	2	12
4	9	19
5	18	28
6	29	39

X=0

The expression on the left is always 10 less than the expression on the right. There is no solution.

19. $x^2 + 5x + 1 = 1 + x(5 + x)$
$Y1 = x^2 + 5x + 1$
$Y2 = 1 + x(5 + x)$

X	Y₁	Y₂
0	1	1
1	7	7
2	15	15
3	25	25
4	37	37
5	51	51
6	67	67

X=0

The expressions are equal for all x-values in the table. The solution is the set of all real numbers.

21. $x^2 - 2x + 6 = 12 - 4x + 2x^2$
$Y1 = x^2 - 2x + 6$
$Y2 = 12 - 4x + 2x^2$

X	Y₁	Y₂
0	6	12
1	5	10
2	6	12
3	9	18
4	14	28
5	21	42
6	30	60

X=0

The expression on the left is always less than the expression on the right. In standard form, the equation is $x^2 - 2x + 6 = 0$. There is no real-number solution.

23. $\frac{1}{2}x^2 - x = 6 - 3x$

$Y1 = \frac{1}{2}x^2 - x$

$Y2 = 6 - 3x$

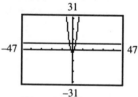

The solutions are –6 and 2.

25. $2 - 0.2x^2 = 0.6x$

$Y1 = 2 - 0.2x^2$

$Y2 = 0.6x$

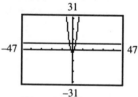

The solutions are –5 and 2.

27. $x^2 - 3 = 6$

$Y1 = x^2 - 3$

$Y2 = 6$

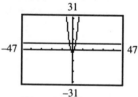

The graphs intersect at (–3, 6) and (3, 6). The solutions are –3 and 3.

29. $x^2 - 3 = 2x$

$Y1 = x^2 - 3$

$Y2 = 2x$

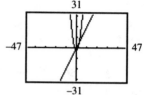

The graphs intersect at (–1, –2) and (3, 6). The solutions are –1 and 3.

31. $x^3 = 4x$

$Y1 = x^3$

$Y2 = 4x$

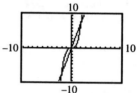

The graphs intersect at (–2, –8), (0, 0), and (2, 8). The solutions are –2, 0, and 2.

33. $x^2 - 3x - 10 = 0$

$Y1 = x^2 - 3x - 10$

$Y2 = 0$

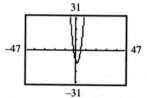

The graphs intersect at (–2, 0) and (5, 0). The solutions are –2 and 5.

35. $x^2 - 2x + 1 = x^2 - 2x - 3$

$Y1 = x^2 - 2x + 1$

$Y2 = x^2 - 2x - 3$

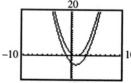

The graphs do not intersect. There is no solution.

37. $x^2 + 1 = 3x^2 + 3$

$Y1 = x^2 + 1$

$Y2 = 3x^2 + 3$

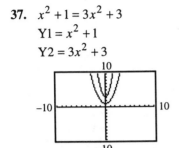

The graphs do not intersect. There is no real-number solution.

39. $x(x + 3) = x^2 + 3x$

$Y1 = x(x + 3)$

$Y1 = x^2 + 3x$

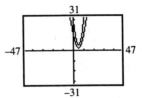

The graphs are the same. The solution is the set of all real numbers.

41. $4x^2 - x^3 = x^2 - 4x$

$Y1 = 4x^2 - x^3$

$Y2 = x^2 - 4x$

The graphs intersect at $(-1, 5)$, $(0, 0)$ and $(4, 0)$. The solutions are -1, 0, and 4.

43. $x^2 - 10x + 30 = \dfrac{1}{2}x^2 - 5x + 15$

$Y1 = x^2 - 10x + 30$

$Y2 = \dfrac{1}{2}x^2 - 5x + 15$

The graphs do not intersect.
There is no real-number solution.

45. $x^3 - 2x^2 + 1 = x^3 - 2x^2 + 9$

$Y1 = x^3 - 2x^2 + 1$

$Y2 = x^3 - 2x^2 + 9$

The graphs do not intersect. There is no solution.

47. $x(x^2 - 3) - 5(x + 1) = x^3 - 8x - 5$

$Y1 = x(x^2 - 3) - 5(x + 1)$

$Y2 = x^3 - 8x - 5$

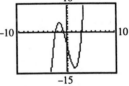

The graphs are the same. The solution is all real numbers.

49. $4x^2 = 9$

$Y1 = 4x^2$

$Y2 = 9$

The graphs intersect at $\left(-\dfrac{3}{2}, 9\right)$ and $\left(\dfrac{3}{2}, 9\right)$. The solutions are $-\dfrac{3}{2}$ and $\dfrac{3}{2}$.

51. $10x^3 - 7x^2 - 4x = 3x - 4$

$Y1 = 10x^3 - 7x^2 - 4x$

$Y2 = 3x - 4$

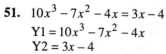

The graphs intersect at $(-0.8, -6.4)$, $(0.5, -2.5)$, and $(1, -1)$. The solutions are -0.8, 0.5, and 1.

53. $x^2 - 0.9x - 10.36 = 0$

$Y1 = x^2 - 0.9x - 10.36$

$Y2 = 0$

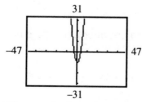

The graphs intersect at $(-2.8, 0)$ and $(3.7, 0)$. The solutions are -2.8 and 3.7.

55. $x^3 + 3.7x^2 = 1.74x + 7.56$

$Y1 = x^3 + 3.7x^2$

$Y2 = 1.74x + 7.56$

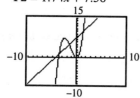

The graphs intersect at $(-3.6, 1.296)$, $(-1.5, 4.95)$, and $(1.4, 9.996)$. The solutions are -3.6, -1.5, and 1.4.

57. $s = -16t^2 + v_0t + s_0$

$0 = -16t^2 + 0t + 40$

$0 = -16t^2 + 40$

$Y1 = 0$

$Y2 = -16x^2 + 40$

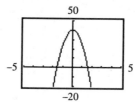

The graphs intersect at about $(-1.58, 0)$ and $(1.58, 0)$. There cannot be a negative amount of time. Therefore, it will hit the ground in approximately 1.58 seconds.

59. $s = -16t^2 + v_0t + s_0$

$0 = -16t^2 - 5t + 40$

$Y1 = 0$

$Y2 = -16x^2 - 5x + 40$

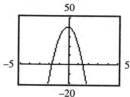

The graphs intersect at approximately $(-1.75, 0)$ and $(1.43, 0)$. There cannot be a negative amount of time. Therefore, the dagger will hit the ground in approximately 1.43 seconds.

61. $s = -16t^2 + v_0t + s_0$

$0 = -16t^2 + 5t + 40$

$Y1 = 0$

$Y2 = -16x^2 + 5x + 40$

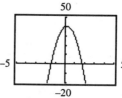

The graphs intersect at approximately $(-1.43, 0)$ and $(1.75, 0)$. Disregard a negative time. It will hit the ground in approximately 1.75 seconds.

63. a. $30 + x$

b. $40 - x$

c. $(30 + x)(40 - x) \doteq 1200 + 10x - x^2$

d. $1100 = 1200 + 10x - x^2$

e. $Y1 = 1100$
$Y2 = 1200 + 10x - x^2$

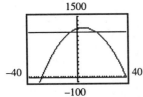

The graphs intersect at approximately
$(-6.18, 1100)$ and $(16.18, 1100)$.
Disregard a negative number of seats.
There should be 16 unsold seats and 24
sold seats to meet this goal.

65. $y = 11.55 + 20.67x - 1.35x^2$

$100 = 11.55 + 20.67x - 1.35x^2$
$Y1 = 100$
$Y2 = 11.55 + 20.67x - 1.35x^2$

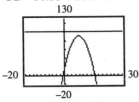

The graphs do not intersect. There are no
number of sales reps that can be assigned
and meet the goal. There is no real-number
solution.

67. Volume = length × width × height
Volume = 2000
height = x
length = $2x$
width = $x + 1$
$2000 = 2x(x + 1)(x)$
$2000 = 2x^3 + 2x^2$
$Y1 = 2000$
$Y2 = 2x^3 + 2x^2$

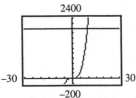

The graphs intersect at about $(9.68, 2000)$.
The dimensions are about
20 in. by 11 in. by 10 in.

10.1 Experiencing Algebra the Calculator Way

Students should verify solutions with the
calculator.

10.2 Experiencing Algebra the Exercise Way

1. $(x + 6)(x + 11) = 0$
$x + 6 = 0$ or $x + 11 = 0$
$x = -6$ $x = -11$
The solutions are -11 and -6.

3. $\left(\dfrac{3}{5}x - \dfrac{9}{20}\right)\left(x + \dfrac{2}{3}\right) = 0$

$\dfrac{3}{5}x - \dfrac{9}{20} = 0$ or $x + \dfrac{2}{3} = 0$

$\dfrac{3}{5}x = \dfrac{9}{20}$ $x = -\dfrac{2}{3}$

$x = \dfrac{3}{4}$

The solutions are $-\dfrac{2}{3}$ and $\dfrac{3}{4}$.

5. $3x(x + 9)(2x - 5) = 0$
$3x = 0$ or $x + 9 = 0$ or $2x - 5 = 0$
$x = 0$ $x = -9$ $2x = 5$
$x = \dfrac{5}{2}$

The solutions are -9, 0 and $\dfrac{5}{2}$.

7. $(7x - 49)(49x - 7) = 0$
$7x - 49 = 0$ or $49x - 7 = 0$
$7x = 49$ $49x = 7$

$x = 7$ $x = \dfrac{1}{7}$

The solutions are $\dfrac{1}{7}$ and 7.

9. $(4x + 3)(2x - 9)(x + 6) = 0$

$4x + 3 = 0$ or $2x - 9 = 0$ or $x + 6 = 0$

$4x = -3$ $2x = 9$ $x = -6$

$x = -\dfrac{3}{4}$ $x = \dfrac{9}{2}$

The solutions are -6, $-\dfrac{3}{4}$, and $\dfrac{9}{2}$.

11. $(0.2x + 6.8)(1.3x - 1.69) = 0$

$0.2x + 6.8 = 0$ or $1.3x - 1.69 = 0$

$0.2x = -6.8$ $1.3x = 1.69$

$x = -34$ $x = 1.3$

The solutions are -34 and 1.3.

13. $0 = x^2 + 10x + 24$

$0 = (x + 6)(x + 4)$

$x + 6 = 0$ or $x + 4 = 0$

$x = -6$ $x = -4$

The solutions are -6 and -4.

15. $x^2 + 33 = 14x$

$x^2 - 14x + 33 = 0$

$(x - 3)(x - 11) = 0$

$x - 3 = 0$ or $x - 11 = 0$

$x = 3$ $x = 11$

The solutions are 3 and 11.

17. $4x^2 + 5x + 24x + 30 = 0$

$(4x^2 + 5x) + (24x + 30) = 0$

$x(4x + 5) + 6(4x + 5) = 0$

$(4x + 5)(x + 6) = 0$

$4x + 5 = 0$ or $x + 6 = 0$

$4x = -5$ $x = -6$

$x = -\dfrac{5}{4}$

The solutions are -6 and $-\dfrac{5}{4}$.

19. $5x^2 + 3x = 8$

$5x^2 + 3x - 8 = 0$

$(5x + 8)(x - 1) = 0$

$5x + 8 = 0$ or $x - 1 = 0$

$5x = -8$ $x = 1$

$x = -\dfrac{8}{5}$

The solutions are $-\dfrac{8}{5}$ and 1.

21. $15x^2 = 35x$

$15x^2 - 35x = 0$

$5x(3x - 7) = 0$

$5x = 0$ or $3x - 7 = 0$

$x = 0$ $3x = 7$

$x = \dfrac{7}{3}$

The solutions are 0 and $\dfrac{7}{3}$.

23. $18x^2 - 3x = 5 - 30x$

$18x^2 + 27x - 5 = 0$

$(6x - 1)(3x + 5) = 0$

$6x - 1 = 0$ or $3x + 5 = 0$

$6x = 1$ $3x = -5$

$x = \dfrac{1}{6}$ $x = -\dfrac{5}{3}$

The solutions are $-\dfrac{5}{3}$ and $\dfrac{1}{6}$.

25. $16x^2 + 72x + 81 = 0$

$(4x + 9)^2 = 0$

$4x + 9 = 0$

$4x = -9$

$x = -\dfrac{9}{4}$

The solution is $-\dfrac{9}{4}$.

27. $4x^2 + 25x + 18 = 5x - 7$

$4x^2 + 20x + 25 = 0$

$(2x + 5)^2 = 0$

$2x + 5 = 0$

$2x = -5$

$x = -\dfrac{5}{2}$

The solution is $-\dfrac{5}{2}$.

29. $b^2 + 7 = 71$

$b^2 - 64 = 0$

$(b + 8)(b - 8) = 0$

$b + 8 = 0$ or $b - 8 = 0$

$b = -8$ $b = 8$

The solutions are -8 and 8.

31. $9z^2 = 25$
$9z^2 - 25 = 0$
$(3z + 5)(3z - 5) = 0$
$3z + 5 = 0$ or $3z - 5 = 0$
$3z = -5$ $\qquad$ $3z = 5$
$z = -\dfrac{5}{3}$ $\qquad$ $z = \dfrac{5}{3}$
The solutions are $-\dfrac{5}{3}$ and $\dfrac{5}{3}$

33. $x^2 + 10x + 20 = 0$
The equation does not factor.

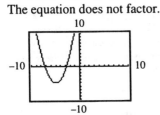

The solutions (the x-coordinates of the intersections) are approximately -7.24 and -2.76.

35. $(x + 1)^2 = 49$
$(x + 1)^2 - 49 = 0$
$[(x + 1) + 7][(x + 1) - 7] = 0$
$(x + 8)(x - 6) = 0$
$x + 8 = 0$ or $x - 6 = 0$
$x = -8$ $\qquad$ $x = 6$
The solutions are -8 and 6.

37. $(x - 3)(x - 2) = 42$
$x^2 - 5x + 6 = 42$
$x^2 - 5x - 36 = 0$
$(x - 9)(x + 4) = 0$
$x - 9 = 0$ or $x + 4 = 0$
$x = 9$ $\qquad$ $x = -4$
The solutions are -4 and 9.

39. $(x - 7)(2x + 3) = 2(x^2 - 5x - 10) - (x + 1)$
$2x^2 - 11x - 21 = 2x^2 - 10x - 20 - x - 1$
$2x^2 - 11x - 21 = 2x^2 - 11x - 21$
$0 = 0$
This is an identity. The solution is the set of all real numbers.

41. $(x + 3)^2 = x^2 + 6x + 11$
$x^2 + 6x + 9 = x^2 + 6x + 11$
$9 = 11$
This is a contradiction.
There is no solution.

43. $x^2 + (x + 3)^2 = 225$
$x^2 + x^2 + 6x + 9 = 225$
$2x^2 + 6x - 216 = 0$
$2(x^2 + 3x - 108) = 0$
$2(x + 12)(x - 9) = 0$
$x + 12 = 0$ or $x - 9 = 0$
$x = -12$ $\qquad$ $x = 9$
The solutions are -12 and 9.

45. $x^3 + 7x^2 - 9x - 63 = 0$
$x^2(x + 7) - 9(x + 7) = 0$
$(x + 7)(x^2 - 9) = 0$
$(x + 7)(x + 3)(x - 3) = 0$
$x + 7 = 0$ or $x + 3 = 0$ or $x - 3 = 0$
$x = -7$ $\qquad$ $x = -3$ $\qquad$ $x = 3$
The solutions are -7, -3, and 3.

47. $18x^3 + 45x^2 - 50x - 125 = 0$
$9x^2(2x + 5) - 25(2x + 5) = 0$
$(2x + 5)(9x^2 - 25) = 0$
$(2x + 5)(3x + 5)(3x - 5) = 0$
$2x + 5 = 0$ or $3x + 5 = 0$ or $3x - 5 = 0$
$2x = -5$ $\qquad$ $3x = -5$ $\qquad$ $3x = 5$
$x = -\dfrac{5}{2}$ $\qquad$ $x = -\dfrac{5}{3}$ $\qquad$ $x = \dfrac{5}{3}$
The solutions are $-\dfrac{5}{2}$, $-\dfrac{5}{3}$, and $\dfrac{5}{3}$.

49. $3x^3 - 3x^2 + 12x - 12 = 0$
$3x^2(x - 1) + 12(x - 1) = 0$
$(x - 1)(3x^2 + 12) = 0$
$(x - 1)3(x^2 + 4) = 0$
$x - 1 = 0$ or $x^2 + 4 = 0$
$x = 1$ $\qquad$ $x^2 = -4$
The solution is 1.

51. a. length = x ft

height = 3 ft

width = $\left(\dfrac{1}{2}x+1\right)$ ft

volume = length × width × height

volume = $x(3)\left(\dfrac{1}{2}x+1\right)=3x\left(\dfrac{1}{2}x+1\right)$

$=\dfrac{3}{2}x^2+3x$

The volume is $\left(\dfrac{3}{2}x^2+3x\right)$ ft^3.

b. $72=\dfrac{3}{2}x^2+3x$

$0=3x^2+6x-144$

$0=(3x-18)(x+8)$

$3x-18=0$ or $x+8=0$

$3x=18$ $\qquad x=-8$

$x=6$

Disregard a negative length. The dimensions are 6 ft by 3 ft by 4 ft.

53. a. height = x in.

width = $(3x)$ in.

length = $(4x+2)$ in.

surface area

$=2x(3x)+2x(4x+2)+3x(4x+2)$

$=6x^2+8x^2+4x+12x^2+6x$

$=26x^2+10x$

The surface area is $(26x^2+10x)$ in^2.

b. $700=26x^2+10x$

$0=26x^2+10x-700$

$0=2(13x^2+5x-350)$

$0=2(13x+70)(x-5)$

$13x+70=0$ or $x-5=0$

$13x=-70$ $\qquad x=5$

$x=-\dfrac{70}{13}$

Disregard a negative height. The dimensions are 15 in. by 5 in. by 22 in.

55. a. base = x ft

height = $(x+4)$ ft

area = $\dfrac{1}{2}x(x+4)=\dfrac{1}{2}x^2+2x$

The area is $\left(\dfrac{1}{2}x^2+2x\right)$ ft^2.

b. $30=\dfrac{1}{2}x^2+2x$

$0=x^2+4x-60$

$0=(x+10)(x-6)$

$x+10=0$ or $x-6=0$

$x=-10$ $\qquad x=6$

Disregard a negative base.
The dimensions are 6 ft base and 10 ft height.

57. height = x ft

base = $(x+6)$ ft

hypotenuse = $(x+12)$ ft

$(x+12)^2=x^2+(x+6)^2$

$x^2+24x+144=x^2+x^2+12x+36$

$0=x^2-12x-108$

$0=(x-18)(x+6)$

$x-18=0$ or $x+6=0$

$x=18$ $\qquad x=-6$

Disregard a negative height.
The measurements are: 18 ft height, 24 ft base, and 30 ft hypotenuse.

59. $V^2=30FS$

$V^2=30(0.40)(243)$

$V^2=2916$

$V=\sqrt{2916}$

$V=54$

The vehicle was traveling at 54 mph.

10.2 Experiencing Algebra the Calculator Way

1. $x^2-133x+4350=0$

$Y1=x^2-133x+4350$

$Y2=0$

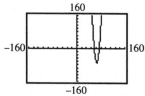

The solutions (the x-coordinates of the intersection) are 58 and 75.

2. $12x^2 - 12x - 3672 = 0$

$Y1 = 12x^2 - 12x - 3672$

$Y2 = 0$

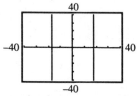

The solutions (the x-coordinates of the intersection) are -17 and 18.

3. $x^2 + 250x + 15,625 = 0$

$Y1 = x^2 + 250x + 15,625$

$Y2 = 0$

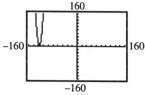

The solution (the x-coordinate of the intersection) is -125.

4. $x^2 - 186x + 8649 = 0$

$Y1 = x^2 - 186x + 8649$

$Y2 = 0$

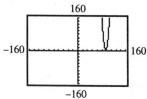

The solution (the x-coordinate of the intersection) is 93.

5. $13x^2 + 195x = 377x + 5655$

$Y1 = 13x^2 + 195x$

$Y2 = 377x + 5655$

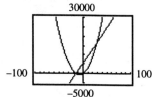

The solutions (the x-coordinates of the intersections) are -15 and 29.

6. Answers will vary.

10.3 Experiencing Algebra the Exercise Way

1. $\sqrt{81} = \sqrt{9^2} = 9$

3. $-\sqrt{100} = -\sqrt{(10)^2} = -10$

5. $\sqrt{3}\sqrt{27} = \sqrt{3 \cdot 27} = \sqrt{81} = 9$

7. $\sqrt{6}\sqrt{21} = \sqrt{6 \cdot 21} = \sqrt{126} = \sqrt{9 \cdot 14} = 3\sqrt{14}$

9. $\sqrt{147} = \sqrt{49 \cdot 3} = 7\sqrt{3}$

11. $-\sqrt{125} = -\sqrt{25 \cdot 5} = -5\sqrt{5}$

13. $\sqrt{\dfrac{36}{49}} = \dfrac{\sqrt{36}}{\sqrt{49}} = \dfrac{6}{7}$

15. $\sqrt{\dfrac{11}{144}} = \dfrac{\sqrt{11}}{\sqrt{144}} = \dfrac{\sqrt{11}}{12}$

17. $\sqrt{\dfrac{64}{13}} = \dfrac{\sqrt{64}}{\sqrt{13}} = \dfrac{8}{\sqrt{13}} \cdot \dfrac{\sqrt{13}}{\sqrt{13}} = \dfrac{8\sqrt{13}}{13}$

19. $-\dfrac{\sqrt{28}}{\sqrt{9}} = -\dfrac{\sqrt{4 \cdot 7}}{3} = -\dfrac{2\sqrt{7}}{3}$

21. $\dfrac{\sqrt{45}}{\sqrt{24}} = \dfrac{\sqrt{9 \cdot 5}}{\sqrt{4 \cdot 6}} = \dfrac{3\sqrt{5}}{2\sqrt{6}} \cdot \dfrac{\sqrt{6}}{\sqrt{6}} = \dfrac{3\sqrt{30}}{2 \cdot 6} = \dfrac{\sqrt{30}}{4}$

23. $-\sqrt{\dfrac{18}{48}}$

$= -\sqrt{\dfrac{9}{24}}$

$= -\dfrac{\sqrt{9}}{\sqrt{24}}$

$= -\dfrac{3}{\sqrt{4 \cdot 6}}$

$= -\dfrac{3}{2\sqrt{6}} \cdot \dfrac{\sqrt{6}}{\sqrt{6}}$

$= -\dfrac{3\sqrt{6}}{12} = -\dfrac{\sqrt{6}}{4}$

25. $x^2 = 144$
$x = \sqrt{144}$ or $x = -\sqrt{144}$
$x = 12$ or $x = -12$
The solutions are ± 12.

27. $a^2 = 13$
$a = \sqrt{13}$ or $a = -\sqrt{13}$
The solutions are $\pm\sqrt{13}$.

29. $q^2 = 98$

$q = \sqrt{98}$ or $q = -\sqrt{98}$

$q = \sqrt{49 \cdot 2}$ $q = -\sqrt{49 \cdot 2}$

$q = 7\sqrt{2}$ $q = -7\sqrt{2}$

The solutions are $\pm 7\sqrt{2}$

31. $2x^2 - 32 = 0$
$2x^2 = 32$
$x^2 = 16$
$x = \sqrt{16}$ or $x = -\sqrt{16}$
$x = 4$ $x = -4$
The solutions are ± 4.

33. $4x^2 - 25 = 0$
$4x^2 = 25$
$x^2 = \dfrac{25}{4}$

$x = \sqrt{\dfrac{25}{4}}$ or $x = -\sqrt{\dfrac{25}{4}}$

$x = \dfrac{5}{2}$ $x = -\dfrac{5}{2}$

The solutions are $\pm\dfrac{5}{2}$.

35. $9x^2 = 2$
$x^2 = \dfrac{2}{9}$

$x = \sqrt{\dfrac{2}{9}}$ or $x = -\sqrt{\dfrac{2}{9}}$

$x = \dfrac{\sqrt{2}}{3}$ $x = -\dfrac{\sqrt{2}}{3}$

The solutions are $\pm\dfrac{\sqrt{2}}{3}$.

37. $x^2 + 4 = 6$
$x^2 = 2$
$x = \sqrt{2}$ or $x = -\sqrt{2}$
The solutions are $\pm\sqrt{2}$.

39. $m^2 + 7 = 5$
$m^2 = -2$
There is no real-number solution because there is no real number whose square is negative.

41. $(x - 5)^2 = 0$
$x - 5 = \sqrt{0}$
$x - 5 = 0$
$x = 5$
The solution is 5.

43. $(z - 7)^2 = 4$
$z - 7 = \sqrt{4}$ or $z - 7 = -\sqrt{4}$
$z - 7 = 2$ $z - 7 = -2$
$z = 9$ $z = 5$
The solutions are 5 and 9.

45. $(4a-3)^2 = 4$

$$4a-3 = \sqrt{4} \qquad \text{or} \qquad 4a-3 = -\sqrt{4}$$
$$4a-3 = 2 \qquad\qquad\qquad 4a-3 = -2$$
$$4a = 5 \qquad\qquad\qquad\quad 4a = 1$$
$$a = \frac{5}{4} \qquad\qquad\qquad\quad a = \frac{1}{4}$$

The solutions are $\frac{1}{4}$ and $\frac{5}{4}$.

47. $x^2 = 1.69$

$$x = \sqrt{1.69} \qquad \text{or} \qquad x = -\sqrt{1.69}$$
$$x = 1.3 \qquad\qquad\qquad x = -1.3$$

The solutions are ± 1.3.

49. $(x+3)^2 - 1 = 3$

$$(x+3)^2 = 4$$
$$x+3 = \sqrt{4} \qquad \text{or} \qquad x+3 = -\sqrt{4}$$
$$x+3 = 2 \qquad\qquad\qquad x+3 = -2$$
$$x = -1 \qquad\qquad\qquad x = -5$$

The solutions are -5 and -1.

51. $2(m-4)^2 - 6 = 12$

$$2(m-4)^2 = 18$$
$$(m-4)^2 = 9$$
$$m-4 = \sqrt{9} \qquad \text{or} \qquad m-4 = -\sqrt{9}$$
$$m-4 = 3 \qquad\qquad\qquad m-4 = -3$$
$$m = 7 \qquad\qquad\qquad\quad m = 1$$

The solutions are 1 and 7.

53. $x^2 + 10x + 25 = 9$

$$(x+5)^2 = 9$$
$$x+5 = \sqrt{9} \qquad \text{or} \qquad x+5 = -\sqrt{9}$$
$$x+5 = 3 \qquad\qquad\qquad x+5 = -3$$
$$x = -2 \qquad\qquad\qquad x = -8$$

The solutions are -8 and -2.

55. $9x^2 - 6x + 1 = 144$

$$(3x-1)^2 = 144$$
$$3x-1 = \sqrt{144} \qquad \text{or} \qquad 3x-1 = -\sqrt{144}$$
$$3x-1 = 12 \qquad\qquad\qquad 3x-1 = -12$$
$$3x = 13 \qquad\qquad\qquad\quad 3x = -11$$
$$x = \frac{13}{3} \qquad\qquad\qquad\quad x = -\frac{11}{3}$$

The solutions are $-\frac{11}{3}$ and $\frac{13}{3}$.

57. $(x-7)^2 - 5 = 1$

$$(x-7)^2 = 6$$
$$x-7 = \sqrt{6} \qquad \text{or} \qquad x-7 = -\sqrt{6}$$
$$x = 7 + \sqrt{6} \qquad\qquad\qquad x = 7 - \sqrt{6}$$

The solutions are $7 \pm \sqrt{6}$.

59. $(2x+1)^2 - 3 = 7$

$$(2x+1)^2 = 10$$
$$2x+1 = \sqrt{10} \qquad \text{or} \qquad 2x+1 = -\sqrt{10}$$
$$2x = -1 + \sqrt{10} \qquad\qquad 2x = -1 - \sqrt{10}$$
$$x = \frac{-1 + \sqrt{10}}{2} \qquad\qquad x = \frac{-1 - \sqrt{10}}{2}$$

The solutions are $\frac{-1 \pm \sqrt{10}}{2}$.

61. $(x+3)^2 - 6 = 6$

$$(x+3)^2 = 12$$
$$x+3 = \sqrt{12} \qquad \text{or} \qquad x+3 = -\sqrt{12}$$
$$x+3 = 2\sqrt{3} \qquad\qquad\qquad x+3 = -2\sqrt{3}$$
$$x = -3 + 2\sqrt{3} \qquad\qquad\quad x = -3 - 2\sqrt{3}$$

The solutions are $-3 \pm 2\sqrt{3}$.

63. $s = -16t^2 + v_0 t + s_0$

$$0 = -16t^2 + 0t + 400$$
$$16t^2 = 400$$
$$t^2 = \frac{400}{16}$$
$$t^2 = 25$$
$$t = \pm\sqrt{25}$$
$$t = 5$$

Use the positive square root. It will take 5 seconds for the balloon to hit the ground.

65. $s = -16t^2 + v_0 t + s_0$

$$5000 = -16t^2 + 0t + 12,000$$
$$16t^2 = 7000$$
$$t^2 = \frac{7000}{16}$$
$$t = \pm\sqrt{\frac{7000}{16}}$$
$$t = \frac{10\sqrt{70}}{4}$$

$t = \dfrac{5\sqrt{70}}{2} \approx 20.9$

Use the positive square root.

It will take $\dfrac{5\sqrt{70}}{2}$ or approximately

20.9 seconds.

67. $a^2 + b^2 = c^2$

$a^2 + (60)^2 = (61)^2$

$a^2 = 121$

$a = \pm\sqrt{121}$

$a = 11$

Use the positive square root.
It is 11 inches high.

69. $a^2 + b^2 = c^2$

$a^2 + (3000)^2 = (5000)^2$

$a^2 = 16,000,000$

$a = \pm\sqrt{16,000,000}$

$a = 4000$

Use the positive square root.
The distance is 4000 ft.

71. $a^2 + b^2 = c^2$

$x^2 + x^2 = (50)^2$

$2x^2 = 2500$

$x^2 = 1250$

$x = \pm\sqrt{1250}$

$x = \pm\sqrt{625 \cdot 2}$

$x = 25\sqrt{2}$

Use the positive square root.

The distance is $25\sqrt{2}$ or approximately
35.36 feet.

73. $C(x) = 0.044x^2 + 4700$

$9000 = 0.044x^2 + 4700$

$4300 = 0.044x^2$

$\dfrac{4300}{0.044} = x^2$

$\pm\sqrt{\dfrac{4300}{0.044}} = x$

$312.61 \approx x$

Use the positive square root.
She manages about 313 accounts.

10.3 Experiencing Algebra the Calculator Way

Original Pizza Diameter	Twice as Large Pizza Diameter
5 inches	7 inches
6 inches	8 inches
7 inches	10 inches
8 inches	11 inches
9 inches	13 inches
10 inches	14 inches
11 inches	16 inches
12 inches	17 inches

10.4 Experiencing Algebra the Exercise Way

1. $x^2 + 6x$

$b = \dfrac{6}{2} = 3$

$b^2 = 3^2 = 9$

We need to add 9.

3. $x^2 - 3x$

$b = -\dfrac{3}{2}$

$b^2 = \left(-\dfrac{3}{2}\right)^2 = \dfrac{9}{4}$

We need to add $\dfrac{9}{4}$.

5. $x^2 + \dfrac{3}{4}x$

$b = \dfrac{3}{4} \cdot \dfrac{1}{2} = \dfrac{3}{8}$

$b^2 = \left(\dfrac{3}{8}\right)^2 = \dfrac{9}{64}$

We need to add $\dfrac{9}{64}$.

7. $x^2 + x$

$b = \dfrac{1}{2}$

$b^2 = \left(\dfrac{1}{2}\right)^2 = \dfrac{1}{4}$

We need to add $\dfrac{1}{4}$.

9. $x^2 - 6x$

$b = -\dfrac{6}{2} = -3$

$b^2 = (-3)^2 = 9$

We need to add 9.

11. $x^2 + 9x$

$b = \dfrac{9}{2}$

$b^2 = \left(\dfrac{9}{2}\right)^2 = \dfrac{81}{4}$

We need to add $\dfrac{81}{4}$.

13. $x^2 + \dfrac{8}{9}x$

$b = \dfrac{8}{9} \cdot \dfrac{1}{2} = \dfrac{4}{9}$

$b^2 = \left(\dfrac{4}{9}\right)^2 = \dfrac{16}{81}$

We need to add $\dfrac{16}{81}$.

15. $x^2 - 14x$

$b = -\dfrac{14}{2} = -7$

$b^2 = (-7)^2 = 49$

We need to add 49.

17. $x^2 + 6x - 20 = 35$

$x^2 + 6x = 55$

$x^2 + 6x + 9 = 55 + 9$

$(x + 3)^2 = 64$

$x + 3 = \sqrt{64}$ or $x + 3 = -\sqrt{64}$

$x + 3 = 8$ $x + 3 = -8$

$x = 5$ $x = -11$

The solutions are -11 and 5.

19. $x^2 - 3x = 28$

$x^2 - 3x + \dfrac{9}{4} = 28 + \dfrac{9}{4}$

$\left(x - \dfrac{3}{2}\right)^2 = \dfrac{121}{4}$

$x - \dfrac{3}{2} = \sqrt{\dfrac{121}{4}}$ or $x - \dfrac{3}{2} = -\sqrt{\dfrac{121}{4}}$

$x - \dfrac{3}{2} = \dfrac{11}{2}$ $x - \dfrac{3}{2} = -\dfrac{11}{2}$

$x = \dfrac{14}{2}$ $x = -\dfrac{8}{2}$

$x = 7$ $x = -4$

The solutions are -4 and 7.

21. $x^2 + \dfrac{4}{7}x + \dfrac{3}{49} = 0$

$x^2 + \dfrac{4}{7}x = -\dfrac{3}{49}$

$x^2 + \dfrac{4}{7}x + \dfrac{4}{49} = -\dfrac{3}{49} + \dfrac{4}{49}$

$\left(x + \dfrac{2}{7}\right)^2 = \dfrac{1}{49}$

$x + \dfrac{2}{7} = \sqrt{\dfrac{1}{49}}$ or $x + \dfrac{2}{7} = -\sqrt{\dfrac{1}{49}}$

$x + \dfrac{2}{7} = \dfrac{1}{7}$ $x + \dfrac{2}{7} = -\dfrac{1}{7}$

$x = -\dfrac{1}{7}$ $x = -\dfrac{3}{7}$

The solutions are $-\dfrac{3}{7}$ and $-\dfrac{1}{7}$.

23. $x^2 + x - 30 = 60$

$x^2 + x = 90$

$x^2 + x + \dfrac{1}{4} = 90 + \dfrac{1}{4}$

$\left(x + \dfrac{1}{2}\right)^2 = \dfrac{361}{4}$

$x + \dfrac{1}{2} = \sqrt{\dfrac{361}{4}}$ or $x + \dfrac{1}{2} = -\sqrt{\dfrac{361}{4}}$

$x + \dfrac{1}{2} = \dfrac{19}{2}$ $x + \dfrac{1}{2} = -\dfrac{19}{2}$

$x = \dfrac{18}{2}$ $x = -\dfrac{20}{2}$

$x = 9$ $x = -10$

The solutions are −10 and 9.

25. $x^2 - 6x = 2$

$x^2 - 6x + 9 = 2 + 9$

$(x - 3)^2 = 11$

$x - 3 = \sqrt{11}$ or $x - 3 = -\sqrt{11}$

$x = 3 + \sqrt{11}$ $x = 3 - \sqrt{11}$

The solutions are $3 \pm \sqrt{11}$.

27. $x^2 + 9x = 1$

$x^2 + 9x + \dfrac{81}{4} = 1 + \dfrac{81}{4}$

$\left(x + \dfrac{9}{2}\right)^2 = \dfrac{85}{4}$

$x + \dfrac{9}{2} = \sqrt{\dfrac{85}{4}}$ or $x + \dfrac{9}{2} = -\sqrt{\dfrac{85}{4}}$

$x + \dfrac{9}{2} = \dfrac{\sqrt{85}}{2}$ $x + \dfrac{9}{2} = -\dfrac{\sqrt{85}}{2}$

$x = \dfrac{-9 + \sqrt{85}}{2}$ $x = \dfrac{-9 - \sqrt{85}}{2}$

The solutions are $\dfrac{-9 \pm \sqrt{85}}{2}$.

29. $x^2 + \dfrac{8}{9}x = 2$

$x^2 + \dfrac{8}{9}x + \dfrac{16}{81} = 2 + \dfrac{16}{81}$

$\left(x + \dfrac{4}{9}\right)^2 = \dfrac{178}{81}$

$x + \dfrac{4}{9} = \sqrt{\dfrac{178}{81}}$ or $x + \dfrac{4}{9} = -\sqrt{\dfrac{178}{81}}$

$x + \dfrac{4}{9} = \dfrac{\sqrt{178}}{9}$ $x + \dfrac{4}{9} = -\dfrac{\sqrt{178}}{9}$

$x = \dfrac{-4 + \sqrt{178}}{9}$ $x = \dfrac{-4 - \sqrt{178}}{9}$

The solutions are $\dfrac{-4 \pm \sqrt{178}}{9}$

31. $x^2 - x - 5 = 0$

$x^2 - x = 5$

$x^2 - x + \dfrac{1}{4} = 5 + \dfrac{1}{4}$

$\left(x - \dfrac{1}{2}\right)^2 = \dfrac{21}{4}$

$$x - \frac{1}{2} = \sqrt{\frac{21}{4}} \quad \text{or} \quad x - \frac{1}{2} = -\sqrt{\frac{21}{4}}$$

$$x - \frac{1}{2} = \frac{\sqrt{21}}{2} \qquad\qquad x - \frac{1}{2} = -\frac{\sqrt{21}}{2}$$

$$x = \frac{1 + \sqrt{21}}{2} \qquad\qquad x = \frac{1 - \sqrt{21}}{2}$$

The solutions are $\dfrac{1 \pm \sqrt{21}}{2}$.

33. $x^2 + x + 10 = 0$

$x^2 + x = -10$

$x^2 + x + \dfrac{1}{4} = -10 + \dfrac{1}{4}$

$\left(x + \dfrac{1}{2}\right)^2 = -\dfrac{39}{4}$

There is no real-number solution because there is no real number whose square is negative.

35. $2x^2 + 6x - 1 = 0$

$2x^2 + 6x = 1$

$x^2 + 3x = \dfrac{1}{2}$

$x^2 + 3x + \dfrac{9}{4} = \dfrac{1}{2} + \dfrac{9}{4}$

$\left(x + \dfrac{3}{2}\right)^2 = \dfrac{11}{4}$

$$x + \frac{3}{2} = \sqrt{\frac{11}{4}} \quad \text{or} \quad x + \frac{3}{2} = -\sqrt{\frac{11}{4}}$$

$$x + \frac{3}{2} = \frac{\sqrt{11}}{2} \qquad\qquad x + \frac{3}{2} = -\frac{\sqrt{11}}{2}$$

$$x = \frac{-3 + \sqrt{11}}{2} \qquad\qquad x = \frac{-3 - \sqrt{11}}{2}$$

The solutions are $\dfrac{-3 \pm \sqrt{11}}{2}$.

37. $3x^2 + x - 7 = 0$

$3x^2 + x = 7$

$x^2 + \dfrac{1}{3}x = \dfrac{7}{3}$

$x^2 + \dfrac{1}{3}x + \dfrac{1}{36} = \dfrac{7}{3} + \dfrac{1}{36}$

$\left(x + \dfrac{1}{6}\right)^2 = \dfrac{85}{36}$

$$x + \frac{1}{6} = \sqrt{\frac{85}{36}} \quad \text{or} \quad x + \frac{1}{6} = -\sqrt{\frac{85}{36}}$$

$$x + \frac{1}{6} = \frac{\sqrt{85}}{6} \qquad\qquad x + \frac{1}{6} = -\frac{\sqrt{85}}{6}$$

$$x = \frac{-1 + \sqrt{85}}{6} \qquad\qquad x = \frac{-1 - \sqrt{85}}{6}$$

The solutions are $\dfrac{-1 \pm \sqrt{85}}{6}$.

39. $x^2 - 14x + 55 = 6$

$x^2 - 14x = -49$

$x^2 - 14x + 49 = -49 + 49$

$(x - 7)^2 = 0$

$x - 7 = 0$

$x = 7$

The solution is 7.

41. $4x^2 - 20x + 30 = 5$

$4x^2 - 20x = -25$

$x^2 - 5x = -\dfrac{25}{4}$

$x^2 - 5x + \dfrac{25}{4} = -\dfrac{25}{4} + \dfrac{25}{4}$

$\left(x - \dfrac{5}{2}\right)^2 = 0$

$x - \dfrac{5}{2} = 0$

$x = \dfrac{5}{2}$

The solution is $\dfrac{5}{2}$.

43. $\frac{1}{2}x^2 + 5x - 2 = 0$

$\frac{1}{2}x^2 + 5x = 2$

$x^2 + 10x = 4$

$x^2 + 10x + 25 = 4 + 25$

$(x+5)^2 = 29$

$x+5 = \sqrt{29}$ or $x+5 = -\sqrt{29}$

$x = -5 + \sqrt{29}$ $x = -5 - \sqrt{29}$

The solutions are $-5 \pm \sqrt{29}$.

45. $\frac{1}{3}x^2 + \frac{1}{9}x - \frac{1}{6} = 0$

$\frac{1}{3}x^2 + \frac{1}{9}x = \frac{1}{6}$

$x^2 + \frac{1}{3}x = \frac{1}{2}$

$x^2 + \frac{1}{3}x + \frac{1}{36} = \frac{1}{2} + \frac{1}{36}$

$\left(x + \frac{1}{6}\right)^2 = \frac{19}{36}$

$x + \frac{1}{6} = \sqrt{\frac{19}{36}}$ or $x + \frac{1}{6} = -\sqrt{\frac{19}{36}}$

$x + \frac{1}{6} = \frac{\sqrt{19}}{6}$ $x + \frac{1}{6} = -\frac{\sqrt{19}}{6}$

$x = \frac{-1+\sqrt{19}}{6}$ $x = \frac{-1-\sqrt{19}}{6}$

The solutions are $\frac{-1 \pm \sqrt{19}}{6}$.

47. $\frac{2}{3}x^2 - 2x - \frac{5}{6} = 0$

$\frac{2}{3}x^2 - 2x = \frac{5}{6}$

$x^2 - 3x = \frac{5}{4}$

$x^2 - 3x + \frac{9}{4} = \frac{5}{4} + \frac{9}{4}$

$\left(x - \frac{3}{2}\right)^2 = \frac{14}{4}$

$x - \frac{3}{2} = \sqrt{\frac{14}{4}}$ or $x - \frac{3}{2} = -\sqrt{\frac{14}{4}}$

$x - \frac{3}{2} = \frac{\sqrt{14}}{2}$ $x - \frac{3}{2} = -\frac{\sqrt{14}}{2}$

$x = \frac{3+\sqrt{14}}{2}$ $x = \frac{3-\sqrt{14}}{2}$

The solutions are $\frac{3 \pm \sqrt{14}}{2}$.

49. Let x = number of cars purchased
$20 - x$ = cost per car

$x(20 - x) = 75$

$20x - x^2 = 75$

$x^2 - 20x = -75$

$x^2 - 20x + 100 = -75 + 100$

$(x-10)^2 = 25$

$x - 10 = \sqrt{25}$ or $x - 10 = -\sqrt{25}$

$x - 10 = 5$ $x - 10 = -5$

$x = 15$ $x = 5$

The number of cars is 15 because the dealer will not sell cars for less than $12.

51. $D(x) = \frac{1}{3}x^2 - 64x + 3100$

$2000 = \frac{1}{3}x^2 - 64x + 3100$

$-1100 = \frac{1}{3}x^2 - 64x$

$-3300 = x^2 - 192x$

$-3300 + 9216 = x^2 - 192x + 9216$

$5916 = (x - 96)^2$

$x - 96 = \sqrt{5916}$ or $x - 96 = -\sqrt{5916}$

$x = 96 + \sqrt{5916}$ $x = 96 - \sqrt{5916}$

$x \approx 173$ $x \approx 19$

Disregard the price over $50. The price should be set at $19.

53. $C(x) = 0.11x^2 + 3.08x + 11$

$55 = 0.11x^2 + 3.08x + 11$

$44 = 0.11x^2 + 3.08x$

$400 = x^2 + 28x$

$400 + 196 = x^2 + 28x + 196$

$596 = (x + 14)^2$

$x + 14 = \sqrt{596}$ or $x + 14 = -\sqrt{596}$

$x = -14 + \sqrt{596}$ $\qquad x = -14 - \sqrt{596}$

$x \approx 10.4$ $\qquad\qquad x = -38.4$

Disregard a negative amount. The contract should be about $10,400,000.

55. Let x = width

$x + 4$ = length

$x(x + 4) = 117$

$x^2 + 4x = 117$

$x^2 + 4x + 4 = 117 + 4$

$(x + 2)^2 = 121$

$x + 2 = \sqrt{121}$ or $x + 2 = -\sqrt{121}$

$x + 2 = 11$ $\qquad\qquad x + 2 = -11$

$x = 9$ $\qquad\qquad\qquad x = -13$

Disregard a negative width. The width is 9 in. and the length is 13 in.

57. Let x = length

$x - 5$ = width

$x(x - 5) = 85$

$x^2 - 5x = 85$

$x^2 - 5x + \dfrac{25}{4} = 85 + \dfrac{25}{4}$

$\left(x - \dfrac{5}{2}\right)^2 = \dfrac{365}{4}$

$x - \dfrac{5}{2} = \sqrt{\dfrac{365}{4}}$ or $x - \dfrac{5}{2} = -\sqrt{\dfrac{365}{4}}$

$x = \dfrac{5 + \sqrt{365}}{2}$ $\qquad x = \dfrac{5 - \sqrt{365}}{2}$

$x \approx 12.1$ $\qquad\qquad x \approx -7.1$

Disregard a negative length. The length is 12.1 ft and the width is 7.1 ft.

59. a. $s = -16t^2 + v_0 t + s_0$

$0 = -16t^2 + 32t + 16$

b. $16t^2 - 32t - 16 = 0$

$t^2 - 2t - 1 = 0$

$t^2 - 2t = 1$

$t^2 - 2t + 1 = 1 + 1$

$(t - 1)^2 = 2$

$t - 1 = \sqrt{2}$ or $t - 1 = -\sqrt{2}$

$t = 1 + \sqrt{2}$ $\qquad t = 1 - \sqrt{2}$

Disregard a negative time. It will take $1 + \sqrt{2}$ seconds.

c. $1 + \sqrt{2} \approx 2.4$

It will take approximately 2.4 seconds.

10.4 Experiencing Algebra the Calculator Way

1. $x^2 + 8x + 15 = 0$

$x^2 + 8x = -15$

$x^2 + 8x + 16 = -15 + 16$

$(x + 4)^2 = 1$

$x + 4 = 1$ or $x + 4 = -1$

$x = -3$ or $x = -5$

$Y1 = x^2 + 8x + 15$

$Y2 = 0$

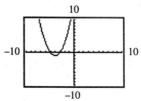

The solutions are -5 and -3.

2. $x^2 + 4x + 7 = 0$

$x^2 + 4x = -7$

$x^2 + 4x + 4 = -7 + 4$

$(x + 2)^2 = -3$

There is no real number solution.

$Y1 = x^2 + 4x + 7$

$Y2 = 0$

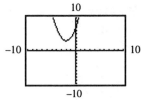

There is no real-number solution.

3. $x^2 + x - 1 = 0$

$x^2 + x = 1$

$x^2 + x + \dfrac{1}{4} = 1 + \dfrac{1}{4}$

$\left(x + \dfrac{1}{2}\right)^2 = \dfrac{5}{4}$

$x + \dfrac{1}{2} = \sqrt{\dfrac{5}{4}}$ or $x + \dfrac{1}{2} = -\sqrt{\dfrac{5}{4}}$

$x = \dfrac{-1 + \sqrt{5}}{2}$ $x = \dfrac{-1 - \sqrt{5}}{2}$

$Y1 = x^2 + x - 1$
$Y2 = 0$

The solutions are approximately −1.618 and 0.618.

4. $2x^2 + 11x + 12 = 0$

$2x^2 + 11x = -12$

$x^2 + \dfrac{11}{2}x = -6$

$x^2 + \dfrac{11}{2}x + \dfrac{121}{16} = -6 + \dfrac{121}{16}$

$\left(x + \dfrac{11}{4}\right)^2 = \dfrac{25}{16}$

$x + \dfrac{11}{4} = \sqrt{\dfrac{25}{16}}$ or $x + \dfrac{11}{4} = -\sqrt{\dfrac{25}{16}}$

$x = -\dfrac{6}{4}$ $x = -\dfrac{16}{4}$

$x = -\dfrac{3}{2}$ $x = -4$

$Y1 = 2x^2 + 11x + 12$
$Y2 = 0$

The solutions are −4 and −1.5.

5. $2x^2 + 6x + 7 = 0$

$2x^2 + 6x = -7$

$x^2 + 3x = -\dfrac{7}{2}$

$x^2 + 3x + \dfrac{9}{4} = -\dfrac{7}{2} + \dfrac{9}{4}$

$\left(x + \dfrac{3}{2}\right)^2 = -\dfrac{5}{4}$

$Y1 = 2x^2 + 6x + 7$
$Y2 = 0$

There is no real-number solution.

6. $8x^2 + 8x - 5 = 0$

$8x^2 + 8x = 5$

$x^2 + x = \dfrac{5}{8}$

$x^2 + x + \dfrac{1}{4} = \dfrac{5}{8} + \dfrac{1}{4}$

$\left(x + \dfrac{1}{2}\right)^2 = \dfrac{7}{8}$

$x + \dfrac{1}{2} = \sqrt{\dfrac{7}{8}}$ or $x + \dfrac{1}{2} = -\sqrt{\dfrac{7}{8}}$

$x = -\dfrac{1}{2} + \dfrac{\sqrt{7}}{2\sqrt{2}} \cdot \dfrac{\sqrt{2}}{\sqrt{2}}$ $x = -\dfrac{1}{2} - \dfrac{\sqrt{7}}{2\sqrt{2}} \cdot \dfrac{\sqrt{2}}{\sqrt{2}}$

$x = \dfrac{-2 + \sqrt{14}}{4}$ $x = \dfrac{-2 - \sqrt{14}}{4}$

$Y1 = 8x^2 + 8x - 5$
$Y2 = 0$

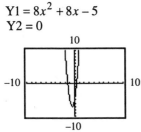

The solutions are approximately -1.435 and 0.435.

10.5 Experiencing Algebra the Exercise Way

1. $x^2 - 12x + 27 = 0$

$$x = \frac{-b \pm \sqrt{b^2 - 4ac}}{2a}$$

$$x = \frac{-(-12) \pm \sqrt{(-12)^2 - 4(1)(27)}}{2(1)}$$

$$x = \frac{12 \pm \sqrt{144 - 108}}{2}$$

$$x = \frac{12 \pm \sqrt{36}}{2}$$

$$x = \frac{12 \pm 6}{2}$$

$$x = \frac{12 - 6}{2} \qquad x = \frac{12 + 6}{2}$$

$x = 3 \qquad\qquad x = 9$

The solutions are 3 and 9.

3. $2x^2 + 3x - 15 = 2x + 6$

$2x^2 + x - 21 = 0$

$$x = \frac{-b \pm \sqrt{b^2 - 4ac}}{2a}$$

$$x = \frac{-1 \pm \sqrt{1^2 - 4(2)(-21)}}{2(2)}$$

$$x = \frac{-1 \pm \sqrt{1 + 168}}{4}$$

$$x = \frac{-1 \pm \sqrt{169}}{4}$$

$$x - \frac{-1 \pm 13}{4}$$

$$x = \frac{-1 - 13}{4} \qquad\qquad x = \frac{-1 + 13}{4}$$

$$x = -\frac{7}{2} \qquad\qquad x = 3$$

The solutions are $-\dfrac{7}{2}$ and 3.

5. $2z^2 + 11z + 5 = 0$

$$z = \frac{-b \pm \sqrt{b^2 - 4ac}}{2a}$$

$$z = \frac{-11 \pm \sqrt{(11)^2 - 4(2)(5)}}{2(2)}$$

$$z = \frac{-11 \pm \sqrt{121 - 40}}{4}$$

$$z = \frac{-11 \pm \sqrt{81}}{4}$$

$$z = \frac{-11 \pm 9}{4}$$

$$z = \frac{-11 - 9}{4} \qquad\qquad z = \frac{-11 + 9}{4}$$

$$z = -5 \qquad\qquad\qquad z = -\frac{1}{2}$$

The solutions are -5 and $-\dfrac{1}{2}$.

7. $8(2p^2 - p + 1) = 7$

$16p^2 - 8p + 1 = 0$

$$p = \frac{-b \pm \sqrt{b^2 - 4ac}}{2a}$$

$$p = \frac{-(-8) \pm \sqrt{(-8)^2 - 4(16)(1)}}{2(16)}$$

$$p = \frac{8 \pm \sqrt{64 - 64}}{32}$$

$$p = \frac{8 \pm \sqrt{0}}{32}$$

$$p = \frac{8}{32}$$

$$p = \frac{1}{4}$$

The solution is $\dfrac{1}{4}$.

9. $x^2 + 3x + 4 = 0$

$$x = \frac{-b \pm \sqrt{b^2 - 4ac}}{2a}$$

$$x = \frac{-3 \pm \sqrt{3^2 - 4(1)(4)}}{2(1)}$$

$$x = \frac{-3 \pm \sqrt{9 - 16}}{2}$$

$$x = \frac{-3 \pm \sqrt{-7}}{2}$$

There is no real-number solution.

11. $-5a^2 + 4a - 7 = 0$

$$a = \frac{-b \pm \sqrt{b^2 - 4ac}}{2a}$$

$$a = \frac{-4 \pm \sqrt{4^2 - 4(-5)(-7)}}{2(-5)}$$

$$a = \frac{-4 \pm \sqrt{16 - 140}}{-10}$$

$$a = \frac{-4 \pm \sqrt{-124}}{-10}$$

There is no real-number solution.

13. $v^2 - 5v + 2 = 0$

$$v = \frac{-b \pm \sqrt{b^2 - 4ac}}{2a}$$

$$v = \frac{-(-5) \pm \sqrt{(-5)^2 - 4(1)(2)}}{2(1)}$$

$$v = \frac{5 \pm \sqrt{25 - 8}}{2}$$

$$v = \frac{5 \pm \sqrt{17}}{2}$$

The solutions are $\dfrac{5 \pm \sqrt{17}}{2}$.

15. $x(x - 4) + 1 = 0$

$x^2 - 4x + 1 = 0$

$$x = \frac{-b \pm \sqrt{b^2 - 4ac}}{2a}$$

$$x = \frac{-(-4) \pm \sqrt{(-4)^2 - 4(1)(1)}}{2(1)}$$

$$x = \frac{4 \pm \sqrt{16 - 4}}{2}$$

$$x = \frac{4 \pm \sqrt{12}}{2}$$

$$x = \frac{4 \pm 2\sqrt{3}}{2}$$

$$x = 2 \pm \sqrt{3}$$

The solutions are $2 \pm \sqrt{3}$.

17. $4x(x + 1) + 6 = 6 - x$

$4x^2 + 5x + 0 = 0$

$$x = \frac{-b \pm \sqrt{b^2 - 4ac}}{2a}$$

$$x = \frac{-5 \pm \sqrt{5^2 - 4(4)(0)}}{2(4)}$$

$$x = \frac{-5 \pm \sqrt{25}}{8}$$

$$x = \frac{-5 \pm 5}{8}$$

$$x = \frac{-5 - 5}{8} \qquad x = \frac{-5 + 5}{8}$$

$$x = -\frac{5}{4} \qquad\qquad x = 0$$

The solutions are $-\dfrac{5}{4}$ and 0.

19. $3d^2 + 10 = 17$

$3d^2 + 0d - 7 = 0$

$$d = \frac{-b \pm \sqrt{b^2 - 4ac}}{2a}$$

$$d = \frac{-0 \pm \sqrt{0^2 - 4(3)(-7)}}{2(3)}$$

$$d = \frac{0 \pm \sqrt{84}}{6}$$

$$d = \frac{\pm 2\sqrt{21}}{6}$$

$$d = \pm \frac{\sqrt{21}}{3}$$

The solutions are $\pm \dfrac{\sqrt{21}}{3}$.

21. $16m = m^2 + 55$

$m^2 - 16m + 55 = 0$

$m = \dfrac{-b \pm \sqrt{b^2 - 4ac}}{2a}$

$m = \dfrac{-(-16) \pm \sqrt{(-16)^2 - 4(1)(55)}}{2(1)}$

$m = \dfrac{16 \pm \sqrt{256 - 220}}{2}$

$m = \dfrac{16 \pm \sqrt{36}}{2}$

$m = \dfrac{16 \pm 6}{2}$

$m = \dfrac{16 - 6}{2}$ or $m = \dfrac{16 + 6}{2}$

$m = 5$ $m = 11$

The solutions are 5 and 11.

23. $(x - 4)(x + 4) = 2(x - 4)$

$x^2 - 2x - 8 = 0$

$x = \dfrac{-b \pm \sqrt{b^2 - 4ac}}{2a}$

$x = \dfrac{-(-2) \pm \sqrt{(-2)^2 - 4(1)(-8)}}{2(1)}$

$x = \dfrac{2 \pm \sqrt{4 + 32}}{2}$

$x = \dfrac{2 \pm \sqrt{36}}{2}$

$x = \dfrac{2 \pm 6}{2}$

$x = \dfrac{2 - 6}{2}$ or $x = \dfrac{2 + 6}{2}$

$x = -2$ $x = 4$

The solutions are –2 and 4.

25. $x^2 - 6.3x + 7.2 = 0$

$x = \dfrac{-b \pm \sqrt{b^2 - 4ac}}{2a}$

$x = \dfrac{-(-6.3) \pm \sqrt{(-6.3)^2 - 4(1)(7.2)}}{2(1)}$

$x = \dfrac{6.3 \pm \sqrt{39.69 - 28.8}}{2}$

$x = \dfrac{6.3 \pm \sqrt{10.89}}{2}$

$x = \dfrac{6.3 \pm 3.3}{2}$

$x = \dfrac{6.3 - 3.3}{2}$ or $x = \dfrac{6.3 + 3.3}{2}$

$x = 1.5$ $x = 4.8$

The solutions are 1.5 and 4.8.

27. $1.8 - 5.6x - x^2 = 0$

$x^2 + 5.6x - 1.8 = 0$

$x = \dfrac{-b \pm \sqrt{b^2 - 4ac}}{2a}$

$x = \dfrac{-5.6 \pm \sqrt{(5.6)^2 - 4(1)(-1.8)}}{2(1)}$

$x = \dfrac{-5.6 \pm \sqrt{31.36 + 7.2}}{2}$

$x = \dfrac{-5.6 \pm \sqrt{38.56}}{2}$

$x = \dfrac{-5.6 \pm 2\sqrt{9.64}}{2}$

$x = -2.8 \pm \sqrt{9.64}$

The solutions are $-2.8 \pm \sqrt{9.64}$

29. $a^2 + 24.01 = 9.8a$

$a^2 - 9.8a + 24.01 = 0$

$a = \dfrac{-(-9.8) \pm \sqrt{(-9.8)^2 - 4(1)(24.01)}}{2(1)}$

$a = \dfrac{9.8 \pm \sqrt{96.04 - 96.04}}{2}$

$a = \dfrac{9.8 \pm 0}{2}$

$a = \dfrac{9.8}{2}$

$a = 4.9$

The solution is 4.9.

31. $1.7z^2 + 1.3z + 5.6 = 0$

$$z = \frac{-b \pm \sqrt{b^2 - 4ac}}{2a}$$

$$z = \frac{-1.3 \pm \sqrt{(1.3)^2 - 4(1.7)(5.6)}}{2(1.7)}$$

$$z = \frac{-1.3 \pm \sqrt{1.69 - 38.08}}{3.4}$$

$$z = \frac{-1.3 \pm \sqrt{-36.39}}{3.4}$$

There is no real-number solution.

33. $x^2 - 11x + 24 = 0$

$b^2 - 4ac = (-11)^2 - 4(1)(24) = 25$

There are two rational solutions because 25 is a perfect square.

35. $a^2 + 12a + 36 = 0$

$b^2 - 4ac = (12)^2 - 4(1)(36) = 0$

There is one rational solution because the discriminant is 0.

37. $z^2 = 4z - 5$

$z^2 - 4z + 5 = 0$

$b^2 - 4ac = (-4)^2 - 4(1)(5) = -4$

There is no real-number solution because the discriminant is negative.

39. $6x^2 - 11x - 7 = 0$

$b^2 - 4ac = (-11)^2 - 4(6)(-7) = 289$

There are two rational solutions because 289 is a perfect square.

41. $7p^2 - 15 = 0$

$b^2 - 4ac = 0^2 - 4(7)(-15) = 420$

There are two irrational solutions because 420 is positive but not a perfect square.

43. $1 - 5x - 4x^2 = 0$

$4x^2 + 5x - 1 = 0$

$b^2 - 4ac = (5)^2 - 4(4)(-1) = 41$

There are two irrational solutions because 41 is positive but not a perfect square.

45. $6.25z - 2.1 = 2.5z^2$

$2.5z^2 - 6.25z + 2.1 = 0$

$b^2 - 4ac = (-6.25)^2 - 4(2.5)(2.1) = 18.0625$

There are two rational solutions because 18.0625 is a perfect square.

47. $0.3x - 2.8 = 1.7x^2$

$1.7x^2 - 0.3x + 2.8 = 0$

$b^2 - 4ac = (-0.3)^2 - 4(1.7)(2.8) = -18.95$

There are no real-number solutions because the discriminant is negative.

49. $A = P(1 + r)^t$

$$6000 = 5000(1 + r)^2$$

$$\frac{6}{5} = 1 + 2r + r^2$$

$$r^2 + 2r - \frac{1}{5} = 0$$

$$r = \frac{-2 \pm \sqrt{(2)^2 - 4(1)\left(-\frac{1}{5}\right)}}{2(1)}$$

$$r = \frac{-2 \pm \sqrt{4 + \frac{4}{5}}}{2}$$

$$r = \frac{-2 \pm \sqrt{4.8}}{2}$$

$$r = \frac{-2 - \sqrt{4.8}}{2} \quad \text{or} \quad r = \frac{-2 + \sqrt{4.8}}{2}$$

$$r \approx -2.095 \qquad\qquad r \approx 0.095$$

Disregard a negative value. The interest rate must be about 9.5%.

51. $R(x) = 6000 + 70x - x^2$

$$7125 = 6000 + 70x - x^2$$

$$x^2 - 70x + 1125 = 0$$

$$x = \frac{70 \pm \sqrt{(-70)^2 - 4(1)(1125)}}{2(1)}$$

$$x = \frac{70 \pm \sqrt{400}}{2}$$

$$x = \frac{70 \pm 20}{2}$$

$$x = \frac{70 - 20}{2} \quad \text{or} \quad x = \frac{70 + 20}{2}$$

$$x = 25 \qquad\qquad x = 45$$

The values of x are 25 and 45.

Answers will vary.

53. $y = 0.086x^2 - 9.34x + 277.78$

$100 = 0.086x^2 - 9.34x + 277.78$

$0 = 0.086x^2 - 9.34x + 177.78$

$x = \dfrac{9.34 \pm \sqrt{(-9.34)^2 - 4(0.086)(177.78)}}{2(0.086)}$

$x = \dfrac{9.34 \pm \sqrt{26.07928}}{0.172}$

$x = \dfrac{9.34 - \sqrt{26.07928}}{0.172} \approx 24.61$ or

$x = \dfrac{9.34 + \sqrt{26.07928}}{0.172} \approx 83.99$

$1950 + 25 = 1975$

The year was 1975.

10.5 Experiencing Algebra the Calculator Way

Equation	Value of Discriminant	Type of Roots	Number of Unlike Roots	Roots
1. $x^2 + 6 = 5x$	1	rational	2	2, 3
2. $9x^2 + 6x = -1$	0	rational	1	$-\dfrac{1}{3}$
3. $2x^2 + 1 = 7x$	41	irrational	2	0.149, 3.351
4. $x^2 + 6x = -10$	−4	not real	2	not real
5. $x^2 = 6 - x$	25	rational	2	−3, 2
6. $5x^2 - 6x = 0$	36	rational	2	0, 1.2
7. $x^2 + 0.36 = 1.2x$	0	rational	1	0.6
8. $1.7x^2 + x + 1.9 = 0$	−11.92	not real	2	not real
9. $1.5x^2 + 1.2x = 3.6$	23.04	rational	2	−2, 1.2
10. $\dfrac{1}{4}x^2 + x = \dfrac{1}{8}$	1.125	irrational	2	−4.121, 0.121
11. $x^2 - \dfrac{1}{6}x = \dfrac{1}{6}$	$0.69\overline{4}$	rational	2	$-0.\overline{3}$, 0.5
12. $\dfrac{1}{5}x^2 + \dfrac{2}{3}x = -\dfrac{7}{8}$	$-0.2\overline{5}$	not real	2	not real

Chapter 10 Review

Reflections

   1–4. Answers will vary.

Exercises

   1. $2x^2 + 5x - 16 = 7x + 8$
   $Y1 = 2x^2 + 5x - 16$
   $Y2 = 7x + 8$

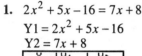

   The solutions are –3 and 4.

   2. $\frac{1}{3}x^2 + 5x + 6 = x - 3$

   $Y1 = \frac{1}{3}x^2 + 5x + 6$

   $Y2 = x - 3$

   The solutions are –9 and –3.

   3. $0.7x^2 - 2.5x + 4.6 = 1.7x + 1.1$
   $Y1 = 0.7x^2 - 2.5x + 4.6$
   $Y2 = 1.7x + 1.1$

   The solutions are 1 and 5.

   4. $x^2 - 4x + 4 = 9$
   $Y1 = x^2 - 4x + 4$
   $Y2 = 9$

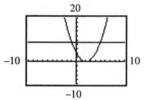

   The graphs intersect at (–1, 9) and (5, 9).
   The solutions are –1 and 5.

   5. $x^2 - 6 = \frac{1}{4}x^2 + 6$

   $Y1 = x^2 - 6$

   $Y2 = \frac{1}{4}x^2 + 6$

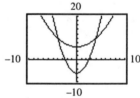

   The graphs intersect at (–4, 10) and (4, 10).
   The solutions are –4 and 4.

   6. $x^2 + 3x - 5 = 7 - 2x - x^2$
   $Y1 = x^2 + 3x - 5$
   $Y2 = 7 - 2x - x^2$

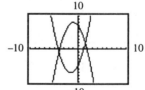

420

The graphs intersect at (–4, –1) and
(1.5, 1.75).
The solutions are –4 and 1.5.

7. $-0.3x^2 + 4x + 5 = 2.2x - 11.5$

$Y1 = -0.3x^2 + 4x + 5$

$Y2 = 2.2x - 11.5$

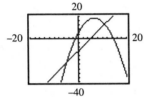

The graphs intersect at (–5, –22.5) and
(11, 12.7).
The solutions are –5 and 11.

8. $x^2 + 8x + 8 = 0$

$Y1 = x^2 + 8x + 8$

$Y2 = 0$

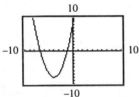

The graphs intersect at about (–6.83, 0) and
(–1.17, 0).
The solutions are approximately –6.83 and
–1.17.

9. $\frac{1}{5}x^2 - 5x + 4 = 8 - \frac{1}{3}x^2$

$Y1 = \frac{1}{5}x^2 - 5x + 4$

$Y2 = 8 - \frac{1}{3}x^2$

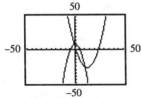

The graphs intersect at about (–0.74, 7.82)
and (10.12, –26.11).
The solutions are approximately –0.74 and
10.12.

10. $s = -16t^2 + v_0 t + s_0$

$0 = -16t^2 + 0t + 96$

$0 = -16t^2 + 96$

$Y1 = 0$

$Y2 = -16x^2 + 96$

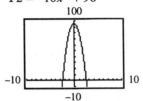

The graphs intersect at about (–2.45, 0) and
(2.45, 0).
Disregard a negative time.
It will reach the ground in approximately
2.45 seconds.

11. $C(x) = 8x^2 + 20x + 450$

$1850 = 8x^2 + 20x + 450$

$Y1 = 1850$

$Y2 = 8x^2 + 20x + 450$

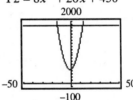

The graphs intersect at about (–14.54, 1850)
and (12.04, 1850).
Disregard a negative number of items.
12 items can be produced.

12. $x^2 - 5 = x + 1$
$x^2 - x - 6 = 0$
$(x - 3)(x + 2) = 0$
$x - 3 = 0$ or $x + 2 = 0$
$x = 3$ $x = -2$
The solutions are –2 and 3.

13. $6x^2 - x - 77 = 0$
$(2x + 7)(3x - 11) = 0$
$2x + 7 = 0$ or $3x - 11 = 0$
$x = -\frac{7}{2}$ $x = \frac{11}{3}$

The solutions are $-\frac{7}{2}$ and $\frac{11}{3}$.

14. $2x + 10 = x^2 - 5x + 2$
$x^2 - 7x - 8 = 0$
$(x - 8)(x + 1) = 0$
$x - 8 = 0$ or $x + 1 = 0$
$x = 8$ $x = -1$
The solutions are -1 and 8.

15. $x^2 - 7x - 60 = 0$
$(x + 5)(x - 12) = 0$
$x + 5 = 0$ or $x - 12 = 0$
$x = -5$ $x = 12$
The solutions are -5 and 12.

16. $3x^2 + 5x - 3 = 1 - 6x$
$3x^2 + 11x - 4 = 0$
$(3x - 1)(x + 4) = 0$
$3x - 1 = 0$ or $x + 4 = 0$
$x = \dfrac{1}{3}$ $x = -4$

The solutions are -4 and $\dfrac{1}{3}$.

17. $6x^2 - 8x = 9x - 12$
$6x^2 - 17x + 12 = 0$
$(3x - 4)(2x - 3) = 0$
$3x - 4 = 0$ or $2x - 3 = 0$
$x = \dfrac{4}{3}$ $x = \dfrac{3}{2}$

The solutions are $\dfrac{4}{3}$ and $\dfrac{3}{2}$.

18. $\dfrac{1}{4}x^2 + \dfrac{3}{2}x - \dfrac{19}{8} = \dfrac{3}{8}x + 2$
$\dfrac{1}{4}x^2 + \dfrac{9}{8}x - \dfrac{35}{8} = 0$
$8\left(\dfrac{1}{4}x^2 + \dfrac{9}{8}x - \dfrac{35}{8}\right) = 8 \cdot 0$
$2x^2 + 9x - 35 = 0$
$(2x - 5)(x + 7) = 0$
$2x - 5 = 0$ or $x + 7 = 0$
$x = \dfrac{5}{2}$ $x = -7$

The solutions are $\dfrac{5}{2}$ and -7.

19. $9x^2 - 49 = 0$
$(3x - 7)(3x + 7) = 0$
$3x - 7 = 0$ or $3x + 7 = 0$
$x = \dfrac{7}{3}$ $x = -\dfrac{7}{3}$

The solutions are $\pm\dfrac{7}{3}$.

20. $s = -16t^2 + v_0 t + s_0$
$0 = -16t^2 + 48t + 16$
$0 = -16(t^2 - 3t - 1)$
$t^2 - 3t - 1 = 0$
The equation does not factor.
$Y1 = x^2 - 3x - 1$
$Y2 = 0$

The graphs intersect at about $(-0.303, 0)$ and $(3.303, 0)$.
Disregard a negative time.
The ball will hit the ground in approximately 3.3 seconds.

21. $\sqrt{25} = \sqrt{5^2} = 5$

22. $-\sqrt{49} = -\sqrt{7^2} = -7$

23. $\sqrt{5}\sqrt{45} = \sqrt{5 \cdot 45} = \sqrt{225} = \sqrt{(15)^2} = 15$

24. $\sqrt{15}\sqrt{21} = \sqrt{315} = \sqrt{9 \cdot 35} = 3\sqrt{35}$

25. $\sqrt{8} = \sqrt{4 \cdot 2} = 2\sqrt{2}$

26. $\sqrt{\dfrac{64}{81}} = \dfrac{\sqrt{64}}{\sqrt{81}} = \dfrac{8}{9}$

27. $\sqrt{\dfrac{13}{25}} = \dfrac{\sqrt{13}}{\sqrt{25}} = \dfrac{\sqrt{13}}{5}$

28. $\sqrt{\dfrac{25}{3}} = \dfrac{\sqrt{25}}{\sqrt{3}} = \dfrac{5}{\sqrt{3}} \cdot \dfrac{\sqrt{3}}{\sqrt{3}} = \dfrac{5\sqrt{3}}{3}$

29. $\sqrt{\dfrac{2}{5}} = \dfrac{\sqrt{2}}{\sqrt{5}} \cdot \dfrac{\sqrt{5}}{\sqrt{5}} = \dfrac{\sqrt{10}}{5}$

30. $\sqrt{\dfrac{15}{21}} = \sqrt{\dfrac{5}{7}} = \dfrac{\sqrt{5}}{\sqrt{7}} \cdot \dfrac{\sqrt{7}}{\sqrt{7}} = \dfrac{\sqrt{35}}{7}$

31. $\sqrt{\dfrac{6}{50}} = \sqrt{\dfrac{3}{25}} = \dfrac{\sqrt{3}}{\sqrt{25}} = \dfrac{\sqrt{3}}{5}$

32. $\dfrac{\sqrt{75}}{\sqrt{3}} = \sqrt{\dfrac{75}{3}} = \sqrt{25} = 5$

33. $4x^2 - 100 = 0$
$4x^2 = 100$
$x^2 = 25$

$x = \sqrt{25}$ or	$x = -\sqrt{25}$
$x = 5$	$x = -5$

The solutions are ± 5.

34. $a^2 + 5 = 9$
$a^2 = 4$

$a = \sqrt{4}$ or	$a = -\sqrt{4}$
$a = 2$	$a = -2$

The solutions are ± 2.

35. $x^2 + 3 = 15$
$x^2 = 12$

$x = \sqrt{12}$ or	$x = -\sqrt{12}$
$x = \sqrt{4 \cdot 3}$	$x = -\sqrt{4 \cdot 3}$
$x = 2\sqrt{3}$	$x = -2\sqrt{3}$

The solutions are $\pm 2\sqrt{3}$.

36. $15 + b^2 = 7$
$b^2 = -8$
There is no real-number solution because there is no real number whose square is negative.

37. $(x - 4)^2 = 9$

$x - 4 = \sqrt{9}$ or	$x - 4 = -\sqrt{9}$
$x - 4 = 3$	$x - 4 = -3$
$x = 7$	$x = 1$

The solutions are 1 and 7.

38. $x^2 + 18x + 81 = 16$
$(x + 9)^2 = 16$

$x + 9 = \sqrt{16}$ or	$x + 9 = -\sqrt{16}$
$x + 9 = 4$	$x + 9 = -4$
$x = -5$	$x = -13$

The solutions are -13 and -5.

39. $(z - 2)^2 + 5 = 7$
$(z - 2)^2 = 2$

$z - 2 = \sqrt{2}$	$z - 2 = -\sqrt{2}$
$z = 2 + \sqrt{2}$	$z = 2 - \sqrt{2}$

The solutions are $2 \pm \sqrt{2}$.

40. $s = -16t^2 + v_0 t + s_0$
$6000 = -16t^2 + 0t + 10,000$
$-4000 = -16t^2$
$250 = t^2$

$t = \sqrt{250}$ or	$t = -\sqrt{250}$
$t = 5\sqrt{10}$	$t = -5\sqrt{10}$

Disregard a negative time.
It will take $5\sqrt{10}$ seconds or approximately 15.8 seconds.

41. $a^2 + b^2 = c^2$
$a^2 + (15)^2 = (35)^2$
$a^2 + 225 = 1225$
$a^2 = 1000$

$a = \sqrt{1000}$ or	$a = -\sqrt{1000}$
$a = 10\sqrt{10}$	$a = -10\sqrt{10}$

Disregard a negative length.
The horizontal distance is $10\sqrt{10}$ feet or approximately 31.6 feet.

42. $s(x) = 28x^2 + 2000$

$3200 = 28x^2 + 2000$

$1200 = 28x^2$

$\dfrac{1200}{28} = x^2$

$x = \sqrt{\dfrac{1200}{28}}$ or $x = -\sqrt{\dfrac{1200}{28}}$

$x = \sqrt{\dfrac{300}{7}}$ $x = -\sqrt{\dfrac{300}{7}}$

$x = \dfrac{10\sqrt{3}}{\sqrt{7}} \cdot \dfrac{\sqrt{7}}{\sqrt{7}}$ $x = \dfrac{-10\sqrt{3}}{\sqrt{7}} \cdot \dfrac{\sqrt{7}}{\sqrt{7}}$

$x = \dfrac{10\sqrt{21}}{7}$ $x = \dfrac{-10\sqrt{21}}{7}$

$x \approx 6.55$ $x = -6.55$

Disregard a negative time.

$1991 + 7 = 1998$

Sales will reach \$3200 million in 1998.

43. $p^2 - 4p - 96 = 0$

$p^2 - 4p = 96$

$p^2 - 4p + 4 = 96 + 4$

$(p - 2)^2 = 100$

$p - 2 = \sqrt{100}$ or $p - 2 = -\sqrt{100}$

$p - 2 = 10$ $p - 2 = -10$

$p = 12$ $p = -8$

The solutions are -8 and 12.

44. $z^2 + 12z = -3$

$z^2 + 12z + 36 = -3 + 36$

$(z + 6)^2 = 33$

$z + 6 = \sqrt{33}$ or $z + 6 = -\sqrt{33}$

$z = -6 + \sqrt{33}$ $z = -6 - \sqrt{33}$

The solutions are $-6 \pm \sqrt{33}$.

45. $2x^2 - 5x - 12 = 0$

$2x^2 - 5x = 12$

$x^2 - \dfrac{5}{2}x = 6$

$x^2 - \dfrac{5}{2}x + \dfrac{25}{16} = 6 + \dfrac{25}{16}$

$\left(x - \dfrac{5}{4}\right)^2 = \dfrac{121}{16}$

$x - \dfrac{5}{4} = \sqrt{\dfrac{121}{16}}$ or $x - \dfrac{5}{4} = -\sqrt{\dfrac{121}{16}}$

$x - \dfrac{5}{4} = \dfrac{11}{4}$ $x - \dfrac{5}{4} = -\dfrac{11}{4}$

$x = \dfrac{5}{4} + \dfrac{11}{4}$ $x = \dfrac{5}{4} - \dfrac{11}{4}$

$x = 4$ $x = -\dfrac{3}{2}$

The solutions are $-\dfrac{3}{2}$ and 4.

46. $x^2 + \dfrac{1}{2}x = 1$

$x^2 + \dfrac{1}{2}x + \dfrac{1}{16} = 1 + \dfrac{1}{16}$

$\left(x + \dfrac{1}{4}\right)^2 = \dfrac{17}{16}$

$x + \dfrac{1}{4} = \sqrt{\dfrac{17}{16}}$ or $x + \dfrac{1}{4} = -\sqrt{\dfrac{17}{16}}$

$x = -\dfrac{1}{4} + \dfrac{\sqrt{17}}{4}$ $x = -\dfrac{1}{4} - \dfrac{\sqrt{17}}{4}$

$x = \dfrac{-1 + \sqrt{17}}{4}$ $x = \dfrac{-1 - \sqrt{17}}{4}$

The solutions are $\dfrac{-1 \pm \sqrt{17}}{4}$.

47. $C(x) = 60 + x(20 + 2x)$

$150 = 60 + x(20 + 2x)$

$150 = 60 + 20x + 2x^2$

$2x^2 + 20x - 90 = 0$

$x^2 + 10x - 45 = 0$

$x^2 + 10x = 45$

$x^2 + 10x + 25 = 45 + 25$

$(x + 5)^2 = 70$

$x + 5 = \sqrt{70}$ or $x + 5 = -\sqrt{70}$

$x = -5 + \sqrt{70}$ $x = -5 - \sqrt{70}$

$x \approx 3.4$ $x \approx -13.4$

Disregard a negative amount. Since x is the number of figurines over 3, a total of 6 figurines can be purchased for \$150.

48. Let $x = $ width

$2x + 8 = $ length

$x(2x + 8) = 90$

$2x^2 + 8x - 90 = 0$

$x^2 + 4x - 45 = 0$

$x^2 + 4x + 4 = 45 + 4$

$(x + 2)^2 = 49$

$x + 2 = \sqrt{49}$ or $x + 2 = -\sqrt{49}$

$x + 2 = 7$ $\qquad$ $x + 2 = -7$

$x = 5$ $\qquad$ $x = -9$

Disregard a negative width. The dimensions are 5 inches by 18 inches.

49. $x^2 - 2x - 63 = 0$

$x = \dfrac{-b \pm \sqrt{b^2 - 4ac}}{2a}$

$x = \dfrac{-(-2) \pm \sqrt{(-2)^2 - 4(1)(-63)}}{2(1)}$

$x = \dfrac{2 \pm \sqrt{4 + 252}}{2}$

$x = \dfrac{2 \pm \sqrt{256}}{2}$

$x = \dfrac{2 \pm 16}{2}$

$x = \dfrac{2 - 16}{2}$ or $x = \dfrac{2 + 16}{2}$

$x = -7$ $\qquad$ $x = 9$

The solutions are -7 and 9.

50. $x^2 = 3x + 3$

$x^2 - 3x - 3 = 0$

$x = \dfrac{-b \pm \sqrt{b^2 - 4ac}}{2a}$

$x = \dfrac{-(-3) \pm \sqrt{(-3)^2 - 4(1)(-3)}}{2(1)}$

$x = \dfrac{3 \pm \sqrt{9 + 12}}{2}$

$x = \dfrac{3 \pm \sqrt{21}}{2}$

The solutions are $\dfrac{3 \pm \sqrt{21}}{2}$.

51. $25x^2 + 1 = 10x$

$25x^2 - 10x + 1 = 0$

$x = \dfrac{-b \pm \sqrt{b^2 - 4ac}}{2a}$

$x = \dfrac{-(-10) \pm \sqrt{(-10)^2 - 4(25)(1)}}{2(25)}$

$x = \dfrac{10 \pm \sqrt{100 - 100}}{50}$

$x = \dfrac{10 \pm 0}{50}$

$x = \dfrac{1}{5}$

The solution is $\dfrac{1}{5}$.

52. $z^2 + \dfrac{7}{20}z - \dfrac{3}{10} = 0$

$20\left(z^2 + \dfrac{7}{20}z - \dfrac{3}{10}\right) = 0 \cdot 20$

$20z^2 + 7z - 6 = 0$

$z = \dfrac{-b \pm \sqrt{b^2 - 4ac}}{2a}$

$z = \dfrac{-7 \pm \sqrt{7^2 - 4(20)(-6)}}{2(20)}$

$z = \dfrac{-7 \pm \sqrt{529}}{40}$

$z = \dfrac{-7 \pm 23}{40}$

$z = \dfrac{-7 + 23}{40}$ or $z = \dfrac{-7 - 23}{40}$

$z = \dfrac{16}{40}$ $\qquad$ $z = -\dfrac{30}{40}$

$z = \dfrac{2}{5}$ $\qquad$ $z = -\dfrac{3}{4}$

The solutions are $-\dfrac{3}{4}$ and $\dfrac{2}{5}$.

53. $x^2 + 2.1x - 10.8 = 0$

$$x = \frac{-b \pm \sqrt{b^2 - 4ac}}{2a}$$

$$x = \frac{-2.1 \pm \sqrt{(2.1)^2 - 4(1)(-10.8)}}{2(1)}$$

$$x = \frac{-2.1 \pm \sqrt{4.41 + 43.2}}{2}$$

$$x = \frac{-2.1 \pm \sqrt{47.61}}{2}$$

$$x = \frac{-2.1 \pm 6.9}{2}$$

$$x = \frac{-2.1 - 6.9}{2} \quad \text{or} \quad x = \frac{-2.1 + 6.9}{2}$$

$$x = -4.5 \qquad\qquad x = 2.4$$

The solutions are –4.5 and 2.4.

54. $y^2 - 5y + 12 = 0$

$$y = \frac{-b \pm \sqrt{b^2 - 4ac}}{2a}$$

$$y = \frac{-(-5) \pm \sqrt{(-5)^2 - 4(1)(12)}}{2(1)}$$

$$y = \frac{5 \pm \sqrt{25 - 48}}{2}$$

$$y = \frac{5 \pm \sqrt{-23}}{2}$$

There is no real-number solution.

55. $x^2 - 10x + 6 = 0$

$$x = \frac{-b \pm \sqrt{b^2 - 4ac}}{2a}$$

$$x = \frac{-(-10) \pm \sqrt{(-10)^2 - 4(1)(6)}}{2(1)}$$

$$x = \frac{10 \pm \sqrt{76}}{2}$$

$$x = \frac{10 \pm 2\sqrt{19}}{2}$$

$$x = 5 \pm \sqrt{19}$$

56. $3a^2 - 4a - 12 = 0$

$$a = \frac{-b \pm \sqrt{b^2 - 4ac}}{2a}$$

$$a = \frac{-(-4) \pm \sqrt{(-4)^2 - 4(3)(-12)}}{2(3)}$$

$$a = \frac{4 \pm \sqrt{160}}{6} = \frac{4 \pm 4\sqrt{10}}{6} = \frac{2 \pm 2\sqrt{10}}{3}$$

57. $x^2 + 2x + 10 = 0$

$$b^2 - 4ac = 2^2 - 4(1)(10) = -36$$

There is no real-number solution because the discriminant is negative.

58. $x^2 + 20x + 55 = 2x - 26$

$$x^2 + 18x + 81 = 0$$

$$b^2 - 4ac = (18)^2 - 4(1)(81) = 0$$

There is one rational solution because the discriminant is 0.

59. $x^2 = 10x + 75$

$$x^2 - 10x - 75 = 0$$

$$b^2 - 4ac = (-10)^2 - 4(1)(-75) = 400$$

There are two rational solutions because 400 is a perfect square.

60. $x = x^2 - 11$

$$x^2 - x - 11 = 0$$

$$b^2 - 4ac = (-1)^2 - 4(1)(-11) = 45$$

There are two irrational solutions because 45 is postive but not a perfect square.

61. $A = P(1 + r)^t$

$$1166.40 = 1000(1 + r)^2$$

$$1.1664 = (1 + r)^2$$

$$1 + r = \sqrt{1.1664} \quad \text{or} \quad 1 + r = -\sqrt{1.1664}$$

$$r = -1 + \sqrt{1.1664} \qquad\qquad r = -1 - \sqrt{1.1664}$$

$$r = -1 + 1.08 \qquad\qquad r = -1 - 1.08$$

$$r = 0.08 \qquad\qquad\qquad r = -2.08$$

Disregard a negative interest. The interest rate would be 8%.

62. $A = P(1+r)^t$

$1200 = 1000(1+r)^2$

$1.2 = (1+r)^2$

$1+r = \sqrt{1.2}$ or $1+r = -\sqrt{1.2}$

$r = -1+\sqrt{1.2}$ $r = -1-\sqrt{1.2}$

$r \approx 0.095$ $r \approx -2.095$

Disregard a negative interest. The interest is about 9.5%.

63. $R(x) = 600 + 26x - 2x^2$

$450 = 600 + 26x - 2x^2$

$2x^2 - 26x - 150 = 0$

$2\left(x^2 - 13x - 75\right) = 0$

$$x = \frac{-(-13) \pm \sqrt{(-13)^2 - 4(1)(-75)}}{2(1)}$$

$$x = \frac{13 \pm \sqrt{469}}{2}$$

$$x = \frac{13 - \sqrt{469}}{2} \quad \text{or} \quad x = \frac{13 + \sqrt{469}}{2}$$

$x \approx -4.33$ $x \approx 17.33$

Disregard a negative amount. For a value of 17.33, the revenue will be approximately $450.

Chapter 10 Mixed Review

1. $-\sqrt{144} = -\sqrt{(12)^2} = -12$

2. $\sqrt{125} = \sqrt{25 \cdot 5} = 5\sqrt{5}$

3. $\sqrt{7}\sqrt{28} = \sqrt{7 \cdot 28} = \sqrt{196} = 14$

4. $\sqrt{\dfrac{121}{225}} = \dfrac{\sqrt{121}}{\sqrt{225}} = \dfrac{11}{15}$

5. $\sqrt{\dfrac{24}{135}} = \dfrac{\sqrt{24}}{\sqrt{135}} = \dfrac{\sqrt{4 \cdot 6}}{\sqrt{9 \cdot 15}} = \dfrac{2\sqrt{6}}{3\sqrt{15}} \cdot \dfrac{\sqrt{15}}{\sqrt{15}}$

$\quad = \dfrac{2\sqrt{90}}{45} = \dfrac{6\sqrt{10}}{45} = \dfrac{2\sqrt{10}}{15}$

6. $\dfrac{\sqrt{63}}{\sqrt{28}} = \sqrt{\dfrac{63}{28}} = \sqrt{\dfrac{9}{4}} = \dfrac{\sqrt{9}}{\sqrt{4}} = \dfrac{3}{2}$

7. $x^2 - 4 = 2x - 1$

$Y1 = x^2 - 4$

$Y2 = 2x - 1$

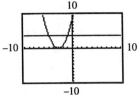

The intersections are $(-1, -3)$ and $(3, 5)$.
The solutions are -1 and 3.

8. $3 - x^2 = -3 - \dfrac{1}{3}x^2$

$Y1 = 3 - x^2$

$Y2 = -3 - \dfrac{1}{3}x^2$

The intersections are $(-3, -6)$ and $(3, -6)$.
The solutions are -3 and 3.

9. $x^2 + 6x + 9 = 4$

$Y1 = x^2 + 6x + 9$

$Y2 = 4$

The intersections are $(-5, 4)$ and $(-1, 4)$.
The solutions are -5 and -1.

10. $15x^2 + 13x - 72 = 0$

$Y1 = 15x^2 + 13x - 72$

$Y2 = 0$

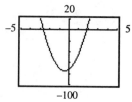

The intersections are

$$\left(-2\frac{2}{3},\, 0\right) \text{ and } \left(1\frac{4}{5},\, 0\right)$$

The solutions are $-2\frac{2}{3}$ and $1\frac{4}{5}$.

11. $2x + 10 = x^2 + 4x + 7$

$\text{Y1} = 2x + 10$

$\text{Y2} = x^2 + 4x + 7$

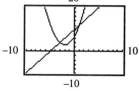

The intersections are $(-3, 4)$ and $(1, 12)$.
The solutions are -3 and 1.

12. $x^2 + 9x + 7 = 0$

$\text{Y1} = x^2 + 9x + 7$

$\text{Y2} = 0$

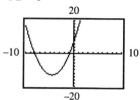

The intersections are about $(-8.14, 0)$ and $(-0.86, 0)$.
The solutions are approximately -8.14 and -0.86.

13. $\frac{1}{6}x^2 - 2x + 7 = 9 - \frac{1}{4}x^2$

$\text{Y1} = \frac{1}{6}x^2 - 2x + 7$

$\text{Y2} = 9 - \frac{1}{4}x^2$

The intersections are about $(-0.85, 8.82)$ and $(5.65, 1.02)$.
The solutions are approximately -0.85 and 5.65.

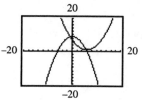

14. $x^2 - 6x - 40 = 0$

$\text{Y1} = x^2 - 6x - 40$

$\text{Y2} = 0$

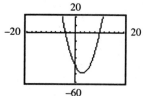

The intersections are $(-4, 0)$ and $(10, 0)$.
The solutions are -4 and 10.

15. $2x^2 - x - 28 = x^2 + 3x + 17$

$\text{Y1} = 2x^2 - x - 28$

$\text{Y2} = x^2 + 3x + 17$

X	Y₁	Y₂
-6	50	35
-5	27	27
-4	8	21
-3	-7	17
-2	-18	15
-1	-25	15
0	-28	17

X= -6

X	Y₁	Y₂
1	-27	21
2	-22	27
3	-13	35
4	0	45
5	17	57
6	38	71
7	63	87

X=1

X	Y₁	Y₂
8	92	105
9	125	125
10	162	147
11	203	171
12	248	197
13	297	225
14	350	255

X=8

The solutions are -5 and 9.

16. $3x^2 - 5x - 10 = 7x + 5$

$Y1 = 3x^2 - 5x - 10$

$Y2 = 7x + 5$

X	Y₁	Y₂
-1	-2	-2
0	-10	5
1	-12	12
2	-8	19
3	2	26
4	18	33
5	40	40

X= -1

The solutions are −1 and 5.

17. $12x^2 + 9x = 8x + 6$

$12x^2 + x - 6 = 0$

$(3x - 2)(4x + 3) = 0$

$3x - 2 = 0$　　or　　$4x + 3 = 0$

$x = \dfrac{2}{3}$　　　　　$x = -\dfrac{3}{4}$

The solutions are $-\dfrac{3}{4}$ and $\dfrac{2}{3}$.

18. $x^2 + 3x - 88 = 0$

$(x + 11)(x - 8) = 0$

$x + 11 = 0$　　or　　$x - 8 = 0$

$x = -11$　　　　　$x = 8$

The solutions are −11 and 8.

19. $5x^2 + 14x - 6 = 8 - 19x$

$5x^2 + 33x - 14 = 0$

$(5x - 2)(x + 7) = 0$

$5x - 2 = 0$　　or　　$x + 7 = 0$

$x = \dfrac{2}{5}$　　　　　$x = -7$

The solutions are −7 and $\dfrac{2}{5}$.

20. $49x^2 - 16 = 0$

$49x^2 = 16$

$x^2 = \dfrac{16}{49}$

$x = \sqrt{\dfrac{16}{49}}$　　or　　$x = -\sqrt{\dfrac{16}{49}}$

$x = \dfrac{4}{7}$　　　　　$x = -\dfrac{4}{7}$

The solutions are $\pm\dfrac{4}{7}$.

21. $5x^2 - 180 = 0$

$5x^2 = 180$

$x^2 = 36$

$x = \sqrt{36}$　　or　　$x = -\sqrt{36}$

$x = 6$　　　　　$x = -6$

The solutions are ± 6.

22. $11 + b^2 = 2$

$b^2 = -9$

There is no real-number solution.

23. $(x - 9)^2 = 16$

$x - 9 = \sqrt{16}$　　or　　$x - 9 = -\sqrt{16}$

$x - 9 = 4$　　　　　$x - 9 = -4$

$x = 9 + 4$　　　　　$x = 9 - 4$

$x = 13$　　　　　$x = 5$

The solutions are 5 and 13.

24. $(p - 7)^2 + 6 = 9$

$(p - 7)^2 = 3$

$p - 7 = \sqrt{3}$　　or　　$p - 7 = -\sqrt{3}$

$p = 7 + \sqrt{3}$　　　　　$p = 7 - \sqrt{3}$

The solutions are $7 \pm \sqrt{3}$.

25. $z^2 + 12z = -33$

$z^2 + 12z + 33 = 0$

$z = \dfrac{-b \pm \sqrt{b^2 - 4ac}}{2a}$

$z = \dfrac{-12 \pm \sqrt{(12)^2 - 4(1)(33)}}{2(1)}$

$z = \dfrac{-12 \pm \sqrt{144 - 132}}{2}$

$z = \dfrac{-12 \pm \sqrt{12}}{2}$

$z = \dfrac{-12 \pm 2\sqrt{3}}{2}$

$z = -6 \pm \sqrt{3}$

The solutions are $-6 \pm \sqrt{3}$.

26. $m^2 + 7m + 12 = 9$

$m^2 + 7m + 3 = 0$

$m = \dfrac{-b \pm \sqrt{b^2 - 4ac}}{2a}$

$m = \dfrac{-7 \pm \sqrt{7^2 - 4(1)(3)}}{2(1)}$

$m = \dfrac{-7 \pm \sqrt{37}}{2}$

The solutions are $\dfrac{-7 \pm \sqrt{37}}{2}$.

27. $b^2 + 2.4b - 4.32 = 0$

$100(b^2 + 2.4b - 4.32) = 0 \cdot 100$

$100b^2 + 240b - 432 = 0$

$b = \dfrac{-240 \pm \sqrt{(240)^2 - 4(100)(-432)}}{2(100)}$

$b = \dfrac{-240 \pm \sqrt{230,400}}{200}$

$b = \dfrac{-240 \pm 480}{200}$

$b = \dfrac{-240 + 480}{200}$ or $b = \dfrac{-240 - 480}{200}$

$b = 1.2$ $\qquad\qquad$ $b = -3.6$

The solutions are –3.6 and 1.2.

28. $q^2 - 7q - 78 = 0$

$(q + 6)(q - 13) = 0$

$q + 6 = 0$ $\qquad$ or $\qquad$ $q - 13 = 0$

$q = -6$ $\qquad\qquad\qquad$ $q = 13$

The solutions are –6 and 13.

29. $x^2 + 20x + 84 = 0$

$(x + 14)(x + 6) = 0$

$x + 14 = 0$ $\qquad$ or $\qquad$ $x + 6 = 0$

$x = -14$ $\qquad\qquad\qquad$ $x = -6$

The solutions are –14 and –6.

30. $2y^2 + 3y - 5 = 0$

$(2y + 5)(y - 1) = 0$

$2y + 5 = 0$ $\qquad$ or $\qquad$ $y - 1 = 0$

$y = -\dfrac{5}{2}$ $\qquad\qquad\qquad$ $y = 1$

The solutions are $-\dfrac{5}{2}$ and 1.

31. $4x^2 + 16x - 2 = 0$

$2(2x^2 + 8x - 1) = 0$

$x = \dfrac{-b \pm \sqrt{b^2 - 4ac}}{2a}$

$x = \dfrac{-8 \pm \sqrt{8^2 - 4(2)(-1)}}{2(2)}$

$x = \dfrac{-8 \pm \sqrt{72}}{4}$

$x = \dfrac{-8 \pm 6\sqrt{2}}{4}$

$x = \dfrac{-4 \pm 3\sqrt{2}}{2}$

The solutions are $\dfrac{-4 \pm 3\sqrt{2}}{2}$.

32. $4 + 8x + x^2 = 0$

$x^2 + 8x + 4 = 0$

$x = \dfrac{-b \pm \sqrt{b^2 - 4ac}}{2a}$

$x = \dfrac{-8 \pm \sqrt{(8)^2 - 4(1)(4)}}{2(1)}$

$x = \dfrac{-8 \pm \sqrt{48}}{2}$

$x = \dfrac{-8 \pm 4\sqrt{3}}{2}$

$x = -4 \pm 2\sqrt{3}$

The solutions are $4 \pm 2\sqrt{5}$.

33. $x^2 = 7x + 2$

$x^2 - 7x - 2 = 0$

$x = \dfrac{-b \pm \sqrt{b^2 - 4ac}}{2a}$

$x = \dfrac{-(-7) \pm \sqrt{(-7)^2 - 4(1)(-2)}}{2(1)}$

$x = \dfrac{7 \pm \sqrt{57}}{2}$

The solutions are $\dfrac{7 \pm \sqrt{57}}{2}$.

34. $36q^2 + 25 = 60q$

$36q^2 - 60q + 25 = 0$

$q = \dfrac{-b \pm \sqrt{b^2 - 4ac}}{2a}$

$q = \dfrac{-(-60) \pm \sqrt{(-60)^2 - 4(36)(25)}}{2(36)}$

$q = \dfrac{60 \pm \sqrt{0}}{72}$

$q = \dfrac{60}{72}$

$q = \dfrac{5}{6}$

The solution is $\dfrac{5}{6}$.

35. $b^2 - 7b + 16 = 0$

$b = \dfrac{-b \pm \sqrt{b^2 - 4ac}}{2a}$

$b = \dfrac{-(-7) \pm \sqrt{(-7)^2 - 4(1)(16)}}{2(1)}$

$b = \dfrac{7 \pm \sqrt{-15}}{2}$

There is no real-number solution.

36. $z^2 + 6z - 55 = 0$

$z = \dfrac{-b \pm \sqrt{b^2 - 4ac}}{2a}$

$z = \dfrac{-6 \pm \sqrt{6^2 - 4(1)(-55)}}{2(1)}$

$z = \dfrac{-6 \pm \sqrt{256}}{2}$

$z = \dfrac{-6 \pm 16}{2}$

$z = \dfrac{-6 - 16}{2}$ or $z = \dfrac{-6 + 16}{2}$

$z = -11$ $\qquad\qquad$ $z = 5$

The solutions are -11 and 5.

37. $4m^2 + 7m = 36$

$4m^2 + 7m - 36 = 0$

$m = \dfrac{-b \pm \sqrt{b^2 - 4ac}}{2a}$

$m = \dfrac{-7 \pm \sqrt{7^2 - 4(4)(-36)}}{2(4)}$

$m = \dfrac{-7 \pm \sqrt{625}}{8}$

$m = \dfrac{-7 \pm 25}{8}$

$m = \dfrac{-7 - 25}{8}$ or $m = \dfrac{-7 + 25}{8}$

$m = -4$ $\qquad\qquad$ $m = \dfrac{9}{4}$

The solutions are -4 and $\dfrac{9}{4}$.

38. $s(t) = -16t^2 + v_0 t + s_0$

$0 = -16t^2 + 0t + 160$

$0 = -16t^2 + 160$

$t^2 = 10$

$t = \pm\sqrt{10}$

$t \approx \pm 3.16$

Disregard a negative time.
The notebook will reach the ground in $\sqrt{10}$ seconds or approximately 3.16 seconds.

39. $C(x) = 5x^2 + 75x + 875$

$3500 = 5x^2 + 75x + 875$

$5x^2 + 75x - 2625 = 0$

$5(x^2 + 15x - 525) = 0$

$x = \dfrac{-15 \pm \sqrt{(15)^2 - 4(1)(-525)}}{2(1)}$

$x = \dfrac{-15 \pm \sqrt{2325}}{2}$

$x = \dfrac{-15 \pm 5\sqrt{93}}{2}$

$x = \dfrac{-15 + 5\sqrt{93}}{2} \approx 16.6$

Disregard a negative value.
The number is about 16 items.

40. $s(t) = -16t^2 + v_0 t + s_0$

$0 = -16t^2 + 64t + 32$

$0 = -16(t^2 - 4t - 2)$

$t = \dfrac{-(-4) \pm \sqrt{(-4)^2 - 4(1)(-2)}}{2(1)}$

$t = \dfrac{4 \pm \sqrt{24}}{2}$

$t = \dfrac{4 \pm 2\sqrt{6}}{2}$

$t = 2 \pm \sqrt{6}$

$t = 2 + \sqrt{6} \approx 4.45$

Disregard a negative value of time.
It will reach the ground in $2 + \sqrt{6}$ seconds
or approximately 4.45 seconds.

41. $s(t) = -16t^2 + v_0 t + s_0$

$4000 = -16t^2 + 0t + 11,000$

$16t^2 = 7000$

$t^2 = 437.5$

$t = \pm\sqrt{437.5}$

$t = \sqrt{437.5} \approx 20.92$

Disregard a negative value of time.
It will take $\sqrt{437.5}$ seconds or
approximately 20.92 seconds.

42. $a^2 + b^2 = c^2$

$(34)^2 + b^2 = (42.5)^2$

$b^2 = 650.25$

$b = \pm\sqrt{650.25}$

$b = \sqrt{650.25}$

$b = 25.5$

The other leg is 25.5 meters.

43. Let $x =$ width

$3x - 6 =$ length

$x(3x - 6) = 144$

$3x^2 - 6x - 144 = 0$

$3(x^2 - 2x - 48) = 0$

$3(x - 8)(x + 6) = 0$

$x = 8$ or $x = -6$

Disregard a negative width. The dimensions
are 8 cm by 18 cm.

44. Let $x =$ length of one leg

$3x - 1 =$ length of other leg

$a^2 + b^2 = c^2$

$x^2 + (3x - 1)^2 = (37)^2$

$x^2 + 9x^2 - 6x + 1 = 1369$

$10x^2 - 6x - 1368 = 0$

$2(5x^2 - 3x - 684) = 0$

$2(5x + 57)(x - 12) = 0$

$5x + 57 = 0$ or $x - 12 = 0$

$x = -\dfrac{57}{5}$ $\qquad$ $x = 12$

Disregard a negative length. The legs
measure 12 in. and 35 in.

45. $A = P(1 + r)^t$

$1323 = 1200(1 + r)^2$

$1.1025 = (1 + r)^2$

$1 + r = \pm\sqrt{1.1025}$

$r = -1 \pm 1.05$

$r = -1 + 1.05 = 0.05$

Disregard a negative value.
The interest rate is 5%.

46. $A = P(1 + r)^t$

$1500 = 1200(1 + r)^2$

$1.25 = (1 + r)^2$

$1 + r = \pm\sqrt{1.25}$

$r = -1 \pm \sqrt{1.25}$

$r \approx -1 + 1.118$

$r \approx 0.118$

Disregard a negative value.
The interest rate is approximately 11.8%.

47. $R(x) = 850 + 35x - 5x^2$

$650 = 850 + 35x - 5x^2$

$5x^2 - 35x - 200 = 0$

$5(x^2 - 7x - 40) = 0$

$$x = \frac{-(-7) \pm \sqrt{(-7)^2 - 4(1)(-40)}}{2(1)}$$

$$x = \frac{7 \pm \sqrt{209}}{2}$$

$$x = \frac{7 + \sqrt{209}}{2} \approx 10.7$$

Disregard a negative value.
A value of about 11 will yield a revenue of about $650.

Chapter 10 Test

1. $x^2 - 4x - 9 = 2x + 7$
$Y1 = x^2 - 4x - 9$
$Y2 = 2x + 7$

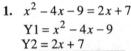

The solutions are –2 and 8.

2. $2x^3 + 9x^2 - 23x - 66 = 0$
$Y1 = 2x^3 + 9x^2 - 23x - 66$
$Y2 = 0$

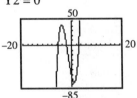

The graphs intersect at (3, 0), (–2, 0), and (–5.5, 0). The solutions are –5.5, –2, and 3.

3. $4(x - 5)^2 + 7 = 71$
$4(x - 5)^2 = 64$
$(x - 5)^2 = 16$
$x - 5 = \sqrt{16}$ or $x - 5 = -\sqrt{16}$
$x - 5 = 4$ $x - 5 = -4$
$x = 9$ $x = 1$
The solutions are 1 and 9.

4. $x^2 - 6x + 13 = 0$

$$x = \frac{-b \pm \sqrt{b^2 - 4ac}}{2a}$$

$$x = \frac{-(-6) \pm \sqrt{(-6)^2 - 4(1)(13)}}{2(1)}$$

$$x = \frac{6 \pm \sqrt{-16}}{2}$$

There is no real-number solution.

5. $2x^2 - 7x + 3 = 2x - 1$
$Y1 = 2x^2 - 7x + 3$
$Y2 = 2x - 1$

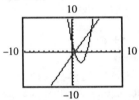

The graphs intersect at (0.5, 0) and (4, 7). The solutions are 0.5 and 4.

6. $x^2 - 5x + 4 = 7 - 3x - x^2$
$Y1 = x^2 - 5x + 4$
$Y2 = 7 - 3x - x^2$

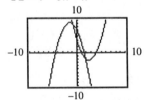

The graphs intersect at about (–0.82, 8.79) and (1.82, –1.79). The solutions are –0.82 and 1.82.

7. $x^2 - 9x + 9 = 7x - 5$
$x^2 - 16x + 14 = 0$
$b^2 - 4ac = (-16)^2 - 4(1)(14) = 200$
There are two irrational solutions because 200 is positive but not a perfect square.

$$x = \frac{-b \pm \sqrt{b^2 - 4ac}}{2a}$$

$$x = \frac{16 \pm \sqrt{200}}{2}$$

$$x = \frac{16 \pm 10\sqrt{2}}{2}$$

$$x = 8 \pm 5\sqrt{2}$$

The roots are $8 \pm 5\sqrt{2}$.

8. $x^2 - 3x + 12 = 0$

$b^2 - 4ac = (-3)^2 - 4(1)(12) = -39$

There are no real solutions because the discriminant is negative.

9. $x^2 - 8x + 3 = 8 + x - x^2$

$2x^2 - 9x - 5 = 0$

$b^2 - 4ac = (-9)^2 - 4(2)(-5) = 121$

There are two rational solutions because 121 is a perfect square.

$$x = \frac{-b \pm \sqrt{b^2 - 4ac}}{2a}$$

$$x = \frac{-(-9) \pm \sqrt{121}}{2(2)}$$

$$x = \frac{9 \pm 11}{4}$$

$$x = \frac{9 - 11}{4} \qquad\qquad x = \frac{9 + 11}{4}$$

$$x = -\frac{1}{2} = -0.5 \qquad x = 5$$

The roots are –0.5 and 5.

10. $x^2 - 25 = 10x$

$x^2 - 10x - 25 = 0$

$b^2 - 4ac = (-10)^2 - 4(1)(-25) = 200$

There are two irrational solutions because 200 is positive but not a perfect square.

$$x = \frac{-b \pm \sqrt{b^2 - 4ac}}{2a}$$

$$x = \frac{-(-10) \pm \sqrt{200}}{2(1)}$$

$$x = \frac{10 \pm 10\sqrt{2}}{2}$$

$$x = 5 \pm 5\sqrt{2}$$

The roots are $5 \pm 5\sqrt{2}$.

11. $x^2 - 2 = x + 4$

$Y1 = x^2 - 2$

$Y2 = x + 4$

a.

X	Y1	Y2
-3	7	1
-2	2	2
-1	-1	3
0	-2	4
1	-1	5
2	2	6
3	7	7

X= -3

The solutions are –2 and 3.

b.

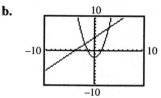

The graphs intersect at (–2, 2) and (3, 7). The solutions are –2 and 3.

c. $x^2 - 2 = x + 4$

$x^2 - x - 6 = 0$

$(x - 3)(x + 2) = 0$

$x - 3 = 0 \qquad$ or $\qquad x + 2 = 0$

$x = 3 \qquad\qquad\qquad\qquad x = -2$

The solutions are –2 and 3.

12. Let x = length of one leg

$x + 4$ = length of other leg

$a^2 + b^2 = c^2$

$x^2 + (x + 4)^2 = (20)^2$

$x^2 + x^2 + 8x + 16 = 400$

$2x^2 + 8x - 384 = 0$

$2(x^2 + 4x - 192) = 0$

$2(x + 16)(x - 12) = 0$

$x = -16$ or $x = 12$

Disregard a negative length. The legs are 12 in. and 16 in.

13. $s = -16t^2 + v_0 t + s_0$

$2600 = -16t^2 + 0 \cdot t + 11,500$

$-16t^2 = -8900$

$t^2 = 556.25$

$t = \pm\sqrt{556.25}$

$t \approx 23.6$

Disregard a negative value of time. Keanu was free-falling for about 24 seconds.

14. $A = P(1+r)^t$

$1573 = 1400(1+r)^2$

$\dfrac{1573}{1400} = (1+r)^2$

$1 + r = \sqrt{\dfrac{1573}{1400}}$ or $1 + r = -\sqrt{\dfrac{1573}{1400}}$

$r = -1 + \sqrt{\dfrac{1573}{1400}}$ $\qquad r = -1 - \sqrt{\dfrac{1573}{1400}}$

$r \approx 0.06$ $\qquad\qquad r \approx -2.06$

Disregard a negative interest rate. The interest rate must be approximately 6%.

15. $R(x) = 100x - 2x^2$

$450 = 100x - 2x^2$

$2x^2 - 100x + 450 = 0$

$x^2 - 50x + 225 = 0$

$x^2 - 50x = -225$

$x^2 - 50x + 625 = -225 + 625$

$(x - 25)^2 = 400$

$x - 25 = \sqrt{400}$ or $x - 25 = -\sqrt{400}$

$x - 25 = 20$ $\qquad x - 25 = -20$

$x = 45$ $\qquad\qquad x = 5$

The revenue will be \$450 if 5 items are sold or if 45 items are sold.

16. Answers will vary.

Chapters 1–10 Cumulative Review

1. $-(-1.5) = 1.5$

2. $-|-1.5| = -|1.5| = -1.5$

3. $\sqrt[3]{\dfrac{64}{216}} = \sqrt[3]{\dfrac{4^3}{6^3}} = \dfrac{4}{6} = \dfrac{2}{3}$

4. $\left(\dfrac{2}{5}\right)^{-3} = \left(\dfrac{5}{2}\right)^3 = \dfrac{5^3}{2^3} = \dfrac{125}{8}$

5. $\left(\dfrac{7}{4}\right)\left(-\dfrac{8}{21}\right) = \left(\dfrac{1}{1}\right)\left(-\dfrac{2}{3}\right) = -\dfrac{2}{3}$

6. $-\dfrac{2}{3} \div \left(-1\dfrac{1}{2}\right) = -\dfrac{2}{3} \div \left(-\dfrac{3}{2}\right)$

$= -\dfrac{2}{3} \times \left(-\dfrac{2}{3}\right) = \dfrac{4}{9}$

7. $(4)(-3) \div (6)(-2) \div 3 = -12 \div (6)(-2) \div 3$

$= (-2)(-2) \div 3 = 4 \div 3 = \dfrac{4}{3}$

8. $12 + 16 \div 4 - \sqrt{2^2 + 4 \cdot 5 - 1^3}$

$= 12 + 16 \div 4 - \sqrt{4 + 20 - 1}$

$= 12 + 16 \div 4 - \sqrt{23}$

$= 12 + 4 - \sqrt{23}$

$= 16 - \sqrt{23} \approx 11.204$

9. $\dfrac{(4^2 - 10) + 2^3}{5 - 1^5} = \dfrac{2(16 - 10) + 2^3}{5 - 1}$

$= \dfrac{2(6) + 8}{4} = \dfrac{12 + 8}{4} = \dfrac{20}{4} = 5$

10. $2^0 x^{-1} y^2 = \dfrac{y^2}{x}$

11. $\left(\dfrac{(2s)^2}{3t}\right)^{-3} = \left(\dfrac{2^2 s^2}{3t}\right)^{-3} = \left(\dfrac{4s^2}{3t}\right)^{-3}$

$= \left(\dfrac{3t}{4s^2}\right)^3 = \dfrac{(3t)^3}{(4s^2)^3} = \dfrac{3^3 t^3}{4^3 (s^2)^3} = \dfrac{27t^3}{64s^6}$

12. $2x - 3(x + 4) + 5 = -x - 12 + 5 = -x - 7$

13. $1.5[2(x + 1) - 3] + [2.5(x + 3) - 4]$

$= 1.5[2x + 2 - 3] + [2.5x + 7.5 - 4]$

$= 1.5[2x - 1] + 2.5x + 3.5$

$= 3x - 1.5 + 2.5x + 3.5$

$= 5.5x + 2$

14. $(2x^3 + 6x^2y - 2xy^2 + y^3) + (3x^3 + 4xy^2 - y^3)$
$= 2x^3 + 3x^3 + 6x^2y - 2xy^2 + 4xy^2 + y^3 - y^3$
$= (2+3)x^3 + 6x^2y + (-2+4)xy^2 + (1-1)y^3$
$= 5x^3 + 6x^2y + 2xy^2 + 0y^3$
$= 5x^3 + 6x^2y + 2xy^2$

15. $(1.2a^2 - 3.6ab + b^2) - (4a^2 + 2.71ab - 3.4b^2)$
$= (1.2a^2 - 3.6ab + b^2) + (-4a^2 - 2.71ab + 3.4b^2)$
$= 1.2a^2 - 4a^2 - 3.6ab - 2.71ab + b^2 + 3.4b^2$
$= (1.2 - 4)a^2 + (-3.6 - 2.71)ab + (1 + 3.4)b^2$
$= -2.8a^2 - 6.31ab + 4.4b^2$

16. $(6.8m^2n)(-2mn^2p)$
$= (6.8)(-2)m^{2+1}n^{1+2}p$
$= -13.6m^3n^3p$

17. $(3a - b)(2a + 4b)$
$= (3a)(2a) + (3a)(4b) + (-b)(2a) + (-b)(4b)$
$= 6a^2 + 12ab - 2ab - 4b^2$
$= 6a^2 + 10ab - 4b^2$

18. $(2x + 3)(2x - 3) = (2x)^2 - 3^2 = 4x^2 - 9$

19. $(2x + 3)^2 = (2x)^2 + 2(2x)(3) + 3^2$
$= 4x^2 + 12x + 9$

20. $\dfrac{-15x^2y^3z}{3xyz} = -5x^{2-1}y^{3-1}z^{1-1} = -5xy^2z^0$
$= -5xy^2$

21. $\dfrac{2m^2n + 4mn - 8n^2}{2m^2n}$
$= \dfrac{2m^2n}{2m^2n} + \dfrac{4mn}{2m^2n} - \dfrac{8n^2}{2m^2n}$
$= 1 + \dfrac{2}{m} - \dfrac{4n}{m^2}$

22. $16a^2 - 25b^2 = (4a)^2 - (5b)^2$
$= (4a + 5b)(4a - 5b)$

23. $x^2 - 2x - 8 = (x - 4)(x + 2)$

24. $3x^2 - 9x - 30 = 3(x^2 - 3x + 10)$
$= 3(x - 5)(x + 2)$

25. $6x^2 + 5x - 4 = (2x - 1)(3x + 4)$

26. $f(x) = -x^2 + 3x - 1$
$f(-2) = -(-2)^2 + 3(-2) - 1$
$f(-2) = -4 + (-6) - 1$
$f(-2) = -11$

27.

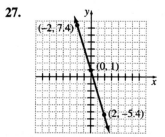

Domain: all real numbers
Range: all real numbers

28. $y = x^2 + 4x + 6$
$a = 1, b = 4, c = 6$
$-\dfrac{b}{2a} = -\dfrac{4}{2(1)} = -2$

x	$y = x^2 + 4x + 6$	y
–2	$y = (-2)^2 + 4(-2) + 6$ $y = 2$	2
–1	$y = (-1)^2 + 4(-1) + 6$ $y = 3$	3
0	$y = 0^2 + 4(0) + 6$ $y = 6$	6

The vertex is (–2, 2); the axis of symmetry is
$x = -2$.

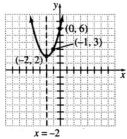

Domain: all real numbers
Range: all real numbers ≥ 2

29. $y = g(x) = -2x^2 + x + 1$
$a = -2, b = 1, c = 1$
$$\frac{-b}{2a} = \frac{-1}{2(-2)} = \frac{1}{4}$$

x	$g(x) = -2x^2 + x + 1$	y
0	$g(0) = -2(0)^2 + 0 + 1$ $g(0) = 1$	1
$\frac{1}{4}$	$g\left(\frac{1}{4}\right) = -2\left(\frac{1}{4}\right)^2 + \frac{1}{4} + 1$ $g\left(\frac{1}{4}\right) = \frac{9}{8}$	$\frac{9}{8}$
1	$g(1) = -2(1)^2 + 1 + 1$ $g(1) = 0$	0
2	$g(2) = -2(2)^2 + 2 + 1$ $g(2) = -5$	–5
$-\frac{1}{2}$	$g\left(-\frac{1}{2}\right) = -2\left(-\frac{1}{2}\right)^2 + \left(-\frac{1}{2}\right) + 1$ $g\left(-\frac{1}{2}\right) = 0$	0

The vertex is $\left(\dfrac{1}{4}, \dfrac{9}{8}\right)$; the axis of symmetry

is $x = \dfrac{1}{4}$.

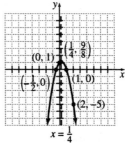

Domain: all real numbers
Range: all real numbers $\leq \dfrac{9}{8}$

30.

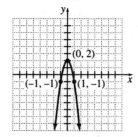

31. There are no restrictions on x, so the domain is the set of all real numbers. The y-value of the function never exceeds 2, so the range is all real numbers less than or equal to 2.

32. The vertical line test shows the relation is a function.

33. The maximum of 2 occurs when $x = 0$. There are no relative minima.

34. The function is increasing on $(-\infty, 0)$ and it is decreasing on $(0, \infty)$.

35. Let $x = 0$.
$y = -3(0)^2 + 2$
$y = 0 + 2 = 2$
The y-intercept is at $(0, 2)$.
Let $y = 0$.
$0 = -3x^2 + 2$
$3x^2 = 2$
$x^2 = \dfrac{2}{3}$
$x = \pm\sqrt{\dfrac{2}{3}} = \pm\dfrac{\sqrt{6}}{3} \approx \pm 0.816$

The x-intercepts are at $\left(\pm\dfrac{\sqrt{6}}{3}, 0\right)$ or approximately $(\pm 0.816, 0)$.

36. $2(x + 3) - 4(x - 1) = x - 2$
$2x + 6 - 4x + 4 = x - 2$
$-2x + 10 = x - 2$
$12 = 3x$
$4 = x$

37. $2.4x - 1.2(2x - 3) = 3.6$
$2.4x - 2.4x + 3.6 = 3.6$
$3.6 = 3.6$
This is an identity. The solution is all real numbers.

38. $x^2 + 2x - 15 = 0$
$(x + 5)(x - 3) = 0$
$x + 5 = 0$ or $x - 3 = 0$
$x = -5$ or $x = 3$

39. $2t^2 + 3t = 4$
$2t^2 + 3t - 4 = 0$
$t = \dfrac{-3 \pm \sqrt{3^2 - 4(2)(-4)}}{2(2)} = \dfrac{-3 \pm \sqrt{41}}{4}$

40. $4x + 5 > 2x + 15$
$2x > 10$
$x > 5$
Interval notation: $(5, \infty)$
Graph:

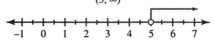

41. $-38x + 14 \ge -18 - 6$
$-38x + 14 \ge -24$
$-38x \ge -38$
$x \le 1$
Interval notation: $(-\infty, 1]$
Graph:

42. From the second equation, $x = 8 - y$.
Substituting $8 - y$ for x in the first equation gives
$3(8 - y) - 4y = 3$
$24 - 3y - 4y = 3$
$-7y = -21$
$y = 3$
$x = 8 - 3 = 5$
The solution is $(5, 3)$.

43.

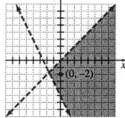

44. $m = \dfrac{y_2 - y_1}{x_2 - x_1} = \dfrac{2 - (-5)}{-4 - 3} = \dfrac{7}{-7} = -1$

$y - y_1 = m(x - x_1)$

$y - (-5) = -1(x - 3)$

$y + 5 = -x + 3$

$y = -x - 2$

45. $2x + 3y = 5$

$3y = -2x + 5$

$y = -\dfrac{2}{3}x + \dfrac{5}{3}$

The line we want has slope $\dfrac{3}{2}$.

$y - y_1 = m(x - x_1)$

$y - 1 = \dfrac{3}{2}[x - (-2)]$

$y - 1 = \dfrac{3}{2}x + 3$

$y = \dfrac{3}{2}x + 4$

46. Let l be the length of the rectangle and w be the width. Since the diagonal measures 20 feet, we have $\sqrt{l^2 + w^2} = 20$ or $l^2 + w^2 = 400$. Substitute $w + 4$ for l in this equation.

$(w + 4)^2 + w^2 = 400$

$w^2 + 8w + 16 + w^2 = 400$

$2w^2 + 8w - 384 = 0$

$w^2 + 4w - 192 = 0$

$(w + 16)(w - 12) = 0$

$w + 16 = 0$ or $w - 12 = 0$

$w = -16$ or $w = 12$

Since length cannot be negative, the width is 12 feet and the length is $12 + 4 = 16$ feet. The rectangle is 12 feet by 16 feet.

47. Let $x =$ the amount of Cameron's bill for the sixth month. Since the average of all six months must be no more than $40:

$\dfrac{35 + 42 + 38 + 50 + 30 + x}{6} < 40$

$\dfrac{35 + 42 + 38 + 50 + 30 + x}{6} < 40$

$\dfrac{195 + x}{6} < 40$

$195 + x < 240$

$x < 45$

Cameron's bill for his sixth month must be less than $45.

48. Let $x =$ number of pounds of 10% nitrogen fertilizer. Since there are 150 pounds in the final mixture, there will be $150 - x$ pounds of 5% fertilizer.

$0.1(x) + 0.05(150 - x) = 0.085(150)$

$0.1x + 7.5 - 0.05x = 12.75$

$0.05x = 5.25$

$x = 105;\ 150 - x = 45$

The chemist must mix 105 pounds of the 10% nitrogen fertilizer with 45 pounds of 5% nitrogen fertilizer.

49. Let x be the number of items produced and sold. The cost of producing x items is $1000 + 0.55x$, while the revenue from selling x items is $1.25x$.

$1.25x \geq 1000 + 0.55x$

$0.70x \geq 1000$

$x \geq \dfrac{1000}{0.7} = \dfrac{10,000}{7} \approx 1428.6$

The company must sell at least 1429 items in order to break even.

50. The position equation is $s = -16t^2 + v_0 t + s_0$. We have to find t when s is 2500 if v_0 is 0 and s_0 is 10,000.

$2500 = -16t^2 + 0t + 10,000$

$2500 = -16t^2 + 10,000$

$16t^2 = 7500$

$t^2 = \dfrac{1875}{4}$

$t = \pm\sqrt{\dfrac{1875}{4}} = \pm\dfrac{25\sqrt{3}}{2}$

Since the amount of time will be positive,
$t = \dfrac{25\sqrt{3}}{2} \approx 21.65$. The skydiver was free-
falling for about 21.65 seconds.